ORE DEPOSITS

A Series of Books in Geology

Editors: James Gilluly
A. O. Woodford

THIRD EDITION

ORE DEPOSITS

Charles F. Park, Jr.
STANFORD UNIVERSITY

Roy A. MacDiarmid
STATE UNIVERSITY OF NEW YORK
AT FREDONIA

W. H. FREEMAN AND COMPANY
San Francisco

Library of Congress Cataloging in Publication Data

Park, Charles Frederick, 1903-
    Ore deposits.

    Includes bibliographies and indexes.
    1.   Ore-deposits.   I.   MacDiarmid, Roy A., 1933-
1966, joint author.   II.   Title.
TN263.P3   1975   553′.1   75-14157
ISBN 0-7167-0272-X

Printed in the United States of America

10   9   8   7   6   5

Contents

Preface

This book presents to the beginning student of economic geology the principles and data basic to understanding the genesis and localization of the metallic ores and the nonmetallic minerals associated with them. It is assumed that the student has at least an elementary knowledge of chemistry and physics, as well as mineralogy, petrology, and structural geology.

Modern civilization's dependence upon minerals makes the search for new ore deposits continually more challenging and difficult. If the future geologist is to meet this challenge he must be given a broad and thorough training; he must develop a blend of ingenuity, imagination, and optimism that will enable him to face the many problems that remain unsolved. It is the purpose of the book to present in an orderly fashion the ideas and principles that will enable the beginning geologist to better understand the challenge as well as the limits of our knowledge.

The physical and chemical characteristics of various types of ore deposits are described and correlated with environments and conditions of deposition. Examples and illustrations have been selected to emphasize structural and chemical controls and to encourage "three-dimensional" thinking. Many of the ideas discussed are controversial and no attempt has been made to settle these controversies. Ideas concerning the genesis of ore deposits are continually changing, hence all sides of the major issues have been presented. Accelerated progress in the science of ore deposits in the past few years has necessitated the preparation of a third edition of this book.

In this edition, as in the first two, the emphasis is on exploration geology for the beginning student, and the development of "exploration thinking." A chapter on the rapidly expanding subject of volcanogenic deposits has been added

(Chapter 17) and descriptions of the Kuroko deposits of Japan have been considerably revised. In order that all of the more common types of ore deposits might be included, a description of the nickel ores at Sudbury, Canada, has been added.

Terminology in the Lindgren classification of ore deposits used in the previous editions has been changed. The term "magmatic" has been largely abandoned in favor of "hydrothermal." Hydrothermal, as commonly used, means simply "hot water," and does not denote any particular origin for either the heat or the water. The parts of the book dealing with fluid inclusions and isotopes have been rewritten and examples given. Progress in both of these fields has been very rapid.

Efforts have been made to keep the book of a length suitable for a one-term or one-quarter lecture course, and this has necessitated the omission of many facets of the science. Nevertheless, many sections of the book have been enlarged and such related subjects as plate tectonics are mentioned briefly.

As with the previous editions, the many helpful comments that have been received from readers are greatly appreciated and have been invaluable in preparing this edition. Special thanks are extended to my colleagues at Stanford University for their many helpful discussions and comments.

February, 1975   *Charles F. Park, Jr.*
                                                    Stanford, California

ORE DEPOSITS

Introduction

Besides, what we are pleased to call the riches of a mine are riches relative to a distinction which nature does not recognize. The spars and veinstones which are thrown out in the rubbish of our mines may be as precious in the eyes of nature, as conducive to the great object of her economy, and are certainly as characteristic of mineral veins, as the ores of silver or gold to which we attach so great a value.

JOHN PLAYFAIR, 1802

Ores are rocks and minerals that can be recovered at a profit. In its strictest sense, *ore* refers only to metals or metal-bearing minerals, but in common usage a few of the nonmetallic minerals, such as sulfur and fluorite, are included. Building stone and industrial materials, such as abrasives, clays, refractories, lightweight aggregates, and salts, are not considered ore; they are classified as industrial rocks and minerals or *economic minerals,* a term that includes both industrial materials and *ore minerals.* Ore minerals are compounds valued for their metal content. Thus, not all minerals containing a given element are classified as ore minerals; for example, most iron silicates, such as fayalite and ferrosilite, are not mined for their iron and therefore are not ore minerals. An ore may be a rock containing veinlets, disseminations, or small amounts of useful minerals or it may be massive ore-bearing material. Although both metallic and nonmetallic minerals are widely distributed in the rocks of the earth's crust, only under exceptional circumstances are they concentrated in amounts sufficient for economic recovery and in a form that permits recovery. Most ore minerals are associated with valueless material called *gangue,* and many ores grade into *protore*—mineralized rock that is too lean to yield a profit. As Playfair stated so well in 1802, the economic value of an ore does not set it apart generically from the worthless pyrite, sericite, calcite, or other gangue and rock with which it is associated. The study of ore deposits is thus a specialized part of the broader field of petrology.

The accelerating growth of the world's population, combined with an improving standard of living throughout the world, is greatly increasing demands for mineral products of all types. These demands will certainly continue to grow. At the same time, the search for ore is becoming more complex; more and more, ore is being sought under cover and at greater depths. In order to obtain sufficient supplies in the future, new ideas and techniques must be devised to supplement the old. Recovery and mining techniques need to be improved so that large bodies of near-surface minerals that are not now economic can be developed. For these reasons the successful economic geologist must develop "exploration thinking." This will require imagination, ingenuity, and a degree of optimism, as well as a thorough knowledge of structural geology, stratigraphy, petrology, and mineralogy, and of how fluids migrate underground. Economic geologists should also be familiar with fundamental techniques in geophysics and geochemistry, as these fields are becoming increasingly useful in the search for buried deposits.

Under what conditions and as a result of what processes are ores formed? What factors lead to concentration of certain elements in one environment and not in another? What causes the localization of ore? Probably the best way to answer these questions, and hence the best way to search for new ore bodies, is to study the structure and genesis of known mineral concentrations and then explore the geologically favorable areas.

Owing to the complex nature of the earth, and because many of the processes involved in ore deposition cannot be observed, the study of ore deposits and ore genesis is not an exact science. Geology, especially the study of ore deposits, is basically a field science. This does not mean that precise laboratory and experimental data are therefore not desirable or useful; it does mean that laboratory data should be viewed critically and used with caution. Laboratory data that is unsupported by field evidence frequently leads to erroneous conclusions. The final test of all theories and hypotheses in geology is their applicability in the field.

1 / The Development of the Modern Theories of Ore Deposition

Early humans were acquainted with ore deposits and worked many of them. They used cinnabar and soft red hematite as pigments long before smelting was devised. Gem stones and native copper, gold, and silver were highly valued as ornaments and as materials for making simple implements. Essentially nothing was understood of the origin of minerals.

In the ancient Greek and Roman civilizations hypotheses about ore genesis were postulated by philosophers, none of whom had firsthand knowledge of ores in the ground. Until the eighteenth century many imaginative theories were advanced to explain the presence of minerals and metals in the earth's crust. Some philosophers envisioned an animated earth that breathed ores or gave off metallic exhalations as a regular function of metabolism. Others considered the ores themselves to be alive, growing from seeds or ripening from base metals into precious metals. A popular conception among such "naturalists" was that the ores existed in the form of a subterranean "golden tree" whose twigs and branches were various kinds of metals and whose roots led down to the center of the earth. Such concepts were not as thoroughly fantastic as they may seem, and many modern ideas can be traced to them. The belief generally held by alchemists in the sixteenth and seventeenth centuries was that ore deposits were generated by celestial powers, such as the sun's rays or planetary "influences." These fanciful ideas were maintained not only by alchemists, but also by some of the world's most renowned thinkers: Aristotle, Pliny, Avicenna, Albertus Magnus, Thomas Aquinas, Robert Boyle, Roger Bacon, and Georges de Buffon. The early hypotheses are of interest to us here principally because their shortcomings emphasize that the study of ore-forming processes must be based upon field observations.

In view of the religious beliefs and superstitions that dominated earlier civilization, it is not at all surprising that at first mysterious forces were called upon to explain and detect ore deposits. Although the ancient Egyptians and Greeks used divination rods for predicting future events, it was not until the fifteenth century that forked twigs were used to locate mineral deposits. The application of the divination rod to prospecting is credited to the Germans of the Harz mining region, who in turn "enlightened" the English (Gardner, 1957). As theories of ore genesis improved, however, so did prospecting techniques, with the result that geophysical methods (de Wet, 1957) and geochemical methods (Hawkes and Webb, 1962; Cameron, 1966) have nearly replaced the hazel stick. Even today, however, the ancient practice of witching (or dousing) for ore deposits is employed by less sophisticated prospectors.

A solid base for the modern theories of ore deposits was formulated in the sixteenth century by Georg Bauer, who is generally known by his Latinized name, Georgius Agricola. Agricola's observations about ore deposits were remarkable for his time, and for this reason he is considered the father of economic geology. He lived in the Erzgebirge region of Saxony and gained an intimate firsthand knowledge of the mines in that district. He wrote many treatises on geology and mining, the most significant of which was his comprehensive and pioneering work, *De Re Metallica* (1556). In spite of the fact that most of Agricola's works were in Latin, and therefore inaccessible to anyone but scholars, his writings gave a significant, leading stimulus to the science of ore deposits, which even today bears the stamp of his influence. Agricola's principal contribution was his attempt to classify ores, for without classification no great scientific progress would be possible. His classification was based upon genesis—whether the deposits were alluvial or *in situ*—and upon form. The classification of *in situ* deposits included, for example, fissure veins, bedded or horizontal deposits, impregnations, stringers, seams, and stockworks; a vein or seam might be straight, curved, inclined or vertical.

According to Hoover and Hoover, translators of *De Re Metallica*, Agricola is responsible for two fundamental principles in addition to his classification of ore deposits. These principles are (1) that ore channels are secondary features, younger than the country rocks, and (2) that the ores are deposited from solutions circulating in these channels. Agricola's work thus marks the transition from speculation to observation. Moreover, this great geologist argued against the use of the hazel twig in prospecting at a time when this technique was first coming into use.

From the time of Agricola until the middle of the eighteenth century, little advancement was made in the study of ore genesis. Nevertheless, a few noteworthy works were published and considerable field knowledge was accumulated. Nicolaus Steno (the Latinized name of Niels Stensen, a Dane who worked in Italy) stands out in the seventeenth century as a scientist of great perception. Responsible for many contributions to general geology, he argued that ores are a product of condensation from vapors ascending through fissures (Steno, 1669)—a remarkably advanced concept that has been retained in modern theory.

During this period, as before, many people throughout the world looked upon mining as dangerous and, what was worse, degrading work. For centuries mining was work for convicts and slaves. The ancient Greeks are believed to have maintained a working force of 20,000 slaves at their silver mines near Laurium. Once taken underground, the slaves rarely saw the light of day again. At night they were chained to the walls; the annual death rate was said to have been about 25 percent of the labor force. Since most mining was done in remote parts of the world and communications were primitive or nonexistent, working in the mines was equivalent to being banished from civilization. Is it any wonder then that for centuries mining was not regarded as worthy of respectable attention?

By the eighteenth century, however, many trained scientists were puzzling over the problems of ore genesis. Most of the early progress was made in Germany, in the Erzgebirge mining district. Henkel (1725, 1727) and Zimmermann (1746) acknowledged the importance of hydrothermal solutions, or vapors of deep-seated origin, which they correctly reasoned should contain dissolved rock materials; they even recognized the products of ore deposition by metasomatism (replacement). In 1749 von Oppel made the important distinction between veins and bedded deposits, the former being cross-cutting features of secondary, open-fissure origin, and the latter being conformably interbedded with the stratified sediments.

It must have been difficult for a student of ore deposits to differentiate between fact and fancy in those days. Not only had the scientific method not been established, but some of the leading scientists were guilty of proposing theories that did not acknowledge all of the field data. The result was a mixture of factual contributions and imaginative ideas, and it is only in retrospect that we are able to distinguish between the two and give credit where it is due. A good example of the sometimes observant, sometimes fanciful scientist is Delius (1770, 1773), who first recognized the superficial alteration of ores by atmospheric agents. He also observed the development of secondary minerals in the alteration zone as well as a zone of supergene enrichment beneath the altered layer. Yet he reasoned erroneously that development of alteration products requires heat from the sun.

A few years later, Charpentier (1778, 1799), a professor at the Mining Academy at Freiberg, Germany, published two small but well-written books. His writings were based upon astute observations made during many years of study in the Freiberg mines. He believed that the veins in these mines were formed by alteration of the country rock; for supporting evidence he showed how some of the veins graded into wall rock. He likened this process to the silicification of wood and, like Henkel and Zimmermann, proposed that one material had been altered to another. On the other hand, Gerhard (1781), a contemporary German writer, considered the veins to be open fissures filled by minerals leached from the country rock. From Charpentier's and Gerhard's writings has come the theory of lateral secretion, which posits that the contents of an ore deposit are leached from the adjacent wall rock by waters that are usually, but not necessarily, of meteoric origin. The theory of lateral secretion was apparently neglected for almost 100 years, and not until Sandberger (1882) published his *Untersuchungen über Erzgänge* (Investigations Concerning Mineral Veins) did the theory become generally known and widely accepted. Sandberger attempted to show that most mineral deposits were formed according to this theory, and his publications profoundly influenced contemporary thought. Actually, the theory has much to commend it; some ore deposits are still believed by careful workers to have formed in this manner.

During the latter part of the eighteenth century two outstanding men of diametrically opposed views dominated geological thought: James Hutton, a Scot, and Abraham Gottlob Werner, a German. Their ideas have had a far-reaching effect on the science of ore deposits. Hutton (1788, 1795), a "plutonist," thought that both igneous rocks and ore deposits were derived from molten magmas at depth and were transported in the liquid state to their present positions. He was impressed by the similarity between metallic ores and the products observed in smelters. Arguing against the idea of ore deposition from hydrothermal solutions, he suggested that ore materials solidified from the molten state, the ore magma having been injected into fissures of tectonic origin. Massive, mixed sulfides were assumed to have an origin similar to that of an igneous rock, as evidenced by the mutually interpenetrating mineral grains. Hutton went too far with his magmatic theory of rocks when he insisted that metals could not have been precipitated from aqueous solutions.

The suggestion that ores are direct magmatic products or are formed as products of differentiation was also advocated by Joseph Brunner (1801), a Bavarian mine official. Scipione Breislak (1811) an Italian geologist, also called upon the process of magmatic segregation to explain how ore minerals became concentrated in definite layers in igneous rocks. The theory remained essentially unchanged until Joseph Fournet revived it about the middle of the nineteenth century. Fournet was an able advocate of the theory of ore magmas, and his ideas, with little modification, were later revived by Spurr (1923). Ore magmas are widely accepted today as the origin of certain types of ore bodies.

In contrast to Hutton, Werner was a "neptunist." He argued that basalts, sandstones, limestones, and ore deposits were formed as sediments in a primeval ocean. In 1791 Werner published a résumé of his ideas in the *Neue Theorie von der Entstehung der Gänge* (New Theory of the Formation of Veins). He pictured veins originating as cracks formed on the bottom of a primeval ocean by slumping or earthquakes; the cracks were subsequently filled with chemical precipitates, as evidenced by the symmetrical banding observed in veins of the Erzgebirge. Werner's personal charm and gift of oratory, as well as his comprehension of mineralogy and local stratigraphy, gained him a devoted following and a reputation as an outstanding teacher and logician. Both directly and indirectly he had great influence on the entire science of geology.

We are dependent upon the students and followers of Hutton and Werner for amplification of the theories of these men. Hutton lacked the ability to present his ideas in a lucid style, and if it had not been for John Playfair (1802), who popularized Hutton's contributions shortly after he died, this great pioneer's influence would not have been so widespread. Probably the best explanation in English of Werner's hypotheses was published by Charles Anderson in 1809.

Heated arguments and discussions between advocates of the plutonist and neptunist theories were conducted by disciples of Hutton and Werner for many years. Observant geologists were quick to discredit both principles as loose generalities. It was readily shown that lavas are not sedimentary formations. Similarly, it was readily shown that minerals, including those that contain metals, are soluble and may be transported in and deposited from aqueous media. Even today, however, this argument between plutonists and neptunists—or, as they are now known, magmatists and syngeneticists—has not been resolved. Some geologists maintain that banded sulfide ores similar to those in the Rammelsberg district, Germany (Fig. 18-13, p. 426) and the extensive "Copper Belt" of Zambia and Zaire are sediments and have the same origin as the enclosing rocks. Other geologists do not accept this theory because parts of the ore bodies show replacement relationships with their wall rocks, and it is not clear how the very large amounts of uncommon ore minerals present could have been concentrated during ordinary sedimentary processes.

Many mines were examined and operated during the early part of the nineteenth century, but comparatively few advances were made in the theories of ore transportation and deposition. Mining methods were improved by technical advances and by the development of such mechanical devices as the hoist. One of the greatest steps forward was the invention of the Cornish pump, which permitted operations to be carried on below the water table. Detailed geological observations were recorded by such careful scientists as J. L. Heim, Alexander von Humboldt, and D'Aubuisson de Voisins. Their observations later provided the basic field data for many theories and conclusions. One result was the demonstration of a spatial relationship between certain types of ore deposits and intrusive rocks. Recognition of this association established the foundation for modern theories of the origin of hydrothermal ores.

It was during this period that geologists began to distinguish between ore deposits of igneous affiliation and those of sedimentary origin. Field data accumulated during the nineteenth century and earlier shows clearly that many large ore deposits are localized in areas of structural complexity and igneous activity, a fact that is used extensively in exploration. Although most evidence that ores are associated with igneous sources is circumstantial and somewhat tenuous, it is nonetheless convincing. Certain types of ore are found so commonly in and around igneous masses of a particular composition, and so seldom elsewhere, that it is natural and reasonable to associate the presence of these ores with definite types of igneous activity. For example, nickel ores are associated with norites and peridotites; disseminated copper deposits are found commonly in monzonite or quartz monzonite stocks; and tin is associated with siliceous plutonic rocks, such as granite and siliceous pegmatite. This close association has strengthened the assumption that a genetic relationship exists between igneous rocks and many ores. The arrangement of minerals in zones around igneous centers likewise suggests that ore-bearing fluids spread from channels that tap deep-seated sources, presumably magma chambers. As the fluids rise and spread through the rocks, they deposit their minerals in favorable structural and stratigraphic environments. Evidence to relate ores to volcanic activity has also been obtained from the studies of fumaroles and hot springs, where many constituents of ore deposits have been recognized, either in the hot fluids or in the wall rocks of the channelways.

In the middle of the nineteenth century Elie de Beaumont made several contributions that are still generally accepted. In addition to his ideas concerning general geology and structural geology, de Beaumont thought that hydrothermal solutions played a part in the formation of ore deposits. Although this suggestion had been made previously, de Beaumont emphasized and developed it. Besides hydrothermal deposits, he recognized and described magmatic segregations and replacement ores formed in the contact metamorphic zones of intrusive igneous rocks. It was during the period of de Beaumont's activity that chemical principles began to influence geologists' ideas about ore genesis, thanks to such men as K. G. Bischof and Bernhard von Cotta.

In the latter half of the nineteenth century many eminent scientists contributed to the theories of ore transportation and deposition. Among the best known were von Cotta, F. Sandberger, and A. W. Stelzner in Germany; A. Daubrée, and later L. de Launay, in France; and F. Posepny in Bohemia; J. A. Phillips in England; J. H. L. Vogt in Norway; and S. F. Emmons in the United

States. The science was coming of age. The accumulation of field data pre-cluded over-generalizing; geologists understood clearly that no single theory would explain the genesis of all ore deposits.

Early in the present century Louis de Launay, J. H. L. Vogt, Waldemar Lindgren, R. Beck, C. R. Van Hise, F. L. Ransome, W. H. Emmons, J. F. Kemp, W. H. Weed, B. S. Butler, and other scientists formulated many of our modern theories. Owing to numerous careful field observations throughout the world, as well as an expanding demand for minerals, economic geology was becoming an important branch of science. It became obvious that struc-ture played a vital role in ore localization by guiding ore-bearing fluids into areas favorable for their deposition. The process of replacement was widely recognized, and knowledge of the more simple chemical changes was gradually extended.

Also in the first decades of the twentieth century the genetic classification of ore deposits was greatly improved. Nearly everyone who worked with ore deposits recognized the need for a systematic arrangement of data in such a way that differences among deposits would be most apparent. Some of the most prominent geologists in the world gathered to discuss this problem; their efforts laid the groundwork for present classification. One symposium (Rickard, 1903) was attended by Waldemar Lindgren, whose endeavors in subsequent years

Waldemar Lindgren

produced the popular form of genetic classification (Lindgren, 1907, 1913, 1922). Lindgren classified deposits according to whether they were products of mechanical or chemical concentration and, if chemical, whether they were deposited from surface waters, from magmas, or within bodies of rock. The major controversy involved the classification of hydrothermal veins, which fit into Lindgren's group of chemical deposits introduced into bodies of rock by igneous processes. Within this group Lindgren included both pyrometasomatic (igneous-metamorphic) and hydrothermal deposits. *Pyrometasomatic deposits* form as high-temperature replacement bodies near the border zones of igneous intrusives. *Hydrothermal deposits*—the ores formed by hot, aqueous solutions—were further subdivided by Lindgren according to temperature and depth of formation. Those formed at great depth and at high temperatures (300° to 500°C) were named *hypothermal;* those formed at intermediate depths and temperatures (150° to 300°C), *mesothermal;* and those formed at shallow depths and relatively low temperatures (50° to 150°C), *epithermal.* By 1933, Lindgren had increased the mesothermal-epithermal temperature boundary from 150° to 200°C, but this change reflected advances in geochemistry rather than a flaw in his classification system. Only two changes in Lindgren's hydrothermal classification have been widely accepted. In 1933 Graton recognized a distinct group of deposits with features that indicate deposition at shallow depths from nearly spent solutions; these he named *telethermal.* Buddington (1935) introduced the *xenothermal* zone, deposits formed at high temperatures and shallow depths. The tendency among American geologists at present is to retain Lindgren's classification but to deny genetic implications to the term "hydrothermal." "Hydrothermal" means simply "hot water," and hydrothermal deposits are not necessarily related to igneous processes.

Proponents of the theory of granitization and of the formation of ore deposits during regional metamorphism have in recent years advanced ideas to explain the origin of various deposits. These geologists reason that during metamorphism the more mobile constituents, including activated meteoric waters and many metallic elements, are "sweated out" of the country rock and migrate along fissures and channels into areas of low pressure and temperature, where the ores are deposited. Even though metamorphism is generally a process of dispersion rather than concentration, the mobile constituents are thought to be concentrated along channelways away from the centers of metamorphism. Thus mobilized, the mineral-bearing metamorphic fluids are believed to act the same as fluids of magmatic origin.

Because sedimentary and residual ores form near and at the surface, where conditions of transportation and deposition may be observed, they are better understood. In spite of this, not all stratabound deposits are sedimentary and the origin of many deposits of this type is controversial (Brown, 1968). Some geologists say these layered deposits are the result of replacement of favorable beds by constituents in ascending hydrothermal fluids. Others say they are of sedimentary origin, and thus are bedded, and that the concentration of ore minerals, unusual in a marine environment, originates from submarine vents, the process of volcanogenesis, or, at certain places, from normal processes of erosion (Oftedahl, 1958).

For many years American economic geologists held the opinion that ore deposits of igneous affiliation are derived from a parent magma differentiated at depth, and that the ore constituents are transported in the uncrystallized fraction. From this uncrystallized fraction is derived the hydrothermal fluid that, mixed with connate or meteoric waters, moves into areas of lower pressure and temperature where structural and environmental conditions are favorable for deposition. Some ores are retained in the magma during differentiation (magmatic segregation deposits); others migrate as magma fractions and fluids. Still other deposits are associated with lavas and related eruptive materials.

In recent years, with the discovery of the Red Sea brines and many detailed and careful studies of isotopes and hydrothermal fluids, the emphasis has changed from igneous to sedimentary and metamorphic processes. Theoretical geochemists have made concerted efforts to examine methods of transportation and deposition of ore minerals. The application of thermodynamics to the problems of ore genesis and mineral stabilities is only now being widely used (Krauskopf, 1967). Also, the role of complexes in transportation is helping to explain problems that were not understood. Laboratory geology is playing a larger role in the understanding of many aspects of ore genesis. Today less emphasis is placed on magmatic processes than was done even a few years ago (Stanton, 1972).

As our understanding of ore genesis slowly changes, so also do field exploration methods. Several types of geochemical sampling (Hawkes and Webb, 1962) and geophysical studies are now standard procedures in exploration. Methods of remote sensing—detection and mapping from aircraft or spacecraft—are gradually becoming better known and may hold promise for the future (Barringer, 1969).

Many of the world's largest and most thoroughly studied mines are still subjects of controversy. Although geology grew around the study of ore deposits, many of the most fundamental problems relating to the transportation and deposition of ore materials remain unsettled. It is more common to find that two or more hypotheses on the origin of a given ore body are being entertained than that its genesis has been satisfactorily unraveled. The debate continues over whether the most famous gold deposits in the world, in the Witwatersrand area of South Africa, are placer or hydrothermal in origin. Many lead-zinc deposits, such as those of Mount Isa, Australia, or the Tri-State district in the Mississippi Valley, are regarded as hydrothermal deposits by some geologists and by others as sedimentary beds composed of chemical precipitates. Some massive sulfide deposits and some oxide deposits have been described as direct products of magmatic differentiation, as replacement ores introduced either in dilute solutions or as gases, or as having been deposited by hydrothermal springs pouring out on a sea floor. Clearly, many fascinating problems remain to be solved.

GENERAL REFERENCES

Adams, F. D., 1934. Origin and nature of ore deposits, an historical study, *Geol. Soc. Amer. Bull.* 45:375-424.

——, 1938. *The Birth and Development of the Geological Sciences,* Baltimore: Williams and Wilkins, pp. 277-328. Reprint, New York: Dover, 1954.

Crook, Thomas, 1933. *History of the Theory of Ore Deposits,* London: Thomas Murby.

Emmons, S. F., 1904. Theories of ore deposition historically considered, *Geol. Soc. Amer. Bull.* 15:1-28.

Geikie, Sir Archibald, 1897. *The Founders of Geology,* London: Macmillan. Reprint (of 2nd ed.), New York: Dover, 1962.

Graton, L. C., 1941. Ore deposits, in *Geology, 1888-1938: Fiftieth Anniversary Volume,* Geological Society of America.

Mather, K. F., and S. L. Mason, 1939. *A Source Book in Geology,* New York: McGraw-Hill.

Zittel, K. A. von, 1889. *Geschichte der Geologie und Paläontologie,* Munich and Leipzig.

REFERENCES CITED

Agricola, Georgius, 1556. *De Re Metallica.* (Engl. tr. by H. C. Hoover and L. H. Hoover, New York: Dover, 1950.)

Anderson, Charles, 1809. *New Theory of the Formation of Veins by Abraham Gottlob Werner,* Edinburgh.

Breislak. Scipione, 1811. *Introduction alla Geologia,* Milan: Stamperia Reale.

Brown, J. S., ed., 1968. Genesis of stratiform lead-zinc-barite-fluorite deposits (Mississippi Valley type deposits), *Econ. Geol. Monogr.* 3.

Brunner, Joseph, 1801. *Neue Hypothese von Entstehung der Gänge,* Leipzig.

Buddington, A. F., 1935. High-termperature mineral associations at shallow to moderate depths, *Econ. Geol.* 30:205-222.

Cameron, E. M., ed., 1966. *Proceedings of the Symposium on Geochemical Prospecting,* Ottawa: Geological Survey of Canada (paper 66-54).

Charpentier, J. F. W., 1778. *Mineralogische Geographie der Chursächischen Lande,* Leipzig.

——, 1799. *Beobachtung über die Lagerstätte der Erze, hauptsächlich aus den sächsischen Gebirgen,* Leipzig.

Delius, C. T., 1770. *Abhandlung von dem Ursprunge der Gebirge und der darinne befindlichen Erzadern,* Leipzig.

——, 1773. *Anleitungen zur der Bergbaukunst nach ihrer Theorie und Ausübung,* Vienna.

Gardner, Martin, 1957. *Fads and Fallacies in the Name of Science,* New York: Dover.

Gerhard, C. A., 1781. *Versuch einer Geschichte des Mineral-Reichs,* Berlin.

Graton, L. C., 1933. The depth-zones in ore deposition, *Econ. Geol.* 28:513–555.

Hawkes, H. E., and J. S. Webb, 1962. *Geochemistry in Mineral Exploration,* New York and Evanston: Harper and Row.

Henkel, J. F., 1725. *Pyritologia oder Kieshistorie,* Leipzig.

———, 1727. *Mediorum Chymicorum Non Ultimum Conjunctionis Primum Appropriatio, etc.,* Dresden and Leipzig.

Hutton, James, 1788. Theory of the earth, *Roy. Soc. Edinburgh Trans.* 1:209–304.

———, 1795. *Theory of the Earth,* Edinburgh.

Krauskopf, K. B., 1967. *Introduction to Geochemistry,* New York: McGraw-Hill.

Lindgren, Waldemar, 1907. The relation of ore deposition to physical conditions, *Econ. Geol.* 2:105–127.

———, 1913. *Mineral Deposits,* New York: McGraw-Hill, pp. 178–188.

———, 1922. A suggestion for the terminology of certain mineral deposits, *Econ. Geol.* 17:292–294.

———, 1933. *Mineral Deposits,* 4th ed., New York: McGraw-Hill.

Oftedahl, Chr., 1958. A theory of exhalative-sedimentary ores, *Geol. Foren. Stockholm Forh.* 80:1–19.

Oppel, F. W. von, 1749. *Anleitungen zur Markscheidekunst nach ihren Anfangsgründen und Aüsubung kürzlich entworfen,* Dresden.

Playfair, John, 1802. *Illustrations of the Huttonian Theory of the Earth,* Edinburgh.

Rickard, T. A., ed., 1903. Ore deposits, a discussion, *Eng. Mining J.* 75:256–258, 476–479, 594–595. Reprint, New York: Engineering and Mining Journal, 1903.

Sandberger, Fridolin, 1822. *Untersuchungen über Erzgänge,* Wiesbaden.

Spurr, J. E., 1923. *The Ore Magmas,* New York: McGraw-Hill.

Stanton, R. L., 1972. *Ore Petrology,* New York: McGraw-Hill.

Steno, N. S., 1669. *De solido intra Solidum Naturaliter Contento,* Florence.

Werner, A. G., 1791. *Neue Theorie von der Entstehung der Gänge, mit Anwendung auf den Bergbau besonders den frebergischen,* Freiberg.

Wet, J. P. de, ed., 1957. *Methods and Case Histories in Mining Geophysics,* Montreal: Canadian Institute of Mining and Metallurgy, 6th Commonwealth Mining and Metallurgical Congress.

Zimmerman, C. F., 1746. *Unteradischen Beschreibung der Meissnischen Erzgebirges,* Dresden and Leipzig: Obersächische Bergakademie.

2 / The Ore-Bearing Fluids

Four stages may be recognized in the formation of ore deposits. These are: (1) the source and character of the ore-bearing fluids; (2) the source of the ore constituents and how they are obtained in solution; (3) the migration of the ore-bearing fluids; and (4) the manner of deposition.

To understand the localization of ore deposits, it is first necessary to understand the nature of the transporting media, which in all cases are liquids or gases. Whether the ores are directly related to magmas, associated with metamorphic processes, or related to groundwaters and sedimentary processes, they are all intimately associated with the movement of fluids. Although hydrothermal fluids may be studied near the surface of the earth, they are likely to be contaminated with admixed groundwaters and to have changed their characteristics through extensive reaction with the wall rocks along their passageways. Little is known of the deeper ore-bearing fluids; their nature must be inferred either from observations of thermal springs, volcanic gases, and other emanations that are the end products of ore-forming processes, or from studies of the ores themselves and the accompanying gangue.

For the purpose of closer investigation, the ore-bearing fluids are divided into four categories: (1) magmas and magmatic fluids; (2) meteoric waters;

(3) connate waters, and (4) fluids associated with metamorphic processes. Any of these fluids may be hot or cold, deep-seated or near-surface. Moreover, if heated and in the liquid state, each category would be considered a *hydrothermal solution* because this term refers to any hot, watery fluid, without regard to origin. If the fluid is in a gaseous form, it is called *pneumatolytic*. Under high pressures the properties of hydrothermal and pneumatolytic fluids are similar. Furthermore, for supercritical fluids under high pressures the distinction between gases and liquids is meaningless. Supercritical fluids or high-pressure gases are therefore commonly (and logically) regarded as simply hydrothermal.

MAGMA AND MAGMATIC FLUIDS

Magma is "naturally occurring mobile rock material, generated within the earth and capable of intrusion and extrusion. . . ." (A.G.I. *Glossary of Geology and Related Sciences,* 1957). Defined less formally, magma is a silicate melt or a mush of liquid and crystals. Solidification of magma produces igneous rocks, the great variety of which, in addition to their diversity of field relations, suggests that the processes by which magmas are generated, transported, and solidified are highly complex. Most magmas are probably not homogeneous in composition; parts may be rich in ferromagnesian constituents, others in silica, sodium or potassium compounds, volatiles, reactive xenoliths, or other substances. Furthermore, the composition of a magma is thought to be constantly changing due to chemical reactions. Magmas are not static; they are not closed systems in which we should expect constant equilibrium.

As a magma cools, it crystallizes and separates into fractions by the processes of differentiation. Metallic elements are concentrated in certain of these facies and may be so abundant locally that the resulting igneous rock constitutes ore. During differentiation the more mafic parts of the magma are enriched in chromium, nickel, platinum, and, in places, phosphorus and other elements. In contrast, concentrations of tin, zirconium, and thorium are found in the silicic facies. Titanium and iron persist throughout the range of composition and are found in all types of igneous rocks.

If a partly crystallized magma is subjected to stresses, the fluid fraction is squeezed off from the residual crystalline mush. This process, known as *filter pressing* (Daly, 1933), is of value in helping to explain the origin of certain ore deposits. Metallic elements may be concentrated in either the crystalline residual mush or in the more fluid molten fraction. If either of these materials is forced into the surrounding rocks, the process is known as *magmatic injection,* and, if ore is present, the product is known as a *magmatic injection deposit.* At Kiruna, Sweden, for example, an injected liquid rich in iron solidified to form one of the world's large iron ore bodies (Geijer, 1931).

Magmas or magmatic fractions that may solidify as ore are called *ore magmas.* Since ore magmas are melts, and not aqueous solutions, they behave like molten rock. This concept was supported by Fournet in the nineteenth century, and more recently by Spurr (1923) and Farmin (1941). Unfortunately, as a result of indiscriminate application of the theory to all types of ore deposits, this idea was largely discarded. After decades of disfavor, however, the theory has again been called upon to explain the emplacement of certain types of ore deposits; for example, Sales (1954) proposed that an ore magma formed the copper sulfide deposits of the Colorado pipe at Cananea, Mexico. Figure 2-1 shows a most spectacular and conclusive example of an ore magma in northern Chile; it is a shallow intrusive and "lava flow" of almost pure magnetite-hematite with minor amounts of apatite (Park, 1961; Ruiz-Fuller, 1965; Rogers 1968). The idea of ore magmas appears to be well established, and there is no theoretical or practical reason why metallic facies of magmas should not exist. The only remaining controversies involve the question of the extent of deposits formed by ore magmas. In short, how important are ore magmas, and do they include sulfide melts and silica melts as well as oxide melts?

The process of crystallization, including differentiation and crystal settling, gradually increases the concentration of the more volatile and fugitive constituents in a magma if the substances have no means of escape. The lighter, volatile fractions, plus the compounds that crystallize at lower temperatures

Figure 2-1
Vertical section in the Laco Sur magnetite-hematite deposit, Chile. The upper surface of the "flow" is about three feet above the hammer. (From Park, 1961.)

than the bulk of the magma, accumulate near the top of the magma chamber. These volatiles and materials of low freezing point are the mother liquors of pegmatites and the hydrothermal or pneumatolytic fluids of magmatic affiliation. They include the more mobile elements present in small but essential amounts in all magmas and probably all rocks.

The mobile elements play an all-important role in the transportation of metals. In general, they are elements of low atomic weight and small ionic radius, though there are some notable exceptions. These elements decrease the viscosity of magmas, lower the freezing points of minerals, and make possible the development of compounds that would not form in a dry melt. They possess great penetrating powers—a factor of considerable significance in the earth's interior. The role the mobile elements play in ore-transportation fluids can best be inferred from the study of ores and the altered rocks associated with ores, from the igneous rocks, and from observations in areas of volcanic and hydrothermal activities.

Water—as a fluid phase—is quantitatively the principal mobile constituent in all magmas and plays a leading part in the transportation of many ores. Estimates of the amounts of water in magmas range from one to eight percent. These estimates have been reached by considering volcanic and metamorphic phenomena as well as by analyzing the water content of volcanic glasses. They emphasize the great difficulties and the many variables encountered in any study of magmas (Daly, 1933; Gilluly, 1937; Morey, 1938). Other important mobile elements are sulfur, chlorine, flourine, boron, phosphorus, carbon dioxide, and arsenic. Micas, clay minerals, zeolites, and amphiboles contain small amounts of chemically bonded water; tourmaline and axinite contain boron; scapolites contain chlorine; and many other common minerals, such as flourite, apatite, and topaz, furnish evidence for the presence of a wide range of readily volatile constituents. Many ores and gangue minerals have trapped liquids and gases, some of which are apparently primary in origin and consequently preserve the mobile materials for observation.

In a series of experiments of particular interest to economic geologists, Goranson (1931) showed that the concentration of the volatile fractions increases as the differentiation of granitic magmas proceeds. He concluded that these fractions, rich in water, have definite solubility limits under specific conditions of temperature and pressure, beyond which they will constitute a separate phase of the magma. Smith (1948) continued the same type of experimentation by slowly cooling an artificial granitic magma that contained an initial two percent of water. At one point on the cooling curve, the magma separated into two immiscible liquids, one of which was mostly water.

Certain mobile constituents, such as chlorine, are most abundant, though not restricted to the mafic differentiates, whereas boron and flourine are most abundant in silicic fractions. Sulfur is one of the most dominant and widespread constituents of ore bodies of igneous affiliation, and elemental sulfur is a common product of volcanic emanations. In spite of its abundance, the role of sulfur (and related elements such as arsenic) in the transportation and precipitation of the metals is but little understood (White, 1968). The sulfide ion concentration in many places is thought to be of fundamental importance in determining the mineral composition of the ores; it may well prove to be one of the controlling factors in both the migration and precipitation of ores.

As crystals grow they may trap some of the gases or liquids in the magma from which they crystallize. If the crystals form at high temperature in a fluid medium, subsequent cooling will permit the fluid to be trapped in the crystals as inclusions, either a gas or two-phase liquid-gas. If the mineral develops below the boiling point of the fluids, the vacuole will be full, or nearly full, of liquid. The assumption is made that the contents of these vacuoles were deposited from the original ore-bearing fluid, and thus offer a clue to the composition and physical state of the ore solutions. By heating the mineral until the liquid expands to fill the vacuole or until a single phase is formed (see Chap. 7), the minimum temperature of the original fluid can be estimated. To determine the composition of the ore-bearing fluids, the minute vacuoles are opened and direct tests are made on the liquid contents. Newhouse (1932) found that sodium and chlorine are the most abundant dissolved substances in liquid inclusions. Many later studies have confirmed this conclusion. If the inclusions truly represent the ore-bearing fluids, as thought, then halides possibly play a much larger role in ore genesis than is ordinarily attributed to them.

Smith (1954) concluded that the compositions of liquid inclusions indicate two classes of ore-containing fluids, one a hydrous silicate melt and the other a watery solution containing abundant dissolved salts or dissolved carbon dioxide. The dissolved salts, as much as 30 percent of the liquids by weight, are primarily chlorides, sulfates, and carbonates of sodium, potassium, and calcium. Smith concluded further that whether the mineralizing solutions were above or below the critical temperature, they must have had the effective densities and solvent properties of liquids because after condensation they nearly filled the vacuoles with liquid. Conversely, some inclusions are all gas, and have no liquid phase. An all-liquid inclusion would be formed in a hydrothermal medium, and an all-gas inclusion would be formed in a pneumatolytic medium. Since both types of inclusions are found, as well as all intermediate combinations of gas and liquid, it is reasonable to conclude that gangue and ore-bearing fluids range from strictly gaseous to strictly liquid (Yermakov, 1957).

Many geologists and geochemists have attempted to estimate concentrations of dissolved metals in hydrothermal fluids, with widely varying results. Direct measurements of metals contained in fluid inclusions generally indicate that the ore-forming solutions contain low concentrations of dissolved metals. However, recent analyses based upon a neutron activation technique have detected concentrations of copper, manganese, and zinc in liquid inclusions from samples of quartz and fluorite that exceed 100 parts per million (ppm) (Czamanske *et al.,* 1963). Such modern techniques of analysis may vastly modify the data on fluid inclusions.

In an interesting discussion of hydrothermal areas being developed as sources of energy, Ellis (1970) states that drilling permits detailed observations of deep-water temperatures and pressures. Concentrations of metals such as lead, zinc, copper, manganese, and iron in high-temperature waters are related directly to salinity. Natural hot brines, such as those of the Salton Sea area, California, contain unusually high concentrations of heavy metals and may produce metal-rich scales in drill pipes. In contrast, most volcanic area thermal waters are dilute salt solutions at 200-300°C with very low concentrations of base metals, silver, and gold. Nevertheless, precipitates from these waters are known to contain a low percentage of antimony, ore grade concentrations of silver and gold, and higher amounts of arsenic, mercury, and thallium. This is a common association in near-surface deposits accompanying Tertiary volcanic activities in the western United States and New Zealand.

For many years economic geologists were unable to explain the very large amounts of watery fluid required to transport the extremely insoluble metallic sulfides. In recent years experimental work with complex compounds, especially those of the chlorides and sulfides, has enabled geologists to offer reasonable explanations of how the metals can be transported in geologically accepted amounts of solutions. There is no longer any doubt but that complexes are adequate to explain this problem.

Calculations on the amounts of water or steam required to transport the metals in individual ore deposits indicate that conveyance as simple dissolved solids would require unrealistic volumes of fluids. The solubility of mercuric sulfide is an extreme example of this fact. The true solubility of HgS at high temperatures is unknown, but between room temperature and 200°C, it varies from about 10^{-15} to 10^{-23} moles/liter, depending upon the pH of the solution (Czamanske, 1959). The solubility increases with increased temperature and with increased acidity; that is, the highest figure given, 10^{-15} moles/liter, refers to a solution at 200°C and a pH of 4 to 5. Since most cinnabar deposits form below 200°C, this temperature should be reasonable for solubility calculations. If HgS were carried in true solution, at 200°C and pH 4, one million times the annual volume of water flowing from the Hudson River would be required to deposit a single ton of cinnabar, provided all the dissolved material could be removed from solution at the site of deposition! At a still lower pH or at a higher temperature, or both, the amount of HgS carried in solution would increase by many orders of magnitude, but would never reach realistic proportions. The mercury must, of course, exist in some form other than a simple solution of HgS in water. Although transportation of mercury as a volatile chloride, a soluble sulfide complex, or a metal vapor will explain the concentrations of HgS found in nature, these mechanisms do not explain occurrences of all metals.

Weissberg (1969) believes that the amount of water present in New Zealand thermal areas is sufficient to account for the formation of many rich gold-silver deposits. The springs have existed for many thousands of years, and even though they carry only minute amounts of metals, near-surface concentration is sufficient to explain many economic deposits at shallow depths.

The formation of complex ions—charged particles consisting of several atoms—increases the solubility of some metals by many orders of magnitude (Barton, 1959). Simple solubilities are measured for the number of single ions or common radicals that go into solution in a given amount of water. However, in the presence of other atoms with which the ion can coordinate, the amount of metal that can enter the solution may well surpass its simple solubility limit; that is, the complex ions may be many times more soluble than the simple ions. Thus, mercury may remain in solution as HgS_2^{-2}, or perhaps as $HgS_2(H_2O)_n^{-2}$, $HHgS_2^-$, and so forth. In any one of these forms the mercury would be in sufficient concentration to account for its transportation in ore-forming solutions of reasonable volumes. Many metals combine as comparatively stable complex ions in the form of sulfides, polysulfides, hydrosulfides, halides, carbonates, hydroxides, oxides, sulfates, and others. The mineral species deposited will depend upon the temperature, pressure, and ratios of ions rather than on the complex by which the metal was transported. Hence, there is no need for a special complex for each mineral formed; a simple sulfide may be deposited from a solution containing multiatomic complex ions. Studies of complex ions have been encouraging and may explain a number of problems concerning ore-bearing fluids, but such ions cannot be considered the panacea for all enigmas of ore genesis; at best, they provide only some of the answers. Nevertheless, the presence of metals as complex ions in solution is especially attractive in light of the general sequence of mineral deposition for most sulfide ores, which is in strong accordance with the relative stabilities of complex ions, but is essentially the reverse of that predicted from simple solubilities (see Chap. 6).

Barnard and Christopher (1966) synthesized chalcopyrite, galena, and sphalerite in the laboratory, using chloride complexes. Because chloride complexes are quite soluble, ores possibly migrate as metallic chlorides and, as Helgeson (1964) points out, the evaluation of different lines of evidence indicates that chloride complexes may be the most important factor in the transport and deposition of ore-forming metals. He showed that the relative stabilities of the chloride complexes of the metals are in the sequence

$$Cu^{++} \quad Zn^{++} \quad Pb^{++} \quad Ag^{++} \quad Hg^{++},$$

which agrees with the commonly observed sequence of deposition of the ore minerals. However, Krauskopf (1967, p. 503) states that the chloride complexes of silver, copper, and mercury do not have the necessary stability to overcome the insolubility of the sulfides, and that in the presence of sulfur the sulfide minerals would be difficult to dissolve.

Some geochemists emphasize the possible role of complexes other than those of chlorine and sulfur. Tyurin (1963) considers the thiosulfate complexes,

$M(S_2O_3)_m{}^{-n}$, to be the most probable form of transportation for metals. He says that the solubility of these complexes is very high; that both sulfide-sulfur and metals would be transported simultaneously; that the complexes are stable between pH 5 and pH 10, and are most stable in neutral or weakly alkaline solution; that changes in pH and Eh may cause precipitation of the complexes; and that the thiosulfate complexes become unstable in a series that corresponds with the zonal distribution and the paragenetic association established by Emmons. Shcherbina (1964) argues against the presence of thiosulfate ions because thiosulfate compounds are unknown among minerals and the S_2O_3 anion could not be detected in either the water of hot springs or in gaseous-liquid inclusions.

Krauskopf (1967) states that a study of hydrosulfide complexes may provide an explanation of how metals can be transported in sulfur-bearing solutions. The main obstacle to metal transport with this complex is the high concentration of H_2S and HS^- required to maintain stability.

Geochemists have argued that the primitive ore-bearing fluids, rich in volatiles, are initially above the critical temperature for water (Niggli, 1929; Fenner, 1933; Bowen, 1933). Under these conditions the ions are closely packed, and, in terms of density, gases behave like liquids; gases may therefore carry considerable amounts of metallic elements. Krauskopf (1957, 1959) demonstrated that gas transport does not adequately explain all metal deposits, but he also showed that many common metals are present in sufficient quantities in magmatic vapors, especially as chlorides, to account for the formation of certain ore deposits. His data however do not make evident the importance of volatility in relation to solubility in little-known supercritical solutions. In addition to simple transport as a gas, it seems likely that solubility effects are significant (Morey, 1957), especially in light of the common-ion effect and the probable importance of complex ions.

Even for deposits that have an obvious igneous affiliation a question exists as to the ultimate or immediate source of the metals. Many plutons are without associated ore deposits; this is generally interpreted to mean that when ore *is* associated with a pluton, it is from magma and not directly from the pluton. In part, the problem of ore genesis is related to the old enigma of where and how magmas originate. Deep in the earth's mantle? Just below the Moho? Within melted sediments? This seemingly intangible question does however have a more practical counterpart, namely, the question of whether the metals in late magmatic fluids are concentrated by simple fractionation processes, or selectively leached from the country rocks or previously solidified igneous rocks through which the ascending fluids must pass. Some geologists have suggested that deuteric alteration of ferromagnesian minerals may release metallic ions in sufficient quantities to form the ore deposits along igneous contacts. The evidence for such a mechanism is especially convincing in the Iron Springs district of southwestern Utah (see Chap. 11), where replacement orebodies of magnetite and hematite exist in limestones that border quartz monzonite porphyry laccoliths (Mackin, 1954; Mackin and Ingerson, 1960). The iron was originally incorporated in hornblende and biotite that crystallized at depth. After intrusion of the magma, the outermost shells of the laccoliths solidified rapidly, undergoing no deuteric alteration; however, the interiors of the plutons contained concentrations of volatiles that leached iron from the nearly rigid crystal mush of ferromagnesian minerals; the volatiles then migrated outward along tension cracks through the peripheral shells. These tension joints are encrusted with magnetite and bordered by iron-deficient quartz monzonite. Exposures are continuous from the deep interior of the laccoliths, through the zone where iron minerals were altered, along magnetite-filled fractures that cut the unaltered border zone of the laccoliths, and into the replacement orebodies at the contacts. The replacement deposits are adjacent to the laccoliths only where there are favorable tension joints. The "deuteric release" hypothesis for the origin of ore deposits is a special case of lateral secretion, the theory proposed by Charpentier in the eighteenth century.

The fact that leaching by deuteric solutions is one mechanism for supplying metals to hydrothermal solutions raises a question that may be fundamental to regional exploration: should we expect the igneous rock associated with ore deposits to be enriched, normal, or deficient in the metals? If the ores are associated with special magmas enriched in the ore metals, we should expect a slight concentration of these ions in the rock minerals; that is, the associated pluton should give anomalously high assays of the metals. If deuteric alteration has leached the metals from the igneous rock, or if favorable magmatic differentiation has selectively concentrated metals in the latest fluid fraction, we should expect the rock to be anomalously low in these components (Ingerson, 1954). Future geochemical studies may ultimately resolve this problem; at present the evidence is conflicting and can be used to support either hypotheses—indeed, both mechanisms are probably operative. Perhaps the processes involved depend upon the depth of intrusion rather than the original composition of the magma. If this were true, injections to shallow depths would favor deuteric activities.

The geological definition of "acidic" and "alkaline" merits discussion. Below pH 7 (neutral), fluids are acidic; above pH 7, they are alkaline, or basic. Unfortunately, geologists have used the terms "acidic," "basic," and "alkaline" in other ways for many years. As a result, the terms "acidic," "basic," and "alkaline" in the literature refer to categories of igneous rocks and to hypothetical reactions (that may not take place) as well as by pH. A few field geologists still categorize materials containing fluorite or barite as acidic, and regard hot springs as alkaline if they have a high calcium or sodium content (even though they may be acidic in terms of pH). Moreover, silicic igneous rocks have been regarded by some geologists as acidic on the erroneous premise that the silica will react with water to form silicic acid, H_4SiO_4, and produce a solution of low pH. Some of these same "acidic" rocks are also called "calc-alkaline" because of the presence of lime, potash, and soda. As a result of this confused chemistry, we have a whole category of igneous rocks with the distinction of being both "acidic" and "alkaline," or "calc-alkaline". The use of this terminology is incorrect; although it must be understood when reading older literature, there is no excuse for continuing it. Geologists should conform to the general chemical definition of acids and bases. An acid is a substance that gives free hydrogen (or hydronium) ions when dissolved in water; a base is a substance that gives free hydroxyl ions when dissolved in water ("alkaline" is essentially synonymous with "basic").

The pH of a solution varies with changes in temperature, pressure, and dissolved solids (Barton, 1959) because the pH is a function of the dissociation constant for water; since this constant varies, the acidity of a solution also varies. The dissociation constant increases slightly with increases in pressure, causing the pH of a neutral solution to drop to slightly less than 7 (Owen and Brinkley, 1941). Conversely, the addition of a dissolved salt brings about a slight decrease in the dissociation constant. The effects of pressure and dissolved solids are minor, amounting to a few tenths of a pH unit at most. Changes in temperature, however, may be highly significant. A pH of 7 is neutral only at about 24°C; at 200°C, neutral pH is about 5.6 (Noyes *et al.*, 1907; Ackermann, 1958). According to Helgeson, the pH of hydrothermal solutions does not deviate markedly from neutrality; hydrothermal solutions that contain high concentrations of chloride probably become weakly acidic as the temperature is decreased from the supercritical region for a given bulk composition (Helgeson, 1964, p. 83–85). The best field evidence and laboratory observations seem to indicate that ore-bearing fluids are nearly neutral and that strong acids and bases are exceptional, though they do exist under special circumstances. Weissberg (1969) has shown that the waters of several thermal spring areas in New Zealand are close to neutral. These waters transport minute amounts of gold, silver, arsenic, antimony, mercury, and thallium, which are concentrated in the muds and sinters. Trace amounts of lead, zinc, and copper have been found in cores obtained from deeper drill holes.

Most crustal rocks are composed either of silicate or carbonate minerals, both of which react with acidic solutions to form bases (the older terminology refers to the silica-rich tectosilicates as "acid" minerals and siliceous rocks as "acidic"). Owing to hydrolysis, any solution in contact with silicate or carbonate minerals will eventually become alkaline. Thus, it is logical to assume that ore-bearing fluids are nearly neutral or basic—at least at pressures close to one atmosphere—otherwise they would react immediately with the wall rocks. Such is the case with ordinary groundwaters, many of which are slightly acidic at the surface and become neutral or alkaline at depth. Yet ore deposition might be due to just such a chemical change. Ore-bearing fluids may be acidic at first; deposition of metals that are soluble in an acid but not in a base would occur when the solutions became neutralized.

The pH of ore-forming fluids can be calculated from thermodynamic considerations if the equilibrium assemblage of minerals is well established. Such calculations have recently become well known and should help in the development of a sound theory of ore-fluid chemistry. Preliminary studies of some common mineral assemblages support the thesis that ore-forming fluids were neutral or slightly alkaline when the ores were being deposited (Barnes and Kullerud, 1961; Krauskopf, 1967; Meyer and Hemley, 1967, Barton and Skinner, 1967; Barnes and Czamanske, 1967; White, 1967). Data on liquid inclusions also suggest that ore solutions are generally neutral or nearly neutral (Gushkin and Prikhid'ko, 1952; Newhouse, 1932; Barton, 1959).

In spite of an abundance of alkaline waters, fumaroles and hot springs may yield acidic, neutral, or alkaline solutions (Zies, 1929; Allen and Zies, 1923; Brannock et al., 1948; White, 1955). The chemical character of fluids near the surface is not a reliable indicator of chemical properties at greater depths because of contamination by surface waters and wall rocks and changes in temperatures and pressures. Specifically, any sulfur present may be oxidized near the surface, thus forming sulfuric acid and lowering the pH.

The acidity of fluids that have traveled long distances through siliceous or calcareous rocks, which are materials that should have raised the pH to neutral or above, is difficult to explain. As these fluids migrate, they react with the wall rocks to form minerals that are stable in an acidic environment. These minerals thus constitute an insulating layer along the channels, and later fluids are protected from the unaltered wall rocks. Accordingly, the solutions can travel long distances with a minimum of reaction. The acidity may be caused by sulfuric acid, which is found in some hydrothermal fluids, and may be generated by the oxidizing reaction of water on sulfur and sulfides carried in the original fluid or present in the rocks through which the fluid moves. The oxidation of sulfur and sulfide minerals, with the production of sulfuric acid, is a widely recognized phenomenon in near-surface environments—especially in the presence of oxygen and water.

Certain minerals form in acidic environments but have a polymorph for less acidic or alkaline conditions. For example, as a result of laboratory experiments, the presence of marcasite is believed to indicate deposition has been substantiated by field observation of marcasite developing in natural springs of low pH. (However, marcasite may form even when partial pressure of S_2 is less than that favorable for the growth of pyrite.)

METEORIC WATERS

Water from the atmosphere is called *meteoric water,* and is especially important in supergene processes. As it sinks into the earth it gradually assumes the temperature of the enclosing rocks; water temperature increases with the depth of circulation. Although exceptions exist, the content of dissolved mineral substances generally increases as water temperature increases. Descending waters gradually assume equilibrium with the enclosing rocks; ascending fluids also tend to reach equilibrium with their environment. Hence, under given conditions the compositions of descending and ascending waters approach each other; in the upper parts of the crust, where these waters can be studied, they are difficult or impossible to distinguish. Heated meteoric waters containing minor amounts of magmatic fluids would be difficult to recognize (since the magmatic fraction would not be detectable), and would probably be regarded as simply meteoric.

Meteoric waters are believed to contain the dominant crustal elements, such as sodium, calcium, magnesium, and the sulfate and carbonate radicals. The more fugitive elements, such as boron and fluorine, are characteristic of juvenile waters (Clarke, 1924). The suggestion has been made that meteoric waters contain lower percentages of the heavy isotopes of sulfur and oxygen than do magmatic waters (Ault, 1959; Ingerson, 1954; Rankama, 1954; White, 1957b; Clayton and Epstein, 1961).

Water trapped in sediments at the time they were deposited is known as *connate water*, a term first applied to the brines flowing into the deeper levels of the Michigan copper mines (Lane, 1908). Connate waters are actually fossil waters. They are widely observed especially in oil field exploration (as the salty edgewaters of many oil accumulations), and a great deal of study has been devoted to them. Most connate waters are abnormally rich in sodium and chlorine, but they also contain considerable amounts of calcium, magnesium, and bicarbonate, and may contain strontium, barium, and nitrogen compounds (White, 1947b, 1968).

Connate waters have little direct relationship to ore-bearing fluids, except where the containing strata are undergoing metamorphism. Here the waters may become heated and activated. When activated, they may become strong solvents of the metals, since they contain large amounts of chlorine. They thus are one source of hydrothermal fluids.

Mine Waters

The only comprehensive studies of mine waters in recent years have been those done on obviously meteoric waters, and little information has been obtained concerning the nature of the ore-bearing fluids (Sato and Mooney, 1960). Although pore moisture remains, groundwater gradually decreases at depth—many deep mines are dry in their lower levels. Most studies of mine waters concentrate on the corrosive effects of these waters on pumps and pipes. Most present mine waters have no relation to the fluids that deposited the ore. Exceptions to this are found in areas of recent volcanism. For example, the hot water in the lower levels of the Comstock Lode, Nevada, may have been in part of volcanic or metamorphic origin. Samples of water from the Comstock Lode, described by Bastin in 1922, apparently were contaminated by sulfuric acid that descended from the zone of weathering. Many mine openings at depths of 2700 feet and less had large flows of water with temperatures up to 70°C. Two analysis of these hot waters are reproduced in Table 2-1. From these analyses it is readily seen that the waters are dominated by sodium and calcium sulfates.

The presence of sodium carbonate waters is commonly reported from deep mines in areas of recent volcanic activity (Lindgren, 1933, p. 53); many geologists consider such waters to have a direct connection with magmatic sources. A sodium carbonate water of this type, from the Homestake gold mine in South Dakota (which is not a region of recent volcanic activity), was described by Noble (1950). The water was taken from a crevice penetrated by a diamond drill hole at a depth of 2,300 feet. Considering the depth of the crevice and the unique proportions of dissolved constituents, Noble reasoned that the water could not be contaminated appreciably by surface waters. Although sodium was abundant in the water, it is practically absent from the country rocks and the ordinary groundwaters in the mine. Noble concluded that the

Table 2-1
Analyses of hot waters from the Comstock lode, Nevada (in parts per million).

Constituent	Union Consolidated mine, 2650-foot level	C. and C. shaft, 2250-foot level
Na	145.0	131.0
K	8.4	53.4
Ca	204.0	100.3
Mg	4.0	5.9
Al	–	1.3
Fe	0.0	6.4
SO_4	752.0	542.6
Cl	12.0	19.0
CO_3	9.3	20.5
HCO_3	0.0	–
OH	7.5	–
SiO_2	61.0	133.4
TOTALS	1203.2	1013.8

Source: Bastin, 1922.

Table 2-2
Analysis of water from the Homestake mine, South Dakota (in parts per million).

Constituent	ppm
Na	428
K	60
Ca	13
Mg	11
SiO_2	11
Al_2O_3 and Fe_2O_3	4
CO_2	392
SO_4	351
S_2O_3	20
H_2S	15
Cl	4
TOTAL	1309

Source: Noble, 1950.

water and its dissolved constituents were probably the products of hydro-thermal mineralization. Table 2-2 presents Noble's analysis of the water. Noble's findings suggest that a comprehensive study of deep mine waters by modern methods might provide valuable information concerning the origin and composition of ore-bearing fluids.

Thermal Springs

Thermal springs, solfataras, and fumaroles have been studied intensively by many geologists, and a growing store of information is applicable to the prob-lems of ore solutions. Among the most recent studies of these phenomena are those of White (1957a, 1957b, 1968), who applied modern geochemical prin-ciples and techniques to the study of Steamboat Springs, Nevada; Salton Sea, California; other areas of thermal activity in the western United States; Provi-dencia, Zacatecas, Mexico, and the recently discovered metal-containing sedi-ments in the Red Sea. On the basis of isotope analyses, whereby the waters of various thermal springs were compared to surface waters of the surrounding areas, White decided that the contribution of direct magmatic material was probably insignificant, although he notes that none of the districts studied is completely understood. He draws five main conclusions: (1) Most ore deposits are formed by complex rather than simple end-member processes. (2) The ore-bearing fluids of base metal deposits are Na-Ca-Cl brines. (3) Brines of similar major element composition may form in at least four ways—magmaticly, con-nately, by solution of evaporates of any dilute water, or by membrane con-centration of dilute meteoric water. (4) In at least three and perhaps five of the districts studied, the ratio of total dissolved metals to dissolved sulfides in the ore fluids was very high since the metals were transported as chloride complexes in the presence of small amounts of sulfide. (5) The density of the ore-bearing brines is normally higher than that of near-surface waters (White, 1968).

Water in regions of recent volcanic activity is commonly characterized by an abundance of sodium chloride, and apparently grades into acidic sulfate-chloride water. Other waters in volcanic areas contain sodium bicarbonate, acid sulfate, and calcium bicarbonate, or mixed sodium and calcium bicar-bonate. White considers the sodium-chloride waters to be the most closely related to magmatic emanations; the other types of thermal waters are sec-ondary products resulting from reactions with wall rocks and changes in the physical environment. White's analyses of sodium chloride waters are given in Table 2-3.

According to White, the volcanic sodium-chloride waters are distinguished from connate and ocean waters by the relatively high content of lithium,

Table 2-3
Analyses of sodium chloride hot spring waters (in parts per million).

	Steamboat Springs, Nevada	Morgan Springs, Tehama County, California	Norris Basin, Yellowstone Park, Wyoming	Upper Basin, Yellowstone Park, Wyoming	Wairakei, New Zealand
Temp. °C	89.2	95.4	84	94.5	>100
pH	7.9	7.83	7.45	8.69	8.6
SiO_2	293	233	529	321	386
Fe	–	–	–	Trace	–
Al	–	–	–	0	–
Ca	5.0	79	5.8	4	26
Mg	0.8	0.8	0.2	Trace	<0.1
Sr	1	10	–	–	–
Na	653	1398	439	453	1130
K	71	196	74	17	146
LI	7.6	9.2	8.4	–	12.2
NH_4	<1	<1	0.1	0	0.9
As	2.7	2.2	3.1	–	–
Sb	0.4	0.0	0.1	–	–
CO_3	0	0	0	66	–
HCO_3	305	52	27	466	35
SO_4	100	79	38	15	35
Cl	865	2437	744	307	1927
F	1.8	1.5	4.9	21.5	6.2
Br	0.2	0.8	0.1	–	–
I	0.1	<0.1	<0.1	–	–
B	49	88	11.5	3.7	26
S_2O_3	–	–	–	2	–
H_2S	4.7	0.7	0	0	1.1
Co_2	–	–	–	–	11
TOTALS	2360	4578	1885	1676	3742

Source: White, 1957a.

fluorine, silica, boron, sulfur, and carbon dioxide in the former, compared to a high content of calcium and magnesium in the latter. Nonvolatile compounds are variably soluble in steam at high pressure (high density steam has solvent properties similar to those of liquid water). In volcanic sodium-chloride waters a high ratio of lithium to sodium and potassium indicates differentiation in a magma and suggests that the alkalies were transported as alkali halides dissolved in liquid or dense vapor. White believes the juvenile fluid is greatly diluted by deeply circulating meteoric water, mixing at a depth of approximately two miles. Where circulation of meteoric water is shallow, the halide-bearing vapors reach low-pressure regions, expand, and precipitate the nonvolatile substances; such a mechanism would remove much of the sodium chloride from the vapor system and produce one of the modified calcium-bicarbonate waters or acidic sulfate waters.

The Valley of Ten Thousand Smokes, an area of fumarolic activity that became active in 1912, provides further information on the nature of hydrothermal fluids (Zies, 1929). The fumaroles formed at the surface of a thick accumulation of hot, rhyolitic pyroclastics. Although this type of igneous hydrothermal system differs from magma crystallizing at depth, it does provide useful data. It is assumed that most of the fumarolic waters were originally meteoric; elements that are not common to rain water were probably derived from the cooling mass of pyroclastic debris. The exhalations of the fumaroles were more than 99 percent steam, at temperatures up to 650°C. Enough HCl, HF, and H_2S were dissolved in the steam to make the exhalations acidic at the surface. Different mineral assemblages formed around the vents as the temperature dropped, the early phase being characterized by magnetite, the later phase by galena and sphalerite. The incrustations contained iron, lead, zinc, molybdenum, copper, arsenic, antimony, tin, silver, nickel, cobalt, thallium, and bismuth; these metals were combined with sulfur, oxygen, fluorine, chlorine, selenium, and tellurium. Zies pointed out that each of these elements would form a volatile compound and that each was transported in the gaseous state.

Recent work by Hewett and his associates (Hewett *et al.*, 1960, 1963) indicates that many hot springs are at least partly volcanic in origin. They deposit manganese oxides and minute but detectable amounts of tungsten, as well as other elements, such as boron, strontium, and fluorine.

FLUIDS ASSOCIATED WITH METAMORPHIC PROCESSES

Under favorable circumstances, connate and meteoric waters enclosed in rocks buried below the surface of the earth may be set in motion and made chemically reactive by heat and pressure accompanying magmatic intrusion or regional metamorphism (Shand, 1943). These are the so-called *metamorphic waters* that many geologists believe are active ore carriers.

It is ordinarily and widely believed that in regional metamorphism metallic and volatile constituents are dispersed rather than concentrated (Taupitz,

1954). In support of this contention, Eskola showed that the palingenetic granites (granites formed by the reconstitution of other rocks) are notably free of metallic constituents. He proposed that they be distinguished from magmatic granites by their general lack of ore deposits (Eskola, 1932). Owing to such reasoning, mining geologists in general have been slow to accept the idea that ore deposits may be produced by regional metamorphic processes. Nevertheless, none of the phenomena usually taken to denote magmatic origin are incompatible with a metamorphic origin. A growing number of geologists attach significance to metamorphic processes, and especially to the action of waters set in motion from buried sediments either by heat from a cooling magma or by regional metamorphism. (Kittl, 1960; Marmo, 1960; Vokes, 1969). For example, in the controversy concerning the origin of the Zambian copper belt in Africa, Garlick (1953) and Davis (1954) believe the ore-bearing fluids that formed these deposits were the same as those from which the enclosing sediments were deposited; that is, they consider the ores to be syngenetic. In contrast, Gray (1959) argues that connate or meteoric waters were activated during regional metamorphism, leaching the ores from the surrounding rocks and concentrating them in areas of reduced pressure or temperature or in areas of reactive wall rock. Sales (1960) disagrees with both ideas, stating that the ore-bearing fluids are directly related to igneous magmas.

Granitization is the process by which nongranitic rocks are converted to rocks of granitic character without passing through a magmatic or liquid stage (Guimarães, 1947; Read, 1948, 1954). During granitization and its associated processes, the volatile and mobile constituents are "activated." They are forced from the rock and migrate toward cooler and, in general, less deformed regions. These volatile and mobile elements include most of the metallic constituents, but water constitutes most of the fluid. The metal-bearing metamorphic water would be the same whether activated by nearby magma or regional metamorphism. Trace-element studies of minerals from different metamorphic facies indicate that certain metals are selectively released during regional metamorphism (DeVore, 1955). Accompanying tectonic processes may provide avenues along which the metals and metamorphic fluids travel, producing a hydrothermal system. The water and its metallic content are believed to move down the metamorphic gradient, in advance of either the regional metamorphism or the intruding magma.

The release of metallic elements during granitization should be verifiable in the field, though the evidence to date is inconclusive. Geochemical studies have not solved the problem; they merely indicate that the original compositions of the sedimentary rocks are highly variable in their content of metals. For example, in a study of the metal content of several fine-grained rocks in New Hampshire Shaw (1954) decided that during regional metamorphism, nickel and copper showed a poorly defined tendency to decrease, whereas elements such as lithium, strontium, and lead increased. Much more careful work is needed before valid generalizations can be drawn. Most evidence advanced in support of a relationship between ore genesis and metamorphic processes is equally applicable to magmatic processes.

The processes of metamorphism are not a convincing explanation of the origin of magmatic segregation deposits, nor are they adequate to explain the introduction of ores into igneous masses, as in some disseminated copper deposits. However, migration of water and the mobile elements in a rock would be normal during either regional metamorphism or intrusion.

One attractive theory is that both igneous and metamorphic processes play significant roles in the activation of volatile constituents in the rocks. Connate waters and groundwaters commonly contain large amounts of soluble salts. Once heated and set in motion, they would become unusually strong solvents and would tend to remove metals from the rocks. Supporting this view is the fact that inclusions in ore minerals generally contain chlorides, indicating that chlorine was present in many of the ore-bearing fluids. These activated and ore-bearing waters might also have combined with fluids released from a magma.

Brown (1965) says that metals, particularly lead, are concentrated in connate waters that are activated by igneous processes or perhaps by metamorphism. The basis for this assertion is the similarity between the lead isotopes of ocean-bottom sediments and those of the youngest lead ores. He contends that the genesis of lead-bearing ores is a process accompanying marine sedimentation and that the principal ore fluid is essentially connate water concentrated by diagenetic processes. Sawkins (1965), in a discussion of Brown's ideas, points out that the data on oceanic isotopes may be misleading because many nodules of manganese that contain lead are of volcanic origin and were not derived from land waste. This may indicate a similarity between the genesis of the youngest lead deposits and volcanic lead.

Goodspeed (1952) contended that some mineral-bearing fluids were derived from the breakdown of hydrous minerals during granitization. He pointed out that clay minerals, which are abundant in geosynclinal prisms, contain about 14 percent water, and that a change into feldspars must release this water, thus providing a potential source of hydrothermal mineralizing solutions.

A study of the Scott magnetite mine, Sterling Lake district, New York, convinced Hagner and his associates that the source of the ore was the pyroxene-bearing amphibolite host rock. They have shown that the percentage of iron in the mafic silicates and in the total rock decreases with proximity to the ore. Much of the amphibolite was replaced by gneiss and pegmatite; the iron thus released migrated to the ore zone to form magnetite. The process resulted from metasomatism accompanying regional metamorphism and granitization (Hagner, Collins, and Clemency, 1963).

Guimarães (1947) emphasized the part played by igneous intrusions rather than metamorphic processes in activating waters. He argued that the mobile elements would move ahead of a slowly advancing intrusive mass, where conditions of temperature and pressure were favorable. Movement of the mobile elements would persist as long as the intrusive continued to advance or until it solidified and cooled.

As a result of many laboratory experiments and studies with blast furnace slags, Sosman (1950) concluded that an intruding magma, not saturated with water, will produce a gradient of both water pressure and water concentration in the surrounding rocks, such that the direction of decreasing pressure and concentration is toward the intrusive, and not, as is commonly assumed, away from it. This conclusion is reinforced by the phenomenon of thermal transpiration, whereby a gas under constant pressure, in a medium having small pores, travels toward the region of higher temperature. The groundwater in sedimentary rocks intruded by a water-deficient magma should therefore travel with a positive temperature gradient toward the intrusive. This process would result in the incorporation of activated groundwaters into the mineral-bearing fluids associated with the magma. Although this scheme seems plausible, field relationships and paragenetic studies indicate that at least the late-stage pneumatolytic and hydrothermal fluids migrate away from intrusives rather than toward them.

Several geologists in Africa strongly support the theory that metamorphic waters are the active ore carriers. After many years of studying the gold deposits of Rhodesia, Macgregor (1951) suggested that gold and other minerals were released as a result of "migmatitic extraction" from the rocks engulfed by magmatic stoping. The waters, he thought, were of igneous or magmatic origin; the metals, however, originated in the country rock. After solution of the metals, the waters migrated upwards and formed epigenetic deposits (deposits of later origin than the enclosing rocks).

Hydrothermal ore-bearing fluids may be largely made up of groundwater or connate water activated by means of igneous intrusions or metamorphism. Such metamorphic waters would combine with fluids emitted from nearby magma, and the metals could come from either the magma or the country rock.

Some of the problems concerning ore-bearing fluids may be solved by the study of samples collected from deep wells that penetrate active hydrothermal systems. The possibility of generating power from wells drilled in thermal spring areas has promoted great interest. For example, a 5,232-foot well drilled for geothermal power near the Salton Sea in California tapped hot brines that at first were thought to represent diluted magmatic waters from an underlying igneous source (White et al., 1963). Later information indicated that meteoric waters were heated at depth and the contained metals leached from surrounding sediments (White, 1968).

These hydrothermal waters contain anomalous concentrations of heavy metals and have no counterparts among known connate or meteoric waters. The temperature exceeds 270°C and may be as high as 370°C. The waters and their residues contain abnormally high amounts of copper, silver, potassium, lithium, antimony, lead, arsenic, boron, beryllium, bismuth, gallium, and gold. A dark, siliceous deposit that accumulated in the discharge pipe contained about 20 percent copper and two percent silver; an estimated five to eight tons of precipitate formed in the pipe within three months. The country rocks appear to be undergoing 'metamorphism, as shown by a mineral assemblage characteristic of the greenschist facies. Groundwaters that originally occupied the pore spaces were displaced by the heavier ore-bearing fluids.

REFERENCES CITED

Ackerman, Th., 1958. The self-dissociation of water from measurements of molar heats of dissolved electrolytes, *Z. Electrochemie* 62:411–419.

Allen, E. T., J. L. Crenshaw, and H. E. Merwin, 1914. Effect of temperature and acidity in the formation of marcasite and wurtzite: a contribution to the genesis of unstable forms, *Amer. J. Sci.* 38:393–431.

Allen, E. T., and E. G. Zies, 1923. A chemical study of fumaroles of the Katmai region, *Nat. Geog. Soc. Contrib. Tech. Pap.*, Katmai series 1(2).

American Geological Institute, 1972. *Glossary of Geology and Related Sciences*, Washington, D.C.

Ault, W. U., 1959. Isotopic fractionation of sulfur in geochemical processes, in *Researches in Geochemistry*, ed. P. H. Abelson, New York: Wiley.

Barnard, W. M., and P. A. Christopher, 1966. Hydrothermal synthesis of chalcopyrite, *Econ. Geol.* 61:897–902.

Barnes, H. L., and G. K. Czamanske, 1967. Solubilities and transport of ore minerals, in *Geochemistry of Hydrothermal Ore Deposits*, ed. H. L. Barnes, New York: Holt, Rinehart and Winston.

Barnes, H. L., and G. Kullerud, 1961. Equilibria in sulfur-containing aqueous solutions, in the system Fe-S-O, and their correlation during ore deposition, *Econ. Geol.* 56:648–688.

Barton, P. B., Jr., 1959. The chemical environment of ore deposition and the problem of low-temperature ore transport, in *Researches in Geochemistry*, ed. P. H. Abelson, New York: Wiley.

Barton, P. B., and B. J. Skinner, 1967. Sulfide mineral stabilities, in *Geochemistry of Hydrothermal Ore Deposits*, ed. H. L. Barnes, New York: Holt, Rinehart and Winston.

Bastin, E. S., 1922. Bonanza ores of the Comstock Lode, Nevada, *U.S. Geol. Surv. Bull.* 735, pp. 57–63.

Bowen, N. L., 1933. The broader story of magmatic differentiation, briefly told, in *Ore Deposits of the Western States* (Lindgren Vol.), New York: American Institute of Mining and Metallurgical Engineers.

Brannock, W. W., P. F. Fix, V. P. Gianella, and D. E. White, 1948. Preliminary geochemical results at Steamboat Springs, Nevada, *Amer. Geophys. Union Trans.* 29:211–226.

Brown, J. S., 1965. Oceanic lead isotopes and ore genesis, *Econ. Geol.* 60:47–68.

Clarke, F. W., 1924. The data of geochemistry, *U.S. Geol. Surv. Bull.* 779, pp. 63–121.

Clayton, R. N., and S. Epstein, 1961. The use of oxygen isotopes in high-temperature geological thermometry, *J. Geol.* 69:447-452.

Czamanske, G. K., 1959. Sulfide solubility in aqueous solutions, *Econ. Geol.* 54:57-63.

Czamanske, G. K., E. Roedder, and F. C. Burns, 1963. Neutron activation analysis of inclusions for copper, manganese, and zinc, *Science* 140(3565):401-403.

Daly, R. A., 1933. *Igneous Rocks and the Depths of the Earth,* New York: McGraw-Hill.

Davis, G. R., 1954. The origin of the Roan Antelope copper deposit of Northern Rhodesia, *Econ. Geol.* 49:575-615.

DeVore, G. W., 1955. The role of adsorption in the fractionation and distribution of elements, *J. Geol.* 63:159-190.

Ellis, A. J., 1970. Present-day hydrothermal systems and mineral deposition, in *Mining and Petroleum Geology,* vol. 2, 9th Commonwealth Mining and Metallurgical Congress.

Eskola, P., 1932. On the origin of granitic magmas, *Mineral. Petrogr. Mitt.* 42:478.

Farmin, R., 1941. Host-rock inflation by veins and dikes at Grass Valley, California, *Econ. Geol.* 36:143-174.

Fenner, C. N., 1933. Pneumatolytic processes in the formation of minerals and ores, in *Ore Deposits of the Western States* (Lindgren Vol.), New York: American Institute of Mining and Metallurgical Engineers.

Garlick, W. G., 1953. Reflections on prospecting and ore genesis in Northern Rhodesia, *Inst. Mining Metall. London Trans.* 63:9-20, 94-106.

Geijer, P., 1931. The iron ores of the Kiruna type, *Geol. Surv. Sweden Ann. Rept.,* ser. C, no. 367.

Gilluly, J., 1937. The water content of magmas, *Amer. J. Sci.* 233:430-441.

Goodspeed, G. E., 1952. Mineralization related to granitization, *Econ. Geol.* 47:146-168.

Goranson, R. W., 1931. The solubility of water in granite magmas, *Amer. J. Sci.* 222:481-502.

Gray, A., 1959. The future of mineral exploration, *Inst. Mining Metall. London Trans.* 68(pt. 2):23-34.

Guimarães, D., 1947. Mineral deposits of magmatic origin, *Econ. Geol.* 42:721-736.

Gushkin, G. G., and P. L. Prikhid'ko, 1952. Chemical composition, concentration, and pH of liquid inclusions in fluorite, *Zap. Vsesoyuz. Mineral. Obshchestuva* 81:120-126 (*Chem. Abstr.,* 1952, vol. 46, col. 10055).

Hagner, A. F., L. G. Collins, and C. V. Clemency, 1963. Host rock as a source of magnetite ore, Scott mine, Sterling Lake, New York, *Econ. Geol.* 58:730-768.

Helgeson, H. C., 1964. *Complexing and Hydrothermal Ore Deposition,* New York: Macmillan.

Hewett, D. F., and M. Fleischer, 1960. Deposits of the manganese oxides, *Econ. Geol.* 55:1-55.

Hewett, D. F., M. Fleischer, and N. Conklin, 1963. Deposits of the manganese oxides: supplement, *Econ. Geol.* 58:1-51.

Ingerson, E., 1954. Nature of the ore-forming fluids at various stages—a suggested approach, *Econ. Geol.* 49:727-733.

Kittl, E., 1960. Lagerstättenbildung und Mobilisierung in Geosynklinalen, *Abh. Deut. Akad. Wiss. Berlin Kl. Bergb. Hüttenw. Montangeol.,* nr. 1, pp. 321-328.

Krauskopf, K. B., 1957. The heavy metal content of magmatic vapor at 600° C, *Econ. Geol.* 52:786-807.

———, 1959. The use of equilibrium calculations in finding the composition of a magmatic gas phase, in *Researches in Geochemistry,* ed. P. H. Abelson, New York: Wiley.

———, 1967. *Introduction to Geochemistry,* New York: McGraw-Hill.

————, 1971. Introductory talk, in *Geochemistry and Crystallography of Sulphide Minerals in Hydrothermal Deposits* (Joint Symposium Vol.: IMA–IAGOD Mtgs. '70), ed. Y. Takéuchi, Tokyo: Society of Mining Geologists of Japan (spec. issue 2).

Lane, A. C., 1908. Mine waters and their field assay, *Geol. Soc. Amer. Bull.* 19:501–512.

Lindgren, W., 1933. *Mineral Deposits,* 4th ed., New York: McGraw-Hill.

Macgregor, A. M., 1951. The primary source of gold, *S. Afr. J. Sci.* 47:157–161.

Mackin, J. H., 1954. Geology and iron ore deposits of the Granite Mountain area, Iron County, Utah, *U.S. Geol. Surv. Mineral Invest. Field Stud. Map* MF 14.

Mackin, J. H., and E. Ingerson, 1960. An hypothesis for the origin of ore-forming fluid, *U.S. Geol. Surv. Prof. Pap.* 400-B, pp. B1–B2.

Marmo, V., 1960. On the possible genetical relationship between sulphide schists and ore, *21st Int. Geol. Congr. Rept.,* pt. 16, pp. 160–163.

Meyer, C., and J. J. Hemley, 1967. Wall rock alteration, in *Geochemistry of Hydrothermal Ore Deposits,* ed. H. L. Barnes, New York: Holt, Rinehart and Winston.

Morey, G. W., 1938. Water in geological processes, *Carnegie Inst. Wash. Publ.* 501, pp. 49–59.

————, 1957. The solubility of solids in gases, *Econ. Geol.* 52:225–251.

Newhouse, W. H., 1932. The composition of vein solutions as shown by liquid inclusions in minerals, *Econ. Geol.* 27:419–436.

Niggli, P., 1929. *Ore Deposits of Magmatic Origin,* tr. H. C. Boydell, London: Thomas Murby.

Noble, J. A., 1950. Ore mineralization in the Homestake gold mine, Lead, South Dakota, *Geol. Soc. Amer. Bull.* 61:221–252.

Noyes, A. A., Y. Kato, and R. B. Sosman, 1907. Hydrolysis of ammonium acetate and ionization of water at 100°, 156°, 218°, and 306°, *Carnegie Inst. Wash. Publ.* 63, pp. 153–235.

Owen, B. B., and S. R. Brinkley, Jr., 1941. Calculation of the effect of pressure upon equilibria in pure water and in salt solutions, *Chem. Rev.* 29:461–474.

Park, C. F., Jr., 1961. A magnetite "flow" in northern Chile, *Econ. Geol.* 56:431–436.

Rankama, K., 1954. *Isotope Geology,* New York: McGraw-Hill.

Read, H. H., 1948. Granites and granites, *Geol. Soc. Amer. Mem.* 28, pp. 1–19.

————, 1954. Granitization and mineral deposits, *Geol. Mijnbouw* 16:95–99.

Rogers, D. P., 1968. The extrusive iron oxide deposits, "El Laco," Chile, *Econ. Geol.* 63:700.

Ruiz-Fuller, C., 1965. *Geología y Yacimientos Metaliferos de Chile,* Instituto de Investigaciones Geológicas, Chile, pp. 245–247.

Sales, R. H., 1954. Genetic relations between granites, porphyries, and associated copper deposits, *Amer. Inst. Mining Eng. Trans.* 199:499–505.

————, 1960. Critical remarks on the genesis of ore as applied to future mineral explorations, *Econ. Geol.* 55:805–817.

Sato, M., and H. M. Mooney, 1960. The electrochemical mechanism of sulfide self potentials, *Geophysics* 25:226–249.

Sawkins, F. J., 1965. Discussion, oceanic lead isotopes and ore genesis, *Econ. Geol.* 60:1083–1084.

Shand, S. J., 1943. *Eruptive Rocks: Their Genesis, Composition, and Classification,* New York: Wiley.

Shaw, D. M., 1954. Trace elements in pelitic rocks, *Geol. Soc. Amer. Bull.* 65:1151–1182.

Shcherbina, V. V., 1964. Problems of the existence of the thiosulfates in hydrothermal solutions [article in Russian], *Geol. Rud. Mestorozhd. Acad. Sci. U.S.S.R.* 6(3):110–112. (Reviewed by E. A. Alexandrov, *Econ. Geol.* 60:645.)

Skinner, B. J., and P. B. Barton, Jr., 1973. Genesis of mineral deposits, *Ann. Rev. Earth Planet. Sci.* 1:183-211.

Smith, F. G., 1948. Transport and deposition of the non-sulphide vein minerals, III: phase relations at the pegmatitic stage, *Econ. Geol.* 43:535-546.

———, 1954. Composition of vein-forming fluids from inclusion data, *Econ. Geol.* 49:205-210.

Sosman, R. B., 1950. Centripetal genesis of magmatic ore deposits, *Geol. Soc. Amer. Bull.* 61:1505.

Spurr, J. E., 1923. *The Ore Magmas,* New York: McGraw-Hill.

Taupitz, K.-C., 1954. Über Sedimentation, Diagenese, Metamorphose, Magmatismus und die Entstehung der Erzlagerstätten, *Chemie der Erde* 17:104-164.

Tyurin, N. G., 1963. The problem of the composition of hydrothermal solutions [article in Russian], *Geol. Rud. Mestorozhd. Acad. Sci. U.S.S.R.* 5(4):24-42. (Reviewed by E. A. Alexandrov, *Econ. Geol.* 59:734.)

Vokes, F. M., 1969. A review of the metamorphism of sulfide deposits, *Earth-Sci. Rev.* 5:99-143.

Weissberg, B. G., 1969. Gold-silver ore grade precipitates from New Zealand thermal waters, *Econ. Geol.* 64:95-108.

White, D. E., 1955. Thermal springs and epithermal ore deposits, *Econ. Geol. (50th Anniv. Vol.),* pp. 99-154.

———, 1957a. Thermal waters of volcanic origin, *Geol. Soc. Amer. Bull.* 68:1637-1657.

———, 1957b. Magmatic, connate, and metamorphic waters, *Geol. Soc. Amer. Bull.* 68:1659-1682.

———, 1967. Mercury and base-metal deposits with associated thermal and mineral waters, in *Geochemistry of Hydrothermal Ore Deposits,* ed. H. L. Barnes, New York: Holt, Rinehart and Winston.

———, 1968. Environments of generation of some base metal ore deposits, *Econ. Geol.* 63:301-335.

White, D. E., E. T. Anderson, and D. K. Grubbs, 1963. Geothermal brine well: mile-deep drill hole may tap ore-bearing magmatic water and rocks undergoing metamorphism, *Science* 139(3558):919-922.

Yermakov, N. P., 1957. Importance of inclusions in minerals to the theory of ore genesis and study of the mineral forming medium [article in Russian]. Tr. E. A. Alexandrov, 1964. *Int. Geol. Rev.* 3(7):575-585.

Zies, E. G., 1929. The Valley of Ten Thousand Smokes, *Nat. Geog. Soc. Contrib. Tech Pap.,* Katmai series 1(4).

3 / Migration of the Ore-Bearing Fluids

Any hypothesis concerning the genesis of ores must explain the migration of large amounts of ore-bearing fluids, which can be magmas, aqueous solutions, or gases. The movement of fluids underground is as significant in ore genesis as it is in the concentration of oil and gas or in the emplacement of dikes and other intrusive masses. In general, migration is controlled by structure, which determines the avenues of permeability. In detail, however, ore solutions may completely permeate the rocks, going around and through the mineral grains. Knowledge of the paths traveled by the ore solution and the mode of ore emplacement is fundamental to the understanding of genesis of an ore deposit.

MIGRATION OF MAGMA

The manner in which ordinary magmas move through rock has been the subject of much discussion in petrology. Although the reasons for their movement are debatable, they do move, and they generally move upward toward areas of lower pressure. Magmas contain gases under pressure, and any release in pressure will allow the gases to expand. Both the process of expansion and the resultant decrease in specific gravity will cause magmas to move upward. Furthermore, tectonic stresses may squeeze the magma or fractions of magmatic differentiates into overlying or adjacent rocks, starting a movement that will be propagated by gas expansion in the areas of lower pressure. Magma may be injected into overlying rocks or force its way between rock layers by actually breaking rocks apart; in fact it is difficult to envisage any other mechanism of emplacement for many sills and dikes. Some magmas are thought to move by means of stoping—a process whereby the magma works its way upward by engulfing blocks of overlying country rock. Presumably, the stoped blocks sink into the magma chamber and are assimilated at depth. Other magmas are believed to move by a sort of migration of heat and highly mobile fluids, whereby rocks around the upper part of the chamber are slowly melted to form part of the magma. As is true of many geologic phenomena, no single theory can explain the emplacement of all intrusive masses. In certain individual plutons the combined effects of forceful intrusion, stoping, and assimilation are evident (Compton, 1955, 1960).

In ore magmas, where the resultant product is relatively pure ore (such as the magnetite at El Laco, Chile), the molten materials cannot have assimilated much of the wall rocks; consequently, the fluids cannot have migrated by dissolving or stoping their way along. It seems likely that most ore magmas move after differentiation at depth and that subsequent directed tectonic pressures cause them to be injected into the adjacent or overlying rocks.

MIGRATION OF FLUIDS AT DEPTH

Ordinarily, permeability and porosity decrease with depth of burial owing to pressure of overlying rocks and cementing action of mineral-laden waters. The lower limit of freely circulating groundwaters varies considerably, and may lie anywhere from a few feet to several thousand feet below the surface, depending upon the nature of the rock. (Many deep mines extend below this limit and, as a result, are dry and dusty in their lower levels; in fact, at some places it is necessary to pipe water down for drilling.) Because of the lack of permeability, some geologists doubt the ability of large quantities of ascending waters or watery ore-bearing fluids to penetrate dense, compact rocks at depth for significant distances. Nevertheless, the preponderance of evidence indicates that solutions in large amounts do move through massive rocks at depth. These fluids transport metallic constituents, and where they are concentrated in traps or modified by chemical reaction, they have a tendency to deposit their loads and form ore deposits. Certainly in crystalline limestones and dolomites the ore-bearing fluids migrated through the rock and replaced parts of it (Fig. 3-1). The mineralizing fluids are generally believed to move around and penetrate the borders of mineral grains, which they corrode and alter, enabling subsequent solutions to pass more readily.

Figure 3-1
Limestone from Eagle Mountains, California. Note the disseminated magnetite.
Natural size.

That fluids under pressure are able to fracture and work their way through rocks has long been an attractive idea. This method of introduction might be likened to the intrusion of dikes, though it is perhaps more complex because the ore fluids are probably less viscous and more mobile than magmas. Field geologists have suggested that ore-bearing fluids can act with sufficient force to keep fissures open, allowing the fluids to circulate freely and permitting time for reaction and deposition (Graton, 1906, p. 60; Spurr, 1906, 1923; Wandke, 1930). Recently, mathematical analyses and model studies have been applied to the problem, with the conclusion that fluid pressures underground can reach significant magnitudes. These studies support the contention that fluids are able to fracture rocks and pass through them to areas of lower pressure (Hubbert and Willis, 1947; Hubbert and Rubey, 1959). Water might serve as a lubricant along the fractures. Such a theory helps to explain the quartz zones commonly found in supposedly impermeable shales and slates—in, for example, some of the gold-bearing quartz in the Southern Piedmont region of the United States (see Fig. 3-2).

Both the relative impermeability of many rocks, such as plutonic masses and shales, and evidence from experimental studies of impermeability suggest that superimposed permeability due to faults and other secondary structures may be more significant in ore transportation and deposition than the original permeability of the rock (Rove, 1947). Nevertheless, ore-bearing fluids have apparently moved through rocks where superimposed permeability is minor or absent.

Anyone who has had an opportunity to examine a large thermal spring area, such as the one at Big Geysers near Healdsburg, California (Fig. 3-3), cannot resist being impressed by the movement of large amounts of hot fluids through

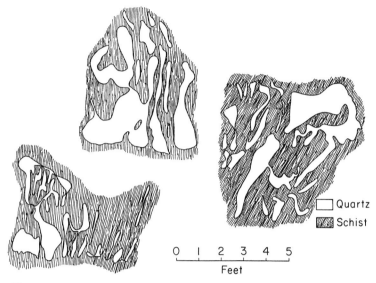

Figure 3-2
Zone of gold-bearing quartz, Laird prospect, Virginia. (From Park, 1936, Fig. 3.)

Figure 3-3
Big Geysers area near Healdsburg, California. The rock was originally Franciscan graywacke and argillite; it has been thoroughly altered to carbonates, chalcedony, and clay minerals. (Photo courtesy of D. A. McMillan, Jr., Thermal Power Company.)

relatively impermeable rocks. At Big Geysers, massive and dense graywackes and argillites are thoroughly altered over an area that exceeds 3,000 acres. The rocks are softened and saturated with steam and hot water. The altered rock is warm at the surface or within inches of the surface, and shallow drill holes release superheated steam under very high pressures; in fact, the area is now being used as a source of electric power. The thermal zone at Big Geysers follows the footwall of a large fault. Although permeability at depth along this fault may localize the path of steam migration, in the exposed area the hot fluids have spread for considerable distances away from the fault and into the footwall. The conclusion is inescapable that ore-bearing fluids are able to move through the densest rocks by working around individual grain boundaries or by other means.

The relative impermeability of massive carbonate rocks has been effectively demonstrated in the Balmat-Edwards district of New York State. In this mine area a drift was driven below and within a few feet of an abandoned, flooded winze. In spite of the pressure exerted by a head of nearly 100 feet of water, the drift remained dry. Similarly, drill holes with 600 feet of hydrostatic head (260 pounds per square inch, or about 17 atmospheres) were plugged where they intersected with mine openings, and no water passed through the limestone around the plugs (Brown, 1948, p. 38). From this evidence Brown concluded that limestones are essentially impermeable to cold watery solutions, even where these solutions are under considerable pressure. Earlier, he had suggested that ores in limestone migrate through heated rocks in a gaseous state (Brown, 1941, 1947). He attributes only a minor role to near-surface water, emphasizing that many deep mines are dry in their lower levels, that the rocks at depth are highly impervious, and that watery fluids would have great difficulty in traversing such material. At Balmat-Edwards, the ore shoots extend for more than 3,000 feet without any great change in character, indicating that the ore-bearing fluids operated over long distances with surprising uniformity. Supposedly, the fluids were guided by openings of microscopic size, probably near the lower capillary limits. According to Brown, the process of ore emplacement reduced rather than enlarged these openings.

Many ideas have been advanced and many experiments attempted in the effort to explain the mechanism by which fluids travel through dense rocks at depth. Maxwell and Verrall (1953) heated specimens of marble, limestone, and travertine under high confining pressures, a treatment that apparently developed permanent expansion and increased permeability. Maxwell and Verrall suggested that this method of expansion may help to explain the permeability to ore-bearing fluids of rocks under high temperature and pressure.

Brown's proposal that such minerals as the metallic sulfides are volatilized directly from the magma and that neither water vapor nor liquid is required for their transportation has not been widely accepted. Krauskopf (1957) has shown that a theory based strictly upon volatility fails to explain ore transport, since several metals have a very low volatility. However, fluids under conditions of high temperatures and pressures can transport metallic ions and operate through minute openings; under these conditions the physical states of liquids and gases are essentially the same. Klinkenberg (1941) determined from experiments that flow rates of gases are slightly higher than those of liquids; the flow rate of a gas through rock is not inversely proportional to viscosity, as is the case with a liquid. Extrapolating the determinations of permeability made on gases to those expected for liquids results in only small error when the medium is unconsolidated sediments. If the medium is a dense carbonate rock, as it is in many ore deposits, the error may be large (Ohle, 1951). Klinkenberg's permeability determinations, in which he used gas, are based upon the theory of slip: where gas is flowing along a solid wall, the layer of gas next to the surface is in motion with respect to that surface—that is, it "slips." Hence the volume of gas flowing through a medium is greater than if no slip occurred. The less dense the gas, the greater the slippage; conversely, the more pressure upon a gas, or the denser the gas, the more nearly it approaches the behavior of a liquid. In the deeper parts of the earth's crust, gases probably behave essentially the same as liquids.

Ohle (1951) separated the permeability values of the rocks in the east Tennessee zinc district into three groups: dolomite, limestone, and "recrystalline" (dolomitic marble, an alteration product of the limestone near mineralized bodies). As a presulfide phase of the mineralization process, some limestone beds were recrystallized and dolomitized, resulting in a significant increase in permeability. The original dolomite and the "recrystalline" are virtually the same in composition, but the replacement orebodies are concentrated in the "recrystalline." Ohle suggests that localization of ore was controlled by permeability, and that the ability of the limestone to recrystallize made it more favorable for replacement than the original dolomite; therefore the replacement orebodies were practically restricted to zones of "recrystalline" within the limestone. Since his calculations from tests showed that under favorable geologic conditions large volumes of aqueous solutions could pass through the dolomitized marble, Ohle concluded that the amount of dilute solution that could permeate these rocks would be sufficient to account for the ore deposits found in them.

As Ohle points out, it is customary to emphasize the relative inefficiency of intergranular flow, compared to flow through open channelways; consequently most geologists underestimate the volume of fluid that will pass through solid rock. Once the permeability of a rock type has been measured, it is possible to calculate with reasonable accuracy the quantity of a given fluid that will move through that rock under given conditions. Ohle's conclusion that the quantity may be adequate to produce large orebodies seems to contradict the conclusions of those who emphasize the impermeable nature of carbonate rocks, especially to the passage of watery solutions at depth.

These contradictory conclusions may reflect different opinions held concerning the depths at which ore-bearing fluids entered the country rocks.

Dolomitized limestones are generally more permeable than the undolomitized. At the Eagle Mine, Gilman, Colorado, Wehrenberg and Silverman (1965) determined that in the immediate area of the mine dolomitization increased the permeability of the Leadville formation by a factor between 10^3 and 10^6. Permeability becomes progressively higher as the ore zones are approached. The shape and extent of the orebodies at Gilman, especially at the manto deposits, were influenced not only by dolomitization, but also by solution channeling.

The difficulty that geologists have explaining the migration of ore-bearing fluids at depth causes them to return continually to the concept of diffusion. Economic geologists define diffusion as a spontaneous movement of particles of molecular or ionic size that causes one substance to become uniformly intermingled with another. Diffusion may take place in the solid, liquid, or gas phase. Water moving through the pores of a rock is therefore not regarded as diffusion, whereas the spread of copper ions through water is. Experimental and geochemical studies offer little evidence in support of the diffusion theory. Even though many experiments indicate that diffusion is unimportant, many field geologists find in this theory an appealing and simple answer to problems of ore transport.

Diffusion of ions or molecules through a liquid phase—for example, through saturated, porous rocks—is not difficult to envisage. It would be an especially efficient mechanism of transport in replacement because it permits the movement of ions both toward and away from the replacement front. The ions diffuse toward the regions of lesser concentration (that is, they move down their own concentration gradients); as a result, the replacing ions migrate toward country rock while the replaced ions move away from it. As the metal ions are deposited at the replacement front, their concentration is automatically lowered, and more ions move in to take their places. Thus, the replacement front grows steadily at the expense of the host until the supply of replacing ions is expended (Holser, 1947).

Diffusion through rocks and solid crystals is less likely as a mechanism of transport because rates of diffusion through these media are much slower than rates through liquids. The migration of ions diffusing through a solid is controlled largely by imperfections in the crystal structure. Indeed, a perfect crystal, maintaining ideal order under all circumstances, does not permit ionic diffusion except at slow rates (Barrer, 1951, p. 247), or permits only very small ions to diffuse. Imperfections may result from foreign ions existing interstitially in the normal structure, thus distorting this structure, or they may result from vacancies in the lattice. Diffusion may take place from one structural imperfection to another, or, in the case of vacant lattice positions, by the migration of holes, *i.e.*, an adjacent ion moves into the vacancy, leaving a hole behind. Thus, diffusion may involve an advancing ion or a retreating vacancy working its way through the crystal structure. The rate of diffusion will, of course, be strongly dependent upon the radius of the particle moving through the crystal. It will also be a function of temperature; near a mineral's melting point, the crystal structure becomes disordered and expanded. Within one or two hundred degrees of a crystal's melting point, diffusion through the crystal may be greatly increased (Holser, 1947; Barrer, 1951).

Diffusion can be measured for several fluids and ions in various media. The rate of diffusion is proportional to the concentration gradient, i.e., the change in concentration with distance, and to a diffusion coefficient, which is a constant for each host material. The concentration gradient depends partly upon the solubility of the substance diffusing, and the rate of diffusion is accordingly a function of both the fluid or ion undergoing diffusion and the medium through which diffusion takes place (Barrer, 1951). Replacement reactions maintain two-way concentration gradients, with the host material diffusing away from the replacement front and the replacing mineral diffusing toward it; deposition of the ore will necessarily reduce the concentration in solution at that point (Duffell, 1937). Indeed, any other mechanism of transfer to and from a replacement front is difficult to imagine.

Field evidence in support of diffusion is found in the trace-element halos in wall rocks near veins. According to diffusion theory, the metal content of wall rocks should increase logarithmically toward the ore deposit (Hawkes and Webb, 1962). Studies of wall-rock aureoles have demonstrated this logarithmic pattern in several areas. For example, Morris (1952) found a logarithmic dispersion of copper, lead, and zinc outward from basemetal veins in dolomite and quartz monzonite of the Tintic district, Utah. Dispersion through dolomite is confined to 10 feet or less from ore, but trace amounts of the heavy metals had moved several times this distance through quartz monzonite. Where fractures permitted fluid flow to modify the diffusion pattern, trace-element aureoles are irregular.

Wehrenberg and Silverman (1965) found that natural systems differ from ideal ones because of the chemical and physical influence of the matrix environment. The application of their data and theory to diffusion in the zinc deposits at Gilman, Colorado, resulted in an estimate of 100,000 years for the period of mineralization.

From calculations of the rate of diffusion through various media, Garrels and Howland (1949) concluded that ore-bearing fluids must be carried along small, closely-spaced fractures, but that diffusion on both sides of a crack will easily account for any massive ore deposits in a reasonable length of geologic time. Even under near-surface conditions—for which many geologists are reluctant to acknowledge a diffusion effect—it can be shown that ions will diffuse significant distances through solid rock. In fact, Garrels and Dreyer (1952) calculated that diffusion of galena through limestone will be at least 300 times more effective than forced flow, even if the diffusion is assumed to take place at only 100°C and the forced flow is given a pressure gradient well beyond that for a reasonable geologic environment.

The process of diffusion is perhaps best illustrated by the well-known phenomenon of exsolution, or unmixing, in ore minerals. For example, Figure 3-4 shows tiny exsolution blebs of chalcopyrite scattered through a specimen of sphalerite. If the specimen is heated to a temperature of 400°C, the chalcopyrite blebs disappear, and the sphalerite seems to be homogeneous. At elevated temperatures the bonds of a crystal are loosened, permitting the entrance of foreign materials, and near 400°C chalcopyrite diffuses and disperses through the expanded structure of sphalerite. Upon slow cooling the bonds tighten again and the chalcopyrite is forced out of the sphalerite structure. As cooling proceeds, the chalcopyrite is arranged in accordance with the atomic structure of the sphalerite, or, where a great excess of chalcopyrite has been exsolved, it may form mineral grains along the boundaries of the sphalerite (Edwards, 1954).

Another common example of diffusion that may indicate considerable migration is in the movement of carbonaceous material during the marmorization

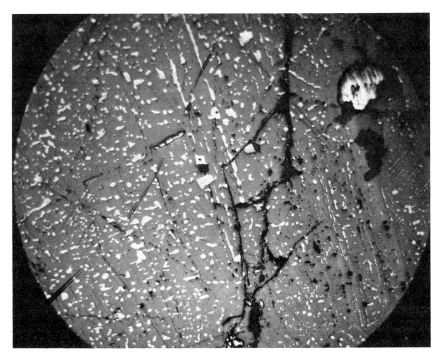

Figure 3-4
Chalcopyrite blebs in sphalerite, Darwin, California. Exsolution texture developed by the separation of chalcopyrite from the sphalerite lattice during slow cooling. ×80.

46

Figure 3-5
Button of carbon ("anthracite")
in partly crystallized dolomite,
Wolf Creek prospect, Metaline
Falls, Washington. ×2.

(marbleization) of limestone. As recrystallization takes place, the grains of calcite become bleached—in fact, most crystalline limestones are white—and the small particles of carbon to which the limestone owes its dark bluish or grayish color are forced out of the calcite structures. In some places the process can be seen to have stopped before completion; the carbon is concentrated as anthracite or graphite buttons within halos of white marble (Fig. 3-5). Where recrystallization has been carried to completion the carbon is entirely dispersed. If carbon migrates in this manner, it is reasonable to suppose that sulfides do also, under favorable conditions of temperature and pressure. The process would probably be aided by the presence of water in the rock pores, which would facilitate crystal growth and help dissolve and transport soluble extraneous materials.

Niggli proposed a mechanism involving the diffusion of a gas phase, whereby considerable amounts of material would be transferred from a cooling magma into the wall rocks. The area overlying a magma chamber would be saturated with vapors evolved from the magma. If the wall rocks are capable of replacement reactions with the vapors, they would act as an absorption apparatus, continually removing the gases and allowing for more "evaporation" from the magma (Niggli, 1929, p. 8–9).

Regarding the efficacy of diffusion in ore genesis, Edwards (1952) emphasized the small scale of migration in exsolution. Even though ore minerals provide the most favorable structures for solid diffusion—far more than the more rigid silicate structures—and even though sulfides form under conditions likely to promote a maximum of diffusion (namely, high temperatures), the linear movement of a given ion during exsolution rarely exceeds a few milli-meters and is generally measurable only in microns. That the exsolved mineral has a larger volume than the host is explained by a corresponding reduction in volume of the residual host mineral, so that although there is separation, there is no migration away from the place of deposition.

The importance of the process of exsolution, however, is not the magnitude of diffusion, but the rate at which diffusion takes place in sulfides. Gill (1960) produced significant diffusion through copper sulfides. In fact, he demon-strated that metal ions, such as ferrous iron, diffuse through sulfides at an appreciable rate (on the order of 1 to 3 millimeters per day) without the aid of hydrothermal solutions and within the temperature range in which many ores have formed. The best results were obtained using CuS; similar, though less striking, results were observed with FeS, PbS, and ZnS.

In contrast to the results obtained by Gill, experimental work has shown that diffusion in silica and the silicates takes place very slowly and over ex-tremely small distances. For example, Verhoogen (1952) attempted to diffuse potassium ions into a quartz crystal and found that, even at a temperature of 500°C, and under the influence of concentration gradient, the ions would travel at the rate of only 110 centimeters per million years.

The migration of fluids through rock is greatly facilitated if the rock is in a state of stress. During mountain-building episodes—when many ores are emplaced—stress is the normal condition. Under a triaxial stress system, fluids tend to migrate along the planes of tension, that is, normal to the axis of min-imum principal stress. Fractures, cleavage planes, and crystal boundaries pro-vide the channelways. Edwards (1954, p. 142) showed that if quicksilver is spread on unstrained brass, it forms only a thin coating, but when the brass is bent, the mercury works through it (largely along grain boundaries) and the brass eventually breaks.

Although the importance of diffusion to ore transport remains unresolved, diffusion in the solid state must be acknowledged as a reasonable possibility under favorable circumstances. Long-distance migration of ore-bearing fluids is undoubtedly controlled by the relatively open channelways provided by fault systems and joint systems. But locally, especially near ore deposits, where free circulation may have been impeded, diffusion of metal ions through liquid and solid media probably contributes to the movement of materials and to the final configuration of ore deposits.

MIGRATION OF FLUIDS AT SHALLOW DEPTHS

The behavior of fluids near the surface of the earth, within the range of drill holes and mine openings, has been carefully studied. Nevertheless, our knowl-edge of the movement of fluids at shallow depths is incomplete (Scheidegger, 1960; Hubbert, 1940, 1953; Muscat, 1937; Meinzer, 1923; Clough, 1936).

The factors to be considered in the near-surface movement of fluids include: the character of the fluid, especially its viscosity and density; the nature of the medium being traversed, especially its porosity and permeability; and the hydraulic head, or liquid pressure. The study can be extremely complex, depending upon the amount of material in solution, the presence or absence of gases, the character and heterogeneity of the rocks invaded, the geologic structure, and the temperature and pressure of the fluids. Many problems encountered in the study of the migration of near-surface fluids are similar to those dealt with in the study of fluids at depth, e.g., the relative roles of gases and liquids and the importance of diffusion.

The simplest and most fundamental law describing the movement of fluids underground is Darcy's law. This law relates the amount of fluid passing through a porous medium to the velocity of fluid movement, the permeability of the fluid-bearing materials, and the hydraulic gradient of the system (Darcy, 1856). Although many modifications covering special conditions have since been stated, Darcy's original statement has proven to be a reasonably correct approximation. The law may be expressed as

$$q = -K \frac{h_2 - h_1}{l}$$

where q = the volume of fluid crossing a unit area in a unit time;
K = a factor of proportionality that depends on the nature of the fluid, geometrical properties of the pore space, and the acceleration of gravity;
$h_2 - h_1$ = the difference in head measured from a standard datum: h_1 being upstream, h_2 being downstream;
l = the length of the column or the distance traversed.

An equivalent expression in physical terms is

$$q_s = k \frac{\mathscr{G}}{\mu} g_s - (1/\mathscr{G}) \frac{\delta p}{\delta s}$$

where q_s = component of q in the direction s;
k = permeability depending on the pore geometry;
$\mathscr{G}$ = fluid density;
μ = fluid viscosity;
g_s = component of gravity in the direction s;
$\dfrac{\delta p}{\delta s}$ = the rate of increase of pressure in the direction s.

This formula, or a modification, is used extensively in determining the laminar flow of water and other liquids through homogeneous permeable media.

Darcy's law, as written, does not apply if the flow is turbulent, that is, rapid enough to cause eddying. In all probability, turbulent flow takes place only in openings of unusually large size, and is of little significance in the passage of ore-bearing solutions.

Hubbert (1953) demonstrated that during migration of fluids consisting of oil, gas, and water, the three components separate and migrate at different rates and in different directions. As we shall see, these findings apply to all fluids at shallow depths. In an oil-gas-water system the interfaces will not be horizontal if one fluid of any pair is in motion in a nonvertical direction. Moreover, the migration paths of gas or oil droplets in moving water are not the same (that is, the impelling forces for oil and gas are not parallel). Thus, the two fluids migrate in divergent directions to different traps or to different parts of the same trap. Since ore-bearing fluids in many environments are thought to consist of both liquids and gases, and since the behavior of such fluids is similar to that of oil-gas-water systems, they tend to separate during migration. Liquids of different densities also will follow divergent paths during migration if they are in separate layers. This phenomenon was demonstrated for salt-water/fresh-water interfaces by Hubbert (1953). From the relationships determined by Hubbert, we should expect migration of hydrothermal solutions along the top of groundwater aquifers and the concentration of these solutions off-center from the crests of structures. Applying hydrodynamics, Thompson (1954) proposed just such a migration path for the mercury-bearing fluids that deposited the ores of the Terlingua district, Texas. Thus, separation of liquid and gas phases may be responsible for zoning of ores around a source (see Chap. 6). It may also explain the presence of monomineralic deposits, such as those of mercury, in which other sulfide minerals are usually either absent or present only in minor amounts.

Whereas diffusion through comparatively dry rocks under near-surface conditions is thought by many geologists to be inefficient for the transportation of ores, diffusion through solutions in saturated rocks is generally considered to be significant. For example, ores are commonly concentrated in permeable layers beneath shales and other relatively impermeable rocks, but in a few places ores thought to have been deposited from ascending waters are found in receptive rocks *above* the impermeable layers. Such an apparently anomalous situation may result from the diffusion of ions through water-saturated rocks (see Fig. 3-6). If a soluble salt is introduced from below into a sequence of alternately permeable and impermeable saturated layers, the salt tends to diffuse through the rocks (owing to the concentration gradient), even though the pore water is not circulating. Since the pore fluids cannot circulate through the strata, the fluid in each layer will attain equilibrious composition with the host rock, and the presence of a chemically receptive stratum directly above the impermeable layer will cause the immediate precipitation of the metals as they emerge from the diffusion zone. The replaced materials will diffuse back through the impermeable stratum.

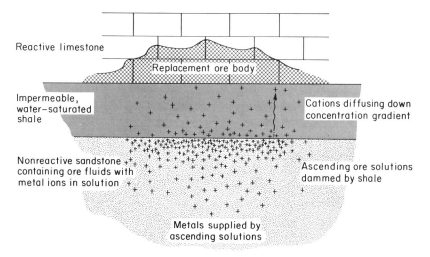

Figure 3-6
Schematic cross section, showing the diffusion of metal ions through a poorly permeable shale. Anions, such as sulfur, would migrate with the metal ions. The ascending ions move down the concentration gradient. Any radical change in chemistry of the country rock, such as the change from shale to limestone, could affect the ore solutions; as a result, metals would be deposited immediately after the solution emerged from the shale.

Some geologists consider the force generated by growing crystals great enough to force open the walls of fissures, enabling fluids to pass. Growing crystals undoubtedly exert a considerable force, as is shown by the fact that ice can split rocks or rupture an automobile crankcase. The force of growing crystals is limited, however, by the internal strength of the crystals; if this limit is exceeded, the crystals will collapse or cease to develop. It is therefore difficult to believe that the force of growing crystals is sufficient to force apart rocks in the depths of the earth. Nevertheless, under light loads, this force may be significant (Taber, 1916, 1926; Boydell, 1926, 1928).

MIGRATION OF METALS IN THE COLLOIDAL STATE

The possibility that colloids may be present in ore-bearing fluids has been considered by many geologists but not widely accepted, perhaps because the behavior of colloids at elevated temperatures and pressures is one of the least-understood phases of physical chemistry (Barton, 1959; Williams *et al.*, 1951; McBain, 1950; Lindgren, 1933; Boydell, 1925, 1927; Lasky, 1930; Liesegang, 1913).

A colloidal system consists of two phases, one of which (the dispersed phase) is diffused in the other (the dispersion medium). Colloidal particles range in size between those in true solution and those in coarse suspension, the general limits being defined at 10^{-7} and 10^{-3} centimeters (Williams *et al.*, 1951, p. 511). The colloidal material may be solid, liquid, or gas, and may be dispersed in any of these same phases. In the study of ore transport, however, we are concerned essentially with solids dispersed in liquid, or possibly gaseous, media. A colloidal system consisting of solid particles dispersed in a liquid is called a *sol.*

Colloidal particles have large surface areas per unit volume; as a result, ions adsorbed on the surface are able to control the behavior of the particles themselves. A given kind of colloidal particle may adsorb cations and behave as a positively charged body, or it may adsorb anions and become negatively charged. Since the particles of a sol all have the same charge, they repel each other and prevent coagulation. Accordingly, if an electrolyte is added to the sol, the colloidal particles become neutralized and flocculate. The dispersed phase is generally made up of molecular clusters, which may be in the form of sulfides, oxides, hydroxides, or other chemical compounds. Most sulfide and organic sols are negative, whereas most oxide and hydroxide sols are positive (but there are some serious exceptions to these generalizations; for example, colloidal silica carries a negative charge). Some sols have particles with a definite electrical charge that is not easily changed; others may be positive, negative, or neutral, depending upon the pH of the dispersion medium. Considering these properties, it is not surprising that various theories have been proposed to explain both the transportation and deposition of ore minerals in colloidal systems.

Whether colloidal metals are transported in hydrothermal systems depends upon whether the sols are stable at high temperatures and pressures, but such colloid stability has not been adequately studied. Colloids are most stable in cool, dilute solutions and may be much more stable in the presence of a second (protective) colloid. Frondel (1938) studied the stability of colloidal gold under simulated hydrothermal conditions and concluded that silica acts as a protective colloid, stabilizing the colloidal gold against electrolytes and against coagulation due to increases in temperature. Unprotected sols (gold) containing no electrolyte coagulated spontaneously at 150° to 250°C, but protected sols (gold and silica) were stable at 350°C.

The pros and cons of the hypothesis that metals migrate in the colloidal state have been debated from both the chemical and geological viewpoints. The difficult problem of explaining the migration of ore-bearing fluids at depth through dense and relatively impermeable materials would seem to be made more difficult by the colloid theory because colloidal particles are larger than ions, atoms, and molecules (Boydell, 1925, 1927). Nevertheless, some geologists argue that certain deep-seated ores were emplaced as colloids; others suggest that ore fluids change from solutions at depth to colloidal sols in near-surface environments (Herzenberg, 1936). Laboratory studies have established that colloids of metals can exist at moderate temperatures and pressures. Further, certain minerals and mineraloids are found in forms that suggest flocculation from a sol.

REFERENCES CITED

Barrer, R. M., 1951. *Diffusion in and Through Solids,* Cambridge Univ. Press.

Barton, P. B., Jr., 1959. The chemical environment of ore deposition and the problem of low-temperature ore transport, in *Researches in Geochemistry,* ed. P. H. Abelson, New York: Wiley.

Boydell, H. C., 1925. Role of colloidal solutions in the formation of mineral deposits, *Inst. Mining Metall. London Trans.* 34:145-337.

———, 1926. A discussion on metasomatism and the linear "force of growing crystals," *Econ. Geol.* 21:1-55.

———, 1927. Operative causes in ore deposition, *Inst. Mining Metall. London Bull.* 227, pp. 1-85.

———, 1928. Metasomatism and the pressure of growing crystals: a discussion, *Econ. Geol.* 23:214-218.

Brown, J. S., 1941. Factors of composition and porosity in lead-zinc replacements of metamorphosed limestone, *Amer. Inst. Mining Eng. Trans.* 144:250-263.

———, 1947. Porosity and ore deposition at Edwards and Balmat, New York, *Geol. Soc. Amer. Bull.* 58:505-545.

———, 1948. *Ore Genesis,* New Jersey: Hopewell.

Clough, K. H., 1936. Study of permeability measurements and their application to the oil industry, *Oil Weekly* 83:27-34.

Compton, R. R., 1955. Trondhjemite batholith near Bidwell Bar, California, *Geol. Soc. Amer. Bull.* 66:9-44.

———, 1960. Contact metamorphism in Santa Rosa Range, Nevada, *Geol. Soc. Amer. Bull.* 71:1383-1416.

Darcy, H., 1856. *Les Fontaines Publiques de la Ville de Dijon,* Paris: Victor Dalmont.

Duffell, S., 1937. Diffusion and its relation to ore deposition, *Econ. Geol.* 32:494-510.

Edwards, A. B., 1952. The ore minerals and their textures, *Roy. Soc. N.S.W. J. Proc.* 85:26-45.

———, 1954. *Textures of the Ore Minerals and Their Significance,* Melbourne: Australasian Institute of Mining Metallurgy.

Frondel, C., 1938. Stability of colloidal gold under hydrothermal conditions, *Econ. Geol.* 33:1-20.

Garrels, R. M., and R. M. Dreyer, 1952. Mechanism of limestone replacement at low temperatures and pressure, *Geol. Soc. Amer. Bull.* 63:325-379.

Garrels, R. M., and A. L. Howland, 1949. Diffusion of ions through intergranular spaces in water-saturated rocks, *Geol. Soc. Amer. Bull.* 60:1809-1828.

Gill, J. E., 1960. Solid diffusion of sulphides and ore formation, *21st Int. Geol. Congr. Rept.,* pt. 16, pp. 209-217.

Graton, L. C., 1906. Reconnaissance of some gold and tin deposits of the southern Appalachians, *U.S. Geol. Surv. Bull.* 293.

Hawkes, H. E., and J. S. Webb, 1962. *Geochemistry in Mineral Exploration,* New York: Harper and Row.

Herzenberg, R., 1936. Colloidal tin ore deposits, *Econ. Geol.* 31:761–766.

Holser, W. T., 1947. Metasomatic processes, *Econ. Geol.* 42:384–395.

Hubbert, M. K., 1940. The theory of ground water motion, *J. Geol.* 48:785–944.

———, 1953. Entrapment of petroleum under hydrodynamic conditions, *Amer. Ass. Petrol. Geol. Bull.* 37:1954–2026.

Hubbert, M. K., and W. W. Rubey, 1959. Role of fluid pressure in mechanics of overthrust faulting: part I, *Geol. Soc. Amer. Bull.* 70:115–166.

Hubbert, M. K., and D. G. Willis, 1957. Mechanics of hydraulic fracturing, *J. Petrol. Tech.* 9(6):158–168.

Klinkenberg, L. J., 1941. The permeability of porous media to liquids and gases, in *Drilling and Production Practice,* American Petroleum Institute.

Krauskopf, K. B., 1957. The heavy metal content of magmatic vapors at 600°C, *Econ. Geol.* 52:786–807.

Lasky, S. G., 1930. A colloidal origin of some of the Kennecott ore minerals, *Econ. Geol.* 25:737–757.

Liesegang, R. E., 1913. *Geologische Diffusionen,* Dresden and Leipzig: Theodor Steinkopff.

Lindgren, W., 1933. *Mineral Deposits,* 4th ed., New York: McGraw-Hill.

Maxwell, J. C., and P. Verrall, 1953. Expansion and increase in permeability of carbonate rocks on heating, *Amer. Geophys. Union Trans.* 34:101–106.

McBain, J. W., 1950. *Colloid Science,* Boston: D. C. Heath.

Meinzer, O. E., 1923. The occurrence of ground water in the United States, *U.S. Geol. Surv. Water-Supply Pap.* 489.

Morris, H. T., 1952. Primary dispersion patterns of heavy metals in carbonate and quartz monzonite wall rocks, Part II; in H. T. Morris and T. S. Lovering, Supergene and hydrothermal dispersion of heavy metals in wall rocks near ore bodies, Tintic district, Utah, *Econ. Geol.* 47:698–716.

Muscat, M., 1937. *The Flow of Homogeneous Fluids Through Porous Media,* New York: McGraw-Hill.

Niggli, P., 1929. *Ore Deposits of Magmatic Origin,* tr. H. C. Boydell, London: Thomas Murby.

Ohle, E. L., 1951. The influence of permeability on ore distribution in limestone and dolomite, *Econ. Geol.* 46:667–706, 871–908.

Park, C. F., Jr., 1936. Preliminary report on gold deposits of the Virginia Piedmont, *Va. Geol. Surv. Bull.* 44.

Park, C. F., Jr., and R. S. Cannon, Jr., 1943. Geology and ore deposits of the Metaline quadrangle, Washington, *U.S. Geol. Surv. Prof. Pap.* 202.

Rove, O. N., 1947. Some physical characteristics of certain favorable and unfavorable ore horizons, *Econ. Geol.* 42:57–77, 161–193.

Scheidegger, A. E., 1960. *The Physics of Flow Through Porous Media,* New York: Macmillan.

Spurr, J. E., 1906. Ore deposits of the Silver Peak quadrangle, Nevada, *U.S. Geol. Surv. Prof. Pap.* 55, pp. 112–128.

———, 1923. *The Ore Magmas,* New York: McGraw-Hill.

Taber, S., 1916. The growth of crystals under external pressure, *Amer. J. Sci.* 41:532–556.

———, 1926. Metasomatism and the pressure of growing crystals: a discussion, *Econ. Geol.* 21:717–727.

Thompson, G. A., 1954. Transportation and deposition of quicksilver ores in the Terlingua district, Texas, *Econ. Geol.* 49:175-197.

Verhoogen, J., 1952. Ionic diffusion and electrical conductivity in quartz, *Amer. Mineral.* 37:637-655.

Wandke, A., 1930. Ore deposition in open fissures formed by solution pressure, *Amer. Inst. Mining Eng. Tech. Pub.* 342. (Also 1931, *Amer. Inst. Mining Eng. Trans.* 96:291-304.)

Wehrenberg, J. P., and A. Silverman, 1965. Studies of base metal diffusion in experimental and natural systems, *Econ. Geol.* 60:317-350.

Williams, J. W., R. A. Alberty, and E. O. Kraemer, 1951. The colloidal state and surface chemistry, in *Physical Chemistry*, Vol. 2, ed. H. S. Taylor and S. Gladstone, New York: Van Nostrand.

4 / Deposition of the Ores

The careful field observer who has a thorough understanding of why an ore deposit is localized in a given environment has a definite advantage in the search for other deposits in similar environments. Whether an ore deposit is a product of igneous, sedimentary, metamorphic, or weathering processes, its localization is generally the result of many factors. One factor, or series of factors, may account for the ore constituent; another for its deposition; still another for the size of the deposit. The ore metals may be supplied by either a differentiating magma or the weathering of rocks at the earth's surface.

Some ores are deposited by gravity; an early formed chromite crystal may settle out of a residual magma or a fragment of gold may settle to the bottom of a layer of agitated sediments. Other ores are deposited because of chemical changes, such as a change in pH, that result from reactions between the ore-bearing solutions and the host rocks. A drop in temperature, pressure, or velocity of the transporting medium, or the admixture of a second solution, may also bring about chemical reactions to cause ore deposition. Localized deposition may in turn depend upon the permeability, structure, brittleness, or chemistry of the rocks. Although the reasons for deposition and the causes of localization are often the same, we rarely know what they are for a given deposit.

GROUND PREPARATION

Many *epigenetic* deposits—ores formed later than the host rock—are restricted to areas that have undergone a favorable premetallization change. Such a change may make the country rock more receptive or more reactive to the ore-bearing solutions; accordingly, the process is known as ground preparation. *Syngenetic* deposits—ores formed at the same time as the host rock—are deposited as part of the original rock and therefore do not require ground preparation.

Ground preparation may take place in several ways. Any process that increases permeability, causes a favorable chemical change, or induces brittleness in the rocks may localize deposition from the ore-bearing fluids. Hence, the type of ground preparation depends upon both the country rock and the preparing agent (heat, fluids, tectonics, or a combination of the three). Silicification, dolomitization, and recrystallization are common examples of ground preparation. The removal of a constituent, as in dedolomitization, may also be a form of ground preparation. Ground preparation at some places is believed to be an early stage of mineralization.

Epigenetic deposits are commonly, though not necessarily, introduced during tectonic activity; many are late-phase products of associated igneous or metamorphic activities. Under these circumstances, ground preparation may be caused by fluids originating from the same source as the ore. For example, dolomitization and ore emplacement may occur sequentially in the country rock around a cooling pluton. Dolomitization tends to make rocks brittle; subsequent movement may shatter the dolomitized rock and produce a clean, permeable breccia that can serve as a porous receptacle for ore deposition.

Much ground preparation is chemical; perhaps the most common reaction is the addition and rearrangement of silica, in the form of either SiO_2 or silicates. One of the common forms of silica is jasperoid, a cryptocrystalline silica that is transported in hydrothermal fluids and replaces country rock (Lovering, 1962). Jasperoid, which abounds at Leadville, Colorado, at Metaline Falls, Washington, and in many other mining districts, forms conspicuous and resistant masses, knobs, and hills along shear zones or faults. It is common in limestone and dolomite, and in some districts the ore is limited to the jasperoid rock. Some jasperoid, hardened by the addition of the fine grained silica, has been fractured and shattered. This jasperoid breccia may contain almost no clay or fine dust. It thus forms favorable sites for fluid circulation because of increased permeability along relatively continuous zones. Not all jasperoid hosts have been brecciated; the silicified rock may have been chemically favorable for ore deposition. Gunning (1948) noted that the ore minerals of the Privateer mine, Vancouver Island, Canada, were concentrated in fissures favored by preore silicification of limy and argillaceous country rocks. The hardness of the silicified materials permitted clean faulting with a minimum of powdering. These rocks were also relatively inert to the ore-forming fluids and were not softened by hydrothermal action, allowing essentially continuous fissuring, unobstructed by gouge (rock powder or paste) or altered wall rock. But the veins narrow strikingly where they cut quartz diorite, because the diorite was powdered during faulting and later altered; both the gouge and the alteration effectively reduced the permeability.

Permeability and brittleness were increased in some districts by simple crystallization of the country rock. Such recrystallization is common near intrusive masses. A pluton may also have preheated the country rock, permitting the passage of hydrothermal solutions that might have precipitated their mineral contents if retained.

The formation of skarn is another type of ground preparation. Skarn is composed mainly of lime-rich silicates produced by the introduction of silicon, aluminum, iron, and magnesium into calcium- and magnesium-rich carbonates. Orebodies are common in skarns and generally are introduced after the skarns have formed. Moreover, skarns are replaced in preference to the unaltered country rock; this suggests that the silicified material is more permeable and more favorable for chemical replacement. It is commonly thought that the skarn is preferentially replaced because it is readily brecciated (and thus rendered permeable), but some unbrecciated skarn zones have been selectively replaced by ore. Accordingly, the common association of ores with skarn zones may not be related to the mechanical influence of ground preparation; instead, it may be a product of physiochemical factors (Titley, 1961).

McDougall (1968) reached the conclusion that lattice defects due to strain in minerals may have been significant in preore ground preparation. Where obvious structural or petrological controls for the emplacement of ore deposits are lacking, the controls may have been inconspicuous concentrations of either lattice defects or above-normal free energy.

Clearly, the formation of ore deposits cannot be studied apart from the formation of associated gangue and alteration products during ground preparation. All aspects of a mineral deposit must be studied as a unit.

DEPOSITION OF
MAGMATIC SEGREGATION DEPOSITS

Magmatic segregation deposits form as a direct result of differentiation. Some are early differentiates that settled through the magma in response to gravity; others are late products of fractionation. Magmatic products of this type migrate and solidify in the same manner as dikes and minor intrusives of ordinary igneous rocks. In fact, they are merely igneous rocks of special economic value. Accordingly, the ores show a wide range of physical and chemical features. Some deposits are conspicuously layered; in others the metals are disseminated through massive igneous rock; and in still others the ore itself is a massive igneous rock.

Good examples of compositional layering of differentiates are found in the Bushveld complex of South Africa and the Stillwater complex of Montana. Layering is especially abundant in mafic or intermediate igneous rocks containing ilmenite, magnetite, or chromite, and may result from any of several processes during differentiation (Wager and Brown, 1967). One widely acknowledged process is crystal settling. As the temperature and pressure of a magma are lowered, crystals begin to form. The composition of these crystals depends, of course, upon the composition of the melt with which the crystals are in equilibrium. In general, the ferromagnesian minerals—including the spinel family—crystallize first. Since these early minerals are more dense than the residual magma, they settle and collect near the base of the igneous chamber. As crystallization proceeds, crystals of progressively different composition develop, producing a layered rock (Bowen, 1928).

Another hypothesis is that layering may be a result of immiscibilities in the magma. Some compounds that are mutually soluble under certain conditions of temperature and pressure are immiscible under other conditions. The heavier liquid settles through the lighter one, thereby producing layering of the liquids. Upon solidification, the igneous mass would contain magmatic segregation layers of a different, though perhaps indistinguishable, type from those formed by crystal settling (Fenner, 1948). Still another suggestion is that layers may be formed by successive injections of the rest magma into a rock. Since the rest magma constantly changes in composition as the solid phases develop, the injections may become parallel layers of different composition within a progressively growing intrusive mass if they are controlled by bedding or structure. It is conceivable, though not probable, that compositional layering could be produced by a single intrusive involving all three of these mechanisms. In such a case crystals would sink, immiscible liquids would develop in the magma chamber, and some of the liquid fraction would be squeezed into the surrounding country rock.

The concentration of ore minerals in a massive igneous rock may be great or small, depending upon the kind of mineral and the degree of differentiation. Moreover, economics may determine that an igneous differentiate constitutes ore because of accessory minerals, such as monazite, cassiterite, or ilmenite. For example, ilmenite is a common accessory mineral in the Tahawas district, Lake Sanford, New York. The Tahawas district was originally prospected for iron, which is much more abundant, but the titanium impurity made smelting costs prohibitive. Now that titanium is a valuable commodity, the ores are worked mainly for the ilmenite, and the iron is a by-product (Killinger, 1942).

Massive ore deposits of direct magmatic affiliation are widespread. They are associated with both mafic and silicic rocks, and include monazite, tin, titanium, iron, or other economic minerals either as early differentiates or late liquid fractions. If the ore mineral settles during differentiation, the residual melt may be squeezed off to leave nothing but the crystal mush; thus, what started out to be a layered deposit would become a relatively homogeneous mass of the solid phase. Similarly, upon solidification the liquid phase may form either a layered or a massive orebody, depending upon the way it is injected into the country rocks.

Most geologists agree that massive sulfides may form directly from a magma; some argue for both an immiscible sulfide liquid phase and a sulfide ore magma. Sales (1954) proposed that the copper orebody of the Colorada pipe, Cananea, Mexico, formed by crystallization of an ore magma injected into the top of a porphyry breccia pipe. Similar hypotheses have been advanced for other massive sulfide deposits.

The unusual rocks known as *carbonatites* are commonly designated as magmatic segregation deposits (Tuttle and Gittins, 1966). A carbonatite is defined as a carbonate-rich rock genetically related to the alkalic rock-forming process and alkalic magma irruption (Pecora, 1956). Carbonatites are associated with either intrusive or extrusive igneous processes. Many carbonatite deposits have been described in the literature, and some have been of great economic value. One of the most interesting deposits is the Oldoinyo Lengai volcano in northern Tanganyika, where several flows of modern times are composed predominantly of sodium carbonate (King, 1965; Dawson, 1966). Other carbonatites, especially those of Africa (as at Palabora in the Transvaal), contain valuable copper deposits. Valuable accessory minerals include niobium (recovered at several places), the rare earth minerals, the flourocarbonates, and the phosphates (Deans, 1966).

In summary, the mechanism of formation of magmatic segregations is either crystal settling or the simple crystallization of an ore-grade fluid phase. The ore may crystallize in place or may be squeezed into the surrounding rocks as a separate magma. Nevertheless, distinguishing the types of magmatic segregation deposits presents problems; moreover, there is a resemblance with some massive replacement deposits. Consequently, just because the physical processes of magmatic segregation are not complex does not mean that the interpretation of a deposit is simple.

STRUCTURAL CONTROLS

Detailed studies of structure are essential in exploration and unquestionably have led to more discoveries of ore than any other approach. This is because the movement of fluids underground is controlled by permeability, a function of both the original character of the rock and of superimposed structure. Permeability and structure are widely understood and used by petroleum geologists. Oil and gas are commonly channeled to the crest of a dome or to an area of overlap or faulting, causing them to be trapped or ponded against an impermeable layer. The laws that govern the migration of oil, gas, and water apply to all fluids underground, including the ore-bearing gases and liquids. Faults or other permeable features, either primary or superimposed, tap supplies of mineral-bearing fluids, allowing them to migrate until they cool and precipitate their mineral content or can react with and replace receptive country rocks.

Serious efforts have been made to classify ore deposits according to their structural control. In general, these efforts have not produced an acceptable classification, largely because any structural or sedimentary feature that permits passage of ore-bearing fluids may contain areas favorable for the precipitation of ore. A particular type of ore deposit may thus be localized along more than one type of structure. Nevertheless, ore concentrations in many districts follow patterns of deposition that are associated predominantly with one type of structure.

The structures and textures that control ore deposition may be primary or secondary (superimposed, such as faults), according to whether they were formed at the same time as the rock mass or later. In certain ore deposits the primary controls dominate; in others superimposed features are the basic controls. Determining a physical control is a fundamental problem in the exploration of any mineralized district.

Primary Features

Primary structures and textures of rocks control the distribution of fluids and hence the localization of ores. Since any textural or structural feature that influences porosity and permeability may control deposition, the variety of primary controls is practically unlimited. A few of the most obvious are (1) permeable (clastic) limestone or dolomite, especially where covered by impermeable cap rocks; (2) reef structures, especially limestones; (3) well-sorted conglomerates that permit easy circulation of ore-bearing fluids; (4) broken and scoreacous tops of lava flows that also permit ready circulation of ore-bearing fluids; (5) permeable sandstones such as channel sands and beach deposits; (6) volcanic domes and piles.

Ohle and Brown (1954) described a sedimentary arch structure of depositional origin that controlled ore trends in the southeastern Missouri lead district. The arch structures apparently were built by currents that formed ridges or bars of limy sediments parallel to their direction of movement. Much of the sediment making up the ridges was clastic shell material, an exceptionally permeable rock compared to the limy muds in adjacent troughs. Ascending aqueous solutions were directed along these permeable arches and restricted from further movement by an overlying bed of impermeable limestone. In this same district other orebodies were controlled by buried knobs and ridges of Precambrian granite against which Cambrian and younger sediments were deposited. The sediments dip in all directions away from the granite highs. Ascending solutions migrated along a basal sandstone that pinches out against the Precambrian granites. Upon reaching the up-dip termination of the sandstone, the ore-bearing fluids entered the overlying carbonate sediments and deposited their ore minerals (James, 1949; Ohle and Brown, 1954).

Brecciation produced by slumping during deposition and compaction of the overlying sediments was yet another control of ore deposition in this district. Differential compaction over the clastic shell ridges led to oversteepening of slopes and to consequent slumping into the basin regions. These submarine landslides produced huge breccia zones that were favorable for the subsequent ingress of ore-bearing solutions. Individual orebodies in the slides range up to more than 6,000 feet in length and contain several million tons of ore (Snyder and Odell, 1958). Superimposed fracture zones have strongly influenced the distribution of ores in southeastern Missouri, but much new ore has been discovered by mapping and following primary sedimentary features, such as the clastics around buried ridges.

Dolomites and dolomitic limestones are ordinarily more permeable and more porous than pure limestones, so that dolomites permit solutions to circulate more readily than do limestones. Many geologists therefore believe that dolomite is more likely to be a host for ore (Hayward and Triplett, 1931). In the Sierra Mojada district, Nuevo León, Mexico, and in many other areas, ores are concentrated in dolomitic rocks, and the purer limestones are barren or low-grade. This is not universally true, however. In the Metaline district of northeastern Washington the ores are in low magnesium limestones beneath impermeable black shale, and underlying dolomites are either barren or only weakly mineralized. Evidently, factors other than permeability must be considered.

The copper ores of the Keweenaw peninsula in northern Michigan furnish excellent examples of hydrothermal deposits that have been channeled into their present positions through permeable conglomerates and broken tops of lava flows (Butler and Burbank, 1929). Concentrations of ore are in conglomerate beds between the flows and in the fragmental, vesicular surface layers of individual flows. These beds were favorable for ore deposition because of extremely high permeabilities.

Permeable strata formed by channel deposits and sands interbedded with siltstones have been mineralized in the Gas Hills uranium district of central Wyoming. The ore deposits are restricted to the coarse-grained facies and are especially concentrated along the valleys of a buried land surface. Apparently, the ore-bearing solutions mixed with ground waters and moved laterally along permeable zones until they were dammed by facies changes or by impermeable shales across the basal unconformity (Zeller, 1957).

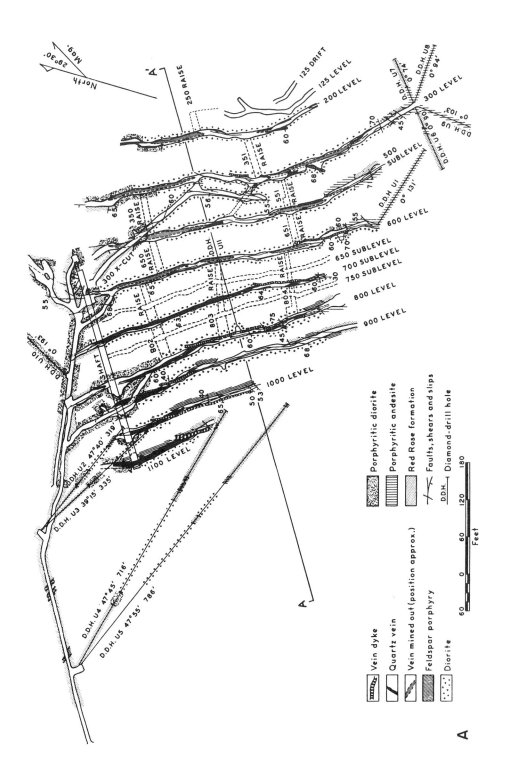

A

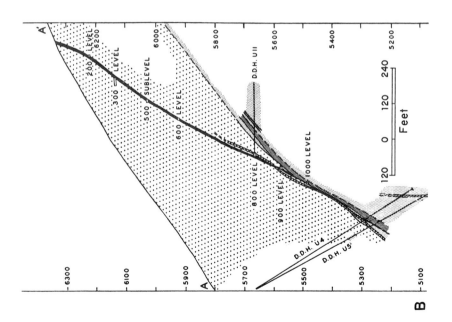

Figure 4-1
Red Rose tungsten mine, Canada.
(A) Plan, showing simple vein;
(B) vertical section AA', normal to vein.
(From Brown, 1957.)

Secondary, or Superimposed, Features

In most epigenetic ore deposits the path of circulation followed by the ore-bearing fluids has been greatly influenced by structures superimposed on the rocks. Faults and folds are probably the most common secondary structures, though breccia zones, pipes, and other features are locally of great significance.

Faults are found nearly everywhere, and many ore deposits are directly related to them. Because fault surfaces are uneven, movement along a fault produces breccia and gouge. A zone of fine-grained gouge frequently hinders circulation of fluids, either along or across a fault. On the other hand, coarse, clean breccia containing a minimum of fines may increase permeability, especially in brittle rocks fractured under light loads (Lovering, 1942). Accordingly, minor faults may be much better hosts for ore solutions than large faults, which are more likely to develop gouge. As a general rule, then, tight gouge-filled fractures are less favorable for ore deposition than more open fractures. As is usual in geology, there are exceptions to this generalization. For example, the Santa Rosa mine in the Huantajaya district of northern Chile is said to have contained silver ore in clay along the side of what seems to be a more permeable quartz vein (Hector Flores Williams, 1958, personal communication). The ores at Butte, Montana are also said to be best where gouge zones are widest.

Veins are tabular bodies, long in two dimensions and short in the third, formed along cracks or fissure zones; fault planes are especially favorable loci. They are either the simple filling of open fissures or replacement masses along a permeable fracture. Veins are classified as *simple* (Fig. 4-1) if they are a single injection, *complex* (Fig. 4-2) if made up of multiple injections along the same fracture, *irregular* if they are of variable thickness, and *anastomosing* or branching (Fig. 4-3) if they are braided or mutually interlacing mineral zones. Vein patterns are further described by qualifying terms, such as *conjugate* (two sets of veins that have the same strike at 90 degrees to each other). Seldom is an entire vein filled with ore; the valuable constituents are usually concentrated in restricted zones, called *ore shoots*. Where many small veinlets are distributed along a tabular zone, the deposit is called a *lead*, a *lode*, or a *fissure zone*. Veins are no longer the predominant type of ore deposit in mining. Their importance has been superseded by irregular replacement deposits, disseminated ores, and segregation and sedimentary materials.

Minor movement along curved fault surfaces causes pinching and swelling, locally separating the footwall from the hanging wall (see Fig. 4-4). Ore-bearing fluids migrate through the more open parts of the fissures (the "swells") and are deflected around the tighter zones (the "pinches"). As a result, rolls or changes in attitude of either the strike or dip of a vein commonly mark the beginning or end of an ore shoot; such irregularities along veins have long been recognized as highly significant in mineral exploration.

Contour maps based upon assay values and dilation maps showing changes in thickness of a vein are of considerable interest and use in exploration; how-

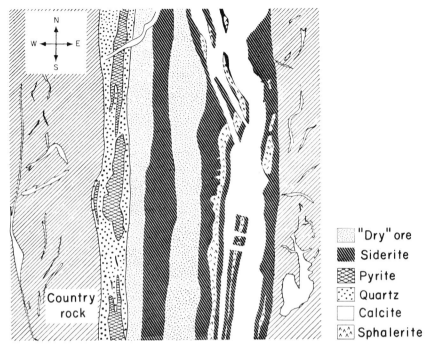

Figure 4-2
Complex vein, generally 10 to 20 cm wide, Příbram, Czechoslovakia. (From Kutina, 1955.)

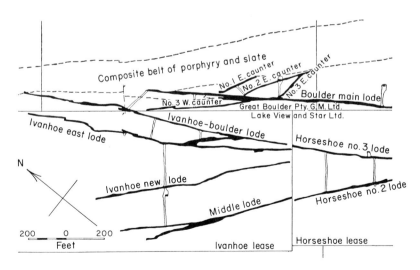

Figure 4-3
Branching vein, West of Boulder Dyke, Kalgoorlie district, Western Australia. (From Finucane, 1948.)

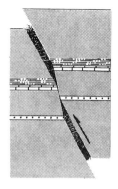
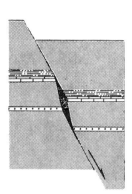

Figure 4-4
Openings caused by reverse and normal movements along veins. Note pinches and swells.

ever, sufficient information is seldom available to make them as complete as desirable. Contour maps are constructed by plotting assay values on a longitudinal section drawn along the vein. The assay data usually represent the entire width of the vein, though they may portray only the richer parts, such as a footwall or a hanging wall streak. Contours are then drawn through points of equal value, the thickness of the vein usually being ignored. Used alone, a contour map of this type has little value, as it shows only the unit worth of the material in the vein.

A more instructive type of contour map is the one showing the analysis of Garnett (1966) at the Geever mine, Cornwall, England. Here the study of tin content and vein structure indicates that spaces formed in the lodes by movement of the walls have been important in controlling the distribution of cassiterite. Variations in the tin content, expressed in terms of lode width, grade, and value (grade × width), are illustrated diagramatically as lode contour diagrams. The arithmetic means of the regular level development samples over constant distances are contoured on longitudinal sections (Fig. 4-5C). The changing structure of the lode is shown by structural diagrams illustrating variations in strikes and dips (Figs. 4-5A and 4-5B). Comparison of these and the lode diagrams allows one to interpret the structural controls imposed upon each of the several phases of cassiterite mineralization.

Other features, such as mineralogy, may also be contoured. Contouring of several features affords a better understanding of the deposit and the distribution of ore, and more efficient underground and surface exploration and development.

At the South Crofty mine, Cornwall, England, a comparison of cassiterite distribution and lode structures reveals that many ore shoots are directly related to changes in dip and strike. Where the shape of the lode and the amount of fault displacement are known, it may be possible to predict the location of wide parts of the lode. At South Crofty the correlation between locations of the ore

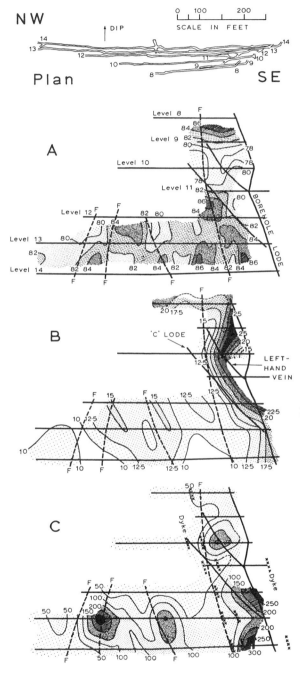

Figure 4-5
Borehole (extension) lode,
Geever Mine, Cornwall,
England, longitudinal section.
(A) Lode dip contour diagram,
in degrees;
(B) lode width contour diagram,
in inches;
(C) lode value contour diagram,
in foot-pounds of SnO_2/ton.
(From Garnett, 1966.)

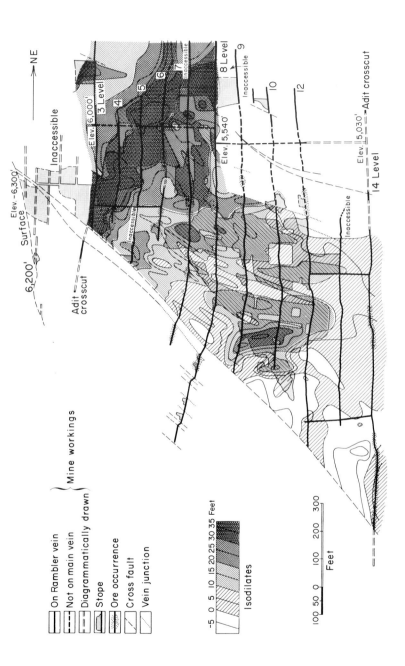

Figure 4-6
Dilation map of the Rambler vein, Slocan district, British Columbia. (From Roscoe, 1951.)

shoots and the predicted locations of wide parts of the lode is sufficiently close to establish that the change in attitude of the lode was a major control in localizing the cassiterite (Taylor, 1966).

When sufficient information is available, dilation maps may be drawn. These are constructed by passing a reference plane below and roughly parallel to the footwall of a vein. Distances from this plane to the footwall are plotted and contours are drawn through points at equal distances from the footwall. Another similar map is made, plotting the distances from the same plane to the hanging wall. Contours based upon the thickness of the vein (the dilatancy) may then be drawn from an overlay of the hanging wall and footwall maps. Each map will show "hills and valleys" where irregularities are present in either wall. If the vein is along a simple fissure, the two contour maps can be adjusted until the contours of one agree closely with the contours of the other. From this, the amount and direction of movement along the fissure can be ascertained, and these, when combined with the amount of dilatancy, enable a projection to be made into undeveloped ground of the widest and more favorable parts of the vein. An interesting and economically rewarding study of this type was made by Roscoe (1951) in the Rambler mine of the Slocan district, British Columbia (see Fig. 4-6).

Pipes or *chimneys*, as the names imply, are bodies that are relatively short in two dimensions and long in the third. The two names are used interchangeably, though efforts have been made to restrict one or the other to rod-shaped deposits having a specifically defined plunge. For example, many geologists restrict the term *pipe* to steeply dipping rod-shaped bodies, and use the word *manto* (Spanish: mantle, or cloak) for flat-lying rod-shaped bodies. However, *manto* is more correctly used to describe beds or flat, bedded, and sheetlike deposits (Prescott, 1915). Pipes are known to change in attitude from nearly vertical to nearly horizontal, so it is desirable to use the term without regard to plunge or dip. Most pipes contain broken rock and accordingly are known as *breccia pipes*.

Many large, valuable orebodies have been mined from pipes. The host brecciated country rock either predates the ore deposition or was formed during mineralization processes. The origin of pipes is one of the most fascinating problems in the field of ore geology, and its study one of the most rewarding. Ore pipes are found in many environments and result from various combinations of processes. Although the genesis of many pipes is unknown, that of others is readily interpreted. Pipes are commonly formed at the intersection of any two tabular features, such as faults, fissures, dikes, bedding, lava flows, or joints. Where the tabular features are faults, brecciation is likely to be most extensive if the fractures intersect at small angles (see Fig. 4-7). Pipes are also formed at the crests of folds, especially where the rocks are highly fractured and where permeability has been increased by strata sliding over each other, leaving areas of reduced pressure at the crests; against igneous contacts, either

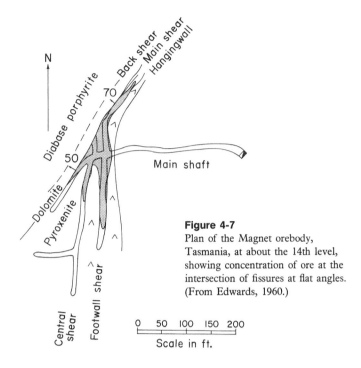

Figure 4-7
Plan of the Magnet orebody,
Tasmania, at about the 14th level,
showing concentration of ore at the
intersection of fissures at flat angles.
(From Edwards, 1960.)

where solutions were channeled along irregularities or where differential movements caused brecciation; in the throats of old volcanoes, and in diatremes, solution channels through carbonate rocks, shoestring sands, along rolls or changes in dip or strike of a vein, and within small cupolas or other igneous bodies (Bryner, 1961; Perry, 1961; Blanchard, 1947; Kuhn, 1941; Emmons, 1938; Butler, 1913; Ransome, 1911; Wagner, 1927). Paul Gemmill (1959, personal communication) studied the ore-bearing pipes in the Bristol mine of eastern Nevada and concluded that some of them resulted from renewed movement along the earlier of two intersecting fractures (see Fig. 4-8). Other pipes at Bristol are aligned along single faults and could not have been formed in this manner (MacDiarmid, 1960). Figure 4-9 shows a pair of pipes thought to have developed in the open zones of pinch-and-swell structures along an irregular fault.

Locke (1926) suggested that mineralization stoping was active in forming ore pipes. This process involves the ascent of ore-bearing fluids through the crust of the earth by stoping their way upward. The undermined and partly dissolved blocks break loose from the roof and settle into the solution cavern, forming a pipe of collapsed breccia. The ore and gangue are then deposited in the brecciated mass. Some of the breccia pipes in the caldera at Cripple Creek, Colorado, appear to have formed as a result of stoping by ascending

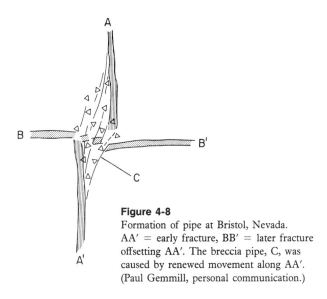

Figure 4-8
Formation of pipe at Bristol, Nevada.
AA' = early fracture, BB' = later fracture
offsetting AA'. The breccia pipe, C, was
caused by renewed movement along AA'.
(Paul Gemmill, personal communication.)

fluids. Many pipes, however, show little or no alteration near their tops, a feature that should be developed where the rocks are partly dissolved.

Diatremes, or volcanic explosion vents, form where gases expand at explosive rates, causing the gases and rock to push violently upwards. The process is probably similar to the two-phase water-steam action in the throats of active geysers, and it may take place where ascending magmas suddenly encounter a porous, water-saturated clastic sediment (Williams, 1936). Numerous well-defined diatreme breccia pipes in northeastern Arizona have been described; they are funnel-shaped, with diameters that range in thickness from 4,000 feet near the top to 500 feet at depth (Hack, 1942). Since the diatremes form in areas of igneous activity and represent highly porous avenues for the escape of hydrothermal fluids, they stand a reasonable chance of becoming loci of pipe-shaped ore deposits.

Actually, this two-step process—development of a diatreme with subsequent hydrothermal activity—is not essential for all "volcanic" ore pipes. The explosive escape of gases from confined magmas and superheated meteoric waters may cause brecciation along a pipelike course, with ore deposition taking place during the waning stages; in this case both the breccia pipe and the metallization are attributed to a single agent. Under conditions of high-velocity gas flow, the rocks tend to expand, shatter, and rotate or become agitated. Continued acceleration of the gas will form streams or avenues of relatively high-velocity gas, which transport the rock fragments upwards. Such a process, known as *fluidization*, has been utilized in large-scale industrial operations for the intimate mixing of finegrained aggregates (Reynolds, 1954).

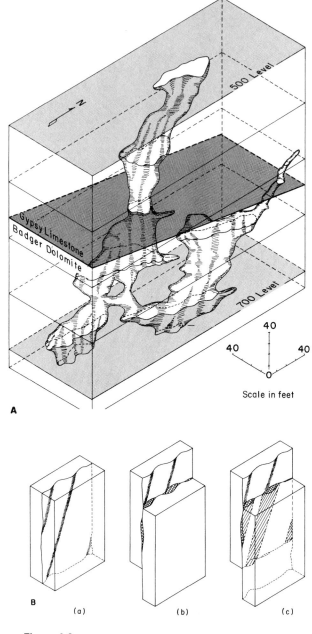

Figure 4-9
(A) Isometric drawing of an ore pipe, Bristol mine, Nevada.
(B) Probable development of breccia pipes at Bristol mine,
Nevada: (a) irregular fault surface before displacement;
(b) after vertical displacement, showing development of
swells: (c) pipes along which collapse breccias would form.

A well-defined example of a breccia pipe in Arizona has been attributed to explosive penetration of superheated steam (Barrington and Kerr, 1961). The pipe, circular in cross section and about 100 feet in diameter, supports a local, 300-foot high, cone-shaped hill known as Black Peak, and extends vertically through the Navajo Sandstone and underlying sediments. Its shape and diameter seem to persist with depth. Surrounding the brecciated axis is a 10-foot vesiculated collar of indurated sandstone and a more extensive, somewhat irregularly shaped zone of hydrothermal alteration. The alteration aureole is elongated parallel to a north-south fracture system, which also controls the orientation of small monchiquite dikes. The pipe breccia consists of subangular to rounded fragments of sandstone, mudstone, shale, and altered monchiquite in a friable sandstone matrix. These breccia fragments range from one to several inches in diameter, though sporadic blocks measure 2 feet across. Alteration of Navajo Sandstone around the pipe grades from bleaching, argillization, cementation, and partial recrystallization near the vesicular collar, to the development of illite from kaolinite in marginal or lower-grade zones. Silica, carbonates, alunite, and minor amounts of iron, manganese, gold, and silver are associated with the pipe.

Striking evidence for a pressurized, pneumatolytic origin of the pipe is the presence of vesiculation and the fact that the well-rounded sand grains are encircled radially with acicular quartz crystals (rather than cryptocrystalline quartz in concentric layers). Moreover, a liquid/solid (as opposed to a gas/solid) process would cause grading and only minor rounding of the breccia particles (Reynolds, 1954). Barrington and Kerr (1961) concluded that the breccia pipe was produced by the turbulent penetration of steam along a joint or joint intersection. They suggested that the brecciation and mixing resulted from a gas/solid fluidization system and that the lack of uranium and copper may be attributed to the solubilities of these metals rather than their absence from the gas. Theoretically, metallization of these metals may have occurred somewhere beyond (above) the section of pipe exposed at Black Peak. A single underlying body of cooling magma would account for the early monchiquite dikes and the later pipeforming steam jet.

A fairly common feature in some mineral districts is the presence of pebble dikes, narrow dike-like bodies that contain pebbles or rounded fragments, brought up from below by explosive activities (Bryant, 1974). Many dikes of this type have been found in Tintic, Utah, and Ouray, Colorado, for example. Geijer (1971) recently discussed the origin of the puzzling "ball ores" of Sweden and concluded that they have the same origin as pebble dikes.

When a sequence of beds is folded, the beds tend to slide over one another and undergo compression on the flanks and dilation along the crests, which are thus areas of low pressure. Fractures develop parallel to the axial planes of the folds and are especially prominent in competent rocks along the crests of anticlines, though they are present to a lesser extent along the troughs of synclines

Figure 4-10
Tightly folded Metaline Limestone adjacent to a
strong fault. Note the thickening of layers at the crest
of the fold and the presence of fractures parallel
to the axial plane. Natural size. (From Park and
Cannon, 1943.)

(see Fig. 4-10). The result of folding and the concomitant fracturing is an increase in permeability in the thickened beds at the crests of folds. Mineralizing
fluids moving along these permeable crests fill the openings and replace favorable rocks nearby. The resultant orebodies are pipe-shaped, and, where situated
above one another, are called *saddle reefs* (see Fig. 4-11). Excellent examples
of saddle reefs are the well-known Bendigo gold field, Victoria, Australia (Fig.
4-12) and the gold measures of Nova Scotia (Fig. 4-13).

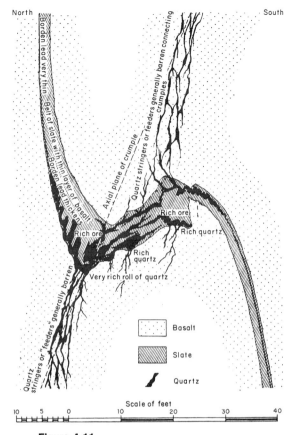

Figure 4-11
Detail of fold, West Lake mine, Mount Uniacke, Canada. (From Malcolm, 1912, Fig. 2.)

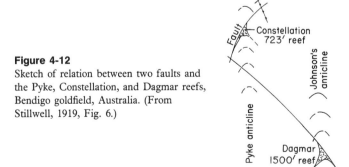

Figure 4-12
Sketch of relation between two faults and the Pyke, Constellation, and Dagmar reefs, Bendigo goldfield, Australia. (From Stillwell, 1919, Fig. 6.)

76

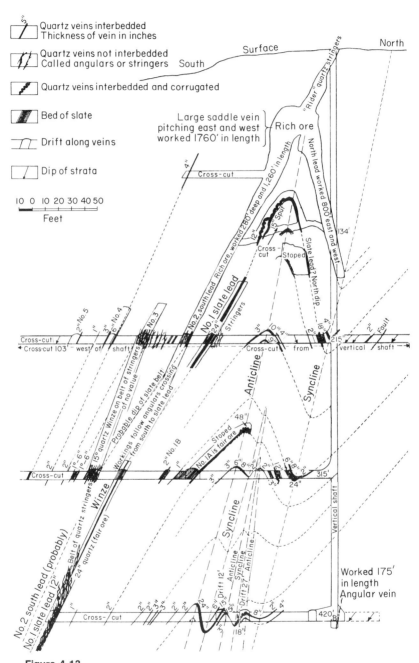

Figure 4-13
Saddle reefs in the Dufferin mine, Salmon River gold district, Nova Scotia. (From Malcolm, 1912, Fig. 9.)

Many iron and gold deposits in the Piedmont region of the southeastern United States are rod-shaped bodies, the long axes of which range in attitude from nearly horizontal to nearly vertical. Some of the rods are as much as 100 feet in cross section and have been mined down the dip of their long axes for more than 1,000 feet. Most, however, are small, especially in the gold-bearing areas. They may be no more than a few inches thick, although at Dahlonega, Georgia, they were so close together that the entire rock was mined at the surface. These orebodies are replacement saddle reefs formed at or near the crests of small, tight folds in schist (see Fig. 4-14). Much money and effort have been spent in an attempt to find vertical or down-dip continuations of the individual rods, on the erroneous assumption that these structures are tabular.

Pipelike orebodies that resemble saddle reefs may form where a fracture intersects a steeply dipping bed. The ore-bearing fluids migrate up the fault or bedding, or both, and the ore localizes along the footwall of the fault in permeable breccias or beneath impermeable strata. The shape of the deposits may suggest a folded structure, and for this reason they are known as *false saddle reefs* (Fig. 4-15).

Small amounts of ore are recovered from *ladder veins* (Fig. 4-16), which, as their name implies, are veins arranged in steplike or ladderlike form. They are ordinarily confined to dikes or competent strata that lie within shales or other incompetent rocks (Grout, 1923). Movement within the shales is taken up by flowage or cumulative displacements along foliation planes, whereas the competent rock tends to fracture. Moreover, ladder veins have been known to form where mineralizing solutions have invaded shrinkage joints in dikes, sills, and lava flows. As the igneous rock cools, it contracts, and cracks develop normal to the planar direction. The Morning Star gold mine at Wood's Point, Victoria, Australia, is a classic example of a ladder vein deposit (Threadgold, 1958). Gold-quartz veins fill a conjugate set of reverse faults developed across a 250-foot dike (Fig. 4-17).

Carbonate rocks below shales are favorable places to find ore, especially if faulting has taken place along the contact. Where a sequence of shales and carbonate rocks has been faulted or folded, the bedding slip is commonly taken up in the more mobile, fissile shales. The movement and drag are partly reflected in the limestones of dolomites, especially near the shale contact, increasing the permeability and permitting easier access of solutions.

Examples of Structural Control

Structurally controlled ore deposits abound. Rather than superficially describe a group of such deposits, we shall discuss three spectacular examples in some detail: The Trepça mine in Yugoslavia, the Tsumeb mine in South-West Africa, and the Rambler vein in the Slocan district of Canada.

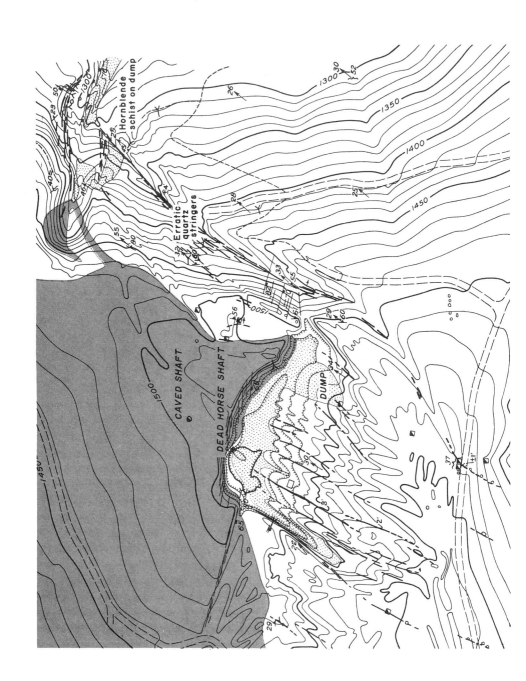

Hornblende schist on dump

Erratic quartz stringers

CAVED SHAFT

DEAD HORSE SHAFT

DUMP

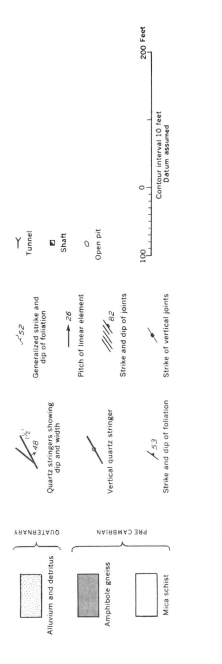

Figure 4-14
Gold-quartz rods formed at the crests of small folds in schist, near Dahlonega, Georgia.
(From Pardee and Park, 1948, Plate 48.)

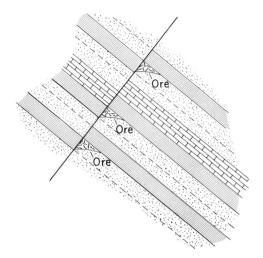

Figure 4-15
Sketch of false saddle reefs.

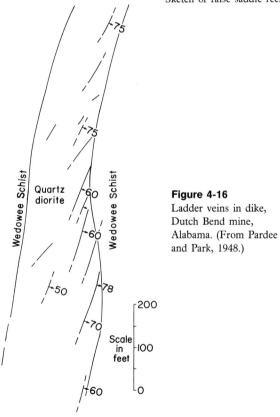

Figure 4-16
Ladder veins in dike,
Dutch Bend mine,
Alabama. (From Pardee
and Park, 1948.)

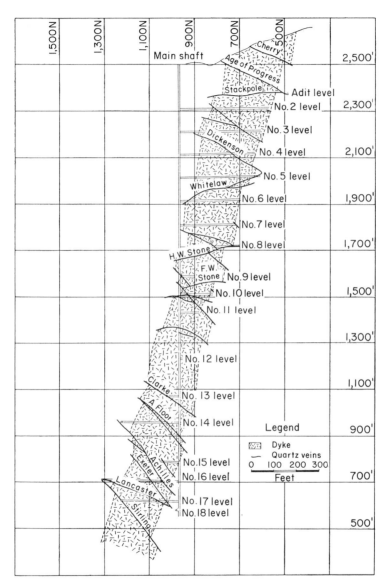

Figure 4-17
Cross section through the main shaft, Morning Star dike, Australia. Fissures are mineralized. (From Clappison, 1953, Fig. 3.)

Trepça Mine, Yugoslavia

The Trepça, or Stari Trg ("Old Market Place"), lead-zinc deposit is in southern Serbia, Yugoslavia, slightly over 200 kilometers south of Belgrade (see Fig. 4-18). The mine area has been studied by Forgan (1948), Schumacher (1950, 1954), and other geologists (Christie, 1950; Brammall, 1930).

Geologic formations consist of the Stari Trg Series, possibly of Ordovician-Silurian age, which forms a complex of schist, phyllite, quartzite, and marmorized limestone. Within this series is one fairly thick bed of pure, recrystallized limestone, called the Mazic Limestone, the principal mineralized stratum in the district. The Stari Trg Series is unconformably overlain and partly obscured by Miocene(?) andesite flows and tuffs, and is intruded by associated igneous rocks ranging from monzonite porphyry to hornblende andesite.

Figure 4-18
The Trepça mining district, Yugoslavia.

The older rocks of the region are sharply folded. An anticlinal structure, along which the mine is developed, plunges about 40°NW (Forgan, 1949). A nearly circular dacite and breccia pipe transects the Stari Trg-Mazic Limestone contact along the crest of the anticline. This pipe is a volcanic plug. At the surface and in the upper parts of the mine the pipe contains a core of dacite surrounded by a mantle of breccia. The dacite decreases in amount with depth and is absent in the lower levels of the mine. The breccia in the upper levels is composed of subangular fragments of the country rock and rounded fragments of dacite. In the deeper levels even the groundmass of the breccia is composed of crushed sedimentary materials.

Ore distribution was controlled both by the structure, which directed the fluid flow, and by the character of the country rock, part of which was reactive to the ore-bearing solutions. The ores are concentrated in the Mazic Limestone around the breccia pipe and along the limestone-schist contact; away from the pipe, the ore gradually thins. The intrusion and the breccia pipe modified the limestone-schist contact, so that in plan the upper contact of the limestone, against the schist and breccia, is M-shaped, with the pipe representing the upper re-entrant of the letter (see Fig. 4-19). The two top points of the M are the most favorable loci for ore, though mineralization does spread laterally in the limestone just below. Within the points of the M, the limestone is completely replaced by ore, from pipe to schist. The orebodies themselves are

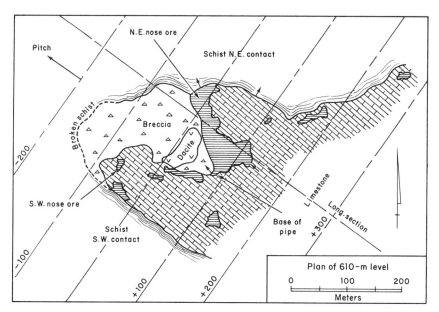

Figure 4-19
Trepça (or Stari Trg) mine, Yugoslavia. Plan of 610-meter level. (From Forgan, 1950.)

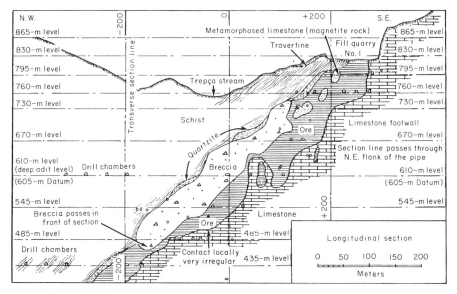

Figure 4-20
Trepça (or Stari Trg) mine, Yugoslavia. Longitudinal section. (From Forgan, 1950.)

pipe-shaped and plunge northwest, following the breccia pipe (see Fig. 4-20).

All significant deformation took place before the ore was deposited, as shown by the relationships between the ore bodies and the breccia pipe, and by the mineralization of the drag folds along the limestone-schist contact. The dacite was intruded along a zone of low pressure, and the jacket of breccia was apparently formed by subsequent explosive activity; thus it is a diatreme breccia. Ore deposition followed consolidation of the pipe, as a late phase of the same igneous activity; hence, mineralization took place in the favorable structures defined by the breccia and the schist-limestone contact (Forgan, 1948). The limestone reacted with the ore-bearing solutions and consequently was replaced in preference to the schist, breccia, and dacite.

The ore is a massive, coarse-grained sulfide mixture of silver-bearing galena, high-iron sphalerite (12 or 13 percent Fe), pyrite, and pyrrhotite, with minor amounts of arsenopyrite, jamesonite ($Pb_4FeSb_6S_{14}$), boulangerite ($Pb_5Sb_4S_{11}$), bournonite ($PbCuSbS_3$), and chalcopyrite. The galena contains recoverable bismuth in addition to the silver. Gangue minerals are minor in most of the ore body; they consist of quartz and the carbonates—calcite, rhodochrosite, siderite, manganosiderite, and dolomite. A skarn type of mineralization is found in parts of the mine and appears to increase with depth. The minerals of the skarn are principally actinolite, hedenbergite, ilvaite [$CaFe_2(FeOH)(SiO_4)_2$], garnet, and magnetite.

According to both Forgan (1948) and Schumacher (1950, 1954), the mineralization changes with depth. Galena remains fairly constant. Sphalerite is relatively more abundant in the upper levels than it is in the deeper workings, and becomes progressively more abundant away from the breccia pipe. Both silver and bismuth show a steady increase with depth. Pyrite and pyrrhotite are present in nearly equal proportions in the upper levels, but at depth the pyrrhotite predominates, especially near the borders of the pipe. Jamesonite is largely confined to the upper levels. Rhodochrosite also decreases in depth, but quartz increases. Skarn is restricted in the upper workings to the contact between limestone and dacite or breccia, but lower in the mine it is intimately mixed with the ore. The textures are coarse in the upper levels but finer-grained below. Along the flanks of the structure, however, the ore at depth is not greatly different from the ore in the upper levels.

These changes in mineralization, following a pattern symmetrical to the breccia pipe, are significant in the history of ore deposition. The mineralizing solutions evidently ascended along the breccia pipe, gradually changing in composition with time and with distance away from the pipe. The deep, near-pipe minerals, such as pyrrhotite and quartz, were deposited early, before the ascending solutions reached the upper levels; the later, relatively spent solutions produced sphalerite, jamesonite, and rhodochrosite. The different mineral assemblages probably reflect changing conditions, from availability of metal ions in time to a drop in temperature and pressure, a change in pH, or a relative increase in the partial pressure of sulfur. Such progressions in ore deposition are common, but Trepça is of special interest because some of the changes are unusual. In most lead-zinc deposits the Zn-Pb ratio increases with depth; at Trepça it decreases, yet the pyrrhotite-pyrite relationship is normal. Probably the most anomalous feature is the restriction of dacite to the upper parts of the breccia pipe; igneous intrusives are normally thought to originate at depth, but there is no sign of dacite low in the Trepça pipe.

Mining at Trepça is difficult because the limestone is highly fractured and contains much water. Ore is approached through inclines in the comparatively impermeable schist. All development in the water-saturated rocks is done behind the protection of isolating concrete dams. According to Christie (1950), the lower levels are intensely hot and uncomfortable. Visible ore reserves in 1950 were said to be between three and four million tons, with a grade of about 6.9 percent lead, 4.2 percent zinc, and 3.8 ounces of silver per ton. Possible reserves amounted to about 10 million tons.

Tsumeb Mine, South-West Africa

Another classic example of an ore pipe is the Tsumeb deposit, recently described by Söhnge (1963) and the staff of the Tsumeb Corporation (1961). The Tsumeb mine is in northern South-West Africa, 235 miles north of

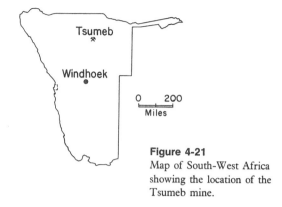

Figure 4-21
Map of South-West Africa
showing the location of the
Tsumeb mine.

Windhoek, the capital (see Fig. 4-21). Precambrian sediments of the Otavi Highland overlie older metamorphic and igneous rocks in the Tsumeb area. Mineralization is confined to a dolomite sequence known as the Tsumeb Stage of the Otavi Formation. This Upper Precambrian dolomite is a fine-grained, compact rock containing shale lenses, chert nodules, stromatolites, limestone beds, and oölitic zones. The sediments were folded into open synclines and anticlines and subsequently intruded by dikes, sills, and pipes of diabase, kersantite, and a rock known locally as pseudoaplite. Both the kersantite and the diabase are postore intrusives, but the pseudoaplite was emplaced before the ore-bearing fluids ascended along the pipe (Schneiderhöhn, 1929, 1931, 1941; Söhnge, 1952, 1963).

The mine is developed along a pseudoaplite pipe that cuts the north limb of a syncline. Petrographically, the pseudoaplite resembles a feldspathic quartzite because it consists of rounded and crushed grains of quartz, microcline, and sodic-plagioclase. No thermal metamorphism is associated with the pseudoaplite, but it is definitely intrusive (Söhnge, 1952, 1963).

The ore pipe, although it is basically an elliptical structure, pinches and swells irregularly and continues from the surface to a depth of at least 4,000 feet (see Figs. 4-22, 4-23, and 4-24). Along its constrictions it consists of a sparse network of thin veins, but it widens abruptly into plan sections of over 10,000 square feet. Its widest section, on the 2390-foot level, measures 600 feet by 250 feet (Söhnge, 1952).

The pipe was formed by fracturing, brecciation, and intrusion of pseudoaplite. Brecciation was restricted to the deeper levels, especially where the pipe cuts the axis of a fold. Reverse faulting along the bedding, which dipped about 45° to the south, caused drag folding, local thrusts across bedding, and brecciation. The pipe is nearly vertical above 700 feet and below 2,000 feet, but the intervening section follows a zone of concentrated bedding-plane faults, thus forming a relatively constricted pipe that plunges about 50° south. The sausage-like intrusive of pseudoaplite was injected after the folding, faulting, and brecciation. It is centrally located in the upper portion of the pipe, but branches and lenses irregularly through the breccia (Tsumeb Corporation Staff, 1961).

The origin of pseudoaplite and its associated ore minerals has been a long-standing enigma. Söhnge (1963) points out that it is neither a simple igneous nor a simple clastic dike—it seems to be a blend of the two. One possibility is that the materials were mobilized by deep-seated paligenesis and the escape of volatiles was made along a previously formed fracture zone. Clastic grains from lower formations would thus be partly dissolved and partly lifted (fluidized) from their original positions. A second hypothesis attributes the pseudoaplite and the ores to the intrusion of a highly volatile fraction of the Karroo magma that is cogenetic with the carbonatites that carry clastic fragments of quartzite derived from below.

Metallization follows the periphery of the pipe, sometimes forming rich pods and veinlike masses as well as large tonnages of disseminated, lowgrade ore. The main ore pipe is a horseshoe-shaped complex of veins with massive ore and intervening low-grade disseminations. High-grade mantos develop where the pipe intersects breccia zones along bedding slippages and project as much as 300 feet both north and south of the pipe. The pipe is widest where it crosses the bedding and narrowest where it parallels the bedding (see Fig. 4-21).

Tsumeb produces lead, zinc, and copper, with accessory germanium, silver, and cadmium, from ore containing such primary minerals as galena, sphalerite, tennantite, chalcocite, bornite, digenite, and enargite, plus subordinate pyrite, chalcopyrite, germanite $[Cu_3(Fe,Ge,Zn,Ga)(S,As)_4]$, reniérite $[(Cu,Fe)_3$ $(Fe,Ge,Zn,Sn)(S,As)_4]$, molybdenite, wurtzite, luzonite (Cu_3AsS_4), greenockite (CdS), and stromeyerite. Numerous products of oxidation are also found in the mine; these include many rare mineral species (Sclar and Geier, 1957; Moritz, 1933).

Individual masses or shoots of ore vary in their ratios of lead, copper, and zinc, but in general, lead is most abundant. Massive sulfide bodies contain up to 60 percent metal. One large sulfide mass contained 26.7 percent lead, 12.4 percent zinc, and 3.6 percent copper; high-grade copper ores have produced as much as 23 percent copper. The disseminated ores form stockworks (networks of fissures) in dolomite and breccia and interstitial specks and veinlets in the pseudoaplite. All rock containing over 3.5 percent metals is considered ore by the Tsumeb Corporation.

Gangue minerals and wall-rock alteration products are not abundant and, where found, consist of quartz, calcite, and dolomite. The wall rock is partially bleached and altered to calcite within the ore pipe, and the pseudoaplite is sericitized.

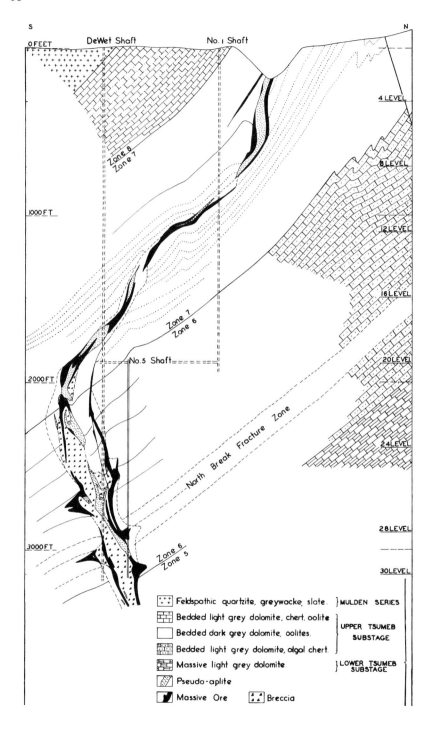

S ... N

OFEET

DeWet Shaft   No.1 Shaft

4 LEVEL

Zone 8
Zone 7

8 LEVEL

1000 FT

12 LEVEL

16 LEVEL

Zone 7
Zone 6

20 LEVEL

No.5 Shaft

2000 FT

North Break Fracture Zone

24 LEVEL

28 LEVEL

3000 FT

Zone 6
Zone 5

30 LEVEL

Feldspathic quartzite, greywacke, slate. } MULDEN SERIES

Bedded light grey dolomite, chert, oolite.
Bedded dark grey dolomite, oolites.
Bedded light grey dolomite, algal chert. } UPPER TSUMEB SUBSTAGE

Massive light grey dolomite } LOWER TSUMEB SUBSTAGE

Pseudo-aplite

Massive Ore   Breccia

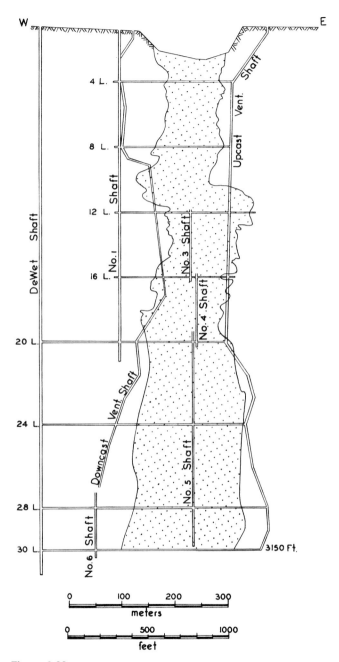

Figure 4-22
The Tsumeb ore pipe. Left, north-south geologic section; above, east-west
geologic section. (From Tsumeb Corporation Staff, 1961.)

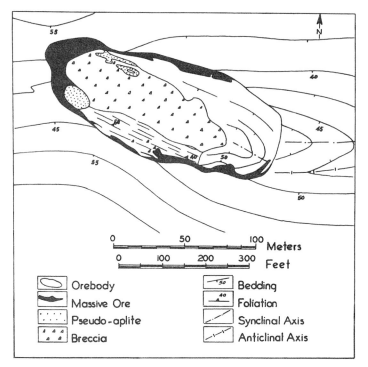

Figure 4-23
The Tsumeb pipe, 26th level, geologic plan.
(From Tsumeb Corporation Staff, 1961, Fig. 5.)

Although minor amounts of sulfides were at one time present, even at the surface, the Tsumeb ores have been oxidized to great depths. Supergene minerals predominate above 1,200 feet. Between 1,200 and 2,500 feet oxidation products diminish strikingly, but the lowest parts of the pipe are thoroughly oxidized. This unusually deep-seated oxidation zone is along permeable horizons in the Tsumeb dolomites, where ground waters migrating along bedding planes and brecciated strata attacked the sulfides even more efficiently than the near-surface waters. Supergene enrichment increased the copper values along upper portions of the pipe, though zinc was concomitantly leached and removed.

Since the first shipments in 1907, Tsumeb has produced over two million tons of metal, of which half or more was lead. The metals were recovered from seven million tons of ore, averaging 15.1 percent lead, 7.0 percent zinc, and 5.9 percent copper. In other words, the ore thus far produced has averaged 28 percent in metals, or eight times the established cutoff percentage—an enviable margin of profit (Pelletier, 1961).

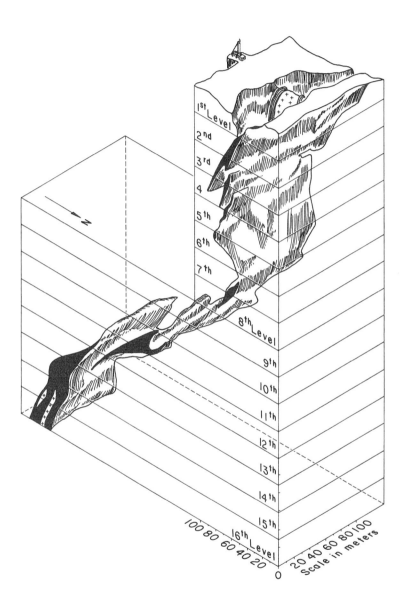

Figure 4-24
The Tsumeb orebody, isometric drawing. (From Schneiderhöhn, 1929, Fig. 9.)

Rambler Mine, British Columbia

The Rambler vein in the Slocan district of southeastern British Columbia (Fig. 4-25) offers a good example of structural control of ore deposition along a fault. The many ore deposits in the Slocan district are nearly all structurally controlled (Cairnes, 1934, 1935, 1948; Ambrose, 1957). The ore-bearing structures include fissures through competent rocks, directional changes along faults, brecciation against dike contacts, and intersections of fissures. For the sake of simplicity, only the Rambler vein (a combination of the first two types) will be described.

The Rambler orebodies (see Fig. 4-26) are shoots in a complex vein that are emplaced along a normal, oblique-slip fault with 40 to 100 feet of displacement determined by the dilation map method. Argillaceous sedimentary rocks of the Slocan Series and granodiorite porphyry sills and dikes are the only rock units near the Rambler mine. The Slocan Series is thought to be Triassic; the granodiorite is Late Cretaceous (Cairnes, 1934).

The Rambler fault cuts the sediments and intrusives at large angles. It varies in strike and dip; where it turns eastward, it widens or steepens, forming openings. The vein is thickest along these changes in attitude, and narrowest along the tight segments where the fault turns westward or flattens.

The Rambler orebodies are composite veins of massive sulfides, small sulfide veinlets, and disseminated sulfides in altered wall rock within the fault zone. The veins, which consist of galena, sphalerite, and pyrite in a subordinate gangue of quartz and siderite, range in thickness from a few inches to as much as 20 feet along major shoots. Argentiferous tetrahedrite and pyragyrite accompany the galena in significant amounts, and recoverable quantities of silver are found in some of the sphalerite and pyrite; thus silver is a major product of the mine.

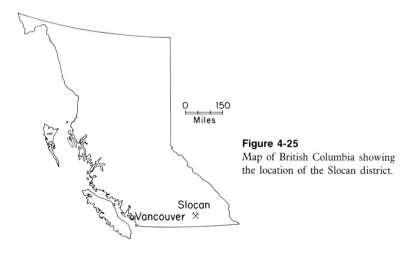

0   150
Miles

Figure 4-25
Map of British Columbia showing the location of the Slocan district.

Slocan
Vancouver

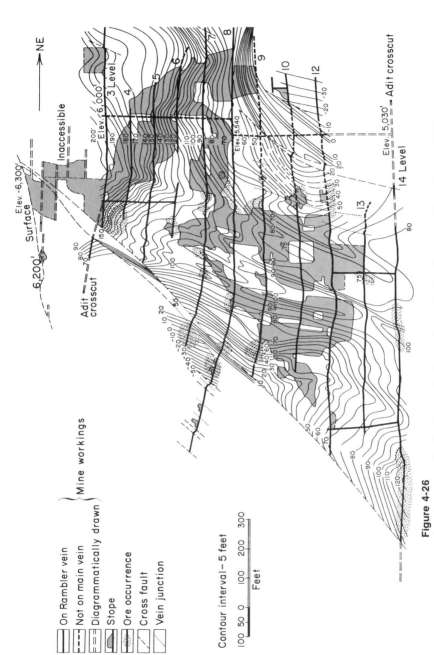

Figure 4-26

Ore shoot along the Rambler fault, Slocan district, British Columbia. (From Roscoe, 1951.)

Roscoe (1951) studied the details of ore distribution along the Rambler fault, and showed that the fault plane was refracted at points of contact between sediment and granodiorite. Wherever the attitude of the granodiorite-Slocan Series contact was favorably oriented with respect to the fault, an opening (or low pressure area) formed along the fault surface. Mineralization took place along such openings, since the orebodies were restricted to the dike swarms and were not found where the fault passed through massive granodiorite or thick, uniform sediments; moreover, the fault was planar except in the dike swarm.

By constructing dilation maps of the vein, Roscoe demonstrated that as fault displacement increased, the central portions of the ore shoots progressively widened. Furthermore, the mineralization was contemporaneous with faulting and spread outward from the central part of dilatant zones. Recurrent small displacements along the Rambler fault were accompanied by deposition of pyrite, sphalerite, and galena, in successive stages. Pyrite was the first mineral to be deposited. As dilation along the ore shoot continued, sphalerite moved into the central zone, replacing the pyrite, which migrated to the fringes. Subsequent movements allowed the galena and silver minerals to take over the central portion while both pyrite and sphalerite spread outward. The structural and paragenetic sequences are clear; the ore shoots grew in size and developed progressive zones of mineralization as faulting created an expanding area of pressure release. As a result, the ore shoots are disk-shaped bodies of composite veins located where the Rambler fault is favorably deflected across contacts between competent and incompetent rocks (see Fig. 4-26). The outside, tapered edge of each disk is predominantly pyrite; the central portion is sphalerite and galena. Where a dilatant area opened rapidly early in the faulting and then widened only slightly later, the shoot is predominantly pyrite with a little sphalerite in the core. Conversely, shoots that dilated late in the faulting contain a high ratio of galena to sphalerite and pyrite. Thus, the configuration and mineral composition of Rambler orebodies strictly depended upon structural events and not upon wall-rock chemistry.

The Rambler mine has been productive since 1895, though it was shut down between 1926 and 1947. Roscoe states that the average metal content of the ore mined has been about 25 ounces of silver per ton, and was 7.5 percent lead, and 5.5 percent zinc.

CHEMICAL CONTROLS

A favorable structural environment, even the presence in this structure of ore-bearing fluids, does not necessarily mean that ore will be deposited. Ore-bearing fluids react continuously with the wall rocks and, like the material they traverse, constantly change in composition. Moreover, factors such as reductions in temperature and pressure may either bring about chemical reactions

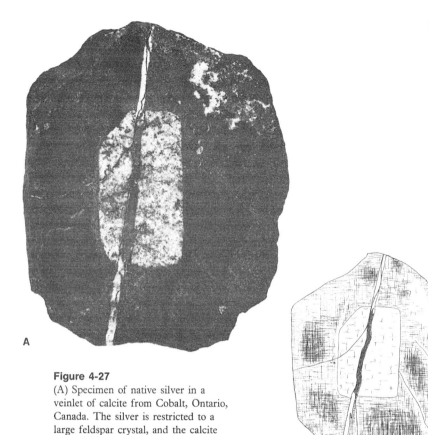

Figure 4-27
(A) Specimen of native silver in a veinlet of calcite from Cobalt, Ontario, Canada. The silver is restricted to a large feldspar crystal, and the calcite is in diabase. Note that the width of the veinlet is nearly constant across the specimen. (B) Sketch of specimen.

A

B

1 in.

or decrease solubilities and thus contribute to the deposition of ore minerals. Geologists are unable to explain why in many places certain beds are mineralized and others, above and below, are barren. These barren beds may have compositions and physical characteristics that appear to be identical with the mineralized strata. Even though the causes of chemical controls are not always clear, their existence can often be demonstrated. For example, Figure 4-27 shows a polished slab of diabase cut by a veinlet of calcite and native silver. The silver in the veinlet is restricted almost entirely to a large feldspar crystal, but calcite fills the veinlet where it passes through diabase. Note that the width of the veinlet is almost constant across the specimen. The silver is thought to have been deposited as a result of chemical reaction between the silver-bearing fluid and the feldspar crystal, rather than as a result of physical differences between the feldspar and the diabase. Although the fracture served as a permeable channel for introducing the fluid into the environment, the silver was deposited only along the chemically receptive part of the vein.

Much ground preparation for ore deposition is chemical. Silicification, dolomitization, and recrystallization are all chemical processes, and even much brecciated ground was made brittle by chemical reaction with earlier solutions.

One explanation of the common localization of ores in carbonate rocks beneath relatively impermeable covers is that ascending ore-bearing fluids are impounded and forced to move laterally into the more permeable carbonate rocks. Since carbonates are both permeable and chemically favorable host rocks, migration through them brought about by forced lateral movement allows ample contact for chemical reaction and precipitation of ore minerals. Carbonates are chemically reactive because they break down readily in the presence of acids and are relatively soluble in water. Just as limestones are preferentially dissolved in humid climates, so they are selectively replaced by mineralizing solutions.

Temperature and pressure are important in ore deposition. The solubilities of many substances increase in direct proportion to the temperature of the solution, so that cooling will precipitate any material whose saturation value has been exceeded. Some ore-bearing fluids may travel away from their source and deposit their metals when they reach a cooler zone, regardless of the host rock. A reduction in pressure may have a similar effect. An ore-bearing fluid flushed into a brecciated zone may deposit its load because of expansion and a resultant loss in pressure. Mechanisms that explain cooling in a hydrothermal system are discussed by Barton and Toulmin (1961) and by Sims and Barton (1962). Besides ordinary heat loss to wall rock, three other processes may be significant: (1) heat exchange in chemical reactions taking place in solution or in wall rock; (2) mixing of solutions with ground water; and (3) adiabatic expansion (including boiling) of the fluid, especially irreversible expansion through a constriction, commonly referred to as *throttling* (Barton and Toulmin, 1961).

Pressure losses would be especially effective in a supercritical fluid because the dissolving power of this phase is directly proportional to the molecular density, which is itself a function of pressure. As the pressure drops, the molecules become less densely packed, and any dissolved matter at the saturation point is precipitated. It has even been suggested that ore deposition may take place along constrictions in channelways because there the pressure decreases as a result of increases in the velocity of passing solutions, in accordance with the Bernoulli principle (Bain, 1936). Furthermore, the solubility of an ore mineral may depend upon the concentration of dissolved volatiles, such as H_2S or CO_2. A reduction in pressure allows these gases to leave the solution, thereby precipitating the ore minerals. The effectiveness of such a mechanism is debatable, but it probably operates in open-fissure deposits. Solutions that ascend through veins will naturally undergo decreases in pressure, and ore deposition may in part depend upon such decreases.

The stability of a solution may be determined by both the conditions of pH and the *oxidation potential* of the environment; a change in either could precipitate the dissolved materials. The oxidation potential (Eh) is a measure of the energy required to add or remove electrons from an element.* The ability of an environment to supply or accept electrons determines the valence of any ions present, and the valence may in turn determine whether or not the ion can remain in solution. For example, iron in simple solution is oxidized to ferric ion. Accordingly, the oxidation potential of the environment determines whether iron will go into solution or be deposited. The energy required for the environment to oxidize or reduce an element is modified by the pH of the solution. Thus, both the oxidation potential and the pH determine whether an ion will remain in solution or precipitate. The ultimate significance of these factors in hydrothermal solutions is unknown, but calculations and laboratory experiments indicate that they are important. Nevertheless, the control of oxidation potential and pH over the deposition of sedimentary ores and the modification of mineral deposits by weathering processes may be predicted with confidence. Thus, favorable environments for deposition of certain metals, such as iron and manganese, can be selected.

Sales and Meyer (1949) pointed out a significant relationship between the composition of the ore at Butte, Montana, and the concentration of sulfur in the mineral veins. A high sulfur ratio seems to have favored the stability of the simple copper and iron sulfides (chalcocite, covellite, and pyrite), whereas low ratios favored the combined copper-iron sulfides (bornite and chalcopyrite). Sulfur combined with the wall-rock iron to form pyrite, hence the ratio of sulfur to sulfide-forming metal ions in the vein solutions decreased with the distance from the source of supply. Yet at any single reference point along a vein the sulfur ratio *increased with time* because the entire system of ore solutions was expanding outward. That is, whereas at any given time the outermost minerals were copper-iron sulfides (because of a low sulfur ratio), subsequent sulfides deposited at the same location were simple sulfides (because the zone high in sulfur was expanding outward from the source area).

Mineral solubilities and the affinities of metallic elements for sulfur are of considerable interest in the study of ore deposition (Kolthoff and Sandell, 1952). Considerable experimental work has been done on metal-sulfur and metal-chloride complexes in an attempt to help explain both transportation and deposition of the metallic minerals (see Chap. 2). Experimental evidence indicates that such complexes play a leading part in ore genesis (Barnes, 1962; Helgeson, 1964; Hemley, 1953; Krauskopf, 1967; Treadwell and Hepenstrick, 1949; Treadwell and Schaufelberger, 1964). Since the solubilities of heavy metals change markedly in response to relatively small changes in composition of solutions, a solution that passes through a fissure may deposit the metals along a restricted portion of the vein where contamination by wall rocks has modified the solutions significantly; the gangue minerals tend to persist beyond these limits (Barton, 1959).

---

*The symbol Eh is used in this book, although E is now more commonly used. Oxidation potential is the same as oxidation-reduction potential.

Certainly in many sulfide deposits the abundance of sulfur in the ore-bearing fluids is a principal factor in controlling the character of mineralization. The formation of the massive pyrite-chalcopyrite deposits at Ducktown, Tennessee, at Yanahara, Japan, and at Rio Tinto, Spain obviously required large amounts of sulfur. Many volcanic areas contain abundant sulfur, which probably was transported as a gas (at least near the surface) and was then deposited either directly as a sublimate or through reaction with wall rocks.

Solfataric activity may account for the great density of native sulfur in the Matsuo deposit in Central Honshu, Japan. A bleached, gypsiferous tuff (Fig. 4-28) within a large crater is estimated to contain at least 160 million tons of ore with an average content of about 35 percent sulfur; an estimate of 50 million tons of native sulfur is reasonable. In addition, large amounts of sulfur are contained in the tremendous quantities of gypsum and in finely divided pyrite, and small amounts of arsenic, bismuth, and antimony sulfides are widely distributed in the ore. The ratio of pyrite to native sulfur is approximately 1:2. If the original tuff at Matsuo had been rich in iron, large bodies of massive pyrite probably would have formed instead of the native sulfur (Fujita, 1954).

The oldest rock in the Matsuo region is Upper Miocene Kitanomatagawa dacite. This is overlain by a Plio-Pleistocene sequence of lavas and sediments that includes the Hachimatai formation, which is the principal host rock of the ores (see Fig. 4-29). The sulfur and iron sulfide deposits are developed best in the augite hypersthene andesite (Takeuchi, Takahashi, and Abe, 1966).

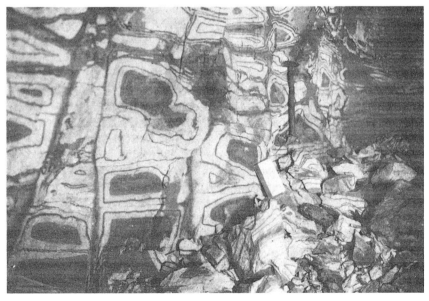

Figure 4-28
Gypsiferous tuff, Matsuo mine, Japan. (Courtesy of the Matsuo Company.)

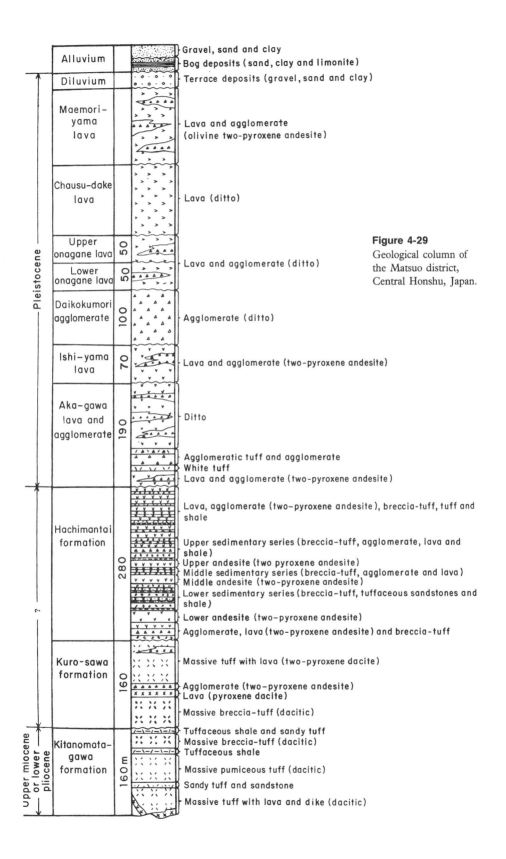

Gravel, sand and clay
Bog deposits (sand, clay and limonite)
Terrace deposits (gravel, sand and clay)

Lava and agglomerate
(olivine two-pyroxene andesite)

Lava (ditto)

Lava and agglomerate (ditto)

Agglomerate (ditto)

Lava and agglomerate (two-pyroxene andesite)

Ditto

Agglomeratic tuff and agglomerate
White tuff
Lava and agglomerate (two-pyroxene andesite)

Lava, agglomerate (two-pyroxene andesite), breccia-tuff, tuff and shale

Upper sedimentary series (breccia-tuff, agglomerate, lava and shale)
Upper andesite (two pyroxene andesite)
Middle sedimentary series (breccia-tuff, agglomerate and lava)
Middle andesite (two-pyroxene andesite)
Lower sedimentary series (breccia-tuff, tuffaceous sandstones and shale)
Lower andesite (two-pyroxene andesite)
Agglomerate, lava (two-pyroxene andesite) and breccia-tuff

Massive tuff with lava (two-pyroxene dacite)

Agglomerate (two-pyroxene andesite)
Lava (pyroxene dacite)
Massive breccia-tuff (dacitic)

Tuffaceous shale and sandy tuff
Massive breccia-tuff (dacitic)
Tuffaceous shale
Massive pumiceous tuff (dacitic)
Sandy tuff and sandstone
Massive tuff with lava and dike (dacitic)

Alluvium
Diluvium
Maemori-yama lava
Chausu-dake lava
Upper onagane lava   50
Lower onagane lava   50
Daikokumori agglomerate   100
Ishi-yama lava   70
Aka-gawa lava and agglomerate   190
Hachimantai formation   280
Kuro-sawa formation   160
Kitanomata-gawa formation   160 m

Pleistocene

Upper miocene or lower pliocene

?

Figure 4-29
Geological column of
the Matsuo district,
Central Honshu, Japan.

The ascending fluids are considered by Takeuchi and his associates to resemble those of present-day solfataras. The fluids have produced ten recognizable alteration zones: (1) silicified (2) iron sulfide ore (3) iron sulfide-sulfur ore (4) sulfur-iron sulfide ore (5) sulfur ore (6) kaolin-alunite-opal (7) kaolin (8) montmorillonite (9) saponite, and (10) quartz. The solutions that caused this alteration consisted of H_2O, H_2SO_4, H_2S, H_2SO_3, CO_2, H_2, N_2, Cl, O_2, sulfates, and other metallic and nonmetallic compounds. Where these ascending acidic fluids mixed with groundwaters, they were oxidized to H_2S-bearing sulfuric acid solutions. When iron was encountered, the reaction was

$$FeO \cdot Fe_2O_3 + H_2SO_4 \rightarrow FeSO_4 + Fe_2O_3 + H_2O \quad (1)$$

$$Fe_2O_3 + 2H_2SO_4 + H_2S \rightarrow 2FeSO_4 + 3H_2O + S \quad (2)$$

$$FeSO_4 + H_2S + S \rightarrow H_2SO_4 + FeS_2 \quad (3)$$

When iron was absent and aluminum was present, the reaction was

$$Al_2O_3 \cdot 2SiO_2 \cdot 2H_2O + 3H_2SO_4 \rightarrow Al_2(SO_4)_3 + 2SiO_2 \cdot H_2O + 3H_2O \quad (4)$$

$$2H_2S + O_2 \rightarrow 2H_2O + 2S \quad (5)$$

$$3H_2S + H_2SO_4 \rightarrow 4H_2O + 4S \quad (6)$$

Takeuchi and Abe (1970) discussed the hydrothermal alteration associated with the sulfur at the Matsuo deposits and they published a table showing the sequence of deposition of the ore minerals and the alteration products (Table 4-1).

Deposition of native sulfur by volcanic or solfataric activity was directly observed at another Japanese locality, the Siretoko-Iosan volcano in Hokkaido (Watanabe, 1940). Molten sulfur was ejected intermittently from a small geyser steam vent. The sulfur flowed down a valley for nearly a mile, reaching thicknesses of more than 16 feet and widths of 65 to 85 feet (see Fig. 4-30). The associated water was very acidic, owing to the presence of sulfuric acid. Much of the water was meteoric, however, and probably leached the sulfur from previously deposited materials in the underlying volcanic agglomerate. That is, the original magmatic fluids may have supplied sulfur in the form of H_2S, meaning that the eruptions of native sulfur were second-cycle products after remobilization. A seasonal and hourly periodicity, reflecting recharge of water, supports this hypothesis.

Certain ore deposits of near-surface or sedimentary origin have been attributed to the action of anaerobic bacteria or other organic processes. Some bacteria reduce the sulfur in sulfates, producing H_2S, which in turn may combine with any metals present in the solution to form metal sulfides. Other bacteria actually store sulfur granules within their cells, possibly forming native

Table 4-1
Paragenesis of alteration and ore minerals in the Matsuo mine, Japan.

	MATSUO MINE					
	Altered zone of sulfur deposit	Sulfur deposit	Iron-Sulfide ore deposit	Altered zone of iron-sulfide ore deposit	Limonite deposit
	Outer Inner			Inner Outer	
Chlorite						
Saponite						
Montmorillonite						
Sericite						
Kaolinite						
Halloysite						
Alunite						
Quartz						
Cristobalite						
Opal						
Sulfur						
Iron-sulfide ore						
Hematite						
Hydrohematite						
Goethite						
Jarosite						
Gypsum						
Barite						
Calcite and siderite						
Leucoxene						
Rutile						
Cinnabar						
Orpiment						
Realgar						
Livingstonite						
Horobetsuite						

(The table cells contain graphical bars and lens-shaped symbols indicating the relative abundance and distribution of each mineral across the deposit zones.)

Figure 4-30
Valley filled with native sulfur from the Siretoko-Iosan volcano, Hokkaido, Japan. (From Watanabe, 1940.)

sulfur deposits or combining with metal ions upon decomposition of the organisms. Oxides or hydroxides of metals, especially iron and manganese, may be deposited directly as a result of the life processes of bacteria. These bacteria are capable of oxidizing the metals. Other bacteria release oxidized metals from organic compounds used for food or simply accumulate the metals mechanically against their mucilaginous sheaths. The importance of these processes to ore depositions is very debatable, and has been a topic of great controversy in the literature (Harder, 1919; Schouten, 1946; Schneiderhöhn, 1923). Many large and important ore deposits have been considered products of bacterial activities; for most deposits, however, a hydrothermal origin cannot be disproved. If metals had been deposited by bacteria at the bottom of stagnant seas (like the present Black Sea), they should have been precipitated as simple minerals in an intimately mixed or rhythmic sequence; no such sulfide deposits are known (Edwards, 1956). Nevertheless, the ability of bacteria to precipitate metal compounds cannot be denied. This mechanism is certainly operative under special conditions, such as those that obtain in bog iron environments, and many geologists attribute the Kupferschiefer in Northern Europe to euxinic conditions. Jensen (1958) suggested that sandstone-type uranium ores owe their precipitation to the reducing action of H_2S generated from sulfates by anaerobic bacteria. Long after the H_2S was produced, according to his hypothesis, uranium-bearing solutions migrating through the sediments encountered the gas, which reduced the uranium from the soluble hexavalent, UO_4^{-2}, to the relatively insoluble tetravalent, UO_2. Accompanying iron and copper ions were deposited as sulfides.

Decaying organic matter has been known to precipitate certain metals in concentrations of economic value. Petrified logs and dinosaur bones with extremely high uranium contents are occasionally found, and coal or carbonaceous material may concentrate such elements as uranium, vanadium, molybdenum, germanium, nickel, titanium, gold, silver, lead, and zinc. Lignites or black shales may contain anomalous concentrations of any of these metals. Some organic matter acts as a reducing agent or supplies sulfide ions, causing the metals to become insoluble. Living organisms also concentrate certain metals, which are requirements of some plants and animals; the mechanism of enrichment in most life forms remains a mystery (Krauskopf, 1955).

A simple factor, possibly significant in ore deposition at shallow depths, is the precipitation of solids from hydrothermal solutions when the latter encounter connate waters or ground waters. Such interstitial waters are generally brackish and many contain large quantities of sulfate ions. Upon merging with these solutions, metal ions might be precipitated as sulfides or sulfates, depending upon the prevailing conditions of pH and oxidation potential. Ransome (1909) suggested that the bonanza gold ores at Goldfield, Nevada, were deposited where ascending ore-bearing solutions encountered descending sulfate waters; he pointed to the widespread development of alunite with the ores as evidence for this mechanism. Similarly, any colloid may be flocculated by ground waters that contain strong electrolytes.

The limestone replacement deposits of lead-zinc in the Pioche district (Fig. 4-31), Lincoln County, Nevada, offer a good example of a chemically controlled zone of deposition. Since operations began in 1869 (Westgate and Knopf, 1932), the district has produced about $100 million worth of ore. The principal ore deposits are in the Pioche Hills, a northwesterly striking range; related ore has been recovered from small mines scattered along the length of the Highland-Bristol Range to the west. Lower and Middle Cambrian rocks make up most of the mountain ranges, though Tertiary lava flows and tuffs cover large areas on all sides of the district. The Cambrian sequence grades from clastics at the bottom to carbonates at the top. A quartz monzonite stock of probable Tertiary age intrudes the Highland-Bristol Range along its west-central edge, and numerous granite porphyry and diabase dikes are found near the mining areas.

The mountain ranges are typical basin-range structures and are thoroughly broken by normal faulting (see Fig. 4-32). The ore-bearing solutions ascended along these faults; in fact most of the high-grade ore mined in the early days came from structurally controlled fissure veins in Lower Cambrian quartzites.

104

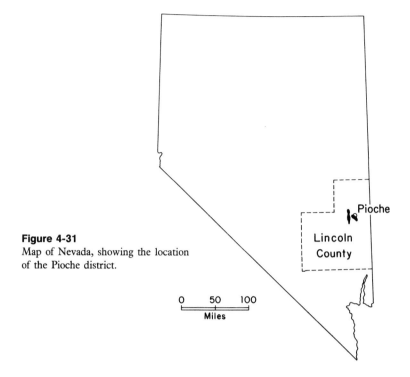

Figure 4-31
Map of Nevada, showing the location
of the Pioche district.

0   50   100
Miles

Where the mineralizing fluids passed through the quartzites and into the over-lying carbonate rocks, deposition was by replacement, generally in the first limestone intersected. Nearly all production since 1924 has come from deposits in these limestone beds.

The principal ore-bearing beds are in the Pioche Shale, of Lower to Middle Cambrian age. This formation contains several thin limestone beds that have undergone selective replacement by ore minerals. Carbonate formations overly-ing the Pioche Shale were also mineralized, but the limestone beds within the shale were the first carbonates met by ascending solutions and accordingly were the most highly mineralized.

The importance of reactive limestone to ore-bearing solutions is strikingly shown by the concentration of ores within the Pioche Shale. The lowermost bed of limestone, known locally as the Combined Metals (or C.M.) Limestone, has produced more than 90 percent of the replacement ores in the Pioche Hills and is also an important ore producer elsewhere. This limestone stratum is as much as 50 feet thick; although the lower half contains most of the ore, in places the whole bed is replaced by sulfides. Numerous replacement orebodies are localized along fractures that acted as channels for the ore-bearing solu-tions. Elongate, "bedded" orebodies, which reach widths of 400 feet, persist along single fractures as much as two miles long, though the C.M. bed (and

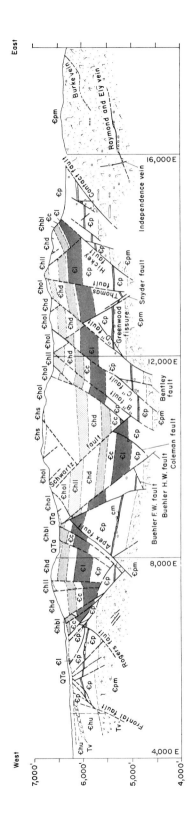

Figure 4-32

East-west cross section through the Pioche Hills, Nevada. Є‍pm = Prospect Mountain Quartzite; Є‍p = Pioche Shale, which includes the C. M. Limestone member (cm), Є‍l = Lyndon Limestone; Є‍c = Chisholm shale; Є‍hbl, Є‍hd, Є‍hll, Є‍hol, and Є‍hs are oldest to youngest members of Highland Peak Limestone (Є‍hu) mapped in this section; Tv = Tertiary volcanic rocks; QTa = Tertiary and Quaternary clastics. (After Park, Gemmill, and Tschanz, 1958.)

hence the ore zone) is offset at many places along this distance by preore cross-faults (Young, 1948).

The "bedded" orebodies, or mantos, consist of an intimate mixture of pyrite, argentiferous galena, and sphalerite. Near the edges of the orebodies mangano-siderite is abundant, suggesting that manganese traveled farther from the source before being deposited; similarly, manganese concentrations have been mined from some of the zones above the Pioche Shale.

At the Bristol mine, 14 miles northwest of Pioche, most of the orebodies are in limestones and dolomites above the Pioche Shale. Nevertheless, drilling has shown that the C.M. Limestone is mineralized with lead and zinc at depth, emphasizing that this bed is exceptionally receptive to ore deposition wherever it is cut by the metal-bearing fissures. In fact, the limestone beds in the Pioche Shale contain replacement deposits of lead-zinc in other Nevada mining districts, such as the Groom district, nearly 100 miles southwest of Pioche (Humphrey, 1945). The fluorite-tungsten-beryllium deposits in the Snake Range, about 75 miles to the north, are also in the same lithologic and stratigraphic position (Whitebread and Lee, 1961).

Where the ore-bearing solutions were confined to quartzite, they deposited their metals in veins. Where these fluids ascended beyond the quartzite without depositing all their metals, they encountered beds of limestone in the overlying Pioche Shale and younger formations. Since the carbonates are more porous than the shale, the solutions were able to move laterally; even more important, they were reactive to the rock and deposited their metals. If localization of the ore had been a direct function of the permeability of the limestone, mineralization would be concentrated along the upper half of the C.M. bed, rather than along its lower half. However, the ore-bearing fluids ascended along structures, forming the orebodies where these structures intersect the limestone beds. Thus, any mineralized fissures found in rocks above the Pioche Shale are strong evidence of the possibility of replacement deposits along the same structures where they intersect the C.M. Limestone at depth.

DEPOSITIONAL TEXTURES

Textures vary among ore deposits, depending upon the nature of the mineralizing fluids, the physical and chemical characteristics of the host rocks, and the mode of emplacement. Texture can be strong evidence for the manner of deposition. In syngenetic deposits—a category including sedimentary, magmatic segregation, and some pegmatite ores—textures reflect the proximity to a source or the rates of crystallization. Furthermore, resorption of early-formed crystals may produce peculiar textures in magmatic segregation ores (see Figs. 4-33 and 4-34), and diagenetic changes may modify the texture of sedimentary deposits. The textures of hydrothermal ores tend to be diversified.

Figure 4-33
Chromite grains in plagioclase. The concavities were produced by partial resorption of early formed chromite. From the Bushveld igneous complex, South Africa. ×20. (Photo by C. M. Taylor.)

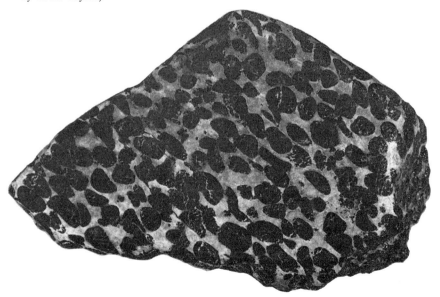

Figure 4-34
Orbicular chromite in altered peridotite. From Greece. Three-fourths natural size. (Photo by C. M. Taylor.)

Interpretation of mineral textures is extremely difficult and often problematical. In spite of the great amount of work done on mineral textures and structures, the causes of textures are poorly understood. One texture may have several interpretations. For example, the texture shown in Figure 3-4, p. 45, has been attributed to the process of exsolution (Schwartz, 1931), replacement (Loughlin and Koschmann, 1942), or, as for chalcopyrite and bornite, the removal of part of the iron from chalcopyrite by hot water or steam (Park, 1931). Brett (1964) called exsolution textures deceiving because they look as if they were formed by replacement. Formerly, such textures were thought to result from crystallization of minerals in eutectic proportions, though this idea is no longer seriously held. Excellent treatises on the textures of ore minerals are available and should be studied by serious students of ore deposits (Edwards, 1952, 1954; Van der Veen, 1925; Ramdohr, 1955; Bastin, 1950; Schneiderhöhn and Ramdohr, 1931; Schouten, 1934).

Ores may be deposited either by open-space filling or replacement. Ores deposited in openings probably followed textural or structural controls; those that replaced preexisting rocks were probably chemically controlled. Accordingly, it is of fundamental importance for the understanding and development of an ore deposit to ascertain whether the minerals originated by replacement or open-space filling, though a deposit formed by either mechanism alone would be an exception—open-space filling is likely to be accompanied by some replacement, and vice versa. Moreover, the textures developed may indicate whether the metals were carried in solution or as colloids. The three so-called types of textures—replacement, open-space filling, and colloidal—will be discussed separately.

Replacement

As defined by Lindgren (1933, p. 91), replacement, or metasomatism, is ". . . the process of practically simultaneous capillary solution and deposition by which a new mineral of partly or wholly differing chemical composition may grow in the body of an old mineral or mineral aggregate." Replacement ordinarily implies little or no change in the volume of the replaced rock, although in some rocks considerable shrinkage or expansion takes place. The process of metasomatism is of great significance in the emplacement of epigenetic ore deposits; many ores are deposited almost entirely in this manner, and nearly all ores show some evidence of replacement. The process is especially characteristic of those deposits formed at high temperatures and under high pressures where open spaces are scarce and where communication with the surface is impeded.

The efficacy of replacement is often astounding. The intimate preservation of plant cells or growth rings in petrified wood is well known, as is the fact that wood—a fibrous substance composed of carbon, hydrogen, oxygen, nitrogen, and minor elements—can be replaced by silica, even though there is no apparent similarity between the two substances. The replacement of one mineral by another may be equally striking and clear; fossils, sedimentary textures, and folded structures are commonly preserved in faithful detail (see Fig. 4-35). A compilation of the minerals that can replace one another indicates that there is practically no limit to the direction of metasomatism. As a bold generalization, it might be stated that given the proper conditions any mineral can replace any other mineral, though natural processes usually make for unilateral reactions. The important factor seems to be the chemical difference between the mineral or rock being replaced and the medium (liquid, gas, or wave of diffusing ions) causing the metasomatism. Hence, merely because quartz is stable at the earth's surface, we cannot conclude that quartz will resist metasomatism; in fact, quartz and the silicates very commonly undergo replacement. In contrast, a fluid that reacts with and replaces limestone may be inert to quartz, or vice versa; as a result, selective replacement may be of the most detailed character. Bastin and his colleagues (1931) proposed the following general rules for replacement: (1) sulfides, arsenides, tellurides, and sulfosalts replace all rock, gangue, and ore minerals (2) gangue minerals replace rock and other gangue minerals but do not commonly replace sulfides, arsenides, tellurides, and sulfosalts (3) high-temperature oxides replace all rock and gangue minerals but are rarely replaced by gangue minerals, and (4) oxides rarely replace sulfides, arsenides, tellurides, and sulfosalts.

Although replacement has been generally recognized and described for many years, the means by which the actual transfer of materials takes place has been the subject of much debate. One fundamental question is how the tremendous volumes of replaced materials are removed by the same solution that deposits the minerals. Presumably, the removed materials are simply transported from

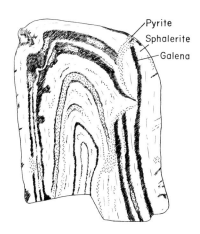

Pyrite
Sphalerite
Galena

Figure 4-35
Sketch of a specimen of ore from the Sullivan mine, British Columbia, showing folded layers of galena, sphalerite, and pyrite. This material has been interpreted both as replacement of a folded sediment by sulfides and, more recently, as a folded sulfide-rich sediment. One-half natural size.

the replacement front by the spent ore solutions by diffusion. The fact that replacement generally takes place volume-for-volume raises another major problem: how to write chemical equations representing electrically and molecularly balanced reactions with equal volumes of solid materials on each side.

Ridge (1949, 1961) attempted to illustrate replacement with equations that balanced molecularly, volumetrically, and electrically. He found that he could write equations acknowledging the known facts of chemistry and geology under near-surface conditions, but could develop only rough approximations for deep-seated reactions. Ridge's work helps to explain the replacement of one sulfide by another and also sheds light on the difficult problem of substituting the sulfur in sulfides for the oxygen in oxides, a process that must take place but seems illogical because of the large size of these ions. As an example of a problem faced by geologists, Ridge described the replacement of covellite by argentite (acanthite), a process likely to take place where surface outcrops are being leached and where deposition is taking place below the water table. For an aqueous, sulfate environment (without regard to the volumes), this reaction may be expressed as

$$2Ag^+ + CuS \rightarrow Ag_2S + Cu^{+2}. \tag{1}$$

According to Ridge, the solid phases would gain nearly 70 percent in volume during this reaction. Of course, such a volume change does not take place in nature. If we assume the oxidation of some sulfur, from cupric sulfide to cupric sulfate, and the concomitant reduction of some iron, from ferric to ferrous, a reasonable reaction may be written

$$17CuS\ (582.08\ \text{Å}^3) + 20Ag^+ + 10SO_4^{-2} + 56Fe^{+3} + 168Cl^- +$$
$$28H_2O \rightarrow 10Ag_2S\ (581.10\ \text{Å}^3) + 17Cu^{+2} + 17SO_4^{-2} +$$
$$56Fe^{+2} + 112Cl^- + 56H^+ + 56Cl^- \tag{2}$$

if the iron is in solution as a chloride, or

$$17CuS + 20Ag^+ + 10SO_4^{-2} + 56Fe^{+3} + 84SO_4^{-2} + 28H_2O \rightarrow$$
$$10Ag_2S + 17Cu^{+2} + 17SO_4^{-2} + 56Fe^{+2} +$$
$$56SO_4^{-2} + 56H^+ + 28SO_4^{-2} \tag{3}$$

if the iron is in solution as a sulfate. The volumes, shown in angstrom units, are the same whether the iron is considered a sulfate or chloride. Since either form of iron is likely under the conditions chosen, it is reasonable to assume that both reactions take place simultaneously. Similarly, in the weathering of a sulfide vein it is not unreasonable to expect fairly acidic conditions, which would be necessary to keep the ferric iron in solution. As written, the volume change in either reaction is only 0.17 percent. The same reaction can be expressed in terms of fewer molecules if a volume change of one or two percent is considered acceptable. Ridge illustrated such a reaction by the equation

$$5CuS\ (171.20\ Å^3) + 6Ag^+ + 3SO_4^{-2} + 16Fe^{+3} + 24SO_4^{-2} + 8H_2O \rightarrow$$
$$3Ag_2S\ (174.33\ Å^3) + 5Cu^{+2} + 5SO_4^{-2} +$$
$$16Fe^{+2} + 16SO_4^{-2} + 16H^+ + 8SO_4^{-2} \qquad (4)$$

for which there is a volume increase of 1.80 percent.

It will be noticed that equations (2), (3), and (4) introduce components on both sides of the equation, which, while taking part in the reaction, leave no solid evidence behind. In nature, there would be no trace of such materials.

The replacement of carbonates by sulfides and of silicates by oxides can be expressed in a similar manner. In some replacement reactions there is even less of a problem because the reactants may not have any ions in common; that is, there will be no chemical reaction that needs balancing. Ridge offers the replacement of limestone by sphalerite as an example; the simple, unbalanced expression is

$$Zn^{+2} + S^{-2} + CaCo_3 \rightarrow ZnS \times Ca^{+2} + CO_3^{-2}. \qquad (5)$$

The equation can be balanced by simple ratios based on the difference between the unit cell volumes of sphalerite and calcite. The resulting hypothetical equation

$$31Zn^{+2} + 31S^{-2} + 20CaCO_3\ (1220.60\ Å^3) \rightarrow$$
$$31ZnS\ (1220.47\ Å^3) + 20Ca^{+2} + 20CO_2^{-2} \qquad (6)$$

shows a volume difference of only 0.01 percent. The only problem for such a replacement is the physical transfer of materials, which pertains for any type of metasomatism and may be a matter of diffusion or practically simultaneous solution and redeposition.

Buerger (1948) pointed out that when the temperature of a mineral is raised, a point is reached at which the atomic structure is disordered but loosely tied together, allowing ions to diffuse through the crystal with ease. The process occurs most readily in sulfides because the tetrahedral coordination of sulfur ions makes for a relatively open crystal structure. Even at moderate temperatures sulfides take up ions from the surrounding solutions; as the ions are passed along, a wave of replacement moves away from the source of supply. The reasoning behind this explanation is like that of Fairbairn (1943), who suggested that minerals having a high packing index (high ratio of ion volume to unit cell volume) are more slowly replaced than loosely packed minerals. We would logically expect replacement to take place more readily in calcite (packing index = 4.0) than in quartz (packing index = 5.2), grossularite (packing index = 6.4), wollastonite (packing index = 5.2), or similar closely packed silicates. Such a relation is shown in tactite ores, in which interstitial calcite is generally selectively replaced before the silicates are attacked.

Limestone replacement under low temperature and pressure was studied by Garrels and Dreyer (1952). They were able to produce under controlled laboratory conditions replacement textures similar to natural ones. Many variables were examined, and the major control of replacement was thought to depend upon the pH of the mineralizing fluid, which in turn controls the solubility of the carbonate host rock. Garrels and Dreyer suggested that dissolving the limestone changes the ore-bearing solutions, causing the ores to be precipitated. Accordingly, they consider the solubility of the host to be the key factor in metasomatism. Ames (1961) proposed that the principal factor is the solubility of the replacement product relative to the solubility of the host rock in the same solution, rather than merely the solubility of the host. In any case, the importance of simultaneous dissolving action and precipitation from solution seems fairly well established.

Garrels and Dreyer also found that the numerous small, closely spaced openings create ideal conditions for replacement and a slightly higher secondary permeability. They concluded that the mineralizing solutions are carried along these zones of secondary permeability, but that the solutions move from the channels to the replacement front mainly by diffusion rather than forced flow.

The amount of replacement that can take place depends upon the amount of time the ore-bearing fluids are in contact with the host; that is, the distance through which ionic diffusion occurs is a function of time. Accordingly, metasomatism is likely to be the most thorough at great depths, where open-space circulation is very limited.

Much has been written about the criteria for recognizing replacement. Pseudomorphs and relict textures are considered diagnostic, but most others are only suggestive and may be formed in other ways. Except for a few diagnostic criteria, it is unwise to base conclusions upon single indications; confidence in any determination is directly proportional to the number of criteria available. A list of twenty of the more reliable criteria is given below. Although the scale of the accompanying illustrations is microscopic, most might be used in the field (Bastin *et al.*, 1931; Schouten, 1934).

1. *Pseudomorphs.* If the form of preexisting mineral is preserved, especially if the internal structure is also discernible, a replacement origin is indisputable.

2. *Widening of a fracture filling to an irregular mass where the fracture crosses certain chemically reactive mineral grains or rock layers.* If a veinlet widens across only one variety of mineral, it suggests that this mineral was receptive to replacement. A mineral vein widening into massive manto deposits along limestone beds is a large-scale example of this.

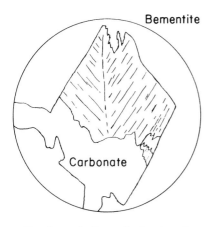

Bementite

Carbonate

Pseudomorph. Bementite (hatchured)
replacing calcite crystal. Olympic
Peninsula, Washington. ×60.

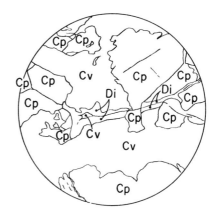

Widening of a fracture filling to an irregular
mass where the fracture crosses certain
chemically reactive mineral grains or rock
layers. Veinlet of digenite (Di) and covellite
(Cv) in chalcopyrite (Cp). Note how the
digenite veinlet pinches out where the
fracture crosses chalcopyrite. Cananea,
Mexico. ×37.5.

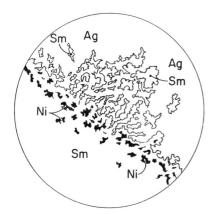

Formation of vermicular intergrowths at
wide places along cracks and at boundaries
of area not related to crystallographic
directions. Preferential replacement of
niccolite (Ni) by native silver (Ag) in intimate
admixture of niccolite in smaltite-chloanthite
(Sm). Cobalt, Ontario, Canada. ×30. (After
an unpublished photo by D. E. Eberlein.)

3. *Formation of vermicular intergrowths at wide places along cracks and at boundaries of areas not related to crystallographic directions.* The vermicular intergrowths may represent an advance wave of replacement, not yet completed. Replacement is not the only mechanism by which vermicular intergrowths are formed; they also develop during crystal growth in a eutectic mixture and by exsolution during the slow cooling of some solid solutions. Such primary vermicular intergrowths are typically related to crystallographic directions, and as a result, only the nonoriented intergrowths can be considered criteria of replacement.

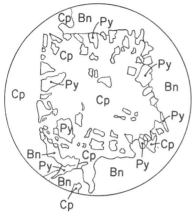

Islands of unreplaced host mineral or wall
rock. Pyrite (Py) cube largely replaced by
chalcopyrite (Cp). Bornite (Bn) borders part
of the grain. Bisbee, Arizona. ×37.5.

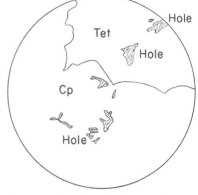

Concave surfaces into the host. Chalcopyrite
(Cp) replacing tetrahedrite (Tet). Coeur
d'Alene, Idaho. ×45.

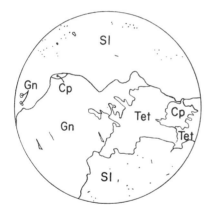

Nonmatching walls or borders of a fracture.
Galena (Gn), tetrahedrite (Tet), and
chalcopyrite (Cp) cutting sphalerite (Sl).
Cananea, Mexico. ×30.

4. *Islands of unreplaced host mineral or wall rock.* Chalcopyrite may replace
pyrite, but if the process is arrested before it goes to completion, remnants
of pyrite remain within the chalcopyrite.

5. *Concave surfaces into the host.* The diffusion of ions at the replacement
front goes on at different rates, so that some parts of the front form concave
reentrants, as if the replacing mineral bit into the host.

6. *Nonmatching walls or borders of a fracture.* If the replacement works outward
from a central fissure, the opposite fronts of replacement should not match
in detail, and may differ radically.

7. *Rims penetrating the crystallographic directions of the host mineral.* Replacement may work outward from any small fissures, including cleavages. For example, galena may be replaced by covellite along directions that are obviously parallel to the cleavages.

8. *Oriented unsupported fragments.* If a piece of the host is completely surrounded by the secondary (replacement) mineral and still maintains its orientation with respect to the host material on the outside, it is practically diagnostic of replacement. (The difficulty lies in proving that the fragment is unsupported.) The fragment may be any size, and the orientation may be proved by crystallographic directions, cleavage, bedding, or foliation.

9. *Selective association.* Since the chemistry is an important factor, some minerals or strata may be selectively replaced while others are left barren. For example, chalcocite may be found with chalcopyrite in preference to pyrite.

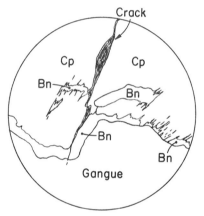

Rims penetrating the crystallographic directions of the host mineral. Bornite (Bn) in chalcopyrite (Cp). Cananea, Mexico. ×40.

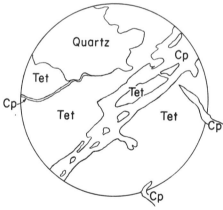

Oriented unsupported fragments. Tetrahedrite (Tet) in chalcopyrite (Cp). Coeur d'Alene, Idaho. ×45.

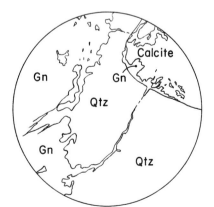

Selective association. Galena (Gn) replacing calcite in preference to quartz. Darwin, California. ×45.

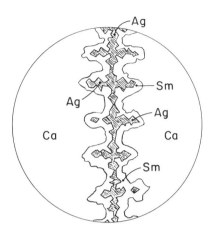

Physicochemical incompatibility between metacrysts and minerals of the host. Silver (Ag) and smaltite (Sm) in calcite (Ca). Cobalt, Ontario, Canada. ×30. (Photo by K. Bhatt.)

10. *Physicochemical incompatibility between metacrysts and minerals of the host.* If the metacrysts normally do not form by the same processes as the host material, replacement origin is suggested. The presence of chemically unrelated substances suggests that metacrysts and host are of different origins, and therefore that the origin of metacrysts is replacement rather than open-space filling. For example, pyrite crystals within calcite are anomalous because these minerals have no ions in common.

11. *Metacrysts transecting original structures.* The presence of a crystal that cuts across bedding or foliation suggests that the structure antedates the metacryst. If the crystal had grown by any process other than replacement, it would have pushed the structure aside.

Metacrysts of pyrite cutting foliation. Mother Lode, California. About nine-tenths natural size.

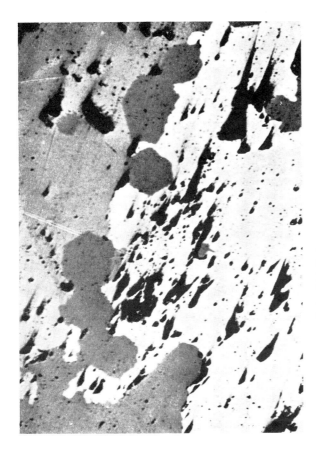

Metacrysts deposited in obvious relation to fractures, cleavage planes, or crystal boundaries. Quartz replacing siderite. Příbram, Czechoslovakia. ×160. (From Kutina, 1963.)

12. *Metacrysts deposited in obvious relation to fractures, cleavage planes, or crystal boundaries.* Since the ore-bearing fluids are introduced along small fractures, the development of crystals by replacement should take place in the wall rock alongside these passageways.

13. *Disparity in size of the metacrysts and the minerals of the host.* Large crystals in a fine-grained groundmass, and vice versa, may indicate that the metacrysts grew independently of the host rock.

14. *Metacrysts deposited along what was clearly an advancing zone of alteration.* If deposition took place by open-space filling, the ore minerals should stop abruptly against the host. Conversely, the replacement may have taken place by the gradual enlargement and merging of metacrysts along the replacement front. Such an advancing zone should be evident from the progressive increase in size of metacrysts and from the completeness of replacement from the wall rock into the ore.

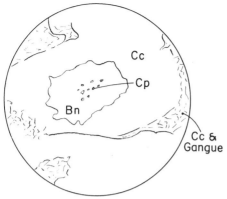

Depositional sequence in which minerals become progressively richer in one constituent. Bornite (Bn) with blebs of chalcopyrite (Cp) in the center, surrounded by chalcocite (Cc) and gangue. Tsumeb, South-West Africa. ×45.

15. *Depositional sequence in which minerals become progressively richer in one constituent.* For example, if polybasite ($Ag_{16}Sb_2S_{11}$) has been acted upon by silver-rich solutions, the replaced material should show a progressive enrichment in silver. A crystal or fragment of polybasite may grade first to argentite (Ag_2S) and then eventually to native silver, reflecting the gradual replacement of antimony and sulfur. An intermediate-stage specimen might have a core of polybasite, rimmed in sequence by argentite and native silver.

16. *Preservation of original structures and textures.* Certain features of sedimentary, igneous, or metamorphic rocks, as well as organic remains, may be preserved pseudomorphically. For example, folded structures or oölites may obviously be preore so their presence in ore is conclusive of replacement.

17. *Doubly terminated crystals.* If a crystal grows within an open cavity, it attaches itself to one wall and develops crystal faces only at the free end. This restriction does not affect crystals growing by replacement, so that doubly terminated crystals may indicate a replacement origin. Naturally, this criterion is limited because doubly terminated crystals may also develop within magmas and by other processes.

Doubly terminated quartz crystal from stibnite vein, Wolf Creek, Fairbanks district, Alaska, containing stibnite crystals. The quartz is said to be a replacement of gangue minerals in vein. ×18. (Collected by P. O. Sandvik; photo by W. J. Crook.)

18. *Gradational boundaries.* Replacement processes may produce either abrupt or gradational contacts between the host rock and the orebody. Since open-cavity filling usually forms abrupt contacts (at least on a microscopic scale), a gradational boundary indicates advancing replacement.

19. *Residual resistant minerals.* Some minerals are stable in the mineralizing solutions and may be left after the surrounding minerals have been replaced. For example, zircon or corundum may be found in the same proportions within the ore body as in a nearby schist; this would support the argument that the ore had replaced some of the schist. The resistant minerals are special types of islands or unreplaced fragments of host rock (see Criterion 4).

20. *No offset along the intersections of fractures.* Movement along a fissure offsets any planar or linear feature that transects the fissure obliquely, but does not offset an intersecting fracture. Both fractures may be replaced and tend to cross one another without any change in course, but only a coincidence of displacement would permit a rematch of intersecting veins that were spread apart and filled.

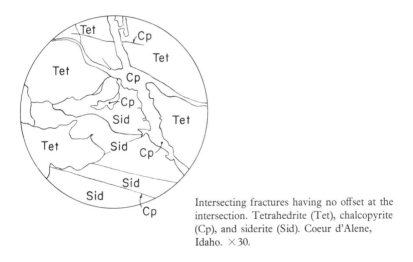

Intersecting fractures having no offset at the intersection. Tetrahedrite (Tet), chalcopyrite (Cp), and siderite (Sid). Coeur d'Alene, Idaho. × 30.

Open-Space Filling

Open-space filling is common in shallow zones where rocks yield by breaking rather than by flowing. The openings in these zones tend to remain open despite pressure from surrounding rock. At these shallow depths the ore-bearing fluids have relatively free circulation; their open connection with the surface permits deposition to be brought about by fairly rapid pressure and temperature changes rather than by prolonged contact with the surrounding rocks, which undergo slow chemical changes. Although the appearance of ores deposited in vugs and open cavities is readily distinguishable from that of replacement ores, the criteria associated with open-space deposition must nevertheless by used with caution because they are sometimes inconclusive (Kutina and Sedlackova, 1961). A list of the common criteria by which open-space filling may be recognized is given below.

1. *Many vugs and cavities.* If veins or orebodies contain many vugs and cavities, these may be interpreted as spaces left by incomplete filling of a larger open space. The inward growth of ore and gangue minerals in an open fissure stops when opposite walls meet. Since growth is not uniform along the fissure, the impeded circulation of the ore-bearing solutions leaves unfilled pockets. (See Figure 4-36a.)

2. *Fine-grained minerals on the walls of a cavity and coarser minerals in the center.* The first crystals that form along the sides of a vein are usually fine-grained because of heat loss to the wall rocks and consequent rapid crystallization. Conversely, crystals forming in the center have more time to develop; these probably form from more dilute, less viscous solutions that allow the ions to move freely. (See Figure 4-36b.)

3. *Crustification.* As ore-bearing solutions change in composition, they deposit different minerals along the walls of a vein or cavity. Early-formed crystals become encrusted with later minerals. The presence of small, euhedral dolomite crystals on top of large, euhedral fluorite crystals is a common example. Some veins have a banded appearance owing to crustification. (See Figure 4-36c.)

4. *Comb structure.* Along the junction of crystals that have grown from opposite walls of a fissure, there is generally an interdigitated, vuggy zone due to the meeting of pointed crystal ends. Because this jagged zone of juncture resembles the outline of a rooster's comb, it is known as comb structure.

A

1 in

B

Figure 4-36
(A) Many vugs left by incomplete vein-filling (quartz). Locality unknown.
(B) Fine-grained quartz grading into coarser crystals in a vug. Locality unknown.
(C) Crustification, comb structure, and open spaces in a veinlet of aragonite through limestone and shale. Locality unknown.

1 in

C

1 in

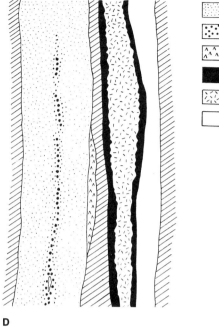

Gangue

Quartz

Sphalerite

Siderite

Ankerite

Calcite

D

Figure 4-36 *(continued)*
(D) Reopening and symmetrical deposition.
Matkobozska vein, Přibram, Czechoslovakia.
(From Kutina, 1957.)
(E) Cockade structure. Quartz in breccia.
Locality unknown.

E

1 in

5. *Symmetrical banding.* Crystals deposited along a cavity or fissure ordinarily grow symmetrically toward the center, in which case the orientation and composition of crystals on opposite sides of a vein are mirror images. As the ore fluid changes in composition, the minerals deposited will differ in composition, forming crustification in a symmetrical pattern inward from the vein walls. Such symmetrical banding may involve many mineralogical and color changes, producing a spectacularly banded vein. (See Figure 4-36*d.*)

6. *Matching walls.* Where a fissure has opened to allow filling, the outlines of opposite walls should match—that is, if the vein material were removed, the wall rocks on opposite sides would fit together like mated pieces of a jigsaw puzzle. The fit across many veins is readily apparent.

7. *Cockade structure.* Mineralization within the open spaces of a breccia (or any other fragmental rock) commonly produces a special pattern of symmetrical banding or crustification known as cockade structure. Each opening is a center for sequential deposition, and the overall rock presents a random pattern of host-rock fragments coated with layers of radiating crystals. (See Figure 4-36*e.*)

8. *Offset oblique structures.* Where a preexisting planar or linear feature intersects a vein obliquely, it is offset at right angles to the vein walls because it was spread apart as the fissure opened. A replacement vein would cause no offset along such a preexisting structure.

Colloidal Deposition

Amorphous minerals, such as opal, neotocite, wood tin, and garnierite, are thought to have been deposited from colloidal solutions. Moreover, it is believed that many cryptocrystalline minerals—chalcedony, some manganese oxides, pyrite, marcasite, pitchblende, and the oxidation products of copper, lead, and zinc (malachite, azurite, chrysocolla, anglesite, cerussite, and smithsonite)—were carried and deposited as colloids that were crystallized shortly after deposition. Some geologists think that gold, especially in the shallow bonanza deposits, traveled in the colloidal state. At the same time, many geochemists consider the existence of colloids in hydrothermal systems highly unlikely.

As a result of his study of colloform sphalerite specimens from many of the world's major geological districts, Roedder (1968) decided that most if not all of these ores grew directly as minute druses of continuously euhedral crystals projecting into an ore fluid. He found that all of the textural features believed to be diagnostic of colloidal deposition were ambiguous, inapplicable, and therefore invalid as criteria of the samples he studied, and perhaps of most other colloform mineral samples as well. However, Roedder did not explain the fact that such minerals as the manganese oxides, garnierite, and opal at times do not reveal any structural pattern under the X-ray; they are noncrystalline.

In a thoughtful discussion of Roedder's paper, Haranczyk (1969) concluded from his extensive studies in the Silesian–Cracovian lead and zinc deposits of Poland that colloform textures might form from both true and colloidal solutions. He proposed the term *hemicolloids* for solutions that are intermediate between true and colloidal ones. The cryptocrystalline, white, claylike form of sphalerite known as brunckite is almost certainly deposited as a colloid. Haranczyk later (1971) clearly showed that some of the second-generation minerals in the Silesian–Cracovian ore deposits of Poland could be of colloidal origin. He states that most of the second-generation sulfides show hemicolloidal textures.

An excellent discussion of the role of colloids in mineral formation was published by Lebedev (1967). He illustrated the independence of crystal structure and exterior mineral form by a stalactite of galena from Raibl, Austria. Ordinary calcite stalactities also may be used to show this feature since they are not uncommonly composed of a single crystal. Crystal structure and exterior form may have nothing to do with mineral transport. Colloids are relatively unstable under ordinary conditions and tend to crystalize as they "age." That a mineral is now crystalline is not proof that it was never a colloid.

Lebedev describes three examples that support his theory of the role of colloids in mineral formation: (1) the Shakh-Shagarla tin deposit, USSR, where botryoidal, wood-tin types of mineral are deposited on small crystals of cassiterite; (2) the Iokun'zh lead-zinc deposits of the Yano-Adychan region of USSR, which have many structures and textures similar to those of the deposits of Upper Silesia; and (3) the Pauzhetka Natural Steam deposits in southern Kamchatka, close to the Koshelev and Kambal'yni volcanoes. His book is well illustrated with photographs that were taken with an electron microprobe and magnified as much as 60,000 times. Mineraloids can be recognized and Lebedev concludes that some minerals in all three of these deposits have passed through a colloidal state.

That some minerals do pass through a colloidal state is firmly established, but the features by which colloidal deposition is recognized are not; most of them may be interpreted in other ways and should be considered only as indicative. The best-available features of colloidal deposition are listed below.

1. *Colloform textures.* Artificial colloids and natural materials that are thought to be of colloidal origin typically develop colloform structures where they extend into open spaces. Hence the presence of colloform textures is strong evidence in favor of colloidal deposition, especially if the botryoids approach sphericity—the shape created by surface-tension phenomena. Care must be taken to distinguish between the truly spheroidal colloform structures and other colloform deposits that develop from precipitation around corners of rock fragments. (See Figure 4-37a.)

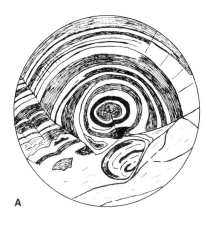

A

Figure 4-37
(A) Colloform texture. Globular sphalerite with concentric banding, considered to be of colloidal origin. Orzel Bialy mine, Katowice, Poland. Natural size. (From Kutina, 1953.)
(B) Shrinkage cracks. Ore from Larap, Philippine Islands. ×87. (Photo by J. E. Frost.)

B

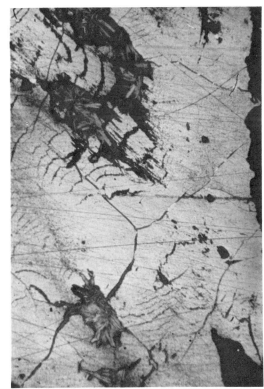

2. *Shrinkage cracks.* Laboratory gels develop shrinkage cracks due to dehydration; similar cracks should be expected in natural colloidal deposits. Their presence suggests colloidal deposition, but most so-called colloids show no signs of shrinkage. (See Figure 4-37*b*.)

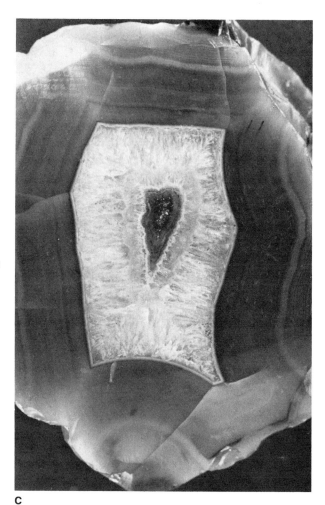

Figure 4-37 *(continued)*
(C) Diffusion banding as
illustrated by an agate. Note
also the pattern of filling in
the central vug. Locality
unknown. ×1.7. (Photo by
W. J. Crook.)

C

3. *Diffusion banding.* Colored bands, or Liesegang rings, may form in a gel if
   an electrolyte is allowed to diffuse into it. Liesegang rings are readily produced in laboratory gels, and similar banding in amorphous or microcrystalline rocks, such as agates, is generally interpreted to indicate a colloidal
   origin. (See Figure 4-37c.)

4. *Absorption of foreign materials causing a variable composition.* Colloids act
   as sponges for many ions because of their electrical charge, and they absorb
   constituents from any surrounding fluids that would otherwise remain in
   solution indefinitely. The presence of rare metals in psilomelane or wad
   may be due to this phenomenon.

5. *Chaotic, noncrystalline structure.* Amorphous minerals or mineraloids are thought to originate as colloids. Since colloidal gels are unstable, they tend to crystallize, so that the amorphous state would not be expected to last indefinitely. Thus, a colloidal precipitate may or may not be amorphous.

6. *Columnar crystals extending in crystallographic continuity through more than one color or compositional zone without interruption.* Colloidal masses that take on colloform shapes generally crystallize into clusters of radiating crystals trending normal to the periphery of the botryoids. Diffusion bands or color variations may also be present in them. Radiating, columnar crystals are able to extend through the diffusion bands because the crystallization was later than, and independent of, the formation of Liesegang rings.

7. *Spheroids.* Some masses of chert contain distinct spherulitic structures, like pisolites. The roundness results from surface tension, which exists in any liquid. If these spheroids were formed during crystallization of a gel, as some geologists maintain (Lindgren, 1925; Lindgren and Loughlin, 1919; Gilluly, 1932; Bastin, 1950), such textures would suggest colloidal deposition, though similar forms may develop in other ways.

A further criterion, used in the past, is the presence of framboidal textures. Clusters of tiny crystals or mineral grains grouped into a spheroidal mass are known as *framboidal spherules* (after *framboise*, the French word for raspberry, which the spherules resemble). The clusters range from about 4 to 50 microns in diameter, and the individual crystals are less than one-tenth the size of the spherule. Pyrite, chalcopyrite, bornite, and chalcocite mineral grains have been reported in spherules. Because framboidal textures have been found in ores thought to be of colloidal origin, they were at first considered indicative of colloidal deposition (Rust, 1935; Bastin, 1950). The framboidal masses were thought to be composite concretions or crystallized globules of gel in which crystallization started simultaneously from many centers. However, later studies showed that framboidal textures have a bacterial origin, the sulfide crystals of the framboids having filled chambers or cells in the organic structures (Love, 1962). The sulfides are probably precipitated out of solution (rather than from colloidal gels) because of the reducing action of H_2S produced by the bacteria (Jensen and Dechow, 1962). Kalliokoski studied the framboidal textures from the Leicester marcasite bed in New York, the Clinton ironstone in New York, and Wabana pyrite beds in Newfoundland. He concluded that the framboids were diagenetic products and that it was highly probable that organic materials and bacteria were necessary for their formation (Kalliokoski, 1966).

Sunagawa and his colleagues in Japan demonstrated that framboidal pyrite can be synthesized in the laboratory from true solutions (Sunagawa *et al.*, 1971). They state that gel phases are unnecessary and that replacement of microfossils likewise is not essential. They also point out that the framboidal pyrite formed at temperatures much higher than previously considered (between 200°–300°C). Framboidal textures can no longer be considered evidence of colloidal process.

Silica gel produced in the laboratory changes first to opal and then to a chalcedonylike material as it "ages" or dries and becomes compact. It undergoes a progressive volume reduction caused by the loss of water and develops characteristic shrinkage cracks. Similar gels in nature should also form shrinkage cracks, but since such cracks are seldom detected, many doubt the existence of natural colloids. Nevertheless, cracks attributed to the shrinkage of hardening colloids have been found in several districts (Gilluly, 1932; Roy, 1959; Kutina, 1952). The absence of dehydration cracks in most deposits of supposed colloidal origin may be due to a gradual accumulation of colloidal material, whereby individual, shell-like layers precipitate and solidify before the subsequent layer is deposited. This mechanism is known as the *Wiegner effect* (Liesegang, 1931; Park and Cannon, 1943). It takes place by means of the coagulation of small colloidal particles around nuclei of larger particles, and may bring about deposition from dilute colloidal systems. The Wiegner effect is significant for deposition or replacement along very small openings because it requires only thin layers of gel at any one time. Accordingly, a large mass of gel may never shrink from dehydration, but a sizable volume of solidified colloid with no visible shrinkage cracks may be built gradually.

Even though the part played by colloids in ore deposition is poorly understood, it seems likely that under near-surface conditions and in watery solutions at comparatively low temperatures the role may be appreciable. Many deposits have been described for which a colloidal origin is probable or in which colloids have played a supporting role. Among the best examples are the lead-zinc deposits of Upper Silesia (see Chap. 15).

An unusual deposit of colloform magnetite was found in an igneous metamorphic environment at the northern end of Vancouver Island, Canada (Stevenson and Jeffery, 1964). The colloform magnetite possessed the textural features usually ascribed to colloidal deposition, and the deposits appeared to have been formed by a type of gel metasomatism of the limestone. The iron, carried in HCl solution, replaced limestone and was precipitated as a colloid during an intermediate state of aggregation that existed between the state of ionic solution and the precipitate.

Hosking (1964) pointed out the presence of opaline gels that are apparently forming in the deposits of Cornwall, England. He also described the tin lodes of St. Agnes, where almost perfect concentrically zoned spherules of "wood tin" are suspended in quartz. These spherules are covered at places with acicular crystals of cassiterite whose long axes are normal to the spherule surfaces. He considers these spherules to have developed as the result of centripetal migration of tin-bearing fluids through a silica gel.

REFERENCES TO SELECTED DISTRICTS SHOWING
STRUCTURAL CONTROL OF ORE DEPOSITS

Balmat-Edwards District, New York

Brown, J. S., 1936*a*. Structure and primary mineralization of the zinc mine at Balmat, New York, *Econ. Geol.* 31:233-258.

————, 1936*b*. Supergene sphalerite, galena, and willemite at Balmat, New York, *Econ. Geol.* 31:331-354.

————, 1941. Factors of composition and porosity in lead zinc replacements of metamorphosed limestone, *Amer. Inst. Mining Eng. Trans.* 144:250-263.

————, 1942. Edwards-Balmat zinc district, New York, in *Ore Deposits as Related to Structural Features*, ed. W. H. Newhouse, Princeton, N.J.: Princeton Univ. Press.

————, 1947. Porosity and ore deposition at Edwards and Balmat, New York, *Geol. Soc. Amer. Bull.* 58:505-545.

Bryant, D. G., 1974. Intrusive breccias, fluidization and ore magmas, in *Mining Yearbook* Colorado Mining Association, pp. 54-58.

Cushing, H. P., and D. H. Newland, 1925. Geology of the Gouverneur quadrangle, *N.Y. State Mus. Bull.* 259.

Lea, E. R., and D. B. Dill, Jr., 1968. Zinc deposits of the Balmat-Edwards district, New York, in *Ore Deposits of the United States* (L. C. Graton-R. Sales Mem. Vol.), vol. 1, ed. J. D. Ridge, American Institute of Mining Engineers.

Newland, D. H., 1916. The new zinc mining district near Edwards, N.Y., *Econ. Geol.* 11:623-644.

————, 1917. Zinc-pyrite deposits of the Edwards district, New York, *N.Y. State Defense Counc. Bull.* 2.

Smyth, C. H., Jr., 1918. Genesis of the zinc ores of the Edwards district, St. Lawrence County, N.Y., *N.Y. State Mus. Bull.* 201.

San Juan Mountains, Colorado

Bastin, E. S., 1923. Silver enrichment in the San Juan Mountains, Colorado, *U.S. Geol. Surv. Bull.* 735-D, pp. 65-129.

Burbank, W. S., 1930. Revision of geologic structure and stratigraphy in the Ouray district of Colorado, and its bearing on ore deposition, *Colo. Sci. Soc. Proc.* 12(6):151-232.

————, 1933*a*. The western San Juan Mountains, *16th Int. Geol. Congr. Guidebook* 19, pp. 34-63.

————, 1933*b*. Vein systems of the Arrastre Basin and regional geologic structure in the Silverton and Telluride quadrangles, Colorado, *Colo. Sci. Soc. Proc.* 13(5):135-214.

————, 1940. Structural control of ore deposition in the Uncompahgre district, Ouray County, Colorado, *U.S. Geol. Surv. Bull.* 906-E, pp. 189-265.

————, 1941. Structural control of ore deposition in the Red Mountain, Sneffels, and Telluride districts of the San Juan Mountains, Colorado, *Colo. Sci. Soc. Proc.* 14(5):141-261.

————, 1951. The Sunnyside, Ross Basin, and Bonita fault systems and their associated ore deposits, San Juan County, Colorado, *Colo. Sci. Soc. Proc.* 15(7):285-304.

Burbank, W. S., E. B. Eckel, and D. J. Varnes, 1947. The San Juan region, in *Mineral Resources of Colorado*, ed. J. W. Vanderbilt, State of Colorado Mineral Resources Board.

Burbank, W. S., and R. G. Luedke, 1968. Geology and ore deposits of the western San Juan Mountains, Colorado, in *Ore Deposits of the United States* (L. C. Graton–R. Sales Mem. Vol.), vol. 1, ed. J. D. Ridge, American Institute of Mining Engineers.

Collins, G. E., 1931. Localization of ore bodies at Rico and Red Mountain, Colorado, as conditioned by geologic structure and history, *Colo. Sci. Soc. Proc.* 12(12):407–424.

Cross, W., and E. S. Larsen, 1935. A brief review of the geology of the San Juan region of southwestern Colorado, *U.S. Geol. Surv. Bull.* 843.

Larsen, E. S., Jr., and W. Cross, 1956. Geology and petrology of the San Juan region, southwestern Colorado, *U.S. Geol. Surv. Prof. Pap.* 258.

Moehlman, R. S., 1936. Ore deposits south of Ouray, Colorado, *Econ. Geol.* 31:377–397, 488–504.

Purington, C. W., 1905. Ore horizons in the San Juan Mountains, *Econ. Geol.* 1:129–133.

Ransome, F. L., 1901. Economic geology of the Silverton quadrangle, *U.S. Geol. Surv. Bull.* 182.

Bawdwin, Burma

Brown, J. C., 1917. Geology and ore deposits of the Bawdwin mines, Burma, *Geol. Surv. India Rec.* 48:121–178.

Clegg, E. L. G., 1941. A note on the Bawdwin mines, Burma, *Geol. Surv. India Rec.* 75(13).

Dunn, J. A., 1937. A microscopic study of the Bawdwin ores, Burma, *Geol. Surv. India Rec.* 72:333–369.

Loveman, M. H., 1917. The geology of the Bawdwin mines, Burma, Asia, *Amer. Inst. Mining Eng. Trans.* 56:170–194.

Roy, S. K., and S. Krishnaswamy, 1935. Notes on the microscopic characters of Bawdwin ores, *Geol. Mining Metall. Soc. India Quart. J.* 7(2):59–69.

REFERENCES CITED

Ambrose, J. W., 1957. Violamac mine, Slocan district, B.C., in *Structural Geology in Canadian Ore Deposits*, vol. 2, Montreal: Canadian Institute of Mining and Metallurgy.

Ames, L. L., Jr., 1961. The metasomatic replacement of limestones by alkaline, fluoride-bearing solutions, *Econ. Geol.* 56:730–739.

Bain, G. W., 1936. Mechanics of metasomatism, *Econ. Geol.* 31:505–526.

Barnes, H. L., 1962. Mechanisms of mineral zoning, *Econ. Geol.* 57:30–37.

Barrington, J., and P. F. Kerr, 1961. Breccia pipe near Cameron, Arizona, *Geol. Soc. Amer. Bull.* 72:1661–1674.

Barton, P. B., Jr., 1959. The chemical environment of ore deposition and the problem of low-temperature ore transport, in *Researches in Geochemistry*, ed. P. H. Abelson, New York: Wiley.

Barton, P. B., Jr., and P. Toulmin III, 1961. Some mechanisms for cooling hydrothermal fluids, *U.S. Geol. Surv. Prof. Pap.* 424-D, pp. 348–352.

Bastin, E. S., 1950. Interpretation of ore textures, *Geol. Soc. Amer. Mem.* 45.

Bastin, E. S., L. C. Graton, W. Lindgren, W. H. Newhouse, G. M. Schwartz, and M. N. Short, 1931. Criteria of age relations of minerals, with special reference to polished sections of ores, *Econ. Geol.* 26:561–610.

Blanchard, R., 1947. Some pipe deposits of eastern Australia, *Econ. Geol.* 42:265–304.

Bowen, N. L., 1928. *The Evolution of the Igneous Rocks*, Princeton Univ. Press. Reprint, 1956, New York: Dover.

Brammall, A., 1930. The Stantrg lead-zinc mine, Yugoslavia, *Mining Mag. London* 42:9-15.

Brett, R., 1964. Experimental data from the system Cu-Fe-S and their bearing on exsolution in ores, *Econ. Geol.* 59:1241-1269.

Brown, A. S., 1957. Red Rose tungsten mine, in *Structural Geology of Canadian Ore Deposits*, vol. 2, Montreal: Canadian Institute of Mining and Metallurgy.

Bryant, D. G., 1974. Intrusive breccias, fluidization and ore magmas, in *Mining Yearbook, Colorado Mining Association*, p. 54-58.

Bryner, L., 1961. Breccia and pebble columns associated with epigenetic ore deposits, *Econ. Geol.* 56:488-508.

———, 1967. Geology of the Barlo mine and vicinity, Dasol, Pangasinan province, Luzon, Philippines, *P.I. Bur. Mines Rept. Invest.* 60.

Buerger, M. J., 1948. The role of temperature in mineralogy, *Amer. Mineral.* 33:101-121.

Butler, B. S., 1913. Geology and ore deposits of the San Francisco and adjacent districts, Utah, *U.S. Geol. Surv. Prof. Pap.* 80.

Butler, B. S., and W. S. Burbank, 1929. The copper deposits of Michigan, *U.S. Geol. Surv. Prof. Pap.* 144.

Cairnes, C. E., 1934. Slocan mining camp, British Columbia, *Can. Geol. Surv. Mem.* 173.

———, 1935. Descriptions of properties, Slocan mining camp, British Columbia, *Can. Geol. Surv. Mem.* 184.

———, 1948. Slocan mining camp, in *Structural Geology of Canadian Ore Deposits*, Montreal: Canadian Institute of Mining and Metallurgy.

Christie, J. J., 1950. Inside Yugoslavia—II, The Trepça mine's great possibilities, *Eng. Mining. J.* (June), pp. 77-79.

Clappison, R. J. S., 1953. The Morning Star mine, Wood's Point, in *Geology of Australian Ore Deposits*, ed. A. B. Edwards, Melbourne: Australasian Institute of Mining and Metallurgy.

Dawson, J. B., 1966. Oldoinyo Lengai—an active volcano with sodium carbonatite lava flows, in *Carbonatites*, ed. O. F. Tuttle and J. Gittins, New York: Wiley.

Deans, T., 1966. Economic mineralogy of African carbonatites, in *Carbonatites*, ed. O. F. Tuttle and J. Gittins, New York: Wiley.

Edwards, A. B., 1952. The ore minerals and their textures, *Roy. Soc. N.S.W. J. Proc.* 85:26-45.

———, 1954. *Textures of the Ore Minerals and Their Significance*, Melbourne: Australasian Institute of Mining and Metallurgy.

———, 1956. The present state of knowledge and theories of ore genesis, *Australas. Inst. Mining Metall. Proc.* 177:69-116.

———, 1960. Contrasting textures in the silver-lead-zinc ores of the Magnet mine, Tasmania, *Neues Jahrb. Mineral. Petrogr. Abh.* 94:298-318.

Emmons, W. H., 1938. Diatremes and certain ore-bearing pipes, *Amer. Inst. Mining Eng. Tech. Pub.* 891.

Fairbairn, H. W., 1943. Packing in ionic minerals, *Geol. Soc. Amer. Bull.* 54:1305-1374.

Fenner, C. N., 1948. Immiscibility of igneous magmas, *Amer. J. Sci.* 246:465-502.

Finucane, K. J., 1948. Ore distribution and lode structures in the Kalgoorlie goldfield, *Australas. Inst. Mining Metall. Proc.* 148-149, pp. 111-129.

Forgan, C. B., 1948. Ore deposits at the Stantrg lead-zinc mine, *18th Int. Geol. Congr. Rept.*, pt. 7, pp. 162-176.

Fujita, Y., 1954. Three-way geophysical method points up huge pyrite deposit, *Eng. Mining J.* (Dec.), pp. 84-88.

Garnett, R. H. T., 1966. Relationship between tin content and structure of lodes at Geever mine, Cornwall, *Inst. Mining Metall. London Trans.* (sec. B) 75:B1-B22.

Garrels, R. M., and R. M. Dreyer, 1952. Mechanism of limestone replacement at low temperatures and pressure, *Geol. Soc. Amer. Bull.* 63:325-379.

Geijer, P., 1971. Sulfidic "ball ores" and the pebble dikes, *Sveriges Geol. Unders. Arsb. Ser. C Avh. Uppsatser* 65(8):1-29.

Gilluly, J., 1932. Geology and ore deposits of the Stockton and Fairfield quadrangles, Utah, *U.S. Geol. Surv. Prof. Pap.* 173.

Grout, F. F., 1923. Occurrence of ladder veins in Minnesota, *Econ. Geol.* 18:494-505.

Gunning, H. C., 1948. Privateer mine, in *Structural Geology of Canadian Ore Deposits,* Montreal: Canadian Institute of Mining and Metallurgy.

Hack, J. T., 1942. Sedimentation and volcanism in the Hopi Buttes, Arizona, *Geol. Soc. Amer. Bull.* 53:335-372.

Haranczyk, C., 1969. Noncolloidal origin of colloform texture, *Econ. Geol.* 64:466-468.

———, 1971. Colloidal transport phenomena of zinc sulfide (brunckite) observed in the Olkusz mine in Poland, in *Geochemistry and Crystallography of Sulphide Minerals in Hydrothermal Deposits* (Joint Symposium Vol.: IMA-IAGOD Mtgs. '70), ed. Y. Takéuchi, Tokyo: Society of Mining Geologists of Japan (spec. issue 2).

Harder, E. C., 1919. Iron depositing bacteria and their geologic relations, *U.S. Geol. Surv. Prof. Pap.* 113.

Hayward, M. W., and W. H. Triplett, 1931. Occurrence of lead-zinc ores in dolomitic limestones in northern Mexico, *Amer. Inst. Mining Eng. Tech. Pub.* 38, class I.

Helgeson, H. C., 1964. *Complexing and Hydrothermal Ore Deposition,* New York: Macmillan.

Hemley, J. J., 1953. A study of lead sulfide solubility and its relation to ore deposition, *Econ. Geol.* 48:113-138.

Hosking, K. F. G., 1964. Permo-Carboniferous and later primary mineralisation of Cornwall and south-west Devon, in *Present Views of Some Aspects of the Geology of Cornwall and Devon,* ed. K. F. G. Hosking and G. J. Shrimpton, Penzance: Royal Geological Society of Cornwall.

Humphrey, F. L., 1945. Geology of the Groom district, Lincoln County, Nevada, *Univ. Nev. Bull.* 39(5).

James, J. A., 1949. Geologic relationships of the ore deposits in the Fredericktown area, Missouri, *Mo. Geol. Surv. Water Resources Rept. Invest.* 8.

Jensen, M. L., 1958. Sulfur isotopes and the origin of sandstone-type uranium deposits, *Econ. Geol.* 53:598-616.

Jensen, M. L., and E. Dechow, 1962. The bearing of sulphur isotopes on the origin of the Rhodesian copper deposits, *Geol. Soc. S. Afr. Trans.* 65(pt. 1):1-17.

Kalliokoski, J. O., 1966. Diagenetic pyritization in three sedimentary rocks, *Econ. Geol.* 61:872-885.

Killinger, P. E., 1942. Report on the titanium mine at Tahawus, New York, *Rocks & Minerals* 17:409.

King, B. C., 1965. Petrogenesis of the alkaline igneous rock suites of the volcanic and intrusive centres of Uganda, *J. Petrology* 6:67-100.

Kolthoff, I. M., and E. B. Sandell, 1952. *Textbook of Quantitative Analysis,* New York: Macmillan, pp. 67-68.

Krauskopf, K. B., 1955. Sedimentary deposits of rare metals, *Econ. Geol. (50th Anniv. Vol.),* pp. 411-463.

———, 1967. *Geochemistry,* New York: McGraw-Hill, pp. 501-504.

Kuhn, T. H., 1941. Pipe deposits of the Copper Creek area, Arizona, *Econ. Geol.* 36:512-538.

Kulp, J. L., W. U. Ault, and H. W. Freely, 1956. Sulfur isotope abundances in sulfide minerals, *Econ. Geol.* 51:139-149.

Kutina, J., 1952. Mikroskopischer und spektrographischer Beitrag zur Frage der Entstehung einiger Kolloidalstrukturen von Zinkblende und Wurtzite, *Geologie* 1:436–452.

——, 1955. Genetische Diskussion der Makrotexturen bei der geochemischen Untersuchung des Adalbert Hauptganges in Pribram, *Chemie der Erde* 17:241–323.

——, 1957. A contribution to the classification of zoning in ore veins, *Univ. Carolina Geol.* 3(3):197–225.

Kutina, J., and J. Sedlackova, 1961. The role of replacement in the origin of some cockade textures, *Econ. Geol.* 56:149–176.

Lebedev, L. M., 1967. *Metacolloids in Endogenic Deposits*, tr. J. B. Southard, New York: Plenum Press.

Liesegang, R. E., 1931. Colloid chemistry and geology, in *Colloid Chemistry*, Vol. 3, ed. J. Alexander, New York: Chemical Catalog Co.

Lindgren, W., 1925. Metasomatism, *Geol. Soc. Amer. Bull.* 36:247–261.

——, 1933. *Mineral Deposits*, 4th ed., New York: McGraw-Hill.

Lindgren, W., and G. F. Loughlin, 1919. Geology and ore deposits of the Tintic mining district, Utah, *U.S. Geol. Surv. Prof. Pap.* 107.

Locke, A., 1926. The formation of certain ore bodies by mineralization stoping, *Econ. Geol.* 21:431–453.

Loughlin, G. F., and A. H. Koschmann, 1942. Geology and ore deposits of the Magdalena mining district, New Mexico, *U.S. Geol. Surv. Prof. Pap.* 200.

Love, L. G., 1962. Biogenic primary sulfide of the Permian Kupferschiefer and marl slate, *Econ. Geol.* 57:350–366.

Lovering, T. G., 1962. The origin of jasperoid in limestone, *Econ. Geol.* 57:861–889.

Lovering, T. S., 1942. Physical factors in the localization of ore, in *Ore Deposits as Related to Structural Features*, ed. W. H. Newhouse, Princeton, N.J.: Princeton Univ. Press.

MacDiarmid, R. A., 1959. Geology and ore deposits of the Bristol silver mine, Pioche, Nevada, Ph.D. dissertation, Stanford Univ., California.

——, 1960. Controls on ore deposition at the Bristol silver mine, Pioche, Nevada, *Geol. Soc. Amer. Bull.* 71:1921.

Malcolm, W., 1912. Gold fields of Nova Scotia, *Can. Geol. Surv. Mem.* 20-E.

McDougall, D. J., 1968. A "lattice defect-free energy" approach to replacement processes in ore deposition, *Econ. Geol.* 63:671–681.

Moritz, H., 1933. Die sulfidischen Erze der Tsumeb-Mine von Ausgehenden bis zur XVI. Sohle (−460 m), *Neues Jahrb. Mineral. Geol. Paläontol.* supp. 67 (sec. A): 118–154.

Ohle, E. L., and J. S. Brown, eds., 1954. Geologic problems in the southeast Missouri lead district, *Geol. Soc. Amer. Bull.* 65:201–222.

Pardee, J. T., and C. F. Park, Jr., 1948. Gold deposits of the Southern Piedmont, *U.S. Geol. Surv. Prof. Pap.* 213.

Park, C. F., Jr., 1931. Hydrothermal experiments with copper compounds, *Econ. Geol.* 26:857–883.

Park, C. F., Jr., and R. S. Cannon, Jr., 1943. Geology and ore deposits of the Metaline quadrangle, Washington, *U.S. Geol. Surv. Prof. Pap.* 202.

Park, C. F., Jr., P. Gemmill, and C. M. Tschanz, 1958. Geologic map and sections of the Pioche Hills, Lincoln County, Nevada, *U.S. Geol. Surv. Mineral Invest. Field Stud. Map* MF 136.

Pecora, W. T., 1956. Carbonatite—a review, *Geol Soc. Amer. Bull.* 67:1537–1555.

Pelletier, R. A., 1964. *Mineral Resources of South-central Africa*, Oxford Univ. Press, pp. 126–129.

Perry, V. D., 1961. The significance of mineralized breccia pipes, *Mining Eng.* 13:367–376.

Pŏsepńy, F., 1873. Die Blei-und galmei-Erzlagerstätten von Raibl, *Jahrb. K. K. Geol. Rachsanstalt* 23:315–420.

Prescott, B., 1915. The main mineral zone of the Santa Eulalia district, Chihuahua, *Amer. Inst. Mining Eng. Bull.* 98:155–198.

Ramdohr, P., 1955. *Die Erzmineralien und Ihre Verwachsungen,* Berlin: Akad. Verlag.

Ransome, F. L., 1909. Geology and ore deposits of Goldfield, Nevada, *U.S. Geol. Surv. Prof. Pap.* 66.

———, 1911. Geology and ore deposits of the Breckenridge district, Colorado, *U.S. Geol. Surv. Prof. Pap.* 75, pp. 144–147.

Reynolds, D. L., 1954. Fluidization as a geological process, and its bearing on the problem of intrusive granites, *Amer. J. Sci.* 252:577–613.

Ridge, J. D., 1949. Replacement and the equating of volume and weight, *J. Geol.* 57:522–550.

———, 1961. Gain and loss of material in a series of replacements, Spec. Paper No. 68, *Geol. Soc. Amer. Abstr. 1961,* pp. 252–253.

Roedder, E., 1968. The noncolloidal origin of "colloform" textures in sphalerite ores, *Econ. Geol.* 63:451–471.

Roscoe, S. M., 1951. Dilation maps, their application to vein-type ore deposits, Ph.D. dissertation, Stanford Univ., California.

Roy, S., 1959. Mineralogy and texture of the manganese ore bodies of Dongari Buzurg, Bhandara district, Bombay State, India, with a note on their genesis, *Econ. Geol.* 54:1556–1574.

Rust, G. W., 1935. Colloidal primary copper ores at Cornwall mines, southeastern Missouri, *J. Geol.* 43:398–426.

Ryznar, G., F. A. Campbell, and H. R. Krouse, 1967. Sulfur isotopes and the origin of the Quemont ore body, *Econ. Geol.* 62:664–678.

Sales, R. H., 1954. Genetic relations between granites, porphyries and associated copper deposits, *Amer. Inst. Mining Eng. Trans.* 199:499–505.

Sales, R. H., and C. Meyer, 1949. Results from preliminary studies of vein formation at Butte, Montana, *Econ. Geol.* 44:465–484.

Schneiderhöhn, H., 1923. Chalkographische Untersuchung des Mansfelder Kupferschiefers, *Neues Jahrb. Mineral. Geol. Paläontol.* supp. 47:1–38.

———, 1929. Das Otavi-Bergland und seine Erzlagerstätten, *Z. Prakt. Geol.* 37:85–116.

———, 1931. *Mineralische Bodenschätze im Südlichen Afrika,* Berlin.

———, 1941. *Lehrbuch der Erzlagerstättenkunde,* Jena: Gustav Fischer, pp. 459–471.

Schneiderhöhn, H., and P. Ramdohr, 1931. *Lehrbuch der Erzmikroskopie,* Berlin: Geb. Borntraeger.

Schouten, C., 1934. Structures and textures of synthetic replacements in "open space," *Econ. Geol.* 29:611–658.

———, 1946. The role of sulphur bacteria in the formation of the so-called sedimentary copper ores and pyritic ore bodies, *Econ. Geol.* 41:517–538.

Schumacher, F., 1950. *Die Lagerstätte der Trepça und Ihre Umgebung,* Belgrade.

———, 1954. The ore deposits of Yugoslavia and the development of its mining industry, *Econ. Geol.* 49:451–492.

Schwartz, G. M., 1931. Textures due to unmixing of solid solutions, *Econ. Geol.* 26:739–763.

Sclar, C. B., and B. H. Geier, 1957. The paragenetic relationships of germanite and reniérite from Tsumeb, South-West Africa, *Econ. Geol.* 52:612–631.

Sims, P. K., and P. B. Barton, Jr., 1962. Hypogene zoning and ore genesis, Central City district, Colorado, in *Petrologic Studies* (Buddington Vol.), ed. A. E. J. Engel, Geological Society of America.

Snyder, F. G., and J. W. Odell, 1958. Sedimentary breccias in the southeast Missouri lead district, *Geol. Soc. Amer. Bull.* 69:899-925.

Söhnge, P. G., 1952. The Tsumeb story: pipe-like, massive-sulphide ore body appears to fill volcanic pipe in depth, *Mining World* 14(6):22-24.

——, 1963. Genetic problems of pipe deposits in South Africa, *Geol. Soc. S. Afr. Trans.* 66:xix-xxii.

Stevenson, J. S., and W. G. Jeffery, 1964. Colloform magnetite in a contact metasomatic iron deposit, Vancouver Island, British Columbia, *Econ. Geol.* 59:1298-1305.

Stillwell, F. L., 1919. The factors influencing gold deposition in the Bendigo goldfield, Australia: part III, *Advis. Counc. Sci. Ind. Bull.* 16.

Sunagawa, I., Y. Endo, and N. Nakai, 1971. Hydrothermal synthesis of framboidal pyrite, in *Geochemistry and Crystallography of Sulphide Minerals in Hydrothermal Deposits* (Joint Symposium Vol.: IMA-IAGOD Mtgs. '70), ed. Y. Takéuchi, Tokyo: Society of Mining Geologists of Japan (spec. issue 2).

Takeuchi, T., and H. Abe, 1970. Regularities of hydrothermal alteration of sulphur and iron-sulphide deposits in Japan, in *Problems of Hydrothermal Ore Deposition*, International Union of Geological Sciences (ser. A, no. 2).

Takeuchi, T., I. Takahashi, and H. Abe, 1966. Wall-rock alteration and genesis of sulphur and iron-sulphide deposits in Japan, *Tohoku Imper. Univ. Sci. Rept.*, ser. 3 (geol.), 9(3):381-483.

Taylor, R. G., 1966. Distribution and deposition of cassiterite at South Crofty mine, Cornwall, *Inst. Mining Metall. London Trans.* 75 (sec. B):B35-B49.

Threadgold, I. M., 1958. Mineralization at the Morning Star gold mine, Wood's Point, Victoria, *Australas. Inst. Mining Metall. Proc.* 185, pp. 1-27.

Titley, S. R., 1961. Genesis and control of the Linchburg orebody, Socorro County, New Mexico, *Econ. Geol.* 56:695-722.

Treadwell, W. D., and H. Hepenstrick 1949. Über die Löslichkeit von Silbersulfid, *Helv. Chim. Acta* 32:1872-1879.

Treadwell, W. D., and F. Schaufelberger, 1946. Zur Kenntnis der Löslichkeit des Quecksilbersulfids, *Helv. Chim. Acta* 29:1936-1946.

Tsumeb Corporation Staff, 1961. Geology, mining methods and metallurgical practice at Tsumeb, *7th Commonw. Mining Metall. Congr. Trans.* South African Institute of Mining Metallurgy, 1:159-179.

Tuttle, O. F., and J. Gittins, eds., 1966. *Carbonatites*, New York: Wiley.

Van der Veen, R. W., 1925. *Mineragraphy and Ore Deposition*, The Hague: B. Naeff.

Wager, L. R., and G. M. Brown, 1967. *Layered Igneous Rock*, San Francisco: W. H. Freeman.

Wagner, P. A., 1927. The pipe form of ore deposit, *Econ. Geol.* 22:740-741.

Watanabe, T., 1940. Eruptions of molten sulphur from the Siretoko-Iôsan volcano, Hokkaidô, Japan, *Jap. J. Geol. Geogr.* 17:289-310.

Westgate, L. G., and A. Knopf, 1932. Geology and ore deposits of the Pioche district, Nevada, *U.S. Geol. Surv. Prof. Pap.* 171.

Whitebread, D. H., and D. E. Lee, 1961. Geology of the Mount Wheeler mine area, White Pine County, Nevada, *U.S. Geol. Surv. Prof. Pap.* 424-C, pp. 120-122.

Williams, H., 1936. Pliocene volcanoes of the Navajo-Hopi country, *Geol. Soc. Amer. Bull.* 47:111-171.

Young, E. B., 1948. The Pioche district, *18th Int. Geol. Congr. Rept.*, pt. 7, pp. 98-106.

Zeller, H. D., 1957. The Gas Hills uranium district and some probable controls for ore deposition, in *Guidebook to Southwest Wind River Basin*, Wyoming Geological Association (12th Ann. Field Conf.).

5 / Alteration and Gangue

The country rocks bordering ore deposits of hydrothermal origin are generally altered by the hot fluids that have passed through them and with which the ores are associated. The alteration is considered to be due as much to the mineralizing processes as the ore itself; indeed, replacement ores are merely commercially valuable products of wall-rock alteration. The nature of the alteration products depends upon (1) the character of the original rock; (2) the character of the invading fluid, which defines such factors as Eh, pH, vapor pressure, and degree of hydrolysis; and (3) the temperature and pressure at which the reactions took place. Wall-rock alteration has been recognized for many years as a valuable tool in exploration; the alteration halos around many deposits are widespread and much easier to locate than the ore bodies. Alteration effects may have reached the surface, or they may be blind and found only in underground workings or in drill holes. Equipment for rapid and precise determination of the mineralogy has become available within recent times, and a great deal of work has been done in an attempt to determine the exact relationships between ores and alteration minerals. The main problems involve such questions as: What types of alteration products are associated with ores deposited at depth? How do they differ from near-surface alterations? What changes are to be expected in the country rocks at various distances from the orebodies? Where in the alteration halos are the ores most likely to be found?

Several types of rock alteration products may be distinguished, and although they are similar in many ways, attempts should be made to separate them. They may result from (1) diagenesis; (2) regional processes; (3) post magmatic processes, such as hydration; and (4) alteration related directly to mineralization.

Gangue minerals in hydrothermal deposits are also direct products of the ore-forming processes. In disseminated copper deposits and in some igneous metamorphic ores, gangue and altered wall rock are the same materials. In some veins the gangue has been derived from the wall rock by alteration processes. In magmatic segregation deposits the gangue and the wall rock may be similar petrologic units, differing only in the percentage of ore mineral present. Gangue minerals may also be useful in prospecting. They generally extend beyond the ore shoots and, as a result, may be used to distinguish mineralized veins from barren structures. Similarly, minor changes in the distribution of gangue minerals may indicate which direction along a vein is likely to lead to ore.

WALL-ROCK ALTERATION

If the wall rocks are unstable in the presence of early, ground-preparing hydrothermal fluids or of ore-forming solutions, they will undergo physical and chemical changes to reach equilibrium under the prevailing conditions. The resulting alteration may be very subtle, such as the incipient hydration of selected ferromagnesian minerals; or it may be very complete, as in the silicification of limestone. The alteration may range from simple recrystallization to the addition, removal, or rearrangement of chemical components. It may take place in advance of emplacement of the ore minerals or during the final waning stages of the hydrothermal activities.

The nature of wall rock alteration by hydrothermal processes offers strong evidence in support of the view that many hydrothermal solutions are neutral or slightly acidic at higher temperatures. Helgeson (1964) states that much hydrothermal alteration is essentially a process of trading H^+ ions for other cations in the rock. The fundamental controls include any process that affects the activities of reacting constituents in solution. Hemley and Jones (1964) list as the most significant controls: (1) reaction with the wall rocks, hence the nature of the rock; (2) changes in the pressure-temperature state of the aqueous phase, such as throttling and boiling, with possible fractionation of volatile components; (3) mixing of hypogene solutions with supergene solutions or groundwaters; and (4) oxidation of H_2S to stronger sulfur acids.

Wall-rock alteration may bring about recrystallization, changes in permeability, and changes in color. Carbonate rocks are characteristically recrystallized along the borders of a vein or near an igneous contact. The recrystallized rock is generally more permeable than the unaltered rock, suggesting that some ores may be deposited following an advance wave of recrystallization. Conversely, argillization may reduce the permeability of a rock, leaving the orebody enclosed within a relatively impermeable shell. Color changes include bleaching, darkening, and aureoles of various colors. Pastel colors are especially prominent around some ore deposits and may form conspicuous leads to ore at the earth's surface. Clay minerals are generally white or light shades of green, brown, and gray, hence argillization may produce a noticeable bleaching effect; even a black basalt may be altered to a white or light-green body of clays and other hydrous minerals. Similarly, the common formation of chlorite or epidote produces a green color.

Since iron is one of the most abundant metals, the alteration products of iron, and especially pyrite, are commonly formed around sulfide ore bodies. Normally pyrite develops wherever sulfur is added to a host rock containing iron-bearing minerals. Even if pyrite is finely disseminated, it may cause a striking color change. For example, the pyritization of a red sandstone or shale containing iron-oxide pigment produces a bleached zone due to the reduction or iron. Conversely, any pyritized rock is likely to be made conspicuous at the earth's surface by oxidation of the iron, to produce a red or red-brown weathered zone. Most of the physical effects produced by wall-rock alteration are brought about by chemical reactions. Thus hydration or reduction may cause bleaching. Studies of wall rocks at progressively greater distances from veins show that the altered rock has undergone complex ion exchange, whereby some constituents are removed, others are added, and still others are merely redistributed. Silicification, carbonatization, argillization, and hydration are typical of the processes that take place in alteration zones—and they may all operate simultaneously. Although generalizations are hazardous because the possibilities are too diverse, certain reactions in specific environments can be expected. For example, water is usually added to the alteration zone, except where the rocks are completely replaced by anhydrous silica, and carbon dioxide is generally removed from carbonate host rocks. Furthermore, certain minerals are to be expected in alteration zones; sericite, quartz, chlorite, pyrite, epidote, zoisite, carbonates, and clay minerals are the most common (Schwartz, 1959), but many others have been formed by hydrothermal alteration.

Conditions of temperature and chemistry usually differ at various distances from a vein, so that different types of alteration are produced simultaneously (Hemley, 1959; Burnham, 1962). In the outer fringes of the alteration zone the ferromagnesian minerals may be slightly hydrated, the interior zone silicified or sericitized, and the intermediate zone argillized. The result is a more or less symmetrical zoning of different alteration products about the central vein. In some deposits such zoning is conspicuous and is an excellent guide to ore. For example, at Casapalca, Peru, the veins that traverse grayish-green porphyry intrusives are outlined sequentially from the veins outward by white, pink, green, and purplish alteration zones (McKinstry and Noble, 1932).

Any hydrothermal solution passing through rock pores or along a fissure is likely to alter the country rocks, whether or not the vein contains ore minerals. Thus, not all alteration zones are guides to ore deposits, and mineralization in any region that has undergone a complex history of hydrothermal activity is complicated by transgressing and sequential zones of alteration associated with both metalliferous and barren phases. On the other hand, the chemistry of each hydrothermal solution may be reflected in details of the wall-rock alteration, slight differences of which indicate the proximity of metal-bearing shoots.

Different kinds of ore deposits are typified by certain alteration minerals and different degrees of alteration. Owing to their depth of burial, the wall rocks around deep-seated ores are relatively impermeable and usually almost as hot as the ore-bearing fluids. Therefore, unless there is a strong chemical contrast between the ore fluids and the country rock, the alteration zone is likely to be thin and inconspicuous. Conversely, hot solutions that invade cool, shallow, permeable rocks typically produce prominent, widespread alteration halos because the country rocks are far out of equilibrium with the fluids. On the following pages the general products of alteration for each type of ore deposit will be described, beginning with magmatic segregations and extending to the shallowest zone.

Magmatic Deposits

Wall-rock alteration around magmatic segregation and magmatic injection deposits is generally inconspicuous and nondiagnostic. The contact zones commonly have assay walls; that is, the valuable constituents gradually decrease in amount, and the gangue minerals increase to a ratio at which the material is below ore grade. In some deposits, however, the contacts between ore and gangue are sharply defined; this is especially true for injection-type deposits that were differentiated at depth prior to emplacement. Wall-rock alteration around injection deposits is likely to be very thin or absent, unless the injection was accompanied or followed by hydrothermal activity. A thin reaction zone may be present in some deposits that have been concentrated by gravity settling, due to a reaction between the ore minerals and the rest melt.

Pegmatites

Contact zones around pegmatites may be either narrow and sharply defined or wide and gradational. Some pegmatites contain rare minerals, and the contact zone around these igneous masses may contain ions of the rare earths, boron, fluorine, beryllium, lithium, and other pegmatitic constituents, producing a varied and unique mineralogy of alteration products. Among the alteration minerals found in and near pegmatites are beryl, monazite, sphene, tantalite-columbite, lithium micas, topaz, zircon, fluorite, and allanite; the most widespread are disseminated feldspars, micas, garnets, and tourmaline (Jahns, 1955). Many pegmatitic fluids are rich in potassium, which may be introduced into the wall rocks in conspicuous amounts; conversion of wall-rock amphiboles to biotite is evidence of this.

Jahns (1946) recorded fine-grained muscovite, plagioclase, potash feldspars, and quartz in the poorly delimited border zones of pegmatites in the Petaca district of New Mexico. Wall rocks that permitted soaking by the pegmatitic fluids were most conspicuously altered; a slabby micaceous quartzite formation showed characteristic alteration for a few feet or even many tens of feet from the pegmatite, extending tongues along bedding or joint planes. Around some pegmatite bodies the zone between true pegmatite and unaltered quartzite consists of a hybrid rock, making the contact between igneous rock and wall rock gradational.

Metacrysts of black tourmaline and garnet are present in the wall rocks of nearly all mica and beryl pegmatites of the Avon district, Idaho. The host-rock mica schists and gneisses were altered to a light-colored aggregate of plagioclase, quartz, muscovite, and schorl, which was formed by the recrystallization of original constituents and accompanying metasomatism by pegmatitic materials. Relict bedding and schistosity are recognized in most places. In general, the alteration zone is less than a foot wide, but at the Muscovite mine it extends about 20 feet (Stoll, 1950).

The Etta spodumene mine near Keystone, South Dakota, is noteworthy in having a pronounced contact zone around the pegmatite. The alteration halo is sharply defined and consists of a friable, fine-grained, sugary rock produced by recrystallization and metasomatism of a fine-grained mica schist. The biotite and muscovite of the schist were altered to a granoblastic and poikilitic assemblage of plagioclase, orthoclase, and microcline. Apatite and tourmaline also were formed in the contact aureole, which has an average width of seven or eight feet and a maximum width of 15 feet (Schwartz and Leonard, 1927).

Pegmatites commonly occur within the parent igneous mass, where the physical and chemical conditions are not amenable to the development of a conspicuous alteration zone. In such cases the contact between pegmatite and host rock is likely to be a gradational change in texture rather than in mineralogy or rock type. Conversely, some pegmatites may represent a gradual and simple change in mineralogy, with or without an increase in grain size over that of the host pluton. For example, in the Kaniksu batholith of northeastern Washington, large, irregular or oval bodies of muscovite-quartz-feldspar rock crop out in the normal biotitic intrusive mass. Although these bleached products are usually of the same grain size as the main intrusive, locally (along fractures and in the centers of the bodies) the grain is coarser and grades into characteristically pegmatitic materials. The most notable feature of these pegmatites is the bleaching and removal of iron minerals from their borders. Biotite is changed to muscovite, and much of the rock shows well-developed myrmekitic intergrowths (Park and Cannon, 1943).

In general, the alteration zones around pegmatites are thin and of little or no value in prospecting. Even the rare conspicuous halos extend only a few feet from the pegmatite. At a few places, for example in the African pegmatites, studies of the minor elements in barren zones indicate the presence of ores (Hornung, 1962).

Igneous-Metamorphic Deposits

Wall-rock alteration products associated with igneous-metamorphic deposits around intrusive masses are extensive in many places and contain conspicuous, diagnostic minerals—garnets (especially grossularite, andradite, and almandite), wollastonite, epidote, pyroxenes (such as hedenbergite, salite, and diopside), amphiboles (principally tremolite and actinolite), ilvaite, idocrase, minerals of the humite group, serpentine, spinels, scapolite, and many others. The character of the alteration products depends in large part upon the invaded materials. For example, the introduction of siliceous fluid into limestone ordinarily results in such calcium-rich minerals as wollastonite and idocrase, whereas the invasion of the same fluid into dolomites forms magnesium-rich minerals, such as serpentine and diopside. The following equations illustrate possible reactions:

$$\underset{\text{limestone}}{CaCO_3} + H_4SiO_4 \rightarrow \underset{\text{wollastonite}}{CaSiO_3} + 2H_2O + CO_2$$

$$\underset{\text{dolomite}}{CaMg(CO_3)_2} + 2H_4SiO_4 \rightarrow \underset{\text{diopside}}{CaMg(SiO_3)_2} + 2CO_2$$

Shales in igneous metamorphic zones tend to develop the characteristic sugary texture of a hornfels; locally, knots of chlorite, andalusite, garnet, or cordierite form; and epidote may be abundant. Volcanic rocks undergo comparable alteration. For example, the abundant andesites and andesitic tuffs of central and northern Chile contain conspicuous epidote near intrusive masses. Carbonate rocks near intrusives are often thoroughly metasomatized to skarn that consists of various silicate and oxide minerals. In some igneous-metamorphic deposits the limestones and dolomites are merely recrystallized to coarsely crystalline marbles that contain no new constituents. During recrystallization the carbonate rocks tend to expel impurities, such as carbon, so that the recrystallized rocks are generally whiter than their unaltered counterparts.

The alteration effects around contacts between igneous and sedimentary rocks range in width from a few feet to more than a mile. In general, the more extensive the metamorphic zone, the more favorable the area is for ore because much heat and hydrothermal fluid have obviously penetrated the rocks. Contact zones that are simply recrystallized and show no other metamorphic effects are said to be "dry" because they seldom contain ore deposits.

Silicification is abundant around igneous metamorphic deposits. The silica may be introduced by hydrothermal fluids or may recrystallize from silica already present in the rock. Shales, which are nearly impermeable and thus not receptive to ore-bearing fluids, may be rendered hard, brittle, and permeable by silicification. Silicified shales are typically aphanitic, but small, doubly terminated quartz crystals can be distinguished in some. Silicification is one form of ground preparation, whereby soft, impermeable, and unfavorable rocks are made more competent and more receptive to the ore fluids and to the deposition of ores.

Hypothermal Deposits

The deepest hydrothermal veins grade into both pegmatites and igneous metamorphic deposits at their lower limits, and into the mesothermal zone above. Accordingly, the alteration products associated with the deeper hydrothermal ores may resemble either pegmatitic or igneous metamorphic alterations, and the upper hypothermal alteration products may grade into mineral suites characteristic of mesothermal deposits. Some deep-seated veins are mineralogically like pegmatites. For example, at the Passagem mine, Brazil, the gangue, and in places the wall rock, contains abundant quartz plus garnets, tourmaline, muscovite, mariposite, phlogopite, biotite, orthoclase, plagioclase, kyanite, and ankerite. Hussak (1898) described the deposit as a pegmatite. Minerals closely akin to those of pegmatites are found in the southern Piedmont region of the United States. Near Dahlonega, Georgia, they are found both in gangue and in the country rock. Within the orebodies the minerals are coarse-grained, decreasing in size laterally into the schists, where it is impossible to pick a precise limit of mineralization (Pardee and Park, 1948, p. 47). Alteration of this type, apparently related to pegmatitic fluids, is seldom identifiable more than a few feet away from ore; it is of little value in exploring for new deposits.

A second type of alteration associated with the deeper hydrothermal deposits is shown by the ores of the Homestake gold mine, South Dakota, and by the Morro Velho gold mine of Brazil. This type grades into the extensive alteration zones characteristic of the mesothermal zone. Ankerite is common, though other carbonates may also be present. The country rocks are commonly silicified and generally contain either sericite or chlorite, or both, but seldom in large amounts. Where both are developed, the sericite is typically closer to the ore deposit than the chlorite. Pyrite, arsenopyrite, and pyrrhotite are widely distributed and locally abundant.

Mesothermal Deposits

Alteration products are most conspicuous and most useful as guides to ore deposits at intermediate and shallow depths, although not all these deposits do show alteration halos. Sericite is the most persistent and abundant alteration mineral in mesothermal deposits; carbonates are also common, especially calcite and dolomite, though siderite, rhodochrosite, and ankerite are conspicuous in places. Chlorite, most characteristic of epithermal deposits, may also form around mesothermal veins; it commonly develops next to sericite, but on the cooler side, away from the ore. Silica is generally present, and in many places abundant. A prevalent feature is jasperoid, the fine-grained silica of hydrothermal origin. Pyrite is another widespread and conspicuous alteration product, and may form euhedral crystals associated with the sericite or, less commonly, with the chlorite. In places, the wall rocks around veins of the mesothermal zone are altered to a fine-grained aggregate of silica, sericite, pyrite, and to a lesser extent feldspars and clay minerals.

The Mother Lode district of California furnishes excellent examples of the alteration products developed in the deeper parts of the mesothermal zone (Knopf, 1929, p. 41-45). In these deposits ankerite, the dominant alteration mineral, abounds in all types of wall rocks. Sericite is probably second in importance—except in serpentine, where mariposite, the chromium mica, is common. The chief characteristic of the alteration zone is the addition of large amounts of carbon dioxide, the quantity of which apparently depends upon the abundance of iron, magnesium, and calcium in the original rock. Other added elements are potassium, sulfur, sodium, and arsenic, represented by sericite, pyrite, and arsenopyrite. Large volumes of silica, more than all the silica in the veins, have been removed from silicates so that it is unnecessary to attribute this compound to the magma. In places the alteration extends 10 feet or more from the edges of the veins, forming a valuable guide to the associated vein.

The conspicuous, extensive alteration zones in and around disseminated copper deposits, which probably form in the mesothermal zone, have been well studied. Unaltered orthoclase in the altered igneous host rock has been widely recognized and is easily confused with original constituents. Alteration feldspars have been described by Gilluly (1946), Schwartz (1947), and Anderson (1950), and are recognized in such districts as Bingham Canyon, Utah; Ely, Nevada; and Bagdad, Arizona, among others. Argillic alteration has been described in the disseminated copper deposit at Castle Dome, Arizona (Peterson *et al.*, 1946), and was recognized by Schwartz at Morenci and San Manuel, Arizona, and at Santa Rita, New Mexico. Sericitic alteration is common to all disseminated deposits, and silica is abundant in places. In many of these deposits the mafic minerals are altered to biotite. Rose (1970) discussed the alteration products of disseminated copper deposits and attempted to distinguish between hypogene and supergene alteration, especially that formed by the action of sulfuric acid. He also emphasized the fact that alteration products frequently are arranged in zones. In studies of the Panguna disseminated copper deposits in Bouganville Island, New Guinea, Fountain (1972) emphasized the close relationship between mineralization and biotite of the alteration zone.

Probably one of the most abundant alteration products bordering ore deposits in silicate and granitic rocks (such as those found at Butte, Montana) is quartz-sericite. A zone of this material may itself be bordered by a zone of argillic products that grade outward into fresh rock. The major compositional change in the rocks is the removal of calcium and sodium, as well as some magnesium, and the addition of chemically equivalent amounts of hydrogen. Aluminum tends to remain unchanged, as does potassium, though where sericite is formed, potassium may be added to the rock. The reaction for the formation of sericite from andesine feldspar may be expressed by the equation (of Hemley and Jones, 1964):

$$0.75Na_2CaAl_4Si_8O_{24} + K^+ + 2H^+ \rightarrow$$
$$\underset{\text{andesine}}{}$$

$$\underset{\text{sericite}}{KAl_3Si_3O_{10}(OH)_2} + 1.5Na^+ + 0.75Ca^{++} + \underset{\text{quartz}}{3SiO_2}$$

Similarly (according to Hemley and Jones, 1964), kaolinite may form in the presence of water by which hydrolysis furnishes available hydrogen.

$$\underset{\text{andesine}}{Na_2CaAl_4Si_8O_{24}} + 4H^+ + 2H_2O \rightarrow$$

$$\underset{\text{kaolinite}}{2Al_2Si_2O_5(OH)_4} + \underset{\text{quartz}}{4SiO_2} + 2Na^+ + Ca^{+2}$$

Calcium plagioclase breaks down before sodium plagioclase, which in turn breaks down before the potassium feldspars. This is reflected in experimental data; higher alkali/HCl ratios are needed for stability of the Ca-Na feldspars. Thus the alteration in a granodiorite, as at Butte, would be complex. The zones would agree with the theory discussed above because the plagioclase sequence would be outside the alkali feldspar sequence and, given a high K^+/H^+ or cation/H^+ ratio, the innermost kaolinite zone would not form. Instead, sericite would develop next to the vein. The release of K^+ with the breaking down of the K-feldspar would automatically create the high K^+/H^+ environment at the reaction site (Hemley and Jones, 1964).

Hemley and Jones also point out that temperatures of alteration can be approximated from the mineral assemblages. For example, kaolinite forms only below about 400°C, and montmorillonite below about 450°C. But the differences in mineralogy may be due to high silica activity or high alkali/H^+ ratios. Thus, whether kaolinite or sericite alteration forms may reflect K^+ rather than epithermal or mesothermal conditions.

Epithermal Deposits

Chlorite is probably one of the most abundant minerals in the wall rocks around epithermal deposits, and in deposits such as the Comstock Lode, Nevada, the enclosing rocks are highly chloritized to a propylite. Sericite is common, though in smaller amounts than in mesothermal deposits; other alteration minerals are alunite, zeolites, chalcedony, opal, calcite, and other carbonates. Argillization may be especially widespread; among the numerous clay minerals, nacrite, dickite, kaolinite, beidellite, illite, and montmorillonite may be present. In shallow deposits alteration products are ordinarily fine-grained, and many are difficult to separate for identification. Probably the most diagnostic minerals are the clays and chlorite. The country rock may be thoroughly altered for several hundred feet from an orebody, making it difficult in such extensive zones to locate the ore within the aureole.

Telethermal Deposits

The alteration products associated with telethermal deposits, such as those of the Mississippi Valley lead-zinc district, are few and inconspicuous. Telethermal deposits are thought to have been formed by warm waters at great distances from their source. They are characterized by calcite, dolomite, marcasite, and cryptocrystalline silica, but these are seldom helpful in exploration.

Hydrothermal Deposits

It is logical to assume that the physical and chemical characteristics of ore-bearing fluids vary outward from the vein into the cooler wall rock, and that the solutions passing through the vein change in both temperature and composition during the history of hydrothermal activity. Such changes are in fact recorded in the different zones of alteration parallel to the contact. A good example is the common association of chlorite, a typical epithermal mineral, outside the sericite zone of a mesothermal vein. Alteration zones ordinarily grade into lower-temperature types wherever ores have been deposited in country rocks that were appreciably cooler than the ore-bearing fluids. A good example is wall-rock alteration associated with xenothermal deposits—ores formed in shallow zones but at high temperatures. These ores are characterized by a mixture of alteration minerals belonging to the entire hydrothermal range, from hypothermal to epithermal.

Studies of the hydrothermal alteration processes of metal-deposition at Steamboat Springs, Nevada, by Schoen and White (1965), indicate that these waters are like some ore-forming fluids. The alteration mineral assemblages along these fractures resemble those at Butte, Montana. Other alteration products, related to depth, include the change of plagioclase to hydrothermal K-feldspar in the upper 300 feet of the spring system and the formation of calcite and albite below 200 feet.

Studies of wall-rock alteration have shown that the old theory of lateral secretion was not entirely fantasy. In most deposits the metals cannot have originated in the adjacent wall rocks, although the silica may have been derived from the wall rocks during hydrothermal alteration. Schmitt (1954) pointed out that in many mineral deposits the altered wall rock contains less silica than the unaltered rock. He thought, as do many field geologists, that the silica was simply rearranged and concentrated locally. Although it may have traveled short distances along the vein, a hydrothermal source is not required; local supplies of silica are generally more than adequate. Thus, the quartz host in the Mother Lode gold veins has been attributed to lateral secretion from the alteration zone (Knopf, 1929); similarly, Boyle (1955) suggested that the quartz in the Canadian Yellowknife veins was secreted from the walls.

The mineralogy of hydrothermal alteration depends not only upon the composition and temperature of the ore-bearing fluids, but also partly upon the composition of the wall rocks. As Schwartz (1950) pointed out, the ore solutions at Butte, Montana, and Bisbee, Arizona, both contained manganese, but only the Bisbee wall rocks were affected by this metal. No manganese minerals were formed in the alteration zone at Butte, where the wall rock is quartz monzonite, though as much as 12 percent manganese (as manganocalcite) was added to the limestone wall rocks at Bisbee. Rhodochrosite and rhodonite are abundant in the Butte gangue and ore, thus proving that the ore solutions contained an appreciable amount of manganese. As a corollary to the fact that the type of alteration depends upon the nature of the wall rock, it should be emphasized that the absence of an element in the alteration zone does not indicate its absence in the ore-bearing fluid.

The relationship of alteration to ore deposits may be obvious, obscure, or nonexistent. Areas subjected to a complex history of hydrothermal activity may retain a complex record of alteration, of which ore deposits may or may not be a part. One fundamental question posed by deposits that show more than one type of alteration is whether the different types are produced by a single continuous hydrothermal phase or a series of separate injections. If the alteration is produced by multiple stages of hydrothermal activity passing along different avenues, the use of alteration zones as prospecting guides is seriously complicated.

A thorough and remarkably clear study by Sales and Meyer (1948, 1949, 1950) of the Butte, Montana veins demonstrated that the alteration products there are an integral part of the mineralization—the alteration minerals were deposited at essentially the same time and by the same fluids as the ore. Successive zones of sericitized and argillized quartz monzonite lie adjacent to every ore-bearing fissure, regardless of its size, attitude, ore mineralogy, or relative age. Except where overlap of the alteration effects between adjacent fissures has eliminated the lower grade zone, the two types of alteration always occupy the same relative positions: sericite is adjacent to the vein, and clay minerals are between the seritized rock and fresh quartz monzonite. Nowhere does the sericite zone grade directly into fresh wall rock. The sequence of alteration is even more striking in detail. The argillic zone is made up of a kaolinitic phase close to the vein and a montmorillonite-rich phase close to the unaltered wall rock. Moreover, the relative proportions of chemical components change systematically from the vein outward, presenting a graphic picture of the hydrothermal alteration process. As shown in Figure 5-1, the lime, soda, and silica contents diminish rapidly toward the vein, and the H_2O content increases; a similar but less striking decrease is also shown by iron and magnesium. These changes reflect the fact that plagioclase was readily hydrolyzed, releasing silica, calcium, and sodium, and that the ferromagnesian minerals were altered to chlorite. Continued alteration leached iron and magnesium from the montmorillonite to form kaolinite. Further leaching finally broke down the orthoclase, supplying potassium for the production of biotite and sericite. The iron and sulfur contents show a marked increase near the vein, because iron leached from the early destruction of ferromagnesian minerals combined with hydrothermal sulfur to form pyrite. Silica released from feldspars was deposited as quartz near the vein. Thus, there was a two-way transfer of ions by diffusion, some ions moving into the wall rock and others moving toward the vein. As long as active circulation continued along the vein, each zone migrated away from the fissures; that is, it grew at its outer edge and simultaneously receded at its veinward edge because of encroachment by the next inner zone. Sales and Meyer concluded that a hydrothermal solution, at a given distance from its magmatic source, causes a specific type of alteration in a reactive wall rock, whereas at a different distance from the source, the same fluid in the same country rock causes a different type of alteration. The various mineralogical and chemical responses to the attack by ore-bearing fluids depend upon continuous changes in the physiochemical environment within the wall rock and not upon a complete change in composition of the hydrothermal fluid. The diffusion gradients in a typical alteration envelop of the Butte type are shown in Figure 5-2.

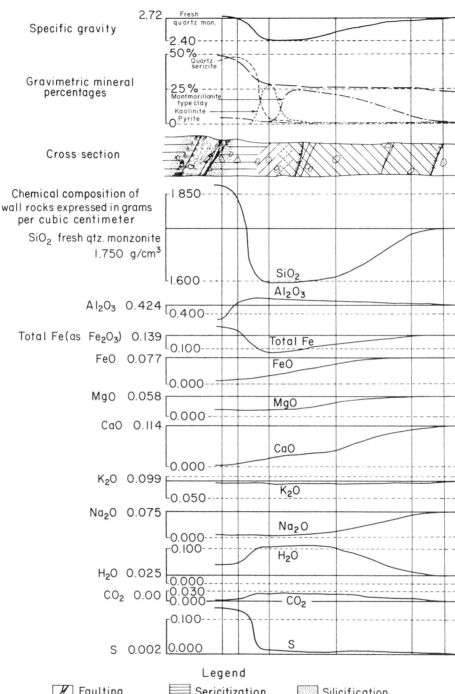

Specific gravity

Gravimetric mineral
percentages

Cross section

Chemical composition of
wall rocks expressed in grams
per cubic centimeter

SiO_2 fresh qtz. monzonite
1.750 g/cm³

Al_2O_3 0.424

Total Fe(as Fe_2O_3) 0.139

FeO 0.077

MgO 0.058

CaO 0.114

K_2O 0.099

Na_2O 0.075

H_2O 0.025

CO_2 0.00

S 0.002

Legend

Faulting   Sericitization   Silicification
Vein material   Kaolinitization   "Green" zone
(Montmorillonite·rich)

Figure 5-1 *(facing page)*
Cross section of a copper-zinc vein at Butte, Montana,
showing the physical, chemical, and mineralogical
changes throughout the zone of alteration. (From
Sales and Meyer, 1948, Fig. 7.)

Bundy (1958) distinguished four wall-rock zones in a detailed study of the
wall-rock alterations in the Cochiti mining district of New Mexico. Represented from the vein outward, the zones are (1) dickite (2) illite-kaolinite (3)
vermiculite-halloysite, and (4) chlorite-montmorillonite. Each mineral was
formed during a continuous process of alteration in the order chlorite, montmorillonite, vermiculite, halloysite, illite, kaolinite, and dickite—a paragenesis that
records the outward growth of the alteration zones very convincingly. Bundy
suggested that the intensity of alteration is a function of time and pH, rather
than temperature and pressure or changes in the composition of vein solutions.

Lovering (1949) reached a different conclusion in a study of the alteration
at Tintic, Utah. He described five stages of hydrothermal alteration, which he
considered mutually independent. The five states are (1) an early barren stage,
during which limestones were dolomitized and volcanics were chloritized; (2)
a mid-barren stage, characterized by argillization; (3) a late barren stage closely
associated with the ore shoots and characterized by jasperoid, barite, pyrite,
and minor chlorite in the sediments (carbonates, quartzites, and shales), plus
allophane, quartz, barite, pyrite, calcite, and minor delessite (iron-rich chlorite)
in the volcanics; (4) an early productive stage closely associated with orebodies
but consisting of only an inconspicuous zone of sericite-hydromica (representing the introduction of potassium), plus minor quartz and pyrite; and (5) the

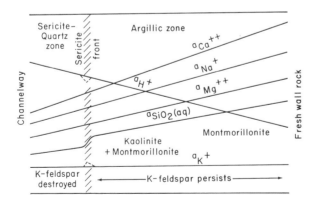

Figure 5-2
Diffusion gradients and zonal relations in a typical alteration
envelope, Butte, Montana. (From Hemley and Jones, 1964.)

productive stage, during which the ore was deposited. According to Lovering's interpretation, each of these stages represents a period of hydrothermal activity separated from the others by appreciable time intervals, and the close association of ore with more than one of these stages reflects a common source of solutions rather than any contemporaneity in genesis.

Although both of these views were reached after a great deal of careful work, neither can be regarded as a completely demonstrated generality. It does seem, however, that a fluid capable of transporting large amounts of ore minerals and reacting with country rocks sufficiently to cause the deposition of these minerals would also have other effects—profound in places—upon the nonmetallic wall-rock constituents. Similarly, it is difficult to believe that the ore-bearing fluids are active only within the narrow confines of their channels; the fluids must permeate the walls and must surely react with some of the mineral constituents. It is possible, however, that the mineral-bearing fluids are injected in surges, each successive surge being of a somewhat different composition, temperature, and pressure. If Lovering is correct and much of the alteration is independent of the ore, then it follows that the chances of finding ore as a result of alteration studies are greatly reduced.* Alteration products and ore would be related only where the fluids used the same paths of migration. Likewise, the spatial and mineralogical relationships between alteration and ore should not be constant. For example, why should orebodies be surrounded by envelopes of silica, sericite, and chlorite in successive zones away from the ore? Some geologists believe that in most mining districts the solutions that deposited the ores were at the same time reacting with the wall rocks and producing the alteration minerals. The injection of hydrothermal fluids in surges is considered exceptional.

It is generally concluded that alteration products in the walls of ore deposits are of great potential aid in the discovery of new orebodies, especially in deposits formed along igneous contacts. Alteration products deserve much more careful attention than has been given them in the past. The identification of fine-grained minerals, such as the clays, is difficult and time-consuming, but with modern techniques (including differential thermal analysis, spectrographic studies, and x-ray equipment) it is now possible to identify the materials with confidence. More precise information about the distribution and relationships of these fine-grained minerals is needed, similar to that obtained by Sales and Meyer, Lovering, and Bundy.

In many outcrops the alteration products are easily recognized and are of wide extent. Elsewhere the surface expression of alteration is small and inconspicuous; possibly only minor amounts of hydrothermal fluids leaked through

---

*Nevertheless, knowledge of the chronology of alteration phases can well be applied to prospecting, as it has by Lovering and others, with impressive success, in the Tintic district (Bush *et al.*, 1960).

a capping rock to reach the surface. In still other areas no surface expression of the alteration is found, but indications may be discovered in drill cores or in underground workings. It is easier, of course, to find a large body of alteration products than it is to locate the smaller orebodies within the envelope of alteration materials. Further study is needed concerning the details of the genetic relationships between the different types of alteration products and the distribution of ores relative to these products.

GANGUE

Gangue consists of all noneconomic minerals in an ore deposit. A mineral considered gangue in one deposit may be regarded as an ore mineral in another. For example, the fluorite in small proportions in the gold ores at Cripple Creek, Colorado, is gangue, whereas the nearly pure fluorite in many districts is ore. Also, a gangue mineral may become an ore mineral as a result of technological advances or market demands.

The gangue minerals are principally silicates and carbonates, with subordinate oxides, fluorides, and sulfates. Other compounds, even sulfides, may be gangue minerals, but are not usually classified as such. Some geologists feel that any sulfide or metallic mineral in a deposit should be classified as an ore mineral, regardless of its economic value. For this reason, a few geologists classify pyrite, usually a gangue mineral, among the ore minerals. Similarly, any opaque mineral included in a petrographic description is apt to be categorically classed as an ore mineral. Such modifications of the terms "ore" and "gangue" are common and easily recognized by informed geologists.

Studies of gangue minerals are of widespread scientific and economic interest. The composition of gangue minerals and the paragenesis of the ore and the gangue must be considered in any hypothesis regarding the nature of ore-bearing fluids and the history of ore deposition. The metallic ore minerals generally replace the nonsulfide minerals, since the ore minerals are normally deposited shortly after the gangue, though the opposite reaction is always possible.

Studies of liquid inclusion are generally easier to perform on the transparent gangue minerals or on such ore minerals as translucent sphalerite because most ore minerals are opaque. The temperatures of ore deposition and the composition of the ore-forming fluids are estimated by examining the gangue, recognizing of course the paragenesis of the ore and gangue. Knowledge of a common origin for the ore and gangue minerals can be especially important for prospecting; if the gangue was produced by the same mechanism as the ore, its presence may be used as a guide to ore-bearing structures or potentially favorable environments. Mining engineers, metallurgists, and geologists must be aware of the physical and textural relationships between ore and gangue minerals in a deposit because the entire milling program—perhaps even the economic criterion dividing ore from subore—is largely a function of the relative densities, magnetic properties, wetting abilities, and intimacy of mixture between the valuable and waste fractions of the rock.

Certain gangue minerals are unique to hydrothermal environments, an association that may be universal or only local. For example, smoky quartz may indicate the presence of uranium because the color is generally due to radioactive bombardment. Similarly, a strong association between dark purple fluorite and economic concentrations of uranium minerals has been observed. An association has also been noted between fluorite and beryllium and between fluorescent varieties of calcite and certain ores. Some calcite fluoresces salmon-pink or red under ultraviolet excitation, and it has been suggested that trace amounts of manganese and possibly a second activator (such as lead) are responsible for this phenomenon. Whatever causes the fluorescence, a striking relationship between orebodies and fluorescent calcite has been recorded at some mines, indicating that the fluorescence in the calcite was formed with the ore (Stone, 1959; MacDiarmid, 1959).

Trace amounts of metal ions can be detected in gangue minerals by spectrographic analyses or chemical assays. Nonmetallic minerals that form in the presence of a metal-bearing fluid can be expected to include these metals in trace amounts; the trace-metals will, in turn, act as guides to ore-bearing structures. Some metals, such as gold, may even reach ore grades as "impurities" within seemingly barren gangue minerals. For example, gold-bearing pyrite would undoubtedly be discarded if the gold content were not discovered by laboratory analysis. At Carlin, Nevada, the gold is so fine grained it is seldom seen. Moreover it cannot be recovered in a pan and the value of the ore must be determined by assay. Most of the gold is only visible under an electron microprobe.

Certain associations between ore and gangue minerals are well known. Quartz is a common gangue mineral in vein deposits; it is especially prominent with gold. The massive white quartz (bull quartz) of many gold deposits characteristically forms conspicuous ridges across the mining district. Because of the common association between bull quartz and gold, early prospectors made a practice of testing all quartz veins for the presence of gold. The wisdom of such a practice becomes obvious when one realizes that a ton of gold ore containing one ounce of gold is high grade. A block of bull quartz about 12 cubic feet in volume will contain about one-tenth of a cubic inch of gold dispersed throughout the entire mass of quartz, or a volume ratio of about one part gold to 32,000 parts quartz.

Another common ore-gangue association is magnetite with apatite. Iron deposits with this combination have been described from all over the world, and include magmatic segregations as well as hydrothermal veins and replacement ores. The veins in any district may range from pure apatite to pure magnetite. In central Chile many small mines have been opened on strong apatite veins, which grade in depth into apatite-bearing magnetite. Fluid immiscibilities in a silica melt, as proposed by Fischer (1950), help explain magnetite-apatite deposits of magmatic segregation origin, but hydrothermal replacement and "lateral-secretion" of deuteric-release ores demand either a modification of this theory or another mechanism.

Well-known but less common associations are fluorite with lead and zinc deposits; barite with lead, silver, and copper ores; tourmaline and topaz with cassiterite veins; and arsenopyrite with tin, tungsten, and gold.

Almost any mineral can be mined for some use if it is found pure and in large volume, with minor gangue. Many gangue minerals commonly found as waste products in metalliferous deposits are the sole or principal ore minerals in other mines. Pyrite is a gangue mineral in most ore deposits, but in localities close to industrial centers, it is mined as an iron ore and for its sulfur content (pyrite is commonly recovered for the manufacture of sulfuric acid).

Barite and fluorite are widespread gangue minerals in hydrothermal deposits. They form under similar conditions and are frequently associated in a single deposit. Wherever either barite or fluorite is concentrated in large deposits, it is likely to be ore. Barite is found both in veins and in massive replacement ores, deposited possibly as a result of the oxidation of a sulfide solution. Fluorite is mined from fissure veins and replacement deposits in many types of rock, but limestone is the most common. The Kentucky-Illinois area, covering 700 square miles, was once the world's largest source of fluorite; most of the ores fill open fissures along faults that pinch and swell, causing the orebodies to vary in thickness from 0 to 60 feet within short distances. A substantial part of the Kentucky-Illinois fluorite was deposited as a replacement of Mississippian limestones (Weller *et al.*, 1952; Gillson, 1960). The principal gangue mineral with both barite and fluorite ores is normally calcite. Another interesting deposit of fluorite-in-limestone is at the Las Cuevas mine in San Luis Potosi, Mexico, where a large body of cavernous fluorite, resembling aragonite, has replaced limestone along a contact with intrusive rhyolite. The main orebody does not crop out, but was discovered by drilling. Although numerous small veinlets of cinnabar are present in the upper workings of the mine, none has been found at depth (Froberg, 1962).

Even quartz is sometimes mined as an ore. Silicon is an increasingly important commodity and is produced by driving the oxygen from quartz. Small, pure deposits of vein quartz are mined for silicon, in preference to the relatively impure varieties present as gangue in metalliferous veins. Other quartz deposits are of economic value because they contain large quartz crystals of optical quality; Brazil is the world's most noteworthy producer of optical quartz.

The study of ore-gangue relationships is especially important to the engineers and geologists who determine how to concentrate ore minerals. The grain size of both ore and gangue will establish how finely the ore must be ground. Relative wetting abilities, specific gravities, and magnetic properties between the ore and gangue minerals limit the mechanisms by which ore can be separated from the gangue after crushing. In many oxidized ores, such as copper carbonates and silicates, the physical contrast between ore minerals and gangue or host rock are so slight that mechanical beneficiation methods are not successful, and the ore must be sent directly to the smelter.

REFERENCES CITED

Anderson, C. A., 1950. Alteration and metallization in the Bagdad porphyry copper deposit, Arizona, *Econ. Geol.* 45:609-628.

Boyle, R. W., 1955. The geochemistry and origin of the gold-bearing quartz veins and lenses of the Yellowknife greenstone belt, *Econ. Geol.* 50:51-66.

Bundy, W. M., 1958. Wall-rock alteration in the Cochiti mining district, New Mexico, *N. Mex. Bur. Mines Mineral Resources Bull.* 59.

Burnham, C. W., 1962. Facies and types of hydrothermal alteration, *Econ. Geol.* 57: 768-784.

Bush, J. B., D. R. Cook, T. S. Lovering, and H. T. Morris, 1960. The Chief Oxide-Burgin area discoveries, East Tintic district, Utah; a case history, *Econ. Geol.* 55: 1116-1147, 1507-1540.

Fischer, R., 1950. Entmischungen in Schmelzen aus Schwermetalloxyden, Silicaten, und Phosphaten, *Neues Jahrb. Mineral. Abh.* 81:315-364.

Fountain, R. J., 1972. Geological relationships in the Panguna copper deposit, Bouganville Island, New Guinea, *Econ. Geol.* 67:1049-1064.

Froberg, M. H., 1962. Geological features of some fluorite deposits in the state of San Luis Potosi, Mexico, *Geol. Ass. Can. Proc.* 14:9-19.

Gillson, J. L., ed., 1960. *Industrial Minerals and Rocks,* 3rd ed., American Institute of Mining Engineers, pp. 364-365.

Gilluly, J., 1946. The Ajo mining district, Arizona, *U.S. Geol. Surv. Prof. Pap.* 209.

Helgeson, H. C., 1964. *Complexing and Hydrothermal Ore Deposition,* New York: Macmillan, p. 85.

Hemley, J. J., 1959. Some mineralogical equilibria in the system $K_2O-Al_2O_3-SiO_2-H_2O$, *Amer. J. Sci.* 257:241-270.

Hemley, J. J., and W. R. Jones, 1964. Chemical aspects of hydrothermal alteration with emphasis on hydrogen metasomatism, *Econ. Geol.* 59:538-569.

Hornung, G., 1962. Wall rock composition as a guide to pegmatite mineralization, *Econ. Geol.* 57:1127-1130.

Hussak, E., 1898. Der goldführende, kiesige Quarzlagergang von Passagem in Minas Geraes, Brasilien, *Z. Prakt. Geol.* 6:345-357.

Jahns, R. H., 1946. Mica deposits of the Petaca district, Rio Arriba County, New Mexico, *N. Mex. Bur. Mines Mineral Resources Bull.* 25.

———, 1955. The study of pegmatites, *Econ. Geol. (50th Anniv. Vol.),* pp. 1067-1070.

Knopf, A., 1929. The Mother Lode system of California, *U.S. Geol. Surv. Prof. Pap.* 157.

Lovering, T. S., 1949. Rock alteration as a guide to ore—East Tintic district, Utah, *Econ. Geol. Monogr.* 1.

Lowell, D. G., and J. M. Guilbert, 1970. Lateral and vertical alteration—mineralization zoning in porphyry ore deposits, *Econ. Geol.* 65:373-408.

MacDiarmid, R. A., 1959. Geology and ore deposits of the Bristol silver mine, Pioche, Nevada, Ph.D. dissertation, Stanford Univ., California.

McKinstry, H. E., and J. A. Noble, 1932. The veins of Casapalca, Peru, *Econ. Geol.* 27:501-522.

Pardee, J. T., and C. F. Park, Jr., 1948. Gold deposits of the Southern Piedmont, *U.S. Geol. Surv. Prof. Pap.* 213.

Park, C. F., Jr., and R. S. Cannon, Jr., 1943. Geology and ore deposits of the Metaline quadrangle, Washington, *U.S. Geol. Surv. Prof. Pap.* 202.

Peterson, N. P., C. M. Gilbert, and G. L. Quick, 1946. Hydrothermal alteration in the Castle Dome district, Arizona, *Econ. Geol.* 41:820-840.

Rose, A. W., 1970. Zonal reactions of wallrock alteration and sulfide distribution at porphyry copper deposits, *Econ. Geol.* 65:920-936.

Sales, R. H., and C. Meyer, 1948. Wall-rock alteration at Butte, Montana, *Amer. Inst. Mining Eng. Trans.* 178:9-35; also *Amer. Inst. Mining Eng. Tech. Pub.* 2400.

———, 1949. Results from preliminary studies of vein formation at Butte, Montana, *Econ. Geol.* 44:465-484.

———, 1950. Interpretation of wall-rock alteration at Butte, Montana, *Colo. School Mines Quart.* 45(1B):261-273.

Schmitt, H. A., 1954. The origin of the silica of the bedrock hypogene ore deposits, *Econ. Geol.* 49:877-890.

Schoen, R., and D. E. White, 1965. Hydrothermal alteration in GS-3 and GS-4 drill holes, main terrace, Steamboat Springs, Nevada, *Econ. Geol.* 60:1411-1421.

Schwartz, G. M., 1947. Hydrothermal alteration of the porphyry copper deposits, *Econ. Geol.* 42:319-352.

———, 1950. Problems in the relation of ore deposits to hydrothermal alteration, *Colo. School Mines Quart.* 45(1B):197-208.

———, 1959. Hydrothermal alteration, *Econ. Geol.* 54:161-183.

Schwartz, G. M., and R. J. Leonard, 1927. Contact action of pegmatite on schist, *Geol. Soc. Amer. Bull.* 38:655-664.

Stoll, W. C., 1950. Mica and beryl pegmatites in Idaho and Montana, *U.S. Geol. Surv. Prof. Pap.* 229.

Stone, J. G., 1959. Ore genesis in the Naica district, Chihuahua, Mexico, *Econ. Geol.* 54:1002-1034.

Weller, J. M., R. M. Grogan, and F. E. Tippie, 1952. Geology of the fluorspar deposits of Illinois, *Ill. Geol. Surv. Bull.* 76.

6 / Paragenesis and Zoning

An ore-bearing fluid gradually changes as it migrates from its source. It reacts with wall rocks, changes its chemical composition, pH and other properties, travels into regions of lower pressure, and loses heat to cooler country rocks. As these physical and chemical changes take place, the ore and gangue minerals approach their respective stability constants and are sequentially deposited; they thus leave a detailed record in time and space of the evolutionary trends in an ore-forming solution.

The chronological order of mineral deposition is known as the *paragenesis* of a deposit; the spatial distribution is known as *zoning*. Paragenesis is determined by mineral studies that commonly focus upon microscopic textural features. Zoning patterns are manifested by mineralogic changes along both vertical and horizontal traverses of mineralized areas. The zones may be defined by differences in mineral species, differences in metallic elements and trace element content, differences in sulfur content, or even subtle differences in the ratios between certain elements. Whichever relationship is used to define a zone, the zoning and the paragenesis are cogenetic—they are merely two aspects of the same phenomenon.

Zoning was defined and contrasted with paragenesis in 1965 by a committee of the International Association on Genesis of Ore Deposits (Kutina, Park, and Smirnov, 1965). Such action became necessary because different meanings were attached to these terms by geologists of different countries. The committee decided that:

> Zoning in ore deposits is any regular pattern in the distribution of minerals or elements in space; it may be shown in a single orebody, in a mineral district, or in a large region. Although zoning is related to the spacial distribution of elements and minerals, both time and space must be considered in the study of zonal phenomena. The term paragenesis, as used in the United States, is the distribution in time, or the sequence of, minerals or elements. Paragenesis, as widely used in Europe, is an association of minerals having a common origin, for example, a tin-tungsten association or paragenesis.

A thorough study of carefully chosen samples made into thin sections and polished sections, in conjunction with detailed field mapping, will allow the geologist to develop a reasonably complete picture of the ore and wall rocks, including changes along the three space coordinates as well as those that took place during emplacement. Details in the character of mineralization—correlated with structure, the character of the wall rocks and their alteration products, the chemical reactions involved, and such factors as permeability and porosity of the host rocks—will lead to a better understanding of the processes of ore genesis and hence to sound exploration for new ore deposits.

PARAGENESIS

Studies of mineral sequences are practically restricted to the microscope; microtextures and microstructures are used to decide the order of mineral deposition, though field relationships between mineralized veins that cut across one another are also valuable clues to mineral paragenesis. In the study of a mine or mineralized district the geologist records the relative ages of each distinguishable mineral pair. He may find, for example, that chalcopyrite always formed earlier than sphalerite, and that sphalerite formed either earlier or at the same time as galena, but that some galena definitely formed later than sphalerite. He may find, further, that pyrite was ubiquitous. Such results would be plotted to show the relative sequence of mineral deposition, and the graph would be a paragenesis diagram similar to the one given in Figure 6-1.

The work of many geologists, studying countless polished sections throughout the world, has established a general sequence of mineral deposition in ore deposits. This sequence is based upon mineral stability ranges and is fairly

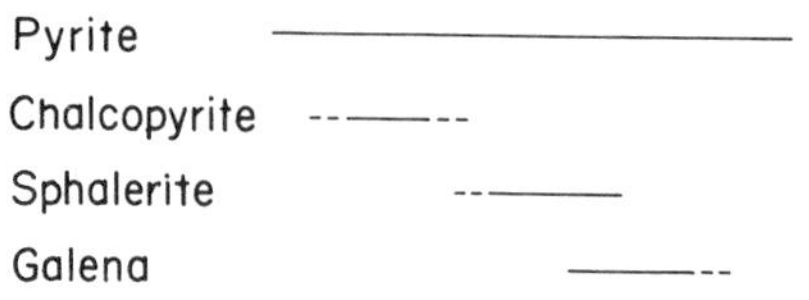

Pyrite	———————————
Chalcopyrite	-- -——— --
Sphalerite	-- -———
Galena	————— --

Figure 6-1
Sample diagram of paragenesis, showing the formation of chalcopyrite before sphalerite and galena, with an overlap in the deposition of sphalerite and galena. Pyrite was deposited with all of the ore minerals. Relative time is indicated from left to right.

constant for most deposits, regardless of the depth and temperature of formation, and regardless of origin—whether magmatic, hydrothermal, pneumatolytic, meteoric, or metamorphic. Since the character of an ore-bearing fluid gradually changes as it moves, different minerals will be formed along different parts of the channel; while one mineral is being deposited under certain conditions at one place or level, other minerals are forming elsewhere under different conditions. Thus, the deposition of one mineral commonly overlaps the deposition of others in both space and time. Slight changes in temperature, pressure, or chemistry of the transporting fluids may also alter the normal course of deposition and cause reversals or interruptions in the process. Furthermore, the paragenesis of each stage in a multiple pulsation system has a chronologic sequence of its own, hence the order of mineral deposition in any two pulses may not agree exactly. For example, a mineralized fissure may be repeatedly opened by tectonic activity (a common phenomenon), allowing discontinuous surges of ore-bearing fluids to enter; since the chemistry of the fluids may change in time, the mineralogy of each surge may be different. Determining the order of deposition is therefore seldom simple, and the true paragenesis can rarely be ascertained from the study of a few samples; it requires an examination of many thin sections and polished sections from samples taken at widely scattered locations within the deposit.

The established mineral sequence, as determined first by Lindgren and later by Edwards, is given in Tables 6-1 and 6-2. The minerals within each group are listed in the order of their deposition, the earliest-formed minerals being listed first. It must be emphasized that although the mineral sequence holds in general, there are many exceptions, reversals, and examples of overlapping deposition.

The agreement between Lindgren's and Edwards' paragenetic models suggests that generalized statements about the sequence of ore deposition may be formulated. We note that the oxides are deposited early; sulfides and arsenides of iron, nickel, cobalt, tin, and molybdenum are generally contemporaneous with or slightly later than the oxides; zinc, lead, silver, and combined copper-iron sulfides are intermediate in the paragenesis and are mixed with, or slightly older than, the copper, lead, and silver sulfosalts; native metals and tellurides are typically late; and antimony and mercury sulfides are the latest.

Table 6-1
Paragenesis in hydrothermal ore deposits, according to Lindgren (1937).

1. Quartz (continued), chlorite, tourmaline, lime-iron silicates, sericite, albite, adularia, barite, fluorite, siderite, rhodochrosite, ankerite, calcite (continued).
2. Magnetite, specularite (sometimes later), uraninite.
3. Pyrite, arsenopyrite, cobalt and nickel arsenides.
4. Cassiterite (sometimes preceding pyrite), wolframite (scheelite), molybdenite(?).
5. Pyrrhotite, pentlandite, chalcopyrite, stannite, bismuthinite(?).
6. Sphalerite, enargite, tennantite, tetrahedrite, chalcopyrite, bornite, galena, chalcocite, stromeyerite, argentite, ruby silver, polybasite, chalcopyrite, lead-silver sulfantimonides, native silver, native bismuth, electrum, tellurides, native gold.
7. Stibnite, cinnabar.

Table 6-2
Paragenesis in hydrothermal ore deposits, according to Edwards (1947, 1952).

Ore minerals

1. Magnetite, ilmenite, chromite, hematite.
2. Cassiterite, tantalite, wolframite, molybdenite.
3. Pyrrhotite, pentlandite, lollingite, arsenopyrite, pyrite, cobalt and nickel arsenides.
4. Chalcopyrite, sphalerite (interchangeable), bornite.
5. Tetrahedrite, galena, lead sulfosalts, silver sulfosalts, native bismuth and bismuthinite, tellurides, stibnite, cinnabar.

Gangue minerals

1. Quartz, tourmaline, topaz.
2. Siderite (often manganiferous), fluorite, calcite, barite, chalcedony.

ZONING

The paragenesis of mineral formation in moving ore fluids produces changes in ore and gangue mineralogy along the course of deposition. Such changes are described as zoning, and are found in sedimentary deposits as well as in magmatic and metamorphic ores. In the ideal event of a radiating hydrothermal or pneumatolytic fluid, changes in chemistry, temperature, and pressure along the fissures result in the deposition of different minerals in concentric zones at increasing distances from the source. Syngenetic deposits, however, may be zoned parallel to a shore line or along a stream channel leading away from the source rock. Any detection of a zonal pattern—epigenetic or syngenetic—is important to economic geology because it helps to predict changes in mineralization as a deposit is developed and mined.

The theory of zoning was probably first stated as a generalization by Spurr (1907), though many workers had previously recognized the phenomenon (de la Beche, 1839; Henwood, 1843; Collins, 1902; Waller, 1904; de Launay, 1900). At the present time zoning is widely recognized and is accepted as a working hypothesis. Yet the causes of zoning are still being debated, and it is difficult to explain why certain deposits are zoned and others are not (Czechoslovak Acad. Sci., 1963).

Zoning in ore deposits is conveniently divided into three intergradational classes, based upon size but independent of origin. These classes are:

1. *Regional zoning*—zoning on a very large scale, as exemplified by the Southern Piedmont region of the southeastern United States and by ore deposits associated with the Sierra Nevada batholith (Park, 1955).

2. *District zoning*—the zoning shown by closely grouped mines, a category which includes the well-known mining districts of Butte, Montana (Sales and Meyer, 1949), Cornwall, England (Davison, 1927; Lilley, 1932), and Bingham, Utah (Peacock, 1948).

3. *Orebody zoning*—changes in the character of mineralization within a single orebody or a single ore shoot (Riley, 1936). Many massive sulfide deposits of volcanogenic origin, such as the Kuroko bodies in the volcanic rocks of Japan, as well as many single orebodies within zoned districts, are in this category.

Because of the great disparity in size between mineralized regions and individual orebodies, confusion often arises in the discussion of zoning. This confusion led Sampson (1936) to suggest that the term "zoning" be limited to districts. But restricting the term in this way further confuses the usage. The addition of adjectives—"regional," "district," and "orebody"—to indicate the relative scale is preferable. This dimensional classification of zoning is not universally accepted by European geologists, some of whom feel that ore deposits usually result from discontinuous surges or pulsations of mineralizing solutions, resulting in a complicated overlapping of zones. They suggest that zoning should be classified according to whether the deposits were developed by a single pulsation or by many (monoascendent or polyascendent fluids), and whether they are in normal or reversed sequence (Kutina, 1957).

For zoned deposits of all scales field observation has shown that groups of minerals were formed in more or less constant sequence from the source of ore fluids outward. By gradually piecing together the field data on relative zoning in individual regions, districts, and orebodies, geologists have been able to construct a theoretical vein system against which any single deposit may be compared. Early observers noted that tin minerals characteristically lie deeper or closer to the source than copper minerals, and that copper minerals, in turn, are inside the silver zone. Expansion of these observations led to the relatively detailed model vein-system in Table 6-3. No single field example includes all of the mineral groups in the model, but the vein system is of value in the study and understanding of zoning. Note the similarity between this system and the models of paragenesis by Lindgren and Edwards.

As would be expected, there are many discrepancies between actual vein systems and the theoretical system. Irregularities and reversals result from rapid deposition and the overlapping of zones, among other factors. Discrepancies and poorly defined zoning are caused by the overlapping of deposits from two or more mineralizing centers, by the retreat or advance of mineralizing

Table 6-3
A reconstructed vein system, from the surface downward.

1. *Barren:* Chalcedony, quartz, barite, fluorite. Some veins carry small amounts of mercury, antimony, or arsenic.

2. *Mercury:* Cinnabar deposits, commonly bearing chalcedony, marcasite. Barite-fluorite veins.

3. *Antimony:* Stibnite deposits, locally passing downward into galena with antimonates. Some carry gold.

4. *Gold-silver:* Bonanza gold deposits and gold-silver deposits. Argentite with arsenic and antimony minerals common. Tellurides and selenides in places. Relatively small amounts of galena, adularia, alunite, with calcite, rhodochrosite, and other carbonates.

5. *Barren:* Most nearly consistent barren zone; represents bottom of many Tertiary precious-metal veins. Quartz, carbonates, and small amounts of pyrite, chalcopyrite, sphalerite, and galena.

6. *Silver:* Argentite veins, complex silver minerals with antimony and arsenic, stibnite, some arsenopyrite; quartz gangue, in places with siderite.

7. *Lead:* Galena veins, generally with silver; sphalerite usually present, increasing with depth; some chalcopyrite. Gangue of quartz and carbonates.

8. *Zinc:* Sphalerite deposits; galena and some chalcopyrite generally present. Gangue is quartz and, in some deposits, carbonates of calcium, iron, and manganese.

9. *Copper:* Tetrahedrite, commonly argentiferous; chalcopyrite present. Some pass downward into chalcopyrite. Enargite veins, generally with tetrahedrite.

10. *Copper:* Chalcopyrite veins, most with pyrite, many with pyrrhotite. The gangue is quartz and, in some places, carbonates and feldspar. Orthoclase and sodic plagioclase not rare, but high-calcium plagioclase very rare. Generally carry precious metals. Uranium; probably main horizon of uraninite.

11. *Gold:* Deposits with pyrite, arsenopyrite, quartz, carbonates, and some with feldspar gangue. Tourmaline. Tellurides not uncommon and at places abundant. Some deposits have zones 10 and 11 reversed.

12. *Arsenic:* Arsenopyrite with chalcopyrite.

13. *Bismuth:* Bismuthinite, native bismuth, quartz, and pyrite.

14. *Tungsten:* Veins with tungsten minerals, arsenopyrite, pyrrhotite, pyrite, chalcopyrite. Tungsten occurs in higher zones in fairly large amounts, but this is the main horizon.

15. *Tin:* Cassiterite, with quartz, tourmaline, topaz, feldspar.

16. *Barren:* Quartz, feldspar, pyrite, carbonates, and small amounts of other minerals.

Source: Emmons, 1936.

centers during one period of deposition, by repeated periods of mineralization in a single area, and by other causes not understood.

Where the minerals of one zone overlap those of another, the deposit is said to be *telescoped*. Near the surface, hydrothermal fluids are subjected to steep temperature and pressure gradients, causing rapid deposition of the ore minerals and a shortening, or telescoping, of the ore zone. At depth the temperature and pressure gradients are gentle; under these conditions deposition takes place slowly, and the separation of minerals is well defined. Telescoping is therefore restricted largely to deposits formed under shallow conditions, where changes in temperature and pressure are rapid (Borchert, 1951).

Where conditions change gradually, as in high-temperature, high-pressure deposits, zoning is generally minor, and is expressed, for example, by variations in the fineness of gold (Pryor, 1923) or in the amounts of minor constituents. Many districts and mines show no recognizable zoning; ore from the lower levels is apparently identical with ore from the upper levels. Probably the best examples of unzoned deposits are some of the deeper gold mines, though any metal mine may lack zoning. At Morro Velho, Brazil, for example, in the bottom workings (at a depth of about 8,000 feet vertically, or about 13,000 feet down the dip) the ore is apparently identical with ore from the upper workings. The Mother Lode district of California is another example of a high-temperature, high-pressure gold district that shows no zoning, either in the district or in individual orebodies. Yet when all the mineral deposits associated with the Sierra Nevada batholith (including those in Nevada) are plotted on a map, they show a regional pattern. Deposits of the California copper belt, as well as other base-metal deposits, are farther from the intrusive complex than are the gold deposits. The tungsten ores of Bishop, Kings Canyon, and other areas, also fit into the zonal pattern; they are closer to the intrusive than are the gold deposits. Analogous regional zoning may occur in Brazil near the Morro Velho deposits, but not enough is known of the area to permit definite conclusions.

The practical application of district zoning may be illustrated in the Comstock district of Nevada, where the distribution of mercury suggests horizontal zoning emanating outward from a central silver-gold area to a mercury area. If similar zoning exists vertically, several mercury anomalies may possibly be underlain by silver-gold mineralization (Cornwall, Lakin, Nakagawa, and Stager, 1967). The same authors indicate that a similar type of exploration might be useful at Tonopah, Nevada.

Regional Zoning—Southern Piedmont

An example of regional zoning is furnished by the ore deposits of the Southern Appalachian-Piedmont province from Washington, D.C., southward to the Coastal Plain of Alabama (Pardee and Park, 1948). In this region the metalliferous deposits are zoned around the central core of the Piedmont region, where both Paleozoic and Precambrian intrusives are abundant. The central part of the region contains deposits of pyrite and gold. East of this central gold-pyrite zone, Mesozoic Coastal Plain sediments cover the evidence; regional zoning is well shown only to the west of the core. Closely associated with the gold ores, but slightly to the west, are deposits of pyrrhotite and pyrite, as at the Gossan Lead, Virginia. Farther westward are pyrrhotite-chalcopyrite ores of the type found at Ducktown, Tennessee; west of the copper zone are the lead-zinc deposits at Austinville, Virginia, and in the Mascot-Jefferson City district, Tennessee. The outermost zone is defined by barite, which has been mined from several areas west of the lead-zinc ores. Figure 6-2 shows the zonal arrangement of these deposits in relation to the intrusive masses. It is unknown whether the ores are genetically associated with the plutonic rocks or whether both the ores and the intrusives are products of metamorphism deep within the Appalachian geosyncline.

These studies of regional zoning have considerable economic value. In the Southern Piedmont, for example, generally favorable sites for the discovery of lead-zinc deposits lie in the area between southwestern Virginia and southwestern Tennessee, as well as north and south of the known limits of mineralization. If there are any mineral deposits in the eastern part of this region, they are buried under Coastal Plain sediments; as geophysical methods are developed and refined, additional ores may be found beneath these materials.

District Zoning—Cornwall-Devon, England

The tin-and-copper district that runs through Cornwall and Devonshire, England, has long been recognized as one of the classical examples of district zoning. The mines have been ably and amply described by many geologists since the early comprehensive report of de la Beche (1839). Yet in spite of the voluminous descriptive literature that pertains to the district, surprisingly little has been published concerning the methods of transportation and precipitation of the ores. The generally accepted hypothesis is that zoning resulted from centrifugal deposition from an igneous center, and that deposition was largely controlled by changes in temperature and pressure. Hosking (1964) states that even though the deposition of a given mineral may have been strongly determined by temperature, it was much more strongly determined by the extent and nature of the fracture systems through which the fluids moved.

Hosking suggested that two thrust-fault zones—running north of east on a path essentially parallel to the long axis of the land mass of southwest England—outline a broad synclinal area that includes the Cornwall-Devonshire district. Between the fault zones the Paleozoic slates, sandstones, grits, limestones, and greenstones are highly faulted and folded (Davison, 1921, 1926,

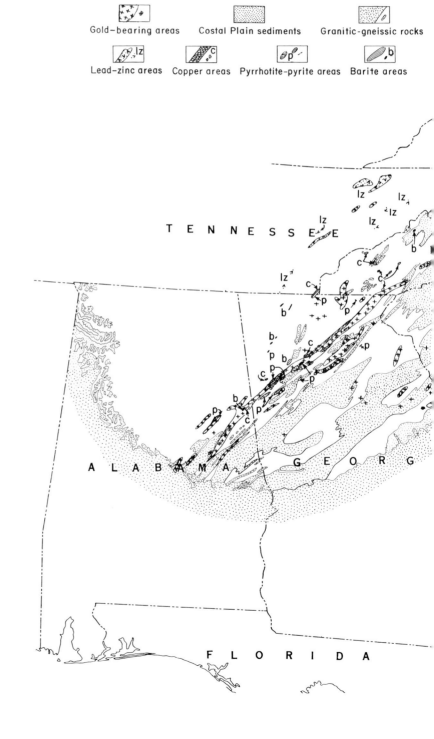

Gold–bearing areas   Costal Plain sediments   Granitic-gneissic rocks

Lead–zinc areas   Copper areas   Pyrrhotite-pyrite areas   Barite areas

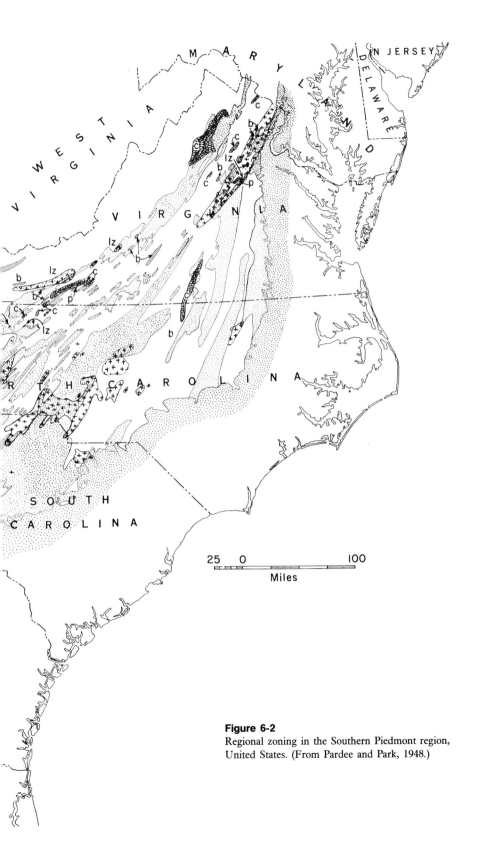

Figure 6-2
Regional zoning in the Southern Piedmont region,
United States. (From Pardee and Park, 1948.)

1927; Dewey, 1925, 1935; Hosking and Shrimpton, 1964). Following deformation a general uplift took place, and the syncline between the fault zones was invaded by granitic magma. Erosion has revealed five stocklike granitic masses and many smaller satellitic intrusive bodies of Permo-Carboniferous age. Skarns were extensively developed around the intrusive bodies, and were followed by shearing parallel to the trough of the syncline. It has long been suspected that the intrusive outcrops were connected at depth (see Fig. 6-3).

Figure 6-3

Map of southwest England, showing the distribution of
the granite outcrops, metamorphic aureoles, lodes, and dikes.
(From Hosking, 1964.)

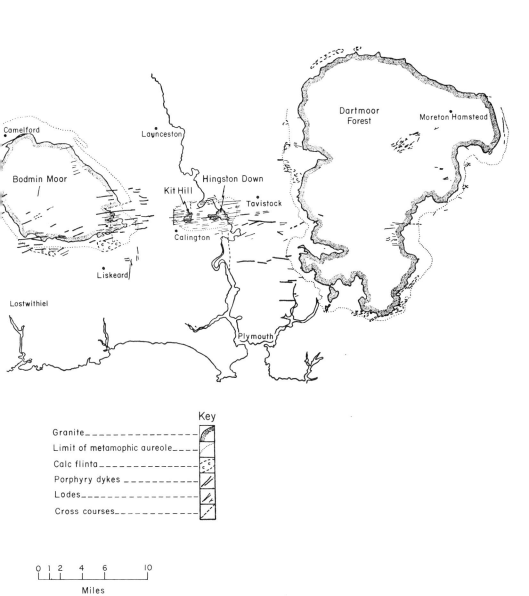

Key

Granite _____

Limit of metamophic aureole ____

Calc flinta _____

Porphyry dykes _____

Lodes _____

Cross courses_____

0 1 2   4   6   10

Miles

Bott and his co-workers (1958) confirmed this suspicion by gravity and other geophysical methods. They showed that the upper surface of the underlying batholith was irregular, that the present outcrops are apexes of the underlying mass, and that the batholith ascended vertically at the south and flowed northward (somewhat like a sill). Many pegmatite dikes and other small intrusive bodies of quartz porphyry, "trap," and other types were developed.

According to Hosking (1964), the mineral deposits of Cornwall are undoubtedly the products of several distinct phases of mineralization that took place between the emplacement of the batholith in Permo-Carboniferous time and the Eocene. Hot chloride-rich springs have been reported in several of the mines and possibly these are the end products of Tertiary mineralization processes, as may also be the gelatinous silica and opal currently being deposited in some of the kaolinized granite of the St. Austell stock. Hosking states, however, that the amount of Mesozoic and Tertiary igneous activity was insignificant compared to that of the earlier times.

MacAlister (1908) said that the intrusions and their accompanying mineralization took place in three gradational stages: (1) intrusion of granitic magma with accompanying thermal metamorphism in the rocks near the intrusive masses; (2) intrusion of quartz porphyry dikes along cleavage planes of the metamorphosed sedimentary rocks; and (3) deposition of the ores in both sedimentary and igneous host rocks. With various refinements, this statement is valid today.

The minerals mined from the Cornish mines were predominantly tin and copper, with lesser amounts of arsenic, tungsten, lead, zinc, silver, uranium, radium, iron, manganese, bismuth, nickel, and cobalt. The region is also well known for its valuable deposits of china clay, an alteration product that is associated with the regional mineralization.

Most of the ores have been recovered from complex bodies, among which lodes or fissures predominate. These lodes traverse both the granite and the adjoining slate. As a rule, they dip away from the granite, though in places the reverse is true (see Fig. 6-4); many of the stronger ones trend nearly east-west, parallel to the grain of the country rock. The east-west lodes are commonly cut by others trending north-south. In addition to the large lodes, swarms of pre-lode, greisen-bordered veins, and others occupy the upper few hundred feet of the apexes of the granite bodies. The lodes are ordinary fault zones that have been opened repeatedly, and along which the mineralizing fluids have ascended. As the success of many miners has shown, the best ores in the lodes are commonly found where the dips are steepest (Garnett, 1966).

The ores are thought to be genetically associated with the granitic intrusions; the veins in the slates at a distance from the main granite masses contain small amounts of granitic materials that intruded with them along the same fissures. The presence of topaz and brown-black tourmaline, which form readily under high temperatures and pressures, also indicates a close association between the

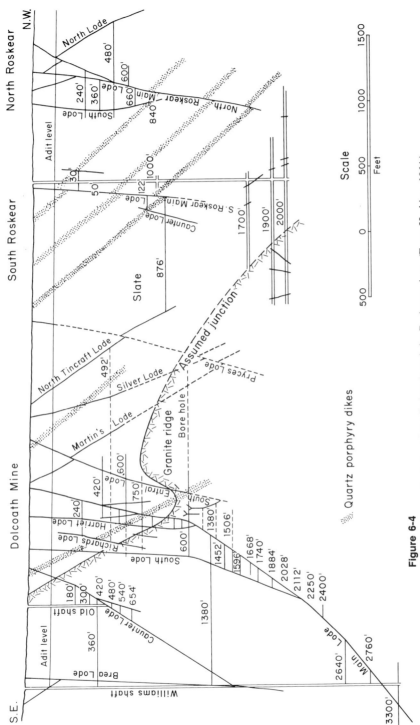

Figure 6-4

A section across the lodes of the Dolcoath and the Roskear mines. (From Hosking, 1964.)

ores and the igneous rocks. Jones (1931) stated that the granite masses are to some extent stanniferous, and that, as far as could be determined, the cassiterite was genetically and intimately related to the granites. Cassiterite appears to be as much an indication of the common parentage of the different igneous plutons as any other mineral common to them.

Alteration products are prominently developed in much of the area. Five types of hydrothermal processes have been recognized: tourmalization, greisenization, chloritization, silicification, and kaolinization. The tourmaline is ordinarily brown-black schorl, but in many lodes the finer needles are blue. In the lower levels of the tin lodes a common rock is composed of tourmaline and quartz. Where the rocks are rich in alumina and lime, axinite and garnet may form in place of the tourmaline. Greisenization is usually restricted to the intrusive masses that are changed to rock composed of quartz and sericite with minor amounts of topaz, tourmaline, and other minerals. Chlorite is abundant around some of the high-temperature ores, where it sometimes forms a chlorite-quartz rock segment. Kaolin is developed principally in the granitic rocks that are so thoroughly altered in places they form some of the world's largest and purest bodies of kaolin. Silification accompanied all other types of alteration; some of the quartz was produced as a byproduct during the breakdown of wall-rock minerals and some was primary silica (Hosking, 1951). Hosking (1964) listed hematitization as another form of alteration. He said that as the lodes in the granite are approached the feldspars become pink; locally, in and near the lodes, a hematite-quartz rock is developed. The alteration clearly varies with time.

Variations in the mineralization, or zoning, of the lodes and their alteration products have been recognized at Cornwall for more than a century, and have frequently been used as a tool in exploration. Changes in mineralization are recognized both along lode strikes and at depth. According to Hosking (1964), the distribution of major, minor, and even trace elements lends strong support to the view that during the major period of ore formation associated with the Permo-Carboniferous granites the lode minerals were deposited at progressively greater distances from the sources. According to Dines (1933, 1956), a zone is more extensive than any below due to the fact that fracturing is more widely developed toward the surface.

Four zones were recognized by Davison (1926). Iron-manganese ores containing small amounts of antimony were deposited farthest from the granitic rocks. Next toward the intrusives are the lead-zinc-silver ores, which in turn lie beyond the copper-arsenic-tungsten minerals. Closest to the intrusives—in fact, extending well into them—are the tin ores. As may be seen in Figure 6-5, many of the zones overlap and locally copper and tin are mined together. In other mines copper and tin are separated by as much as several hundred feet of barren rock. Before the separation of minerals into zones was understood, the presence of the thick layer of barren rock led to the abandonment of many mines that produced copper in their upper levels. These mines were later re-opened as the barren zone was penetrated and tin ore was encountered at depth (see Table 6-4).

Field data accumulated subsequent to Davison's work indicate that the mineral zones are not parallel to the granite contact, though there seems to be no objection to Davison's model for the relative sequence of deposition. Deeper mining has shown that the ore zones reflect the granite contacts but are less steeply inclined (see Fig. 6-6). Consequently, the tin zone is confined to the granite only at the peaks of the stocks; on the flanks it may lie entirely within the metamorphosed sediments (Dines, 1933; Hosking, 1951). Presumably, the zones represent isogeothermal surfaces, which compromise in attitude between the horizontal ground surface and the inclined granite-slate contact.

The iron-manganese ores are mostly oxidation products such as limonite, hematite, manganese oxides, and the hypogene carbonates, siderite and rho-dochrosite. A small amount of rhodonite has also been identified. Studies of the paragenesis indicate that these minerals were the last of the ores to form.

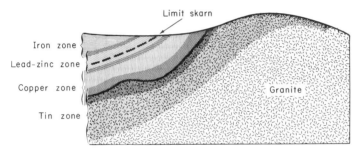

Figure 6-5
Sketch of Cornish mineral zones. (From Davison, 1926.)

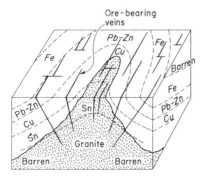

Figure 6-6
Block diagram of the arrangement of ore zones relative to granite-slate contact, Cornwall, England. (From Hosking, 1951.)

Table 6-4.
A generalized mineral paragenesis of the mineral deposits of southwest England.

Gangue	Zone	Ore-minerals	Economically important elements	Composition of wolframite and sphalerite	Trace element data*	Wall-rock alteration
Quartz · Felspar · Mica · Tourmaline · Chlorite · Hematite · Fluorite · Chalcedony · Barite · Dolomite · Calcite (ranges shown by bars)	7.	Barren: (pyrite).	Fe. Sb. *(Mesothermal and Epithermal Lodes. Generally at right-angles to granite ridges.)*			Skarn-type rocks derived from greenstones and calc. sediments.
	6.	Hematite. Stibnite. Jamesonite. Tetrahedrite. Bournonite. Pyrargyrite? Siderite. Pyrite: (marcasite).				Silicified slate.
	5b.	Argentite. Galena. Sphalerite.	Ag. Pb. Zn.		Bi and Sn decrease in galena.	Silicified granite. Kaolinized granite (?).
	5a.	Pitchblende. Niccolite. Smaltite. Cobaltite. (Native bismuth: bismuthinite?)	U. Ni. Co. Bi.			Hematitized slate. Hematitized granite. Chloritized slate: peach.
	4.	Chalcopyrite. Sphalerite. Wolframite: (scheelite). Arsenopyrite. Pyrite.	As ← W ← Cu ← *(Hypothermal Lodes. Generally parallel to granite ridges and dykes.)*	Fe increases in wolframite and decreases in sphalerite.	In, Mn, and Sn decrease, and Ge and Ga increase, in sphalerite.	Chloritized granite: peach. Tourmaline slate hornfels: capel.
	3.	Chalcopyrite: (stannite). Wolframite: (scheelite). Arsenopyrite. Cassiterite: (Wood tin).	Sn			Tourmalinized granite: capel.
	2.	Wolframite: (scheelite). Arsenopyrite: (molybdenite?). Cassiterite.				Quartz-sericite slate hornfels.
	1.	Cassiterite. Specularite. *(Veins often in granite cusps.)*				Greisenized granite.
	Greisen bordered veins	Arsenopyrite. Stannite. Wolframite, Cassiterite. Molybdenite.				
	Pegmatites	Arsenopyrite, Wolframite. Cassiterite. Molybdenite. *Earliest Minerals*				

*See El Shazly, E. C., Webb, J. S., and Williams, D., Trace elements in sphalerite, galena and associated minerals from the British Isles (*Trans, Inst. Min. Metall.*, v. 66, p. 241-271).

Note: Some of the components of Zones 3 and 4 may in part pre-date the cassiterite of Zones 1 and 2.

This outer zone is beyond the limits of recognized metamorphism caused by igneous activity and is of only minor economic value.

The lead-silver ores originated slightly later than the tin and copper-bearing lodes and are slightly older than the iron-manganese lodes. They are not as numerous as the tin and copper ores of the deeper zones. Most of these lead-zinc-silver deposits are beyond the outside edges of the metamorphic aureoles, but locally they extend into the metamorphosed ground. Their minerals include argentiferous galena, pyrargyrite, jamesonite, sphalerite, stibnite, bournonite, tetrahedrite, argentite, pyrite, quartz, dolomite, fluorite, chalcedony, barite, and minor amounts of uranium, nickel, bismuth, and cobalt.

Deposition of the copper-arsenic-tungsten ores took place during and after the formation of the deeper tin minerals. The copper-arsenic-tungsten zone itself is differentiated into an upper chalcopyrite-rich region and a lower arsenopyrite-wolframite region. Most of the copper-rich lodes extend from near the borders of the granites into the metamorphic aureoles, but not beyond the limits of thermally metamorphosed slates. Gangue minerals are quartz, chlorite, tourmaline, fluorite, pyrite, muscovite, and jasper. The chlorite is restricted to the upper zones; tourmaline is found only at depth; small amounts of cassiterite are recognized in the deepest parts of the copper-arsenic-tungsten lodes.

Tin belongs to the deepest of the recognized zones; it is generally found from depths of about 4,000 feet within the granite to a short distance above the granite in the metamorphosed slate. The principal ore mineral is cassiterite, accompanied by minor amounts of other metallic minerals from the copper-arsenic-tungsten zone. Throughout the district there is a well-defined tendency for the tin content to increase in all lodes at depth. The character of the gangue also changes. Chlorite, so common in the intermediate zones, disappears at depth; in its place is a fine-grained, dense, blue quartz crowded with tiny black tourmaline crystals. In many mines the blue quartz is brecciated and has been recemented by coarse-grained cassiterite and tourmaline.

Although zoning is apparent in many of the mines and in the district as a whole, there are still complications. Northward-dipping lodes containing sulfides, arsenides, and tungstates are cut locally by later, northward-striking lodes containing tourmaline and cassiterite. Accordingly, it is inferred that the temperatures increased during this period of mineral deposition, since cassiterite and tourmaline are thought to have formed at slightly higher temperatures than the sulfides, arsenides, and tungstates. (However, numerous other hypotheses, based upon physical-chemical reasoning, could be advanced.) Such a reversal of the normal sequence of mineral formation naturally leads to confusion. Another complicating factor is that many of the lodes have been subjected to repeated movements, brecciation, and subsequent infilling, with changes in composition of both the gangue and the metalliferous minerals. In spite of the complications, however, the zoning at Cornwall is well established and has been used successfully as a tool in exploration.

District Zoning—Tonopah, Nevada

The Tonopah district in Nevada is a good example of how subtle the effect of zoning may be. The Tonopah mines are closely spaced and do not encompass as large an area as the Cornwall district. Nevertheless, the zoning is defined by the deposits as a whole rather than by single orebodies.

Silver was discovered at Tonopah at the beginning of the twentieth century—late enough so that the U.S. Geological Survey was able to witness the development of the district and record the geologic and mineralogic details with accuracy. Numerous reports were published by some of the foremost geologists in the United States (Spurr, 1905, 1915; Burgess, 1909; Locke, 1912; Bastin and Laney, 1918; Nolan, 1935). From these and later studies, geologists have developed a three-dimensional picture of the mineralization.

The ore at Tonopah is mined principally for silver, though gold is also recovered. Faults provided avenues for mineralizing fluids, and the deposits are mainly replacement lodes along these fissures. The host rock is a confusing sequence of altered and faulted Tertiary volcanics ranging in composition from rhyolite to trachyte and andesite, and including lava flows, dikes, and pyroclastics. The most common ore minerals are acanthite, electrum, polybasite, and pyrargyrite, which are in a gangue of quartz, barite, and a pinkish carbonate mineral (rhodochrosite or dolomite?). Two principal types of alteration have been defined in the district: a central zone of quartz-sericite-adularia, and a surrounding envelope of chlorite-carbonate alteration, both of which are superimposed upon early, district-wide albitization (Nolan, 1935).

As a result of detailed field studies, Nolan (1935) determined that the ratio between silver and gold varies systematically, matching the configuration of the ore horizon as well as the pattern of alteration. Nolan contoured the upper and lower limits of mineralization over the district and found that the productive zone describes a somewhat symmetrically domed shell that persists without interruption across formation contacts and is only locally modified across faults. Furthermore, the higher-temperature alteration, represented by quartz-sericite-adularia, coincides with the central part of this domed shell, and the chlorite-carbonate alteration describes a peripheral zone. A plot of silver-gold ratios over the contour and alteration map (see Fig. 6-7) shows relatively low ratios near the top of the dome, compared to higher silver-gold ratios around the outside; silver tended to travel farther than gold. The symmetrical pattern of Nolan's compilation, along with the zoning of alteration products and the distribution of silver-gold ratios, led him to suggest that the lower and upper surfaces of the productive zone represent the approximate isothermal lines for temperature limits of ore deposition, theoretically reflecting a deeper hydrothermal source. This hypothesis is supported by the fact that deviations from a truly symmetrical dome correspond to major fault zones, which the mineralizing fluids could ascend most readily.

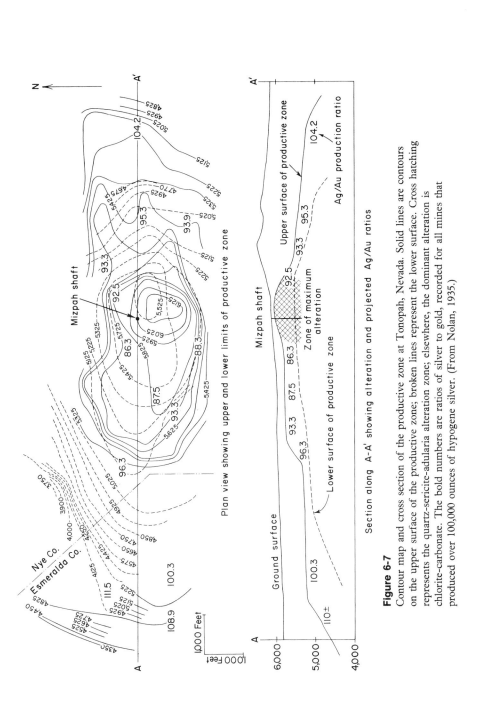

Figure 6-7

Contour map and cross section of the productive zone at Tonopah, Nevada. Solid lines are contours on the upper surface of the productive zone; broken lines represent the lower surface. Cross hatching represents the quartz-sericite-adularia alteration zone; elsewhere, the dominant alteration is chlorite-carbonate. The bold numbers are ratios of silver to gold, recorded for all mines that produced over 100,000 ounces of hypogene silver. (From Nolan, 1935.)

Recent work with oxygen isotopes on the rocks and ore-bearing materials from the Tonopah district indicate that both the hydrothermal alteration and the mineralization were produced by heated groundwaters that had undergone very little or no O^{18} enrichment by the addition of magmatic waters (Taylor, 1973).

Orebody Zoning—Red Mountain, Colorado

The best examples of orebody zoning are among the deposits formed at shallow depths, where temperatures and pressures change rapidly. Many of the deposits are in regions of comparatively recent volcanic activity. They are typically of Tertiary age and, in general, were formed at depths only slightly below their present locations.

The Red Mountain district, in the Silverton volcanic series of the San Juan Mountains, between Ouray and Silverton, contains excellent examples of orebody zoning. The volcanic rocks of Red Mountain are intruded by many plugs and pipes of breccia, porphyritic latite, and porphyritic rhyolite. Fracturing in and around these pipes is intense, and the area contains many weakly mineralized fissures. Large bodies of the rocks are impregnated with finely divided pyrite and hydrothermal alteration products, especially clay minerals, diaspore, alunite, and quartz.

Active mining in the Red Mountain district was restricted to the last decades of the nineteenth century. Typical orebodies were vertical chimneylike masses of roughly elliptical outline, bordered by fractures and fault planes. Within the main brecciated and mineralized pipes there were subsidiary chimneys, fissures, and irregular bodies of ore. Most, but not all, of the chimney orebodies were surrounded by envelopes of silicified country rock, which in turn were enclosed within highly argillized zones (Ransome, 1901; Burbank, 1941, 1947).

Orebody zoning was especially well developed in the Guston and Yankee Girl mines (Ransome, 1901), though it was recognized in many other properties of the Red Mountain district. Just below the weathered outcrop, the main orebody of the Guston mine contained galena and possibly some tetrahedrite, which assayed 50 to 60 percent lead, 30 to 40 ounces per ton of silver, and only a trace of gold. This ore continued to a depth of about 115 feet, where high-grade stromeyerite abruptly increased the silver values. Appreciable amounts of galena were found to a depth of about 290 feet, but below this level the lead content gradually diminished. With increasing depth, stromeyerite and ruby silvers, associated with pyrite and chalcopyrite, characterized the rich ore of the mine.

The best and most abundant ore was recovered between the fifth and sixth levels (288 and 378 feet below the surface); select carloads assayed up to 15,000 ounces of silver per ton, and the general ore ran 12 percent copper and from

1/10 to 3 ounces of gold per ton. At a depth of about 500 feet the orebody was cut by a fault, below which the content of silver was appreciably lower and the ore was harder and more compact. Mineralization down to the lowest level, at a depth of nearly 1300 feet, consisted largely of (low-grade) silver-bearing pyrite and some bornite, enargite, chalcopyrite, and barite; stromeyerite was not present in quantity. Some of the best ore in this fault block was found on the ninth level, about 680 feet beneath the surface. It consisted of bornite carrying 125 to 450 ounces of silver and $\frac{1}{4}$ to $\frac{1}{2}$ ounce of gold per ton, as well as 25 to 50 percent copper; chalcopyrite with 15 to 75 ounces of silver per ton, up to one ounce of gold per ton, and 8 to 15 percent copper; and massive pyrite that assayed up to 15 ounces of silver per ton. The most striking changes below the fault were a gradual increase in the ratio of pyrite to copper and silver minerals and an increase in the gold content. Ore on the ninth level and below carried up to 29 ounces of free gold per ton; much of the gold was in barite. The deepest levels contained large masses of lowgrade pyrite with sporadic nodules of bornite, chalcopyrite, and barite.

In general, the zoning pattern was one of gradual and overlapping changes from galena to stromeyerite, to bornite, to chalcopyrite, and finally to pyrite, considering the most prominent mineral in each zone; also, the gold content increased with depth. These zones did not grade smoothly from one to the next, however, and complications were not uncommon. Pyrite had the greatest vertical range; chalcopyrite extended above bornite but was important only within or below the bornite zone; the upper limit of stromeyerite was abrupt rather than gradational; and the highest values of silver and gold were associated with the greatest concentrations of copper. Nevertheless, zoning in the Guston mine was clearly evident when the ore was being extracted. Although the vertical changes in mineralogy were indisputable, not all of the zoning was vertical, as evidenced by horizontal sections of the high-grade ore between the fifth and sixth levels, where the main ore was stromeyerite and galena, enveloped in a low-grade pyrite-chalcopyrite aureole.

THEORIES OF SEQUENTIAL DEPOSITION

The reasons for time and space arrangement in mineral deposits have been among the most controversial theoretical factors in the study of ore deposits. Zoning and mineral paragenesis have been correlated at one time or another with the densities of fluids, atomic weights of the metals, mineral hardnesses, free energies of mineral formation, volatilities, metal-sulfur ratios, electrode potentials, and many other properties. They have been explained as resulting from changes in temperature and pressure that influence solubilities, and recently as deposition of minerals in regular sequences during the breakup of various complex ions, e.g., sulfide and chloride complexes. Any hypothesis must explain the general mineral sequence given by Lindgren, Edwards, and Emmons. And, as in most geologic controversies, more than one hypothesis will probably be valid under various circumstances; discrepancies in the theoretical ore sequence may be the result of either a relative abundance of elements in the ore fluid or differences in local geologic conditions.

According to Stôces (1934) and Brown (1948), as a general rule the latest minerals to form in an ore deposit are those farthest from the hydrothermal source, and the earliest minerals to form are those existing at depth under conditions of high temperature and pressure. Aside from the fact that the temperature of ore-bearing solutions should decrease with time, this observation does not explain the paragenesis of minerals at any one location, nor does it explain zoning. In the past, local zoning was explained mainly in terms of mineral solubilities; the most insoluble minerals are supposed to have been deposited nearest the source, and the more soluble, farther away. But the observed zoning does not agree with known mineral solubilities.

Although there are notable exceptions, the zonal distribution of minerals is generally the reverse of what would be expected from solubilities, and the younger minerals are typically the least soluble in a paragenetic series. For example, the copper and iron minerals commonly deposited in many central zones are more soluble than the far-removed cinnabar. Similarly, galena is typically younger and farther from the center of control than is the more soluble sphalerite. The manganese-lead relationship would be an exception to this inverse-solubility distribution because manganese is more soluble than lead, yet the minerals of manganese are generally deposited beyond the galena zone. As an explanation of these relationships, Magnée (1932) suggested that the solubilities reverse at elevated temperatures, but laboratory experiments later discredited this hypothesis (Verhoogen, 1938). Newhouse (1928) proposed that the chronologic order of deposition reflected the sequence of removal from the watery rest melt in differentiating magmas; the least soluble would be the last to be removed from the magma and consequently would be the latest to arrive at the mineral deposit. The geochemistry of this process seems a little dubious, but if the metals are supplied by deuteric leaching from a nearly crystallized pluton, it may help explain mineral paragenesis in some cases. In replacement deposits deposition may take place in an inverse order of solubilities, the least soluble mineral being deposited at the expense of more soluble minerals (Newhouse, 1928).

Owing to the evidence against simple mineral solubilities as an explanation of the paragenesis and zoning of mineral deposits, various alternative hypotheses have been proposed. A popular explanation is that sequential deposition may be at least partly related to mineral volatilities, which match the zoning sequence more closely than the solubilities. This hypothesis is based on the observation that minerals are formed roughly in the order of decreasing vapor pressure (assuming that the metals travel mainly as sulfides, oxides, and chlorides). Holland (1959) demonstrated that most mineral assemblages are in equilibrium with S_2-O_2-CO_2 gas at the time of deposition, without implying that these gases are the transporting agent. The concept of gaseous transport cannot explain all deposits, however, because abnormally high temperatures would be required to account for the deposition of some minerals (Krauskopf, 1957). Furthermore, it seems difficult to explain the gaseous replacement and removal of nonvolatile host rocks, such as chert and limestone.

Correlations between mineral sequences and atomic weights or ion percentages have been the basis of some plausible hypotheses on zoning and paragenesis. Newhouse (1928) pointed out the relationship between atomic weights of the metals and their chronologic order of deposition—the heavier elements are generally the youngest in any given deposit. A similar correlation can be shown for the oxides, sulfides, arsenides, and antimonides, which seem to form in the order of increasing atomic number of the anion. The implication of this observation is that the lighter elements are more readily expelled from the magma chamber. Brown (1950) prefers to relate zoning and paragenesis to the specific gravity of the ore minerals, rather than to the atomic weight of the elements, and his correlation is impressive. He suggests that the ore fluids are stratified within the magma chamber according to specific gravity and geochemistry, just as the melt in a blast furnace is layered into slag, matte, and speiss. Escape as volatiles would remove the light silicates and oxides (slag) first and the progressively heavier sulfides and native metals (matte and speiss) later. Similarly, fluids of different densities and those containing both liquids and gases might separate into fractions during migration; such separation might produce different types of mineralization around the source and thus would be a possible explanation of zoning.

Bandy (1940) suggested that the fundamental control over paragenesis is a progressive change in the weight percentage of the anion radical. According to Bandy's hypothesis, each successive oxide mineral has a higher weight percentage of the anion radical than the next older mineral. The sequence determined in this manner is essentially the same as that given by both Lindgren and Edwards, with a few minor exceptions. The sulfide relationship may be stated in another way, considering the metal ions rather than the anion radical. Accordingly, the percentage of metal in the sulfide minerals seems to be inversely proportional to the relative ages; late sulfides contain more metal ions than earlier ones. As a result, the hardness of ore minerals may be shown to follow a paragenetic pattern; the younger, metal-rich sulfides are softer than the older sulfides.

The order of mineral deposition in zoning has been related to the electrode potentials of the elements (Butler and Burbank, 1929; Butler, 1956). The electrode potentials are thought to have a direct and controlling influence on oxidation-reduction reactions and thus to have a vital bearing on ore deposition. Arranged in the order of their electrode potentials, the elements show a periodicity and a definite and striking relation to their distribution in primary ore minerals.

Zoning on a small scale, especially in replacement ore bodies, may be due in part to differential diffusion because the diffusion rate decreases with increasing ionic radius. The small ions should diffuse farther than the large ones (Holser, 1947). However, a comparison of ionic sizes with the apparent distances of diffusion in replacement ores seems to refute this hypothesis; the sequence is more nearly the reverse of what would be expected. Perhaps metals migrate to the replacement front as hydrated ions, in which event the effective ionic radius would be inversely proportional to the radius of the metal ions (Edwards, 1956).

Magnée (1932) suggested that precipitation of sulfides is conditioned by a complex equilibrium between H_2O, H_2S, HCl, HF, CO_2, and the oxides, sulfides, chlorides, and fluorides of the metals. The sulfides are the least soluble; in the presence of excess H_2S differential precipitation should be governed by the solubilities of the sulfides in water. According to this hypothesis, the concentration of sulfides in the magma-derived solution would be proportional to the relative abundance of the metals. Slight changes in concentration would be equivalent to large changes of temperature in determining when and where crystallization takes place. Zonal distribution would result from a single solution in the course of its migration from the magma.

From a study of the Mississippi Valley lead-zinc deposits, in which galena was deposited beyond sphalerite, Garrels (1941) concluded that concentrated chloride solutions could have been responsible for the zoning. Many workers think that chloride ion complexes furnish the most reasonable explanation for the nonsolubility order of mineral precipitation. Garrels and Dreyer (1952) suggested that a common constituent of the ore solution, such as a complex bisulfide, would make the metal sulfides soluble. In the presence of these complex ions, simple metal sulfides would be expected to precipitate in the order of their activity products.

Recently, complex ions have been called upon to explain hypotheses concerning problems in ore transport, deposition, paragenesis, and zoning. The study of complex ions is receiving concentrated attention, and they must be considered in any problem involving the geochemistry of ore fluids. Studies by Barnes (1962, 1967), as interpreted by him, show that the sequence of mineral deposition as predicted by the stabilities of covalently bonded complexes matches the observed paragenesis and zoning of ores, both in detail and on a large scale. The relative stabilities of complex ions can be determined using thermodynamics, provided we consider metals of the same valence for a single type of complex. Using this approach, Barnes found that cobalt, iron, nickel, tin, zinc, copper, lead, and mercury, in that order, should be increasingly stable (increasingly soluble as a complex) within any single type of anion complex. Furthermore, his calculations indicate that chalcopyrite and sphalerite should be closely related, and that chalcopyrite should be zoned nearer the center of mineralization. These thermodynamic considerations imply a wide gap between lead and mercury, the latter being more distantly removed from the magmatic source. Since this sequence refers only to the simple sulfides, iron occurs as pyrrhotite (or troilite), and tin, if included, would probably occur as stannite. Comparison of this theoretical sequence with those given by Lindgren, Edwards, and Emmons shows a remarkable match—in fact, as Barnes suggests, too remarkable to be fortuitous. However, any natural sequence of ore minerals is more complicated than the divalent sulfide series outlined above because each nondivalent mineral belongs to its own, unrelated paragenetic series. Also, referring to orders of complex stabilities in terms of the cations involved invites oversimplification. This is especially true for temperatures above 25°C, since the effect of temperature on the entropy of association may differ among complexes in a given family (Helgeson, 1964, pp. 6–7). Hence, Barnes' sequence may be valid for only a single complex and may not hold for any single type of anion complex.

Sales and Meyer (1949, 1950) noted a relationship at Butte, Montana, that is possibly a product of complexing sulfide ions. They found that a high sulfur-to-metal ratio in solution appears to favor deposition of the simple copper and iron sulfides, chalcocite and pyrite, whereas a relatively lower sulfur ratio favors the combined copper-iron sulfides, bornite and chalcopyrite. Covellite and digenite are also more prevalent in the high-sulfur environment, as their compositions would suggest. The sulfur concentration did not determine the actual sulfur content of each mineral; the critical variable was the fugacity, or partial pressure. Thus, pyrite accounts for most of the sulfur in the chalcocite-pyrite zone. At Butte, the sulfur-metal ratios decrease outward from the center of mineralization, describing a district zoning with the sulfur-rich minerals in the center and at depth. The sulfur ratio determined the nature of the minerals formed and in turn was reduced by deposition of the early sulfur-rich ores. Wall-rock alteration reduced the sulfur ratio further because destruction of ferromagnesian minerals in the quartz monzonite supplied iron that combined with sulfur from the vein fluids to form pyrite.

Sosman (1950) made the novel suggestion that zoning reflects a migration of metals toward rather than away from an intrusive. He argues that a magma undersaturated with water should imbibe water from the surrounding sediments. A gradient of both water pressure and water concentration would thus be established toward the intrusive. Moreover, the process of thermal transpiration should contribute to this gradient; that is, a gas under constant pressure, in a medium having small pores, will travel up the temperature gradient (toward the cooling magma) and down a chemical potential gradient. Most field evidence does not support this hypothesis. Paragenetic studies of ore suites from many districts indicate that, with few exceptions, minerals closest to the central zone are deposited before those farther away. In view of this evidence, it is difficult to contend that fluids migrate toward the magma. Most ore-bearing fluids probably migrate away, thus showing that most magmas are saturated with water.

District zoning is not always described in terms of intrusive igneous rocks. For example, Brown (1936) suggested that the zoning of Austinville, Virginia, is related to faulting rather than to a magmatic center. Pryor (1923) found that zoning is shown by an increase in the fineness of gold with depth in the famous Kolar deposits of India.

The causes of regional zoning are probably similar to those of district and orebody zoning, but the time interval involved in regional zoning is likely to be much longer. For example, in a region undergoing tectonic disturbances and the intrusion of a batholith or a complex of nested batholiths, the rate at which the igneous masses are differentiated, intruded, and cooled is of tremendous significance. The intrusions possibly involve long periods of time, during which the overlying rocks are slowly elevated, engulfed, or metamorphosed. Crystallization and differentiation of the magma cause the volatile fraction and rest melt to change radically from the earliest stages until the latest, ore-forming stages. Similarly, fluids entrapped in the country rock may be activated and set into motion. The escape paths of the fluids may differ at successive stages of intrusion or metamorphism. If the batholith pushes its way up slowly, the overlying rocks may be elevated into a slowly rising dome, with the formation of both radial and concentric faults. The history of fissuring may therefore depend upon the evolution of a domal structure; in turn, the zoning of ore deposition may be patterned according to the progressive order of faulting (Wisser, 1960). The zoning established during one stage of intrusion or metamorphism may be complicated or effectively masked by other ores deposited at a later stage. If the intrusion or metamorphism takes place during more than one geologic period, the zoning may also be of two or even more geologic ages. In this case, ores of different periods may have the same source, and fit together in a regionally zoned arrangement.

Some regional zoning seems to be independent of igneous activity, and may thus be related to metamorphic processes active near the bases of eugeosynclines. As metamorphism proceeds in the deeper wedge of the geosyncline, residual fluids are activated that should dissolve the most readily soluble materials. The fluids then migrate upward and outward into locations where the metallic salts are precipitated (Wells, 1956). Under these conditions, the fluids migrate and behave as though they originated in a magma. Metamorphic zoning, based upon the intensity of metamorphism and mineralogic rearrangement of chemical constituents, has been described in many areas and is universally accepted by petrologists. Original metallic components of highly metamorphosed rocks are generally considered to have been dispersed rather than concentrated during metamorphism, although this is not always true. Conditions near the edges of the most intensely metamorphosed zones could be such that metallic sulfides could be precipitated and preserved.

The lead-zinc ores of the Åmmeberg district in central Sweden have been attributed to metamorphism, and are zoned (on a district scale) about the center of metamorphic intensity (Magnusson, 1950). In sequence from the highest to the lowest metamorphic facies, the rock types include migmatitic gneisses; a complex unit of banded gray leptite containing lenses of metamorphosed limestone, biotite gneiss, skarn minerals, and ore deposits; and a barren red leptite (see Fig. 6-8). The ores and skarn minerals appear to have formed by the same process, and pegmatites in the ore zones indicate a similar genetic relationship. The sulfide minerals include pyrrhotite, sphalerite, and galena, which are zoned in sequence away from the migmatite front. A traverse away from the gneiss would encounter, successively: pyrrhotite, wollastonite-

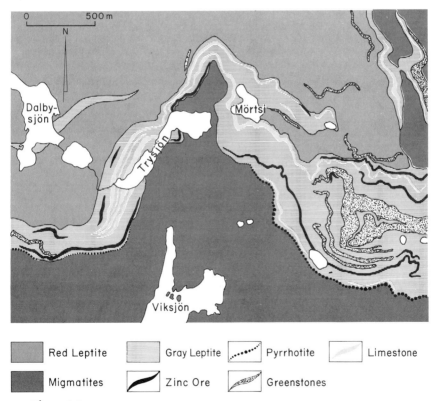

| | Red Leptite | | Gray Leptite | | Pyrrhotite | | Limestone |
| | Migmatites | | Zinc Ore | | Greenstones | | |

Figure 6-8
Geologic map of the Åmmeberg district, Sweden. (From Magnusson and Johansson, 1950, Fig. 65.)

diopside skarn, pyroxene-garnet-hornblende-mica skarn with pyrrhotite, sphalerite-rich ore, and finally galena-rich ore. All of the economic concentrations of lead and zinc are near the migmatite front, and the intensity of sulfide mineralization decreases gradually into the lower-grade metamorphic facies. Magnusson (1950) concluded that the migmatitic gneisses were formed by palingenesis during intense metamorphism, driving the sulfide solutions, skarn-forming solutions, and some pegmatitic solutions ahead of the migmatite front, where they were deposited in receptive rock formations and along favorable structures.

Zoning is also recognized in deposits that are thought to be of sedimentary origin, wherein minerals are deposited at varying depths relative to the shoreline. Overlap might be caused by advance or a retreat of this shoreline. Such an explanation was advanced by Davis (1954) to explain variations in the composition of the ores in the Roan Antelope deposits of Zambia. This explanation has also been applied to the Lake Superior iron formations, where hematite was deposited close to the shore, whereas deposits of magnetite, siderite, and pyrite are found toward deeper water (James, 1954).

The biochemical aspects of sulfide precipitation were discussed briefly by Temple (1964), who stated that the sulfides of different metals would have different rates of settling, particularly under the stagnant conditions of sulfate-reducing basins. This, together with the complexities of precipitation of soluble, colloidally suspended, or variously adsorbed metals, indicates a possible origin for the form of zoning exemplified by the layered sulfide deposits of Zaire and Zambia, in which the copper-iron ratio decreases outward from the shoreline. Temple believes that the aspects of microbial sulfate reduction are adequate, both qualitatively and quantitatively, for the syngenetic theory.

REFERENCES CITED

Bandy, M. C., 1940. A theory of mineral sequence in hypogene ore deposits, *Econ. Geol.* 35:359–381, 546–570.

Barnes, H. L., 1962. Mechanisms of mineral zoning, *Econ. Geol.* 57:30–37.

Barnes, H. L., and G. K. Czamanske, 1967. Solubilities and transport of ore minerals, in *Geochemistry of Hydrothermal Ore Deposits,* ed. H. L. Barnes, New York: Holt, Rinehart and Winston.

Bastin, E. S., and F. B. Laney, 1918. The genesis of the ores at Tonopah, Nevada, *U.S. Geol. Surv. Prof. Pap.* 104.

Beche, H. T. de la, 1839. *Report on the Geology of Cornwall, Devon, and West Somerset,* London, pp. 283–289.

Borchert, H., 1951. Die Zonengliederung der Mineralparagenesen in der Erdkruste, *Geol. Rundsch.* 39(1):81–94.

Bott, M. H. P., A. A. Day, and D. Masson-Smith, 1958. The geological interpretation of gravity and magnetic surveys in Devon and Cornwall, *Phil. Trans. Roy. Soc. London* 251:161-191.

Brown, J. S., 1948. *Ore Genesis*, New Jersey: Hopewell.

————, 1950. An alternative to the hydrothermal theory of ore genesis, *18th Int. Geol. Congr. Rept.*, pt. 2, pp. 37-44.

Brown, W. H., 1936. A quantitative study of the zoning of ores at the Austinville mine, Wythe County, Virginia, *16th Int. Geol. Congr. Rept.*, pt. 1, p. 459.

Burbank, W. S., 1941. Structural control of ore deposition in the Red Mountain, Sneffels, and Telluride districts of the San Juan Mountains, Colorado, *Colo. Sci. Soc. Proc.* 14(5):178-209.

————, 1947. Red Mountain district, Ouray County, in *Mineral Resources of Colorado*, ed. J. W. Vanderbilt, State of Colorado Mineral Resources Board.

Burgess, J. A., 1909. The geology of the producing part of the Tonopah mining district, *Econ. Geol.* 4:681-712.

Butler, B. S., 1956. Mineralizing solutions that carry and deposit iron and sulfur, *Amer. Inst. Mining Eng. Trans.* 205:1012-1017.

Butler, B. S., and W. S. Burbank, 1929. Relation of electrode potentials of some elements to formation of hypogene ore deposits, *Amer. Inst. Mining Eng. Trans. Yearb.*, pp. 341-356.

Collins, J. H., 1902. Notes on the principal lead-bearing lodes of the west of England, *Roy. Geol. Soc. Cornwall Trans.* 12:713.

Cornwell, H. R., H. W. Lakin, H. M. Nakagawa, and H. K. Stager, 1967. Silver and mercury geochemical anomalies in the Comstock, Tonopah, and Silver Reef districts, Nevada-Utah, *U.S. Geol. Surv. Prof. Pap.* 575-B, pp. B10-B20.

Cronshaw, H. B., 1921. The structure and genesis of some tin lodes occurring in the Camborne district of West Cornwall, *Inst. Mining Metall. London Trans.* 30:408-467.

Czechoslovak Academy of Sciences, 1963. *Symposium on the Problems of Postmagmatic Ore Deposition*, Prague: Czechoslovak Acad. Sci.

Davis, G. R., 1954. The origin of the Roan Antelope copper deposit of Northern Rhodesia, *Econ. Geol.* 49:575-615.

Davison, E. H., 1921. The primary zoning of Cornish lodes, *Geol. Mag. London* 58:505-512.

————, 1926. *Handbook of Cornish Geology*, Royal Geological Society of Cornwall, pp. 65-77.

————, 1927. Recent evidence confirming the zonal arrangement of minerals in the Cornish lodes, *Econ. Geol.* 22:475-479.

Dewey, H., 1925. The mineral zones of Cornwall, *Geol. Ass. Proc.* 36:107-135.

————, 1935. Copper ores of Great Britain, in *Copper Resources of the World*, vol. 2, 16th International Geological Congress.

Dines, H. G., 1933. The lateral extent of ore shoots in the primary depth zones of Cornwall, *Roy. Geol. Soc. Cornwall Trans.* 16:279-296.

————, 1956. The metalliferous mining region of southwest England, *Gt. Brit. Geol. Surv. Mem.*

Edwards, A. B., 1947. *Textures of the Ore Minerals*, Melbourne: Australasian Institute of Mining and Metallurgy.

————, 1952. The ore minerals and their textures, *Roy. Soc. N.S.W. J. Proc.* 85:26-46.

————, 1956. The present state of knowledge and theories of ore genesis, *Australas. Inst. Mining Metall. Proc.* 177, pp. 69-116.

Emmons, W. H., 1936. Hypogene zoning in metalliferous lodes, *16th Int. Geol. Congr. Rept.*, pt. 1, pp. 417-432.

Garnett, R. H. T., 1963. Local mineral zoning in Geevor mine, Cornwall, in *Symposium on the Problems of Postmagmatic Ore Deposition,* vol. 1, Prague: Czechoslovak Academy of Sciences.

———, 1966. Relationship between tin content and structure of lodes at Geevor mine, Cornwall, *Inst. Mining Metall. London Trans.* 75(sec. B):B1–B22.

Garrels, R. M., 1941. The Mississippi Valley-type lead-zinc deposits and the problem of mineral zoning, *Econ. Geol.* 36:729–744.

Garrels, R. M., and R. M. Dreyer, 1952. Mechanism of limestone replacement at low temperatures and pressure, *Geol. Soc. Amer. Bull.* 63:325–379.

Helgeson, H. C., 1964. *Complexing and Hydrothermal Ore Deposition,* New York: Macmillan, pp. 6–7.

Henwood, W. J., 1843. The metalliferous deposits of Cornwall and Devon, *Roy. Geol. Soc. Cornwall Trans.* 5:213–224.

Holland, H. D., 1959. Some applications of thermochemical data to problems of ore deposits, I: stability relations among the oxides, sulfides, sulfates and carbonates of ore and gangue metals, *Econ. Geol.* 54:184–233.

Holser, W. T., 1947, Metasomatic processes, *Econ. Geol.* 42:384–395.

Hosking, K. F. G., 1951. Primary ore deposition in Cornwall, *Roy. Geol. Soc. Cornwall Trans.* 18:309–356.

Hosking, K. F. G., and G. J. Shrimpton, eds., 1964. *Present Views of Some Aspects of the Geology of Cornwall and Devon,* Penzance: Royal Geological Society of Cornwall.

James, H. L., 1954. Sedimentary facies of iron-formation, *Econ. Geol.* 49:235–293.

Jones, W. R., 1931. Discussion on "Genesis of ores in relation to petrographic processes," *Brit. Ass. Advan. Sci. Rept. Centenary Mtg.,* pp. 387–389.

Krauskopf, K. B., 1957. The heavy metal content of magmatic vapor at 600° C, *Econ. Geol.* 52:786–807.

Kutina, J., 1957. A contribution to the classification of zoning in ore veins, *Univ. Carolina Geol.* 3(3):197–225.

Kutina, J., C. F. Park, Jr., and V. I. Smirnov, 1965. On the definition of zoning and on the relation between zoning and paragenesis, in *Symposium on the Problems of Post-magmatic Ore Deposition,* vol. 2, Prague: Czechoslovak Academy of Sciences.

Launay, L. de, 1900. Les variations des filons métallifûres en profondeur, *Rev. Gen. Sci. Pures Appl.* 11:575–588.

Lilley, E. R., 1932. Geology and economics of tin mining in Cornwall, England, *Amer. Inst. Mining Eng. Tech. Pub.,* class, I, no. 41, pp. 5–10.

Lindgren, W., 1937. Succession of minerals and temperatures of formation in ore deposits of magmatic affiliation, *Amer. Inst. Mining Eng. Trans.* 126:356–376.

Locke, A., 1912. The geology of the Tonopah mining district, *Amer. Inst. Mining Eng. Trans.* 43:157–166.

MacAlister, D. A., 1908. Geological aspect of the lodes of Cornwall, *Econ. Geol.* 3:363–380.

Magnée, I. de, 1932. Essai d'explication physico-chimique du phénomùne de la répartition zonaire des minerais d'origine magmatique, *Soc. Géol. Belg. Ann.* 55(*Mem.*): M17–M35.

Magnusson, N. H., 1950. Zinc and lead deposits of central Sweden, *18th Int. Geol. Congr. Rept.,* pt. 7, pp. 371–379.

Newhouse, W. H., 1928. The time sequence of hypogene ore mineral deposition, *Econ. Geol.* 23:647–659.

Nolan, T. B., 1935. The underground geology of the Tonopah mining district, Nevada, Univ. Nev. Bull. 29(5).

Pardee, J. T., and C. F. Park, Jr., 1948. Gold deposits of the Southern Piedmont, *U.S. Geol. Surv. Prof. Pap.* 213.

Park, C. F., Jr., 1955. The zonal theory of ore deposits, *Econ. Geol. (50th Anniv. Vol.)*, pp. 226-248.

Peacock, H., 1948. An outline of the geology of the Bingham district, *Mining & Metall.* 29:533-534.

Pryor, T., 1923. The underground geology of the Kolar gold field, *Inst. Mining Metall. London Trans.* 33:95-135.

Ransome, F. L., 1901. Economic geology of the Silverton quardrangle, Colorado, *U.S. Geol. Surv. Bull.* 182.

Riley, L. B., 1936. Ore-body zoning, *Econ. Geol.* 31:170-184.

Sales, R. H., and C. Meyer, 1949. Results from preliminary studies of vein formation at Butte, Montana, *Econ. Geol.* 44:465-484.

————, 1950. Interpretation of wall-rock alteration at Butte, Montana, *Colo. School Mines Quart.* 45(1B):261-273.

Sampson, E., 1936. Zonal distribution of ore deposits, *16th Int. Geol. Congr. Rept.*, pt. 1, p. 461.

Sosman, R. B., 1950. Centripetal genesis of magmatic ore deposits, *Geol Soc. Amer. Bull.* 61:1505.

Spurr, J. E., 1905. Geology of the Tonopah mining district, Nevada, *U.S. Geol. Surv. Prof. Pap.* 42.

————, 1907. A theory of ore deposition, *Econ. Geol.* 2:781-795.

————, 1915. Ore deposition at Tonopah, Nevada, *Econ. Geol.* 10:713-769.

Stôces, B., 1934. How to determine probable changes in primary mineralization with increasing depth, *Econ. Geol.* 29:93-95.

Taylor, H. P., Jr., 1973. O^{18}/O^{16} evidence for meteoric-hydrothermal alteration and ore deposition in the Tonopah, Comstock Lode, and Goldfield mining districts, Nevada, *Econ. Geol.* 68:747-764.

Taylor, R. G., 1966. Distribution and deposition of cassiterite at South Crofty mine, Cornwall, *Inst. Mining Metall. London Trans.* 75 (sec. B):B35-B49.

Temple, K. L., 1964. Syngenesis of sulfide ores: an evaluation of biochemical aspects, *Econ. Geol.* 59:1473-1491.

Verhoogen, J., 1938. Thermodynamic calculation of the solubility of some important sulphides, up to 400° C, *Econ. Geol.* 33:34-51, 775-777.

Waller, G. A., 1904. Report on the Zeehan silver-lead mining field, *Tasmania Geol. Surv. Bull.*, p. 24.

Wells, F. G., 1956. Relation entre gîtes minéraux et géosynclinaux, *Rev. Ind. Minerale* (numero spécial 1R, Jan.): 95-107.

Weston-Dunn, J. A., 1923. Some relations between metal content, lode filling, and country rock, *Econ. Geol.* 18:443-473.

Wisser, E., 1960. Relation of ore deposition to doming in the North American Cordillera, *Geol. Soc. Amer. Mem.* 77.

7 / Geothermometry and Isotopic Studies

The temperatures and pressures at which ores are deposited range from very high at depth to atmospheric at the surface. Placer deposits and other sedimentary ores form under atmospheric conditions. Veins, pegmatites, and magmatic segregation deposits may form at depths of as much as several miles and at temperatures above 500°C, and at times even above 1,000°C. Lindgren used temperature and pressure of formation as the basis for his classification of ore deposits; continued use of this classification depends upon acquisition of better data for determining conditions of deposition. As indicated by studies of paragenesis and zoning, mineral assemblages are themselves indicators of the temperatures and pressures of deposition. However, mineral assemblages are only indicative; in order to establish more precise limits of deposition, more accurate information is needed.

Knowledge of temperature and pressure of mineral deposition is based mainly upon laboratory studies of ores and accompanying gangue minerals. Several methods of study are in common use (Ingerson, 1955a, 1955b; Seifert, 1930; Kullerud, 1959; Roedder, 1967; Kelly and Turneaure, 1970), but all must be used with great care and caution. Most methods have proven difficult to apply or are of limited usefulness and accuracy; one of the principal difficulties is the correlation of temperature and pressure findings with an established paragenesis. The study of fluid inclusions is proving to be the best method of determining temperatures of deposition; in conjunction with the study of isotopes, it is beginning to provide a reasonably good picture of the nature and conditions of fluids at time of ore deposition.

There are practical as well as theoretical reasons for studying the temperature of ore deposition and the character of the depositing fluids. Aside from the fact that the genesis of a mineral deposit cannot be understood until the conditions of deposition and the nature of the fluid involved are determined, it is important for exploration purposes to know if the mineralizing fluid was hot or cold, and if hot, whether or not it cooled quickly or slowly. A deposit precipitated at a high temperature would be more likely to persist at depth than one that formed at a low temperature near the surface. If the depositing solutions cooled slowly, the vertical range of ore would probably be greater than where they cooled rapidly. As a logical generalization, it might be stated that deep, high-pressure thermal gradients are more gradual than shallow, lower-temperature ones. Consequently, compared to shallow deposits, ores deposited under deep-seated, high-temperature conditions will show little change over long distances.

Similarly a knowledge of pH, Eh, and other geochemical properties relating to the character of the ore bearing fluids is helpful in determining environments favorable for deposition.

STUDIES OF FLUID INCLUSIONS

Fluid inclusions have probably received more study in recent years than any other method for determining temperature of deposition of ore minerals. Such studies are proving to be consistent and reasonable, and as a result are becoming more and more useful. In addition to temperature and pressure of deposition, fluid inclusions furnish considerable information as to the chemical character of the fluid. Recently, many ore deposits have been subjected to exhaustive studies using fluid inclusion techniques, sometimes in conjunction with isotope studies (Kelly and Turneaure, 1970; Takenouchi and Imai, 1971; Imai, Takenouchi, and Kihara, 1971; Sillitoe and Sawkins, 1971; Robinson and Ohmoto, 1973).

The theory behind fluid inclusion studies is simple (Newhouse, 1933; Yermakov, 1957; Roedder, 1967). It is assumed that vacuoles in the minerals were formed when the minerals were precipitating and that the contents of these vacuoles were residues from the original solutions. The fluids in these vacuoles are thought to have been homogeneous before cooling, when they were separated into gases, liquids, and solids.

The study of inclusions is easier and quicker in transparent minerals than in opaque minerals; nearly all recent work has been done with transparent minerals. Inclusions may contain only liquid, liquid and gas, or in some, liquid, gas, and solids (see Figure 7-1). When inclusions are heated to the temperature of homogenization, where the minor or subordinate phases disappear, the lower limit for temperature of formation of the fluid is reached (see Figure 7-2).

Three types of fluid inclusions are distinguished: primary, pseudo-secondary, and secondary. Numerous careful studies have developed criteria for recognizing the three types, and the dependability of the results from fluid-inclusion studies rests directly upon the successful determination of the type of inclusion. *Primary inclusions* are those isolated with no clear relationship to any structure that would permit the escape or entry of either gas or liquid. *Pseudo-secondary inclusions* are commonly aligned in streaks; they are best described as lenticular clusters where the inclusions show no relationship to structures in which leakage could take place. *Secondary inclusions* are in zones or streaks that approach or touch fractures, either open or sealed, in which leakage might be possible; secondary inclusions must be avoided in analysis, since results obtained from them are unreliable.

To study fluid inclusions, small chips or thin sections of the transparent mineral are placed on a heating stage under the microscope. The temperature is then raised until homogenization takes place. This, becomes a minimum point on the geothermometer.

The degree of salinity of the fluid in the inclusion may be determined by placing a chip containing an inclusion on a freezing stage and lowering the temperature until freezing of the liquid phase. The degree of salinity is then calculated from this temperature, assuming that the salinity is caused by halite (NaCl). Halite is the most widely distributed, abundant, and easily determined of the solid minerals in inclusions. In addition to halite, at least two dozen other minerals have been reported in inclusions.

Although skepticism concerning the value of fluid inclusion studies is still expressed, so much work has been done and the results so consistent and reasonable that most scientists now accept the results. Ratios of liquids to gases and solids are ordinarily recorded during observation, and these ratios are found to be comparatively constant through wide ranges of inclusions in a given deposit. This would not be true if leakage took place, in which case a wide variation in composition would be expected. Secondary inclusions and leakage can ordinarily be determined by careful study (Smith, 1954; Smith *et al.,* 1950; Ingerson, 1955a; Richter and Ingerson, 1954; Roedder, 1960).

The decrepitation method applied to fluid inclusions was formerly widely used, but has now been largely abandoned. The principle of this method is to heat the inclusion until the vacuole explodes or decrepitates. This is supposed to give an idea of the upper limit of temperature of formation, but the determination is unreliable because the point of decrepitation depends to a large extent upon the strength of the containing mineral.

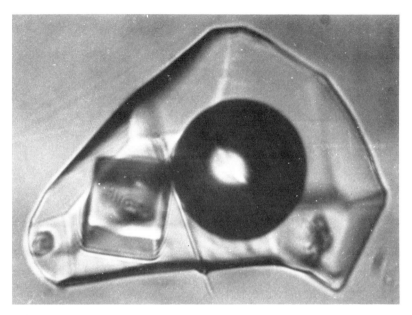

Figure 7-1
Primary inclusion in quartz with large halite cube and two unidentified daughter salts. (From Kelly and Turneaure, 1970.)

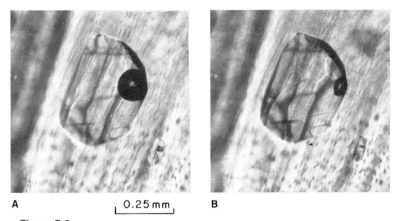

A
| 0.25 mm |
B

Figure 7-2
Fluid inclusion at (A) room temperature and (B) 130°C. The bubble within the liquid is much smaller at the higher temperature and would disappear completely above 140°C, the temperature of homogenization in this case. The bubble appears dark due to total reflection along its sides. The specimen shown is fluorite from the Deardorff mine, Cave-in Rock district, Illinois. (Photos by Bruce M. Harrison.)

SYNTHESIS OF MINERALS

A great deal of work has been done on the synthesis of minerals, although much of it was conducted with dry components, under conditions that do not approach those found in nature. Probably the most complete treatise on this subject is Doelter's *Handbuch der Mineralchemie* (1912-1931). A review of the literature was made by Morey and Ingerson in 1937.

Ordinarily, in geothermometric studies of synthesis the components of a mineral are mixed in stoichiometric proportions and are heated in a closed container. The lowest temperature at which a mineral will form under constant pressure is the accepted temperature for application to geothermometry; that is, a minimum temperature is established for that mineral at each specified pressure. In many instances the resulting products are similar to the natural minerals, and such synthetic products have been of special use as chemically pure materials for experimental use (Gaudin, 1938; Gaudin and Dicke, 1939). Unfortunately, natural minerals are rarely chemically pure substances; their structures contain small amounts of foreign elements, the presence of which almost certainly must have influenced the conditions of formation. Interpreting the results of synthetic studies is further complicated by the fact that the conditions of mineral genesis in nature depend in part upon the presence of fugitive constituents, such as water, and the presence of the fluxes or other mineralizers, which may never appear as components of the minerals. Consequently, most geologists are reluctant to accept experimental work based on synthetic minerals, especially if the minerals were formed in dry systems.

DETERMINATION OF MELTING POINTS

The temperature (at constant pressure) at which a mineral will melt is assumed to mark the upper limit of stability for that mineral (Birch *et al.*, 1942). Actually, melting points are helpful in establishing the maximum temperatures of mineral genesis, but they are of little value in determining the actual conditions of formation. When the melting point is low, it is a meaningful clue to the temperature of formation; when it is high, it does not even provide an order of magnitude for the temperature. For example, an assemblage of minerals that formed with orpiment (As_2S_3) could not have been deposited above 310°C, whereas a pure corundum deposit could theoretically form at any temperature below 2,050°C (Ingerson, 1955a). Again, the presence of small amounts of extraneous materials in crystals may alter the conditions under which synthetic and natural crystals melt; for example, Figure 7-3 illustrates the profound effect of water on the melting curve of SiO_2.

The melting temperature for individual minerals is less meaningful than the melting temperature for combinations of minerals. Mineral combinations

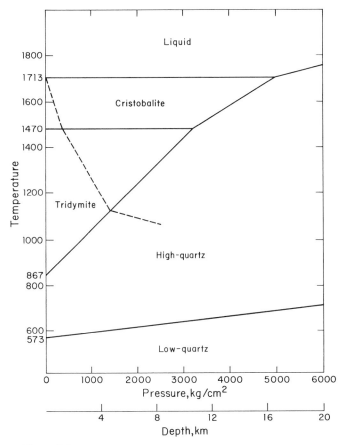

Figure 7-3
Pressure-temperature stability fields of the SiO_2 polymorphs. Solid
lines are for dry systems; broken line is liquidus under water vapor
pressure. (From Mason, 1958.)

melt at lower temperatures than any of their component minerals; the *eutectic
point* for mineral pairs is the minimum melting point for the combination.
Thus, orpiment with smithite ($AgAsS_2$) will melt at 280°C, even though
smithite is stable up to 426°C and orpiment melts at 310°C (Roland, 1966).
Furthermore, the presence of realgar with the orpiment lowers the temperature
maximum to 298°C, which is the eutectic temperature for As_2S_3-AsS. Another
example is argentite and proustite, which melt individually at 838°C and
495°C, respectively; their mixture melts at 469°C. The Ag_2S-As_2S_3 system
is a good example of the effects of the relationships because it includes three
eutectics: Ag_2S + proustite at 469°C, proustite + smithite at 409°C, and
smithite + orpiment at 280°C (Roland, 1966).

DETERMINATION OF INVERSION POINTS
AND STABILITY RANGES

Certain minerals undergo internal crystallographic changes at definite temperatures and pressures. Such changes or inversions are commonly recognizable and have been used in the study of ores (Ramdohr, 1931). The temperature or combination of temperature and pressure that causes the inversion from one polymorph to another should mark the minimum and maximum conditions of formation, depending upon whether the high-temperature polymorph or low-temperature polymorph is being considered. For some polymorphous groups the inversion temperatures are well established or vary within predictable limits; but the stability ranges for many polymorphs are highly controversial, and may depend upon impurities contained in the crystal structure (Buerger, 1961). Inconsistencies in laboratory results suggest that inversion temperatures in nature may cover wide ranges for some mineral groups. Inversion points also depend upon pressure (see Fig. 7-3), since polymorphous inversions involve density (lattice-volume) changes. The presence of water vapor under pressure does not affect the inversion temperatures of sulfides; it may affect the melting point, but not the polymorphic assemblages. If H_2O or OH^- and H^+ are part of the phases, then the temperature of polymorphic change will probably be affected by the water pressure or the pressure of any volatile that makes up part of the phase. This also holds for reactions between phases, e.g., pyrite $\rightleftharpoons$ pyrrhotite $+$ liquid (Kullerud, 1959).

In the Ag_2S system, argentite (isometric) is the stable form at high temperatures, and acanthite (monoclinic) is stable at low temperatures, the inversion temperature being about $180°C$ (Roy et al., 1959). The inversion point for this system seems to be a reliable geothermometer because it does not vary significantly with changes in pressure. The reliability of such a geothermometer may depend upon high purity of minerals and complete stoichiometry; for Ag_2S, the transition temperature between acanthite and argentite is not measurably affected by the likely presence of Pb and Bi as impurities (Van Hook, 1960) or Se (Roy et al., 1959).

In contrast to the acceptance of the argentite-acanthite geothermometer, the validity in geothermometry of the systems Cu-S and Zn-S has been widely debated. Wurtzite has been recognized as the stable form of ZnS above $1,020°C$, and sphalerite the stable polymorph at lower temperatures. Unfortunately, stability is a relative property, hence the "unstable" form may develop under special conditions. In their study of ZnS, Allen and his colleagues (1914) discovered that the hexagonal (wurtzite) crystal forms at low temperatures if the mineralizing solution is acidic. Although the inversion point is always high, it may be significantly lowered by the substitution of iron for zinc in the crystal lattice (Kullerud, 1959; Palache et al., 1944). Thus, the presence of wurtzite may indicate either high temperatures or an acidic environment; for this reason its value in geothermometry is extremely limited. Since, however, the low-temperature form cannot develop above the inversion temperature, the presence of sphalerite (as long as it is not a pseudomorph of wurtzite) indicates deposition below 1,020°C. The pyrite-marcasite system shows a similar relationship (Allen *et al.*, 1914).

The system Cu-S is one of the most controversial polymorphous systems. According to Posnjak and his colleagues (1915) and Merwin and Lombard (1937), chalcocite occurs in two polymorphs: a high-temperature, isometric form and a low-temperature, orthorhombic form. The inversion temperature was reported by Posnjak to be about 91°C, a low figure that should be of interest to the copper mining industry because it should serve to differentiate hot (ascending) from cold (descending) fluids. Relict cleavage remains after chalcocite has passed through the inversion point from the high-temperature to the low-temperature structure, and this cleavage can be recognized under the microscope (see Fig. 7-4).

Buerger (1941) questioned the validity of the chalcocite inversion point. He heated powdered mixtures of Cu_2S (chalcocite) and CuS (covellite) and took x-ray diffraction photographs at controlled temperatures during his experiments. The system Cu_2S-CuS was found to contain three compounds at room temperature: chalcocite, digenite (Cu_9S_5), and covellite. Upon heating, the crystal structures of orthorhombic chalcocite become disordered at 78°C, reaching complete disorder at 105°C. Above 105°C the Cu_2S has a hexagonal structure, rather than the isometric form, as previously thought. The orthorhombic (low-temperature) chalcocite dissolves up to eight atomic percent of Cu_2S or CuS, and hexagonal (high-temperature) chalcocite dissolves only about two percent of these compounds. Digenite, which is Cu_9S_5 (or $4Cu_2S \cdot CuS$) at low temperature, was regarded by Buerger as the material that, prior

Figure 7-4
Triangular etch pattern in chalcocite. Bn = Bornite. ×405. (From Brummer, 1955.)

to his work, had been called the high-temperature "isometric" form of chalcocite. Above 78°C digenite is capable of increasing solid solution of Cu_2S and CuS, and on slow cooling, which results in the return to an ordered crystallographic structure, this solution unmixes into either an intergrowth of chalcocite and digenite or an intergrowth of digenite and covellite (Buerger and Buerger, 1944). Complications result from the fact that chalcocite is actually a crystalline solution of the Cu_2S-CuS system (Eitel, 1958). Apparently the orthorhombic polymorph is stable only below the 78° to 105°C transition temperature, and the hexagonal form is stable above 105°C (Kullerud and Yund, 1960). An isometric form of Cu_2S does exist, but it is stable only above 425°C. This high-temperature polymorph can actually be considered as digenite with extra copper in solid solution because the amount of Cu admitted by the digenite structure is directly proportional to temperature. At room temperature digenite is roughly Cu_9S_5, but as Cu is taken into solution, the Cu:S ratio gradually increases from 9:5 until it reaches 10:5 ($Cu_{10}S_5$, or Cu_2S) at 425°C (Roseboom, 1960). Digenite is isometric but changes to another isometric form at 78°C. Likewise the presence of the mineral djurleite, $Cu_{1.96}S$, may further confuse things.

Buerger's work, plus refinements by many later investigators, confirms the fact that Cu_2S has three forms: cubic above 430 ± 10°C, hexagonal between 430° and 102°C, and orthorhombic below 102°C (Kullerud, 1964; Kullerud and Yund, 1960; Morimoto and Kullerud, 1966). However, as indicated above, the use of Cu-S minerals as geothermometers is not as reliable as was previously thought, and many past determinations must now be questioned. Future studies will have to determine if the minerals used are truly isometric, hexagonal, or orthorhombic, and whether or not exsolution textures are involved.

Chalcopyrite is said to invert from the isometric form to the tetragonal form at 525°C (a negative temperature correction must be applied for increased pressure). Twinned chalcopyrite usually results from this inversion and can, with fair confidence, be assumed to represent high temperature deposition (Kullerud, 1959).

The nonmetallic minerals have also been used to determine inversion temperatures. Quartz has been studied as much as any gangue mineral. Upon cooling, it changes at 573°C from a high-temperature form to a low-temperature form, maintaining the habit of the original polymorph. The inversion temperature may be strongly modified by the chemical environment (Keith and Tuttle, 1952), but with reservation it can be used to estimate the temperatures at which pegmatites and high-temperature veins are deposited. Furthermore, above 870°C the stable form of silica is tridymite, and above 1470°C it is cristobalite, although there are pressure-controlled modifications of these figures and the presence of water vapor changes the entire configuration (see Fig. 7-3).

DETERMINATION OF EXSOLUTION POINTS

High temperatures tend to promote disorder in a mineral structure, and under these conditions other elements are readily absorbed and retained. Upon cooling and the development of a more ordered structure, the extraneous materials may be forced out. They tend to accumulate along cleavage surfaces or crystallographically controlled directions in small blebs and blades. This development of intergrowth structures is called *exsolution* or unmixing, and has been used to a limited extent to indicate temperature points on the geologic thermometer as applied to ore deposits. Unfortunately, wide variations in results have been reported, and determinations based upon exsolutions must be viewed with skepticism.

According to Brett (1964), exsolution is effected by nucleation and subsequent growth of exsolved particles in a previously homogeneous phase. The mechanism is complex and is influenced by temperature, pressure, concentration, purity, structure, and grain size of the solid solution, the coefficients of diffusion, and the stress distribution within the crystal. Lamellar, lens-shaped, varied types of myrmekitic, mutual boundary, and rimming textures have all been ascribed to exsolution processes. Of these, the lamellar textures are considered by Brett to be the most indicative of exsolution.

The process of exsolution has been studied for many years in metallurgy and mineralogy, therefore the conditions under which many of the more common examples form are well known (Schwartz, 1942; Buddington *et al.*, 1955). The usefulness of this method is limited in that it tells us only the point at which order merges into disorder; it does not necessarily indicate the temperatures at which solidification or replacement takes place. Figure 7-5 shows a common example of this phenomenon. When a specimen of bornite is heated, the chalcopyrite blades eventually disappear and the bornite assumes a homogeneous appearance. If the temperature is lowered slowly, the blades reappear.

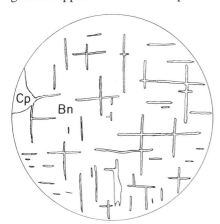

Figure 7-5
Blades of chalcopyrite in bornite. A typical exsolution texture. Brownson mine, San Juan County, Utah. ×192.

Hence, we assume the original mineral was once at temperatures and pressures above those at which unmixing takes place.

Exsolution is more rapid in minerals with low melting points, or, stated in another way, diffusion is faster in minerals with low melting points. Hence, quenching the solid solution will give usable results only for those minerals with high melting points. Heating a mineral pair to produce a homogeneous solid solution will reflect the minimum temperature only if the kinetics allow rapid diffusion and only if the mixture was a true exsolution product at the start.

The exsolved phase forms lamellae when it is isostructural, or nearly so, with the host. Where surface tension on the lamellae is too great, the lamellae tend to break up and form droplets or blebs. The presence of impurities can also have a strong affect on the rates of exsolution. Thus, the absence of exsolution textures cannot be said to imply a definite temperature of formation.

The use of an exsolution solvus curve is illustrated in Figure 7-6. Starting with a homogeneous crystal of composition X at a temperature above the solvus curve, and upon lowering the temperature to a point T_0, exsolution will begin. That is, T_0 is the minimum temperature for the exsolution pair A and B. If cooling is stopped or the size of the lamellae is restricted by the kinetics of lamellar development, at T_1, there will be exsolution lamellae of $A_{12}B_{88}$ in a host of $A_{85}B_{15}$. Continued cooling or arrested growth of lamellae with nucleation beginning at new centers would produce, at T_2, exsolution lamellae of A_3B_{97} in $A_{95}B_5$, the phase that previously was $A_{85}B_{15}$, and lamellae of $A_{95}B_5$ in A_3B_{97}, the earlier $A_{12}B_{88}$ lamellae.

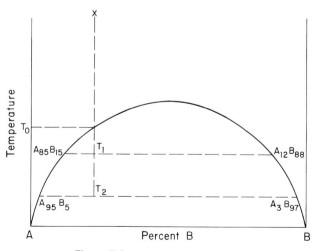

Figure 7-6
Hypothetical exsolution solvus curve.

Exsolution relationships are common in ore minerals, and they include a wide range of temperatures (Edwards, 1954). Examples include such combinations as blebs of pyrrhotite or chalocopyrite in sphalerite and lamellae or crystallographic intergrowths of ilmenite in hematite (and vice versa); ilmenite or hematite in magnetite; chalcocite or chalcopyrite in bornite; covellite or stromeyerite in chalcocite; and bornite or cubanite in chalcopyrite (see Fig. 7-7).

Exsolution in sulfide minerals is influenced by sulfur vapor pressure and as pointed out by Kelly and Turneaure (1970), wide variations in temperatures of exsolution of sulfides are possibly the result of the failure to consider the effects of geologic time.

STUDIES OF MINERAL
TEXTURES AND HABITS

Mineral textures and habits have been used by Edwards (1954) to indicate the temperatures of mineral deposition. He likened these textures to similar textures found in smelter products, and pointed out that they disappear upon annealing (or recrystallization). For example, colloform bands, herringbone texture, and columnar crystals in native copper have been interpreted to indicate repeated deposition at low temperatures from solutions of slightly varying composition; at higher temperatures the copper recrystallizes and assumes a texture similar to that obtained by annealing. Edwards also pointed out that textures such as the oölites in hematite are considered to be diagnostic of near-surface conditions, and do not form at high temperatures.

Drugman (1932) said that fluorite crystals have a form characteristic of their temperature of deposition. Light colored octahedral crystals are produced at relatively high temperatures; the more deeply colored cubic crystals are formed at lower temperatures.

DETERMINATION OF THE ELECTRICAL
CONDUCTIVITY OF MINERALS

Assuming that high-temperature crystals have fewer structural defects than low-temperature crystals, and that crystallographic imperfections retard electrical conductivity, Smith (1947) concluded that the electrical properties of conductive minerals should be a measure of their crystallization temperatures. Accordingly, Smith calibrated pyrite as a geothermometer by recording its thermoelectric potentials over the hydrothermal temperature range. Applications of this technique have given inconsistent and conflicting results, suggesting that the perfection of crystal growth does not depend upon temperature

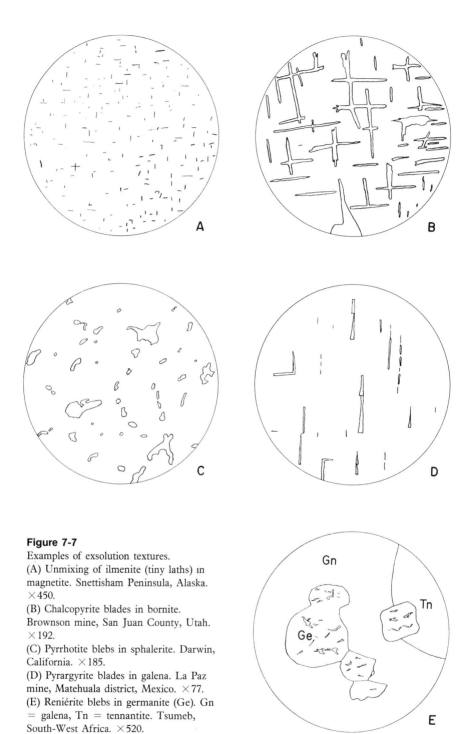

Figure 7-7
Examples of exsolution textures.
(A) Unmixing of ilmenite (tiny laths) in magnetite. Snettisham Peninsula, Alaska. ×450.
(B) Chalcopyrite blades in bornite. Brownson mine, San Juan County, Utah. ×192.
(C) Pyrrhotite blebs in sphalerite. Darwin, California. ×185.
(D) Pyrargyrite blades in galena. La Paz mine, Matehuala district, Mexico. ×77.
(E) Reniérite blebs in germanite (Ge). Gn = galena, Tn = tennantite. Tsumeb, South-West Africa. ×520.

alone (Smith, 1948; Ingerson, 1955a; Lovering, 1958; Hayase and Otuska, 1952). Future studies may resolve the discrepancies and establish pyrite (or some other common sulfide) as a reliable geologic thermometer.

DETERMINATION OF THE CONDITIONS
NECESSARY FOR IONIC SUBSTITUTIONS

For many years field geologists have recognized that sphalerite in igneous metamorphic deposits and in other high-temperature deposits contained more iron that did deposits of lower-temperature zones. This led to the development of the sphalerite geothermometer, in which ZnS was in equilibrium with pyrite and pyrrhotite (Kullerud, 1953, 1959; Barton and Kullerud, 1957; Skinner *et al.*, 1959). Sphalerite has been used in the study of many mineral deposits; in general, the results closely correspond with results obtained by other methods. Barton and Toulmin (1966) and Boorman (1967) claim there are errors in this method and that results obtained with the sphalerite geothermometer may be invalid. Nevertheless, field evidence indicates that in the presence of excess pyrite and pyrrhotite high-temperature sphalerites are generally darker in color and higher in iron than are the low-temperature sphalerites.

Hexagonal pyrrhotite in equilibrium with pyrite has also been used in geothermometry; the curve of the pyrite-pyrrhotite-vapor assemblage is stable from 325° to 743°C, at which point the pyrite melts incongruently (Kullerud, 1959). This method may however be inaccurate in the presence of copper.

Where magnetite forms in the presence of excess titanium—as evidenced by associated ilmenite—the percentage of TiO_2 in the magnetite is largely a function of temperature, though the composition of the hydrothermal or magmatic system is a modifying factor (Buddington *et al.*, 1955). The titaniferous magnetite is of possible use in geothermometry at temperatures between 550° and 1,000°C.

DIFFERENTIAL THERMAL ANALYSIS

Geologists have attempted several other methods for determining temperatures of deposition. One of these, especially applicable to clay minerals or other fine-grained substances, is differential thermal analysis. This technique allows the temperature of exothermic or endothermic reactions to be recorded while a mineral is heated at a constant rate. A differential thermal analysis will establish the temperature at which a mineral gives up its chemically bonded water: Theoretically, no mineral is stable under conditions above which its water of crystallization is expelled from the crystal structure.

AN EXAMPLE OF APPLIED GEOTHERMOMETRY

The tin and tungsten deposits of the eastern Andes of Bolivia were studied for many years by Turneaure and his associates (Turneaure, 1935, 1960, 1971; Turneaure and Welker, 1947). Most of the metals produced in the region have come from vein systems associated with porphyry stocks of intermediate composition or from altered sedimentary rocks. In addition to the large vein systems, many of which are complex, narrow veins, sheeted zones, stockworks, and breccia lenses have been productive.

Hydrothermal alteration appears to have preceded emplacement of the ores in the veins. Chlorite, sericite, kaolinite, alunite, and quartz are the most common gangue and alteration products; tourmaline is present in the tungsten and silver-tin veins and wall rocks. Ore minerals include cassiterite, bismuthinite, native bismuth, pyrrhotite, stannite, marcasite, pyrite, siderite, wolframite, franckeite, teallite, and many others.

Field mapping indicated that mineralization was long continued and that an unusual range of temperatures was involved. Temperatures were thought to vary from 400° to 500°C in the early stages to about 100°C during the later stages. The deposits were also described as having formed in a near-surface environment characterized by rapid changes in both temperatures and pressures.

In order to check the character of the ore-bearing fluids and the temperature of deposition of the ores against the background of this extensive field work, a careful laboratory study of the minerals of tin and tungsten from the Eastern Andes of Bolivia was undertaken by Kelly and Turneaure (1970). Many methods of study were attempted, and the results show clearly the useful contributions that can be made by thorough studies of fluid inclusions.

The Bolivian tin and tungsten ores contain at least minor amounts of most of the sulfide minerals that have been used previously as geothermometers. Kelly and Turneaure applied several of the standard techniques in an attempt to evaluate temperature as a factor in zoning and paragenesis. After a great deal of effort they decided that most of the methods attempted did not reflect initial depositional temperatures. For example, sphalerite that displayed polytypism (sphalerite-wurtzite) could not be correlated with variables of geologic interest. The iron content of the sphalerite reflected equilibrium with pyrrhotite or pyrite, but the significance of this in terms of the temperature was questionable.

Other geothermometers, such as sulfide invariant points, high-low quartz inversion point, melting of native bismuth, and sulfide exsolution temperatures were closely examined. These methods were unsatisfactory when used alone, but they did serve to check results from the fluid-inclusion studies, which were by far the most productive and reliable of the methods tried.

The studies of fluid inclusions showed highly systematic trends of salinity and depositional temperatures during formation of the tin and tungsten ores. Tin-bearing fluids of the early stage were complex sodium chloride brines with low carbon dioxide content. During precipitation of cassiterite and early quartz, salinities reached values as high as 46 weight percent, but the later ore fluids were more dilute and gradually approached fresh water in the closing stages of mineralization.

Temperatures of deposition increased from about 300° to 530°C in the early stages but then later declined to less than 70°C. The bulk of the sulfides was deposited as temperatures declined to about 400° to 260°C. The hypogene mineral sequence is attributed chiefly to the cooling of vein fluids, but little is known of the wall-rock alteration and its influence on the paragenesis.

Boiling in the early vein fluids was indicated by the fluid inclusions in the quartz and cassiterite; it increased the salt content and forced the CO_2 into the vapor phase. Early boiling favored precipitation of the quartz and cassiterite and helps explain the restricted vertical range of very high grade tin ores in several shallow deposits. Evidence of boiling suggests that the ores were deposited at depths no greater than 3,000 meters, a limit that is compatible with that based upon stratigraphic reconstruction.

The data obtained are consistent with a single prolonged period of mineralization for the tin and tungsten deposits throughout Bolivia. A magmatic source is indicated for the mineralizing NaCl brines and their contained metals, but the continued cooling and dilution of these brines is thought to reflect gradual influx of meteoric water into the magmatic hydrothermal system.

STABLE ISOTOPE STUDIES

A great deal of time has been devoted in recent years to the study of stable isotopes, including those of lead, sulfur, oxygen, hydrogen, carbon, and strontium (Hoefs, 1973). The primary objective of the studies is to determine the source and composition of hot-spring waters (ore-bearing fluids), the source of the components, the age of the mineral deposit, and at times to help determine temperature of deposition. Although isotope data have as yet been of little value in exploration, as more information accumulates the studies are finding increasing usefulness.

For example, Krauskopf (1967) pointed out that one of the most promising ways in which to study the origin of hot springs is by means of a comparison of the ratios of the heavy isotopes of oxygen and those of hydrogen. In many hot springs these ratios are identical with the ratios of ordinary surface waters and are considerably different from the ratios of water extracted from granite. The conclusion has been drawn from such studies that most hot spring waters (or, as they have been described, nearly spent ore-bearing fluids) resulted from the heating of surface waters, though minor amounts of igneous source materials may have been added.

Sakai and Matsubaya (1973) studied the thermal-spring waters along the coasts of Japan and concluded from the ratios of heavy oxygen to heavy hydrogen that these thermal waters are largely mixtures of the local meteoric waters and sea waters. They also determined that the sulfur and oxygen isotopes in the sulfates of the thermal waters were the same as those in sea water.

Taylor (1973) stated that each of the waters of the earth's crust—which he classified as magmatic, metamorphic, oceanic, connate, and meteoric—has different values of heavy oxygen and heavy hydrogen. The study of these heavy isotopes and their ratios is of greater value in determining the origin of the waters than the temperature of formation.

Many of the earliest isotope studies were done on lead. Theoretically, rock lead (^{204}Pb) was formed at the same time as the earth and has remained unchanged throughout time. Other leads, ^{206}Pb and ^{207}Pb, are generated during the decomposition of uranium, ^{235}U and ^{238}U, and from the decomposition of thorium, ^{232}Th. Leads generated from uranium and thorium are formed at uniform and unchangeable rates, no matter what the temperature and other conditions. The ratios of these radiogenic leads to the original rock lead should therefore permit estimates of the age of the mineral deposit being studied.

Using this method, Doe (1973) concluded that leads from the Coeur d'Alene district, Idaho, are dominantly of Precambrian age, rather than Cenozoic age as previous geologic field mapping had indicated. Doe also pointed out that the lead-isotope ratios may vary with and be parallel to zoning in the ore deposits, thus serving as a tool for exploration.

Kuo and Folinsbee (1973) found that the lead isotope ratios in minerals of the Ross River area, Yukon Territory, Canada, varied at nearly uniform rates away from the central Ross River mining area, where the deposits contain more than 3.5 million tons of lead. Peripheral deposits become smaller and more radiogenic with increasing distance from Ross River.

Probably more has been published about sulfur isotopes than about any other isotope. Unfortunately, sulfur isotope ratios vary widely and are dependent upon the state of oxidation and reduction, the pH of the solution, temperature, and the ionic bond (Sakai, 1968). As a result of these and other variations that are only now beginning to be understood, the interpretation of sulfur isotope ratios has been at times difficult and controversial (Ault and Kulp, 1960; Jensen, 1957; Sakai, 1957).

Of the four known isotopes of sulfur, ^{32}S, ^{33}S, ^{34}S, and ^{36}S, only ^{34}S and ^{32}S are abundant enough to be studied. The others are generally present only in trace amounts (Jensen, 1967). Ratios of ^{34}S/^{32}S are always referred to standard ratios obtained from the study of troilite in meteorites of the Canyon Diablo fall in Arizona.

The heavy isotope of sulfur ^{34}S tends to become concentrated in molecules during oxidation, while the lighter isotope is higher in sulfur formed during reducing conditions. For example, the lighter isotope becomes concentrated during the reduction (by bacteria or inorganic processes) of sulfates to sulfides. Likewise, the heavier isotope is concentrated during oxidation of sulfide to the sulfate. Some geologists attach considerable importance to the activities of bacteria.

According to Field (1973), sulfides from the hydrothermal deposits of the western American Cordillera are isotopically similar to meteoric sulfur, regardless of the type of deposit, geologic age, or location. Field concluded that the sulfur in these deposits was derived from a deep-seated source and that contributions of heavy sulfur from sea water or evaporite was negligible.

As a result of isotopic, fluid-inclusion, and chemical studies, Heyl and his associates (1973) concluded that the ore fluids in the Mississippi Valley lead deposits were heated oilfield brines, and that both lead and sulfur came largely from a crustal source.

The heavy isotope of sulfur ^{34}S may also be present in varying amounts in a paragenetic sequence of metallic sulfides in a deposit. For example, the heavy isotope is most abundant in pyrite and the amounts decrease progressively in sphalerite, chalcopyrite, and galena (Stanton, 1972).

Studies of the sulfur isotope ratios at the Quemont mine in the Noranda district of northwestern Quebec, Canada, indicated two effects. One was related to the paragenetic sequence, the other was related to a depth effect attributed to the melting of the ores during intrusion of post-ore granites (Ryznar, Campbell, and Krause, 1967).

Stanton and Rafter (1966) studied the sulfur-isotope ratios of twelve stratiform ore deposits and concluded that each deposit was characterized by a narrow spread of values. They concluded that the sulfur was derived neither from migrant hydrothermal solutions nor from the H_2S of normal biological sulfate reduction in seawater. Rather, the ratios in the metallic sulfides were induced by enrichment of the heavier isotope in the metallic sulfide phase or by direct combination with "heavy" sulfur. Stanton and Rafter emphasize that their conclusions do not apply to all types of stratiform sulfide accumulations. As expected, there is an overlap between biogenic and hydrothermal isotopes.

Vinogradov (1964) studied the isotopic composition and origin of the volcanic sulfur in volcanic exhalations in Kamchatka. He concluded that the sulfur in the sulfides deposited by these exhalations resulted from biological or chemical reduction of sulfates. He concluded also that the sulfur supplied to the volcano was derived from groundwater circulating at great depth. His data do not indicate that juvenile sulfur was contributed to the volcanic exhalations.

Of the three known isotopes of oxygen, ^{18}O, ^{17}O, and ^{16}O, only ^{18}O and ^{16}O are abundant enough to be useful in isotope studies. Oxygen isotope analyses are generally referred to the isotopic composition of average sea water as a standard of zero. Isotope analysis of mineral assemblages, followed by calculation of the temperatures of formation from known calibration curves, and then the calculation of the $^{18}O/^{16}O$ ratios of water in equilibrium with the assemblages at their temperatures of formation are used to determine the oxygen isotope composition of a hydrothermal fluid (Taylor, 1967).

Studies have shown that $^{18}O/^{16}O$ ratios are temperature dependent and can be used as geothermometers for hydrothermal mineral pairs (Clayton and Epstein, 1961); the ratios generally pertain to the gangue minerals and cannot be applied directly to the geothermometry of sulfides. A study at the Gilman mine, Colorado, showed increasing $^{18}O/^{16}O$ ratios in dolomite going away from ore, reflecting a progressive decrease in temperature (Engle et al., 1958). The most that can be obtained from such data is the relative temperature of formation.

Other isotopes that have proven to be useful include carbon ^{13}C and ^{12}C. During the process of photosynthesis, these two isotopes are sufficiently separated so that the organic carbon is enriched in ^{12}C relative to ^{13}C (Krauskopf, 1967).

According to Rye and Ohmoto (1973), the studies of carbon and sulfur isotopes that have made the greatest contributions to knowledge of ore genesis are those that have been coupled with detailed geologic, mineralogic, and geochemical investigations. Under conditions of equilibrium, amounts of ^{13}C and ^{34}S in hydrothermal minerals are determined by temperature, pH, fugacities, and the isotope composition of the ore-bearing fluids. Many chemical and physical processes are accompanied by isotopic fractionation of the elements.

THERMOLUMINESCENCE

Thermoluminescence is "frozen-in phosphorescence" or the ability of a substance to emit light when heated. The light is produced by energy released as electrons escape from high-energy levels to more stable, low-energy levels. Each sample will give a characteristic "glow curve" that shows the intensity of luminescence as a function of temperature. Once a sample has been heated, its thermoluminescence is permanently drained because the electrons remain at the lower energy levels; the sample will not glow when reheated unless it has been reenergized by radioactive bombardment (Daniels et al., 1953; Ingerson, 1955b). Lovering (1958) found that the limestones at Gilman, Colorado, gave glow peaks at 235° and 330°C. The low-temperature peak was generally absent from dolomitized materials, but the 330°C peak was preserved everywhere except close to the ore. The thermoluminescence data thus seem to indicate that the wall rocks were heated above 235°C but not above 330°C—

figures that agree with the temperatures determined by liquid-inclusion thermometry.

Actually, the thermoluminescence of a sample is destroyed if it is heated to a temperature below that at which maximum luminescence is produced. Consequently, the absence of the 235°C peak at Gilman need not imply that the wall rocks were heated above 235°C; this peak could have been removed at temperatures as low as 150°C (MacDiarmid, 1963). Lovering suggested that these temperatures represent the premetallization phase of dolomitization and do not reflect the wall-rock temperatures attained during ore deposition. Within 40 feet of ore, even the 330°C peak was suppressed, due to heating effects of the post-dolomitization ore-bearing fluids. McDougall (1966, 1970) pointed out that all thermoluminescence crystals are to some degree semiconductors whose thermoluminescence capabilities increase with increases in the number of lattice defects and in the number of electrons trapped at such defects.

REFERENCES CITED

Allen, E. T., J. L. Crenshaw, and H. E. Merwin, 1914. Effect of temperature and acidity in the formation of marcasite and wurtzite: a contribution to the genesis of unstable forms, *Amer. J. Sci.* 38:398-431.

Ault, W. U., and J. L. Kulp, 1960. Sulfur isotopes and ore deposits, *Econ. Geol.* 55:73-100.

Barton, P. B., Jr., and G. Kullerud, 1957. Preliminary report on the system FeS-ZnS-S and implications regarding the use of the sphalerite geothermometer, *Geol. Soc. Amer. Bull.* 68:1699.

Barton, P. B., Jr., and P. Toulmin, III, 1966. Phase relations involving sphalerite in the Fe-Zn-S system, *Econ. Geol.* 61:815-849.

Birch, F., J. F. Schairer, and H. C. Spicer, 1942. Handbook of physical constants, *Geol. Soc. Amer. Spec. Pap.* 36.

Boorman, R. S., 1967. Subsolidus studies in the ZnS-FeS-FeS$_2$ system, *Econ. Geol.* 62:614-631.

Brett, R., 1964. Experimental data from the system Cu-Fe-S and their bearing on exsolution textures in ores, *Econ. Geol.* 59:1241-1269.

Brummer, J. J., 1955. The geology of the Roan Antelope ore body, *Inst. Mining Metall. London Trans.* 64 (pt. 6):257-318.

Buddington, A. F., J. Fahey, and A. Vlisidis, 1955. Thermometric and petrogenetic significance of titaniferous magnetite, *Amer. J. Sci.* 235:497-532.

Buerger, M. J., 1961. Polymorphism and phase transformations, *Fortschr. Mineral.* 39(1):9-24.

Buerger, M. J., and N. W. Buerger, 1944. Low-chalcocite and high-chalcocite, *Amer. Mineral.* 29:55-65.

Buerger, N. W., 1941. The chalcocite problem, *Econ. Geol.* 36:19-44.

Clark, L. A., and G. Kullerud, 1963. The sulfur-rich portion of the Fe-Ni-S system, *Econ. Geol.* 58:853-885.

Clayton, R. N., and S. Epstein, 1961. The use of oxygen isotopes in high-temperature geological thermometry, *J. Geol.* 69:447-452.

Daniels, F., C. A. Boyd, and D. F. Saunders, 1953. Thermoluminescence as a research tool, *Science* 117(3040):343–349.

Deicha, G., 1955. *Les Lacunes des Cristaux et Leurs Inclusions Fluides,* Paris: Masson.

Doe, B. R., 1973. Lead isotopes, ore genesis, and ore prospect evaluation [abstr.], *Econ. Geol.* 68:1206.

Doelter, C. A., 1912–1931. *Handbuch der Mineralchemie,* 4 vols., Dresden and Leipzig: T. Steinkopff.

Drugman, J., 1932. Different habits of fluorite crystals, *Mining Mag. London* 23:137–144.

Edwards, A. B., 1954. *Textures of the Ore Minerals and Their Significance,* Melbourne: Australasian Institute of Mining and Metallurgy, pp. 1–31.

Eitel, W., 1958. Structural conversions in crystalline systems and their importance for geological problems, *Geol. Soc. Amer. Spec. Pap.* 66, pp. 25–27.

Engel, A. E. J., R. N. Clayton, and S. Epstein, 1958. Variations in the isotopic composition of oxygen in the Leadville limestone (Mississippian) of Colorado, as a guide to the location and origin of its mineral deposits, in *Symposium de Exploracion Geoquimica,* vol. 1, 20th International Geological Congress.

Field, C. W., 1973. Sulfur isotope abundances in hydrothermal sulfate-sulfide assemblages of the American Cordillera [abstr.], *Econ. Geol.* 68:1206.

Gaudin, A. M., 1938. Identification of sulfide minerals by selective iridescent filming, *Amer. Inst. Mining Eng. Tech. Pub.* 912.

Gaudin, A. M., and G. Dicke, 1939. The pyrosynthesis, microscopic study and iridescent filming of sulfide compounds of copper with arsenic, antimony and bismuth, *Econ. Geol.* 34:49–81, 214–232.

Hayase, K., and R. Otuska, 1952. Study on pyrite, part I: on the electrical properties of pyrite, *J. Geol. Soc. Jap.* 58:133–143.

Heyl, A. V., G. P. Landis, and R. E. Zartman, 1973. Isotopic evidence for the origin of Mississippi Valley-type mineral deposits: a review [abstr.], *Econ. Geol.* 68:1207.

Hoefs, J., 1973. *Stable Isotope Geochemistry,* New York: Springer-Verlag.

Imai, H., S. Takenouchi, and T. Kihara, 1971. Fluid inclusion study at the Taishu mine, Japan, as related to geologic structure, *Soc. Mining Geol. Jap. Spec. Issue* 3, pp. 321–326.

Ingerson, E., 1947. Liquid inclusions in geologic thermometry, *Amer. Mineral.* 32: 375–388.

———, 1955a. Methods and problems of geologic thermometry, *Econ. Geol. (50th Anniv. Vol.),* pp. 341–410.

———, 1955b. Geologic thermometry, *Geol. Soc. Amer. Spec. Pap.* 62, pp. 465–488.

Jensen, M. L., 1957. Sulfur isotopes and mineral paragenesis, *Econ. Geol.* 52:269–281.

———, 1967. Sulfur isotopes and mineral genesis, in *Geochemistry of Hydrothermal Ore Deposits,* ed. H. L. Barnes, New York: Holt, Rinehart and Winston.

Keith, M. L., and O. F. Tuttle, 1952. Significance of variation in the high-low inversion of quartz, *Amer. J. Sci. (Bowen Vol.),* pp. 203–280.

Kelly, W. C., and F. S. Turneaue, 1970. Mineralogy, paragenesis and geothermometry of the tin and tungsten deposits of the eastern Andes, Bolivia, *Econ. Geol.* 65:609–680.

Krauskopf, K. B., 1967. *Introduction to Geochemistry,* New York: McGraw-Hill.

Kullerud, G., 1953. The FeS-ZnS system: a geological thermometer, *Norsk Geol. Tidsskr.* 32:61–147.

———, 1959. Sulfide systems as geological thermometers, in *Researches in Geochemistry,* ed. P. H. Abelson, New York: Wiley.

———, 1964. The Cu-Fe-S system, *Carnegie Inst. Wash. Year B.* 63 (1963–64), pp. 200–202.

Kullerud, G., and R. Yund, 1960. Cu-S system, *Geol. Soc. Amer. Bull.* 71:1911–1912.

Kuo, S.-L., and R. E. Folinsbee, 1973. Lead isotopes in mineral deposits spacially related to the Tintina Trench, Yukon Territory (Canada) [abstr.], *Econ. Geol.* 68:1207-1208.

Lovering, T. G., 1958. Temperature and depth of formation of sulfide ore deposits at Gilman, Colorado, *Econ. Geol.* 53:689-707.

MacDiarmid, R. A., 1963. The application of thermoluminescence to geothermometry, *Econ. Geol.* 58:1218-1228.

Mason, B., 1958. *Principles of Geochemistry*, 2nd ed., New York: Wiley.

McDougall, D. J., 1966. A study of the distribution of thermoluminescence around an ore deposit, *Econ. Geol.* 61:1090-1103.

————, 1970. Relative concentrations of lattice defects as an index of the temperature of formation of fluorite, *Econ. Geol.* 65:856-861.

Merwin, H. E., and R. H. Lombard, 1937. The system Cu-Fe-S, *Econ. Geol.* 32:203-284.

Morey, G. W., and E. Ingerson, 1937. The pneumatolytic and hydrothermal alteration and synthesis of silicates, *Econ. Geol.* 32:607-761.

Morimoto, N., and G. Kullerud, 1966. Polymorphism on the Cu_5FeS_4-Cu_9S_5 join, *Z. Kristall.* 123:235-254.

Newhouse, W. H., 1933. The temperature of formation of the Mississippi Valley lead-zinc deposits, *Econ. Geol.* 28:744-750.

Palache, C., H. Berman, and C. Frondel, 1944. *Dana's System of Mineralogy*, 7th ed., vol. 1, New York: Wiley.

Posnjak, E., E. T. Allen, and H. E. Merwin, 1915. The sulphides of copper, *Econ. Geol.* 10:491-535.

Ramdohr, P., 1931. Neue Beobachtungen über die Verwendbarkeit opaker Erze als "Geologische Thermometer," *Z. Prakt. Geol.* 39:65-73, 89-91.

Richter, D. H., and E. Ingerson, 1954. Discussion—some considerations regarding liquid inclusions as geologic thermometers, *Econ. Geol.* 49:786-789.

Robinson, B. W., and H. Ohmoto, 1973. Mineralogy, fluid inclusions, and stable isotopes of the Echo Bay U-Ni-Ag-Cu deposits, Northwest Territories, Canada, *Econ. Geol.* 68:635-656.

Roedder, E., 1960. Fluid inclusions as examples of the ore-forming fluids, *21st Int. Geol. Congr. Rept.* pt. 16, pp. 218-229.

————, 1967. Fluid inclusions as samples of ore fluids, in *Geochemistry of Hydrothermal Ore Deposits*, ed. H. L. Barnes, New York: Holt, Rinehart and Winston.

Roland, G. W., 1966. Phase relations and geologic application of the system Ag-As-S, Ph.D. dissertation, Lehigh Univ., Bethlehem, Pa.

Roseboom, E. H., Jr., 1960. HIgh-temperature x-ray studies in the system Cu-S, *Geol. Soc. Amer. Bull.* 71:1959.

Roy, R., A. J. Majumdar, and C. W. Hulbe, 1959. The Ag_2S and Ag_2Se transitions as geologic thermometers, *Econ. Geol.* 54:1278-1280.

Rye, R. O., and H. Ohmoto, 1973. Carbon and sulfur isotopes and ore genesis: a review [abstr.], *Geol. Soc. Amer. Abstr. 1973 Mtg.*, p. 790.

Ryzner, G., F. A. Campbell, and H. R. Krause, 1967. Sulfur isotopes and the origin of the Quemont ore body (Canada), *Econ. Geol.* 62:664-678.

Sakai, H., 1957. Fractionation of sulfur isotopes in nature, *Geochim. Cosmochim. Acta* 12:150-169.

————, 1968. Isotopic properties of sulfur compounds in hydrothermal processes, *Geochem. J.* 2:29-49.

Sakai, H., and O. Matsubaya, 1973. Isotope geochemistry of the thermal waters of Japan and its implication to Kuroko deposits [abstr.], *Econ. Geol.* 68:1209.

Schwartz, G. M., 1942. Progress in the study of exsolution in ore minerals, *Econ. Geol.* 37:345-364.

Seifert, H., 1930. Geologische Thermometer, *Fortschr. Mineral Kristall. Petrogr.* 14: 167-291.

Sillitoe, R. H., and F. J. Sawkins, 1971. Geologic, mineralogic, and fluid inclusion studies relating to the origin of copper-bearing tourmaline breccia pipes, Chile, *Econ. Geol.* 66:1028-1041.

Skinner, B. J., P. B. Barton, Jr., and G. Kullerud, 1959. Effect of FeS on the unit cell edge of sphalerite: a revision, *Econ. Geol.* 54:1040-1046.

Smith, F. G., 1947. The pyrite geo-thermometer, *Econ. Geol.* 42:515-523.

————, 1948. The ore deposition temperature and pressure at the McIntyre mine, Ontario, *Econ. Geol.* 43:627-636.

————, 1954. Discussion—some considerations regarding liquid inclusions as geologic thermometers, *Econ. Geol.* 49:331-332.

Smith, F. G., and W. M. Little, 1953. Sources of error in the decrepitation method of study of liquid inclusions: discussion, *Econ. Geol.* 48:233-238.

Smith, F. G., P. A. Peach, H. S. Scott, A. D. Mutch, G. D. Springer, R. W. Boyle, and W. M. M. Ogden, 1950. "Pneumatolysis" and the liquid inclusion method of geologic thermometry: a reply, *Econ. Geol.* 45:582-587.

Stanton, R. L., 1972. *Ore Petrology,* New York: McGraw-Hill.

Stanton, R. L., and T. A. Rafter, 1966. The isotopic constitution of sulfur in some stratiform lead-zinc sulfide ores, *Mineralum Deposita* 1:16-29.

Takenouchi, S., and H. Imai, 1971. Fluid inclusion study of some tungsten-quartz veins in Japan, *Soc. Mining Geol. Jap. Spec. Issue* 3, pp. 345-350.

Taylor, H. P., Jr., 1967. Oxygen isotope studies of hydrothermal mineral deposits, in *Geochemistry of Hydrothermal Ore Deposits,* ed. H. L. Barnes, New York: Holt, Rinehart and Winston.

————, 1973. The application of oxygen and hydrogen isotope studies to problems of hydrothermal alteration and ore deposition [abstr.], *Econ. Geol.* 68:1210.

Turneaure, F. S., 1935. The tin deposits of Llallagua, Bolivia, *Econ. Geol.* 30:14-60, 170-190.

————, 1960. A comparative study of major ore deposits of central Bolivia, *Econ. Geol.* 55:217-254, 574-606.

————, 1971. The Bolivian tin-silver province, *Econ. Geol.* 66:215-225.

Turneaure, F. S., and K. K. Welker, 1947. The ore deposits of the eastern Andes of Bolivia: the Cordillera Real, *Econ. Geol.* 42:595-625.

Van Hook, H. J., 1960. The ternary system $Ag_2S-Bi_2S_3-PbS$, *Econ. Geol.* 55:759-788.

Vinogradov, V. I., 1964. The isotopic composition and origin of volcanic sulfur [article in Russian], *Geol. Rud. Mestorozhd. Acad. Sci. U.S.S.R.* 6(3):7-15. (Reviewed 1965 by E. A. Alexandrov, *Econ. Geol.* 60:640-641.)

Yermakov, N. P., 1957. Importance of inclusions in minerals to the theory of ore genesis and study of the mineral forming medium (tr. 1961 by E. A. Alexandrov), *Int. Geol. Rev.* 3(7):575-585.

8 / The Classification
of Ore Deposits

The study of ore deposits requires the examination of great numbers and types of mineral districts; their similarities and differences must be noted and described. Grouping together deposits with similar characteristics facilitates description and, it is hoped, permits generalizations concerning genesis and ore localization. To be useful, a classification must be as simple as possible, especially so if it is to be used in the field during mine examination and mapping. Many attempts have been made to classify ore deposits since Agricola's first rough efforts; however, most have been abandoned, largely because they were cumbersome and restrictive and could not be applied in the field.

Various types of ore deposits grade into each other and their genetic boundaries cannot be precisely defined; classification must therefore be flexible. For example, in many mining districts such as Butte, Montana, mineralization in the outer parts of the area was deposited at relatively low temperatures and pressures, while mineralization in the center of the district was formed at somewhat higher temperatures and pressures. It is impossible to classify the district under only one category; however, the district is usually placed in the category that applies to the majority of the ore.

Whereas past attempts at classification have emphasized form, texture, and the mineral content and associations of ore deposits, more modern classifications are developed around the theories of genesis and environment of deposition. Early on it was recognized that a clear distinction could be made between certain types of sedimentary ores and others associated with igneous processes. Further breakdown into types has been difficult because geologists cannot agree among themselves as to the origin of many deposits. No universally acceptable classification of ore deposits has been proposed and at present three systems are in common use. Europeans favor Niggli's volcanic-plutonic classification and Schneiderhöhn's ore-association classification. The most widely used scheme in the United States is Lindgren's depth-temperature classification. These three classifications were each developed during the early part of this century when vein types of ore deposits were the most common. Such deposits as the massive sulfides associated with volcanic piles, disseminated copper and molybdenum deposits, and stratiform deposits of the Mississippi Valley type, while known, were of far less economic value than now. At the same time, they were of far less scientific interest. Modern studies have made available a great deal of information and necessitate revision and modernization of the earlier classifications.

Niggli (1929) grouped the epigenetic ores into *volcanic,* or near-surface, and *plutonic,* or deep-seated. The plutonic deposits are divided into hydrothermal, pegmatitic-pneumatolytic, and orthomagmatic subgroups, depending upon whether the ores formed from liquids or gases, or as direct crystallization products within the magma. Final classification is based upon chemical and ore-mineral associations.

An outline of Niggli's classification is given in Table 8-1. It can be seen that this system categorizes deposits on the basis of their genesis and mineralogy. For example, it distinguishes between volcanic gold deposits and plutonic gold deposits or between hydrothermal copper ores and pneumatolytic copper ores. Fundamentally, this classification differs little from Lindgren's, and most criticisms applied to Niggli's classification can also be applied to Lindgren's. But since high-pressure fluids above the critical point are neither gases nor liquids, the pneumatolytic-hydrothermal distinction is an artificial one. It defies field application because a mineral deposit that is formed from gaseous transported materials cannot be distinguished from one formed from liquid transported materials.

Schneiderhöhn (1941) classified ore deposits according to (1) the nature of the ore fluid; (2) the mineral associations; (3) a distinction between deep-seated and near-surface deposition; and (4) the type of deposition, host, or gangue. The significant category in this classification is the second group, mineral associations. Schneiderhöhn proposed a detailed list of typical mineral associations, categorized according to the types of ore, host, and gangue found in each. The principal categories of his classification are reproduced in Table 8-2.

Table 8-1
Niggli's classification of ore deposits.

I. Plutonic, or intrusive
  A. Orthomagmatic
    1. Diamond, platinum-chromium
    2. Titanium-iron-nickel-copper
  B. Pneumatolytic to pegmatitic
    1. Heavy metals-alkaline earths-phosphorus-titanium
    2. Silicon-alkali-fluorine-boron-tin-molybdenum-tungsten
    3. Tourmaline-quartz association
  C. Hydrothermal
    1. Iron-copper-gold-arsenic
    2. Lead-zinc-silver
    3. Nickel-cobalt-arsenic-silver
    4. Carbonates-oxides-sulfates-fluorides

II. Volcanic, or extrusive
  A. Tin-silver-bismuth
  B. Heavy metals
  C. Gold-silver
  D. Antimony-mercury
  E. Native copper
  F. Subaquatic-volcanic and biochemical deposits

Table 8-2
Schneiderhöhn's classification of ore deposits.

I. Intrusive and liquid-magmatic deposits

II. Pneumatolytic deposits
  A. Pegmatitic veins
  B. Pneumatolytic veins and impregnations
  C. Contact pneumatolytic replacements

III. Hydrothermal deposits
  A. Gold and silver associations
  B. Pyrite and copper associations
  C. Lead-silver-zinc associations
  D. Silver-cobalt-nickel-bismuth-uranium associations
  E. Tin-silver-tungsten-bismuth associations
  F. Antimony-mercury-arsenic-selenium associations
  G. Nonsulfide associations
  H. Nonmetallic associations

IV. Exhalation deposits

Schneiderhöhn's system is popular in Europe and is advocated by many Americans. Noble (1955) argues that it is the best genetic classification because mineral associations represent metal associations in the ore-forming fluids. Although the schemes of Schneiderhöhn and Lindgren have fundamental similarities, they differ in emphasis. Under Schneiderhöhn's system, a deposit that does not fit one of the given ore-mineral associations or its subdivisions is readily categorized merely by formulating a new group or subdivision. The success of this system for field use, however, is inversely proportional to the number of major groups needed to accommodate all ore deposits; that is, each new category needed weakens the classification. A more detailed example of Schneiderhöhn's system (for Group IIIA) is as follows:

III. Hydrothermal deposits
    A. Gold and silver associations
        1. Hypabyssal suite (deep-seated)
            a. Katathermal (equivalent to hypothermal) gold-quartz veins
            b. Gold-bearing impregnation deposits in silicate rocks
            c. Gold-bearing replacement deposits in carbonate rocks
            d. Mesothermal gold-lead-selenium deposits
        2. Subvolcanic suite (near-surface)
            a. Epithermal propylitic gold-quartz veins and silver-gold veins
            b. Epithermal gold-telluride veins
            c. Epithermal gold-selenium veins
            d. Alunitic gold deposits
            e. Epithermal silver deposits

Lindgren introduced his classification system (Table 8-3) in 1913; it is used today almost in its original form (Lindgren, 1933). Terms such as "tele-thermal" (Graton, 1933) and "xenothermal" (Buddington, 1935) have been added. Ridge (1968) recognized the need for revision, though he retained Lindgren's basic principles.

Lindgren's system is considered the best for use in the field. One modification that seems essential is a deemphasis of the role of magma. Accordingly, in Table 8-3 the source of heat is unspecified. The term "hydrothermal" means simply "hot water," and does not imply magmatic association. Such a change in the system of classification was suggested by K. L. Williams and conforms well with the findings of fluid-inclusion and isotopic studies.

The temperature and pressure designations in Lindgren's scheme are at best only approximate, and subject to constant modification. For example, although most metallization in mesothermal deposits takes place possibly between 300° and 200°C, early and late ore deposition may transgress these limits.

Table 8-3
Lindgren's classification of ore deposits, modified.

I. Deposits produced by chemical processes of concentration, temperatures and pressures vary between wide limits.
  A. In magmas, by processes of differentiation.
    1. Magmatic deposits proper, magmatic segregation deposits, injection deposits. Temperature 700° to 1500°C; pressure very high.
    2. Pegmatites. Temperature, very high to moderate; pressure very high.
  B. In bodies of rocks.
    1. Concentration effected by introduction of substances foreign to the rock (epigenetic).
      a. Origin dependent upon the eruption of igneous rocks.
        i. Volcanogenic, deposits associated usually with volcanic piles. Temperatures 100° to 600°C; pressures moderate to atmospheric.
        ii. From effusive bodies. Sublimates, fumaroles. Temperature 100° to 600°C; pressure moderate to atmospheric.
        iii. From intrusive bodies. (Igneous metamorphic deposits). Temperature probably 500° to 800°C; pressure very high.
      b. By hot ascending waters of uncertain origin, possibly magmatic, metamorphic, oceanic, connate, or meteoric.
        i. Hypothermal deposits. Deposition and concentration at great depths or at high temperature and pressure. Temperature, 300° to 500°C; pressure very high.
        ii. Mesothermal deposits. Deposition and concentration at intermediate depths. Temperature 200° to 300°C; pressure high.
        iii. Epithermal deposits. Deposition and concentration at slight depth. Temperature 50° to 200°C; pressure moderate.
        iv. Telethermal deposits. Deposition from nearly spent solutions. Temperature and pressure low; upper terminus of the hydrothermal range.
        v. Xenothermal deposits. Deposition and concentration at shallow depths, but at high temperatures. Temperature high to low; pressure moderate to atmospheric.
      c. Origin by circulating meteoric waters at moderate or slight depth. Temperature up to 100°C, pressure moderate.
    2. By concentration of substances contained in the geologic body itself.
      a. Concentration by dynamic and regional metamorphism. Temperature up to 400°C; pressure high.
      b. Concentration by ground water of deeper circulation. Temperature 0° to 100°C; pressure moderate.
      c. Concentration by rock decay and residual weathering near surface. Temperature 0° to 100°C; pressure moderate to atmospheric.
  C. In bodies of water.
    1. Volcanogenic. Underwater springs associated with volcanism. Temperatures high to moderate; pressure low to moderate.
    2. By interaction of solutions. Temperature 0° to 70°C; pressure moderate.
      a. Inorganic reactions
      b. Organic reactions
    3. By evaporation of solvents.
II. Deposits produced by mechanical processes of concentration. Temperature and pressure moderate to low.

The criteria commonly used to classify a deposit according to depth and temperature are discussed in following chapters. None of these criteria are infallible; owing to the complexity and variability of the factors involved, minerals that normally form in one zone also form in other places, at higher or lower temperatures and pressures. Depositional zones are characterized by certain associations of ore and gangue minerals (Schneiderhöhn, 1941; Niggli, 1929), as well as by the presence of certain wall-rock alteration products. Some minerals, such as quartz, feldspar, and pyrite, have wide ranges of stability and persist from the deepest to the shallowest zones. Ore textures may indicate the depth-temperature environment; for example, geologists believe that the fine-grained rhythmic banding of some gold ores develops under near-surface conditions. Oölitic textures and colloidal deposits are also generally attributed to low temperatures and pressures. Although knowledge of geologic thermometry is helpful, it must be used with great caution.

Ore deposits shown to be the result of granitization will also fit Lindgren's classification because the *ultimate* origin of the metals is not considered—each category is defined by environmental conditions at the time and place of final deposition.

Despite its apparent simplicity and breadth, Lindgren's classification, has elicited numerous criticisms, particularly the category "in bodies of rock" (IB). In the light of modern knowledge, especially that obtained from fluid-inclusion and isotopic studies, many of these criticisms are justified. In order to remain useful, any system of classification must be able to reflect new findings—it must be revisable. To criticize Lindgren's system only because it fails to classify a particular deposit is no more valid than to argue that a granodiorite batholith cannot be so classified because it contains some quartz monzonite. As long as the classification is understood, modernized, and its limits recognized, it will continue to be meaningful and useful.

Lindgren's zones are commonly based upon the minerals present without much regard to their stability ranges. Thus, pyrrhotite and arsenopyrite are often regarded as high-temperature minerals, which they may indeed be. Yet we now know that pyrrhotite may exist in monoclinic and hexagonal form, and conditions for deposition may vary widely. Many minerals that Lindgren considered to form at high temperatures are now known to form also at lower temperatures, especially through long periods of geologic time. As data on phase equilibria and stability of minerals accumulate, and as geothermometry is refined, the mineral assemblages typical of a given depth zone will undoubtedly be modified. Many deposits will probably be reclassified, and the temperature limits of the hypothermal, mesothermal, and epithermal zones refined and redefined.

Further criticism has been leveled at the Lindgren system because of the uncertainty of exactly what is meant by "high," "medium," and "low" when applied to temperatures and pressures. Refinement of these terms is badly needed. Likewise, as temperatures and pressures of a hydrothermal fluid change, the ores in a single orebody (or even in a hand specimen) may represent more than one of the categories in the table. This is a serious classification problem, but seems unavoidable in the attempt to depict the complex ore-forming processes. Temperature and pressure in the hypothermal zone ordinarily remain high throughout ore deposition, but in shallow zones low-temperature minerals may be superimposed upon veins of a higher temperature. Similarly, it is not uncommon to discover that a high-temperature mineral assemblage has been fractured and subsequently permeated by a cooler ore fluid. In practice such a deposit is assigned to a particular Lindgren zone according to its dominant mineral assemblage.

Originally, Lindgren's classification regarded pressure and temperature as dependent variables, but they may in fact vary independently. Buddington (1935) pointed out nine possible categories for depth-temperature zones; these represented each combination of high, intermediate, and low values of temperature and pressure. Although high temperatures near the surface are entirely reasonable near a shallow pluton, it is perhaps unreasonable to expect near-surface conditions at great depths. Buddington introduced the term *xenothermal* to represent high-temperature deposits that formed close to the surface. It is not practical to distinguish all nine types of deposits in the field, but *xenothermal* is now widely used.

Lindgren's classification does not take into consideration the chemistry of the wall rocks that are unquestionably a fundamental factor in ore deposition. Theoretically this should make no difference because depth and temperature are the only parameters used. But a problem arises where ores are precipitated prematurely with respect to their normal depth-temperature environment. Some igneous metamorphic deposits may, by depth-temperature criteria, fit into Lindgren's hypothermal zone, or even his mesothermal zone (Stone, 1959, p. 1,028), the only difference being that the deposit may lie within a carbonate host rather than a less reactive country rock. These differences suggest that Schneiderhöhn's classification is more suitable where ore deposition has been controlled by chemical differences between the hydrothermal fluids and the wall rocks, whereas Lindgren's classification is preferred where ore deposition has been controlled by pressure and temperature.

Schmitt (1950a) noted that gradations between zones are not as common as one would expect; gradations are most notably absent between epithermal and mesothermal zones, and between pegmatites and hypothermal deposits. The lack of these gradations has never been adequately explained. Nonetheless, a few examples of such gradations are known, and it is likely that as our knowledge of the physicochemical conditions controlling ore deposition increases, a satisfactory explanation will be advanced. Schmitt (1950b) also proposed a classification chart with the ordinate and abscissa defined by the factors of depth and temperature. This method, although more precise, requires information not readily obtainable in the field.

Although Lindgren's classification is the standard in the United States, most deposits are also classified by metal content, form, replacement versus fissure-filling, and so forth. For example, a disseminated copper deposit is only one of many possible types in the mesothermal category; a lead-zinc ore pipe may be hypothermal, mesothermal, or epithermal, depending on the specific deposit.

Lovering (1963) introduced the terms *diplogenetic* and *lithogenetic* in the classification system. "Diplogenetic" is proposed for deposits that are partly syngenetic and partly epigenetic. Although the term refers primarily to time, in a sense it also refers to space; it carries no implication of the source of the constituents or the method of formation. An example of a diplogenetic deposit given by Lovering is one in which the syngenetic cation unites with the epigenetic anion to replace limestone with fluorite. Lovering's second term, "lithogenetic," is applied to the mobilization of elements from a solid rock and their transportation and deposition elsewhere. Lithogenetic deposits could be derived through the action of magmatic, metamorphic, or meteoric fluids.

Lindgren's genetic classification of mineral deposits is closely related to zoning and paragenesis. Theoretically, the depth-temperature zones may correspond to actual mineralogic zones, as at Cornwall, England. Similarly, the high-intensity zones correspond to the earliest paragenetic phase of a district. Zoning, paragenesis, and genetic classification are all expressions of the same phenomena and cannot properly be divorced from one another. This is strong evidence in favor of Lindgren's classification scheme.

REFERENCES CITED

Buddington, A. F., 1935. High-temperature mineral associations at shallow to moderate depths, *Econ. Geol.* 30:205-222.

Czamanske, G. K., 1959. Sulfide solubility in aqueous solutions, *Econ. Geol.* 54:57-63.

Graton, L. C., 1933. The depth-zones in ore deposition, *Econ. Geol.* 28:513-555.

Holland, H. D., 1957. Thermochemical data, mineral associations, and the Lindgren classification of ore deposits, *Geol. Soc. Amer. Bull.* 68:1745.

———, 1959. Some applications of thermochemical data to problems of ore deposits, I: stability relations among the oxides, sulfides, sulfates and carbonates of ore and gangue metals, *Econ. Geol.* 54:184-233.

Lindgren, W., 1913. *Mineral Deposits,* New York: McGraw-Hill.

———, 1933. *Mineral Deposits,* 4th ed., New York: McGraw-Hill.

Lovering, T. S., 1963. Epigenetic, diplogenetic, syngenetic, and lithogene deposits, *Econ. Geol.* 58:315-331.

Niggli, P., 1929. *Ore Deposits of Magmatic Origin,* tr. H. C. Boydell, London: Thomas Murby.

Noble, J. A., 1955. The classification of ore deposits, *Econ Geol. (50th Anniv. Vol.),* pp. 155-169.

Ridge, J. D., 1968. Changes and developments in concepts of ore genesis—1933 to 1967, in *Ore Deposits of the United States* (L. C. Graton-R. Sales Mem. Vol.), vol. 2, ed. J. D. Ridge, American Institute of Mining Engineers.

Schmitt, H. A., 1950*a,* Uniformitarianism and the ideal vein, *Econ. Geol.* 45:54-61.

————, 1950*b.* The genetic classification of the bed rock hypogene mineral deposits, *Econ. Geol.* 45:671-680.

Schneiderhöhn, H., 1941. *Lehrbuch der Erzlagerstättenkunde,* Jena: Gustav Fischer.

Stone, J. G., 1959. Ore genesis in the Naica district, Chihuahua, Mexico, *Econ. Geol.* 54:1002-1034.

9 / Magmatic Segregation Deposits

Ore deposits formed by concentrated fractions of magmatic differentiates were recognized and named before Lindgren developed his classification system (Vogt, 1894). The term *magmatic segregation deposit* is now applied to all ore deposits, exclusive of pegmatites, that are direct crystallization products of a magma. They usually form in deep-seated intrusive bodies, but have been known to form in sills and even in lava flows. A magmatic segregation deposit may constitute the entire rock mass or a compositional layer, or may be defined by the presence of valuable accessory minerals in an otherwise normal igneous rock. The ore minerals may be early or late fractionation products, concentrated by gravitative settling, immiscibility, or filter pressing; and they may remain in place or be injected as an ore magma into a previously solidified pluton or the surrounding country rock. The possibility that liquid separation is active in magmatic segregation ores was advanced by Fischer (1950), who synthesized a magnetite-apatite fluid as an immiscible fraction of a silicate melt.

Certain ore minerals are characteristic of specific igneous rocks; others show no consistent affiliations. The ores commonly found with mafic rocks are chromite, ilmenite, apatite, and platinum; those with igneous rocks of intermediate composition are magnetite, hematite, and ilmenite; and those associated with siliceous rocks are magnetite, hematite, and such accessories as zircon, monazite, and cassiterite. Many associations are even more restrictive; for example, chromite is closely associated with peridotite and dunite, or with serpentine derived from these ultramafic rocks (Thayer, 1946, 1969). This tendency for an ore-host rock association is one of the strongest lines of evidence advanced by the proponents of magmatic segregation as an ore-forming process.

The textures of ore minerals in magmatic segregations are essentially the same as the textures of the including rock (Newhouse, 1936; Ramdohr, 1940; Uytenbogaardt, 1954). For example, chromite in peridotite ordinarily has the same textural characteristics as the rest of the rock, and grains of magnetite in a fine-grained syenite are smaller than those in a coarser-grained syenite. Conversely, many accessory minerals, such as zircon and monazite, form euhedral crystals, suggesting that they crystallized early in the differentiation process (though this is not everywhere clear, and some euhedral crystals may actually be late products).

Minerals formed early in magmatic differentiation will not be in complete equilibrium with the melt during its later stages. Many of these early formed minerals are partly resorbed and show rounded faces and other effects of late magmatic and deuteric activity (see Fig. 9-1). Such corrosive effects may obscure the textures of magmatic segregation deposits, making it difficult to distinguish between these deposits and deposits of hydrothermal origin. Thus, a debate has arisen over whether certain deposits were produced by magmatic segregation or by hydrothermal processes. In reality, the two processes probably grade into one another. Singewald (1917) concluded, for example, that many iron ores, especially titaniferous iron ores, are late-stage magmatic differentiates whose concentration is directly related to the action of mobile fluids. The process advocated by Singewald had been proposed earlier by de Launay, who suggested that as crystallization abates, crystal fractionation and hydrothermal activity must overlap, with an increase in the relative proportion of the fluids. Thus, there would be a gradation from the magmatic stage to the hydrothermal stage. If de Launay and Singewald are correct, as is now commonly assumed, then such corrosive effects as those shown in Figure 4-34 are to be expected in magmatic segregation deposits, especially in those formed during later stages of differentiation.

Microscopic studies of magmatic sulfide ores led Tolman and Rogers (1916) to conclude that the more mobile fluids play an important role in magmatic segregation. They emphasized that early crystallization requires the concomitant squeezing out or displacement of the residual fluid, a process that is not only mechanical but also due to gaseous extraction.

The gradation between magmatic segregation and hydrothermal activity is also manifested by wall-rock alteration effects, which are generally absent from early differentiates but may be clearly developed around the products of

Figure 9-1
Thin section of ilmenite-anorthosite from Allard Lake, Canada. Note the narrow reaction rim around the ilmenite, left of center in the figure. Crossed nicols. ×10.5.

late fractionation. It should be noted, however, that wall-rock alteration is not characteristic of magmatic segregation deposits in general. Even the presence of abundant fluids—as with late magmatic products—does not entail much alteration because there may be little difference between ore and wall-rock temperatures.

Proponents of the magmatic segregation origin for large sulfide masses say that sulfide grains abound in many igneous rocks and are generally considered to be original accessory constituents of the rock. Similarly, the association of many nickel sulfarsenide deposits with norite and related mafic rocks indicates a close genetic connection, and probably means that the nickel-bearing fluids did not travel far from their magmatic sources.

The common association of nickel sulfides with ultramafic rocks has led to the discovery in recent years of the Thompson orebodies in northern Manitoba, Canada, and of the Kambalda deposits in Western Australia (Woodall and Travis, 1970). The Kambalda deposits are in Archaen rocks of the Western Australian Precambrian shield. They were discovered in January, 1966, following the recognition of a nickeliferous limonitic gossan. Geologic mapping shows that this gossan is at the lower contact of an ultramafic body that lies between two metabasalts. The lower contact of the ultramafic body crops out as a domal structure with a strike length of 13 miles. Induced polarization, magnetometer, and geochemical soil surveys were used to prospect the contact zone. Most of the ore is at the base of the ultramafic in contact with metabasalt, though some is in lenses within the ultramafic. Structural irregularities in the basal contact appear to constitute local ore controls. Seven orebodies are known either on or close to this contact.

According to Ewers and Hudson (1972), the profitable Lunnon Shoot in the Kambalda district has a footwall of pillow basalt. Above this is a narrow zone of magnetite-rimmed chromite euhedra in silicates, massive ore with 95 percent sulfides, matrix or disseminated ore with 60-20 percent sulfides, and hydrated and carbonated ultramafic rocks in the hanging wall. Ewers and Hudson considered that the data were consistent with the ore having been emplaced at elevated temperatures as an ultramafic mush containing olivine and chromite crystals, magnesia-rich silicate melt, and immiscible sulfide-oxide liquid. Gravity separation resulted in the ore zone and the hanging wall represents the original crystal mush with minor amounts of sulfides. Pyrite, concentrated near the top of the massive ore was the first subsolidus sulfide mineral to concentrate, at about 400°C, and cobalt was preferentially concentrated in this pyrite.

The ore at Kambalda consists of both massive and disseminated mineralization. Where both are present, the disseminated mineralization overlies the massive. The primary mineralization is a pyrrhotite-pentlandite assemblage with subordinate cobaltiferous pyrite and minor amounts of chalcopyrite. A supergene zone, characterized by the development of violarite as a useful ore mineral, lies between the gossan and the primary sulfides.

Singewald noted in 1917 that the sulfides are deposited along the peripheries of their associated igneous masses, and at places penetrated the wall rock. At least some of the sulfides formed later than the rock-forming silicates, which they replaced without reaction rims (Tolman and Rogers, 1916). Singewald interpreted this relationship to mean that the sulfide-bearing agency also removed the dissolved silicates, indicating the presence of active fluids. Wall-rock alteration around some massive sulfide deposits is further evidence of the presence of these reactive fluids (Singewald, 1917). The relative mobility of sulfide minerals supports the idea that large masses of sulfides, although related to magmatic processes, nonetheless tend to move with the volatile or fluid fraction away from the crystallizing magma.

Probably the best examples of magmatic-segregation ores are found among the chromite deposits, though in places these are thought to be hydrothermal. Wide field experience and the examination of many polished sections of chromite ores led Sampson (1929) to conclude that much chromite crystallizes at a very early magmatic stage, but that considerable amounts may remain in the residual melt or even pass into a highly aqueous solution capable of considerable migration. Both Fisher (1929) and Sampson (1931) recognized three classes of chromite: early magmatic, late magmatic, and hydrothermal. These three categories are widely accepted and apply not only to chromite ores but also to other magmatic segregation minerals, such as ilmenite and magnetite.

Sampson's interpretation implies that the character of the magma that crystallizes chromite is the same whether the mineral is deposited early by crystal settling or later by hydrothermal fluids. Field evidence corroborates this supposition. For example, at the Lambert mine near Magalia, California, small pods of chromite are associated with similar pods of nearly pure albite in a serpentine-peridotite host. The concentration of these minerals along a zone of shearing suggests that they were deposited by hydrothermal processes, but their confinement to the ultramafic rocks implies a genetic relationship between the ore and the host rock. Apparently, the ore-bearing fluids moved only short distances through the ultramafics.

Geijer (1967) gives an excellent review of the mineralogical and textural properties of the magnetite-hematite ores of magmatic affiliation, especially those with a high percentage of apatite. Although he emphasizes the magmatic character of the ores, Geijer states emphatically that the consolidation of these iron ores is not the same as that of a normal igneous rock. He attributes the differences to the role of volatile substances, and says that the form of emplacement in many places is forceful injection and not a filling of preexisting open fissures.

Many magnetite deposits contain coarse-grained apatite, the crystals of which vary in length up to several centimeters. The apatite is commonly perpendicular to the walls of the fractures, though it may be separated from the walls by thin layers of magnetite. Where fragments of country rock are present in magnetite, apatite is likely to be normal to the walls of the fragments (Geijer, 1967). Scapolite has been reported from the ore at Laco, Chile (Ruiz, 1964) and is present in several other deposits of possible magmatic affiliation. According to Geijer, the presence of scapolite does not mean that the ores cannot be magmatic in origin. Hornblende, actinolite, and in a few places tremolite are common companions of magnetite orebodies.

CHROMIUM DEPOSITS

In the Moa district of northeastern Cuba (Fig. 9-2) several chromite deposits are clear-cut products of magmatic segregation. These refractory chromite deposits have been studied in detail by Guild (1947), who described them as sacklike bodies of massive, nearly pure chromite in serpentinized ultramafics (see Fig. 9-3). The orebodies are in an extensive ultramafic complex, which is covered by a broadly domed lateritic slope along the northern range of Oriente Province. Several chromite-bearing districts continue along an east-west belt of serpentinized ultramafics; only Moa is described here because it is more thoroughly understood and less complexly deformed than the other districts.

Figure 9-2
Map of Cuba, showing the location of the Moa district.

Rapid headward erosion of the short but vigorous streams has cut deep, V-shaped valleys, destroying part of the lateritic surface. Between the streams this well-preserved old surface rises at a gradient of three or four degrees from sea level on the north to an altitude of nearly 3,000 feet near the crest of the range. The climate is humid, and thick vegetation considerably handicaps geologic mapping. Chromite bodies project from the canyon walls at a few places, supplying abundant float to the stream beds. Although the deposits have been known since late in the nineteenth century, no exploration work was done until World War I, which motivated a small amount of prospecting. Real development did not begin until 1940; during the next seven years more than 500,000 tons of refractory chromite were produced, averaging about 33 percent Cr_2O_3 (Guild, 1947).

The chromite deposits of the Moa district are often cited as an example of an injected mush of crystals, interstitial liquid, and chromite blocks. The primary structures are largely preserved; many probably formed during consolidation of the mass after intrusion, but there is strong evidence that differentiation of the chromite took place at depth (Guild, 1947).

The serpentine massif originally consisted of peridotite with subordinate dunite, chromitite, and pyroxenite. It contains some less mafic facies, such as gabbro, troctolite, and anorthosite, reflecting variable proportions of olivine, pyroxene, plagioclase, and chromite. With the exception of pyroxene and chromite, all combinations and gradations of these minerals are found. The ultramafic rocks are almost completely serpentinized, and the feldspathic rocks are altered to chalky aggregates of hydrous minerals, including antigorite (?), zoisite, edenite, chlorite, and zeolites.

A basement complex made up of chlorite schist, metamorphosed volcanic tuffs and lavas, thin impure marbles, and intrusive diorite is exposed south of the Moa district and is thought to represent the host into which the ultramafics were intruded. The serpentines show no trace of the intense metamorphism that affected these older rocks, and no diorite has been found in the serpentine.

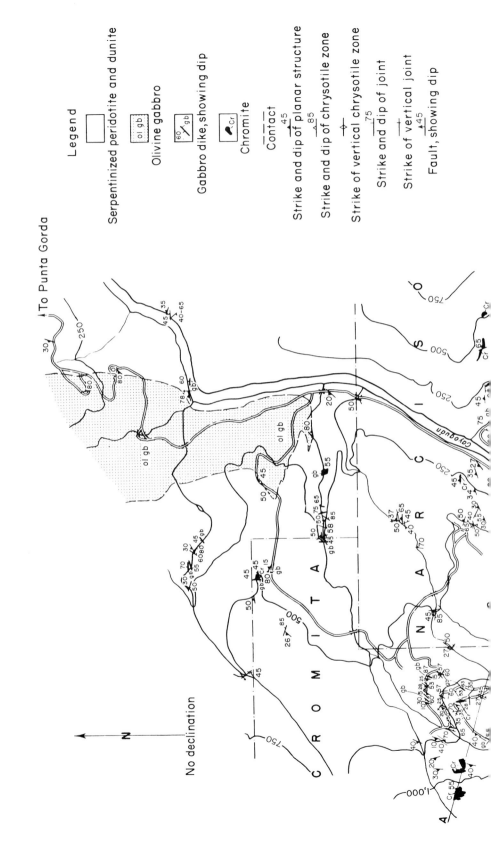

Legend

Serpentinized peridotite and dunite

ol gb Olivine gabbro

60 gb Gabbro dike, showing dip

Cr Chromite

Contact

45 Strike and dip of planar structure

85 Strike and dip of chrysotile zone

75 Strike of vertical chrysotile zone

Strike and dip of joint

45 Strike of vertical joint

45 Fault, showing dip

To Punta Gorda

No declination

N

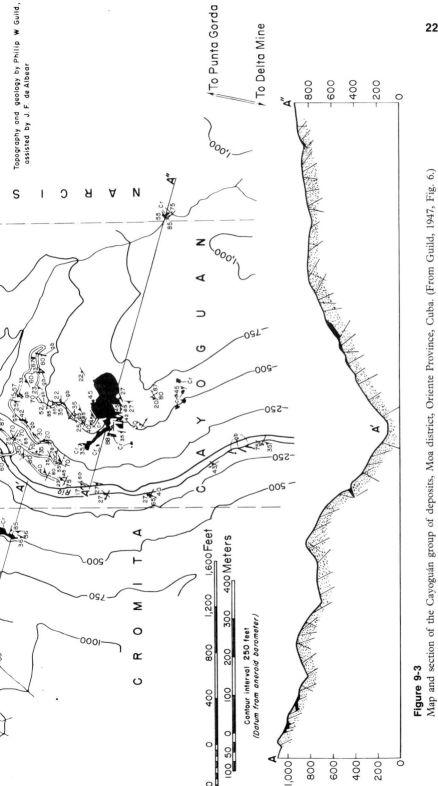

Figure 9-3
Map and section of the Cayoguán group of deposits, Moa district, Oriente Province, Cuba. (From Guild, 1947, Fig. 6.)

Overlying the ultramafics to the west are Upper Cretaceous conglomerates (which contain cobbles of serpentine) and Tertiary deposits (see Fig. 9-4).

The chromite concentration in the ore varies. In some deposits chromite is subordinate to the silicate minerals; but the best deposits contain only a small percentage of gangue minerals, and average ore carries about five percent silicates by weight. The most common gangue minerals are serpentinized olivine and alteration products of the feldspars. Low-grade ores contain individual grains of anhedral chromite disseminated in dunite, whereas the high-grade ores, which constitute most of the commercial deposits, appear almost massive and contain only minor amounts of interstitial silicates. Most of the chromite is fine- to medium-grained with a seriate texture, but a few grains are measured in centimeters, and crystals up to six inches across have been found.

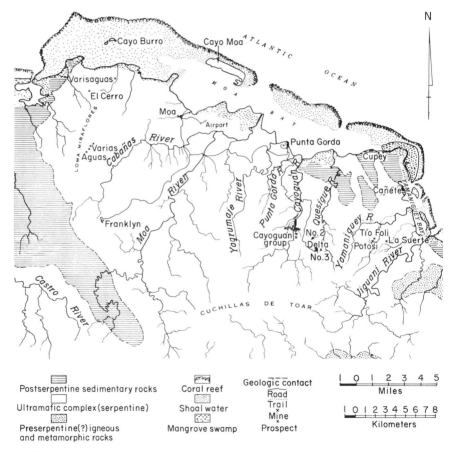

Figure 9-4

Map of the Moa chromite district, Oriente Province, Cuba. (From Guild, 1947, Fig. 2.)

Weathering has had little effect on the chromite ores, but some high-temperature, deuteric alteration products were formed. In one of the small deposits, a pale green chlorite replaced euhedral chromite, leaving only skeletal crystals of the chromite grains. This chlorite is not chromiferous, suggesting that chromium ions are mobile under certain conditions and capable of diffusing away from the reaction front. The pink, chrome-bearing chlorite, kämmererite, is also encountered in many of the deposits. A more ubiquitous alteration product is uvarovite-grossularite garnet, which fills joints in ore or forms reaction rims between pyroxene and chromite.

Each chromite deposit is surrounded by a dunite jacket ranging in thickness from one foot or less to tens of feet. The chromite-dunite pods are in peridotite, hence there is a general sequence from chromite to dunite to peridotite in all orebodies. At any single place the chromite-dunite and dunite-peridotite contacts are virtually parallel. Fault contacts are more common than undisturbed igneous contacts, however; in fact, the association of chromite with dunite is so well established that faults are plotted where drilling encounters peridotite adjacent to ore (Guild, 1947).

The chromite deposits are tabular to lenticular, and range from thin streaks a few feet long to much larger masses; some are elongated down the dip and others along the strike. Guild (1947) described one deposit as 700 feet long (down dip), 320 feet wide (along strike), and 90 feet thick. The contacts are generally sharp, but other ore bodies grade outward through alternating parallel layers of variable silicate content in which the proportion of chromite decreases outward from the deposit to the country rock. Still other deposits are composed entirely of layered material and have no central concentration of chromite.

Fortunately, the internal structures of the ultramafic complex have not been obliterated by later deformation so that the rock fabric still reflects the processes of crystallization and intrusion. Compositional layering and flat mineral grains give the rocks a planar fabric. The most prominent layering is that of the chromite bodies themselves. Feldspathic lenses and streaks or layers of olivine in the chromite are all parallel to the contacts. Another planar structure is formed by the orientation of flat pyroxene grains in the peridotite, generally, but not everywhere, parallel to the compositional layering. It is especially noticeable on weathered outcrops.

The ultramafic complex is intricately jointed and faulted. Joints are encountered every few inches or feet, and all orientations can be found. No systematic pattern of joints can be determined in the massive orebodies, but the peridotite shows two sets of joints that seem related to the planar structures and the processes of intrusion. The principal joints strike parallel to the foliation and dip at a large angle to this planar fabric. These "cross joints" are thought to define planes normal to the direction of rock flowage. A second set of prominent joints, called *longitudinal joints* (Guild, 1947), dips nearly vertically and strikes normal to the foliation. Although many joints are simple fractures in otherwise homogeneous rocks, some cross joints have been filled with gabbro, and some of the longitudinal joints are characterized by chrysotile veins.

Faulting is so complex that it is impossible to reconstruct the original sizes and shapes of some orebodies. As a rule, the faults parallel the prominent joint directions. Normal faults predominate, but reverse faults are also found. The overall picture is one of tension in the dip direction of the ore deposits. Many faults have been filled by gabbro dikes, and some of the dikes themselves have been slickensided by renewed movement or offset by cross-cutting faults. These dikes are more numerous and more coarsely grained in the chromite masses than in the other rocks.

As with most ultramafic complexes, details of differentiation, intrusion, and solidification of the Moa complex are not completely understood. It is known, however, that the Moa district did not undergo extreme deformation after the emplacement of ultramafics; as a result, many primary features have been preserved as evidence of the processes involved. Any feasible theory of origin must (1) explain the compositional layering in the ore, dunite, and peridotite; (2) explain why foliation in the peridotite is generally parallel to the layering but oblique elsewhere; (3) relate the joint systems to the planar structures; (4) explain why the two principal joint systems are filled with different materials; and (5) acknowledge the predominance of gabbro dikes in the ore deposits.

Guild (1947) claimed that the ore was a product of early magmatic differentiation that formed discrete lodes before the ultramafics ascended to their present position. He argued that there could not have been sufficient time for compositional layering and coarse crystals to develop after intrusion. Moreover, pyroxene crystals were oriented during growth by the intrusive forces, and the paragenesis of chromite, olivine, and pyroxene implies that the ore should have been solid at these temperatures. This is shown by the presence of dunite dikes in the chromitite, which shows that the orebodies were rigid enough to fracture even during the olivine stage. The presence of plagioclase as a gangue mineral in deposits enclosed within completely feldspar-free dunite envelopes suggests further that the ore crystallized at a different temperature than the olivine. The orebodies are clustered in a narrow zone in the ultramafic complex, implying intrusion after solidification. Both the parallelism of the orebodies and their layered structures suggest that they were oriented during flow. Two of the larger orebodies, the Cayoguán and Narciso-Cromita deposits, have similar dimensions and almost identical compositions (see Table 9-1), a distinction that cannot be made for any other pair of deposits. This implies that they may be fragments of a single mass that was broken and separated during intrusion.

Guild (1947) concluded that the ultramafic complex probably resulted from the differentiation of a basaltic (?) magma. The mafic fraction, which consists

Table 9-1
Composition of chromite from the Cayoguan, Narciso, and
Potosí deposits, Moa district, Cuba.

Component	Cayoguan	Narciso	Potosí
Cr_2O_3	38.03	38.26	39.62
Al_2O_3	30.96	30.30	24.96
Fe_2O_3	2.59	2.55	5.21
FeO	10.36	10.84	14.00
MgO	17.43	17.22	14.60
MnO	0.03	0.09	0.15
CaO	0.02	0.30	0.20
TiO_2	0.06	0.19	0.44
SiO_2	0.24	0.52	0.80
H_2O	0.10	0.12	0.02
P_2O_5	0.06	0.06	–
S	none	trace	–
TOTALS	99.88	100.45	100.06
Ratio Cr/Fe	2.64	2.56	1.86

Source: Guild, 1947.

of olivine, chromite, and interstitial liquid, was subsequently intruded into the overlying rocks during a period of severe downbuckling. Guild divided the entire history into four overlapping stages: magmatic differentiation, intrusion, complete consolidation, and serpentinization. A summary of this sequence is reproduced in Table 9-2.

TITANIUM DEPOSITS

The expanded use of titanium in recent years has encouraged a corresponding increase in the development of titaniferous ores. Most of these ores are titanium-bearing magnetites and hematites, or intimate exsolution mixtures of these minerals with ilmenite. Most are thought to be of magmatic segregation origin.

Several careful studies of the magnetite-ilmenite deposits near Lake Sanford, New York, indicate that these ores are differentiates of an anorthosite-gabbro magma (Buddington, 1939; Balsley, 1943; Stephenson, 1948). The ores are thought to have formed from residual segregations of magnetite-ilmenite liquid that was partly trapped in the crystal interstices and partly forced into the still-plastic rock. Material grading from gabbro containing less than 10 percent magnetite-ilmenite to almost pure magnetite-ilmenite rock is found. The

Table 9-2

Summary of proposed mode of origin of the ultramafic complex, Moa district, Cuba.

Stage	Processes	Results
Predifferentiation	—Homogeneous basaltic(?) magma—	
Differentiation and segregation	Crystallization of olivine, anorthite, and chrome-spinel	—
	Sinking of crystals, increase of Cr_2O_3 content through reaction of spinel with liquid to form olivine and anorthite	Crystal mush of olivine, chromite, and liquid Chromite concentrated in one or more layers
Intrusion	Downbuckling of crust into "tectogene"	—
	Intrusion of crystal mush into rocks of crust	Layered, sill-like(?) complex of chromite masses in mush of olivine crystals with interstitial pyroxenic and gabbroic liquid
	Start of crystallization of pyroxene	Formation of peridotite
	Laminar flow	Perfection of layering of peridotite, dunite, and chromitite
	Fracturing of chromite masses	—
	Injection of interstitial feldspathic liquid into joints in ore	Gabbro dikes in ore
	Reaction of chromite with pyroxenic liquid to form olivine	Dunite envelopes around ore
	Local turbulence in flow lines	Deviations of layering from normal attitude
Consolidation	Orientation of pyroxene by flow and stresses accompanying consolidation	Foliation of peridotite
	Shearing parallel foliation	Offset of pyroxene layers
	End of crystallization of pyroxene	—
	Inward stress from walls of intrusion chamber	Cross joints
	Injection of gabbroic liquid into joints	Gabbro dikes in peridotite
	Replacement of gabbro in ore by late solutions	Gabbro pegmatite
	Replacement of peridotite by feldspathic solutions	Replacement "dikes" and zones in peridotite
Serpentinization	Continued stress	Longitudinal joints
	Reaction with residual(?) water- and silica-rich solutions	Alteration of olivine and some pyroxene to serpentine
	Shrinkage(?), fissure filling by deposition from serpentine solutions	Chrysotile zones and veinlets

Source: Guild, 1947.

magmatic segregation interpretation has been challenged by Gillson (1956); he argued that the ores are related to pneumatolytic replacement of the anorthosite-gabbro. He offered convincing evidence that all rock types in the area were produced by alteration of anorthosite along faults and fracture zones, and he concluded that the metallization was a late phase of this alteration.

Unusually large ilmenite-hematite deposits have been developed at Allard Lake, Quebec, Canada (Fig. 9-5). The deposits have been interpreted by several geologists as magmatic segregations (Dearden, 1958; Hammond, 1951, 1952). They are in a large mass of anorthosite and anorthosite-gabbro that invades Precambrian metamorphics and Paleozoic limestones. The anorthosites are in turn invaded by granite. Coarse-grained ilmenite containing exsolved hematite makes up the ore, which forms irregular lenses, narrow dikes, large sill-like masses, and various combinations of these forms. Most of the ore contains 32 to 35 percent TiO_2. Typical analyses of ore from Allard Lake are given in Table 9-3.

The titanium ores throughout the region are confined to the light gray, medium- to coarse-grained anorthosites. The Lac Tio deposit, discovered in 1946, is said to be the largest body of titanium ore of its type in the world. This somewhat triangular-shaped orebody has a surface area of 134 acres and contains an estimated 125 million tons or more of ore (Hammond, 1952).

A system of fractures, composed of two steeply dipping sets of joints and faults that trend approximately at right angles to each other, forms a prominent regional pattern throughout much of the anorthosite surrounding the Lac Tio deposit. The more conspicuous set of normal faults strikes nearly north-south. The deposits are displaced by the faults, but the ore appears to be independent of these structures.

The Lac Tio ore contains crystal aggregates of thick, tabular ilmenite grains up to 10 mm across and 2 mm thick. Minor amounts (generally about five percent) of plagioclase, pyroxene, biotite, pyrite, pyrrhotite, and chalcopyrite make up the interstitial material. Microscopic examination reveals an intimate exsolution texture of hematite in the ilmenite. This is remarkable in that there

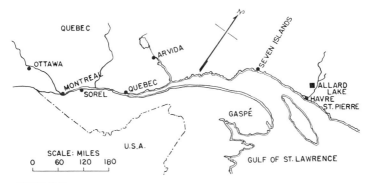

Figure 9-5
Map of the Allard Lake district, Quebec, Canada. (From Hammond, 1952.)

234

Table 9-3
Analyses of titanium ore from Allard Lake, Quebec, Canada.

Constituent	Sample 1	Sample 2	Sample 3
TiO_2	34.8%	36.0%	34.4%
Fe	38.0	42.8	39.1
S	0.36	0.40	0.39
P_2O_5	0.004	0.010	0.012
Cu	0.037	0.12	0.14
V	0.22	0.21	0.21
Mn	0.08	0.08	0.10
Ni	0.03	0.01	0.02
Co	0.014	0.013	0.019

Source: Hammond, 1952.

are exsolution lamellae of ilmenite in the hematite lamellae and further exsolutions of hematite in the ilmenite lamellae (progressive cooling). The ilmenite contains up to 25 percent of intergrown hematite, and the ilmenite-hematite mixture is so fine grained that grinding cannot separate the two minerals.

Widespread accessory ilmenite and commercial concentrations of ilmenite in the anorthosite indicate a genetic relationship between ilmenite and anorthosite. The host rock has been altered only locally, and except for the presence of biotite and the sulfides, no evidence has been found to indicate the presence of the usual mineralizers associated with ores of hydrothermal origin. Dikes and veinlets of ilmenite in anorthosite, as well as inclusions of anorthosite in the orebodies, attest to the younger age of the ore. Both Hammond (1952) and Dearden (1958) suggest that the ilmenite and the anorthosite are differentiates of the same parent magma, the large deposits of commercial ore representing a late segregation. Gillson (1960) diametrically opposes this view and says the ore is a product of pneumatolytic replacement.

Small amounts of vanadium, generally 0.1 or 0.2 percent, are found in both the Lake Sanford and Allard Lake deposits. Vanadium is common in many iron and titanium ores, and appears especially in ores of magmatic affiliation. Vanadium is seldom recovered from these ores, but they may be a source of large amounts in the future. During World War II vanadium was recovered from slag as a by-product of the German iron-and-steel industry (Fischer, 1946). According to Vaasjoki (1947), vanadium is concentrated with magnetite rather than ilmenite probably because vanadium and ferric iron ions are nearly the same size. However, Hutton (1945) states that the differences in ionic size between V^{+3} or Fe^{+3} and Ti^{+4} are not large enough to cause such a selective concentration of vanadium. He favors the hypothesis that the vanadium becomes concentrated in residual magmatic liquids after the ilmenite crystallizes and consequently is more available during the later growth of magnetite.

IRON DEPOSITS

Probably the best-known and most productive magmatic segregation iron deposit in the world is the high-grade ore at Kiruna in northern Sweden (Fig. 9-6). It has been studied intensively over a period of about 50 years, and most geologists who have examined the area agree that the ores are magmatic segregations (Stutzer, 1907; Geijer, 1910, 1931a, 1931b, 1960; Geijer and Magnusson, 1952; Schneiderhöhn, 1958; Frietsch, 1973).

The Kiruna district contains several iron deposits, the largest of which is the Kiirunavaara mine. Kiirunavaara is a sill-like orebody between Precambrian syenite porphyry in the footwall and quartz porphyry in the hanging wall. Overlying the quartz-bearing porphyry is the Lower Hauki complex, consisting of highly altered flows and silicified tuffs(?), and the Vakko Series (or Upper Hauki complex) of sediments. All of these rock units strike north or northeast and dip 50° to 60° east (see Fig. 9-7). The orebody is intrusive between the syenite and quartz-porphyry, extending for over three miles along the strike. It averages about 250 feet in thickness, except at its narrow northern end, which extends over half a mile under a lake. Production began in 1903, and by 1959 had totaled 235 million metric tons averaging 62 percent iron—estimated to be about one-eighth of the available ore (Geijer, 1960).

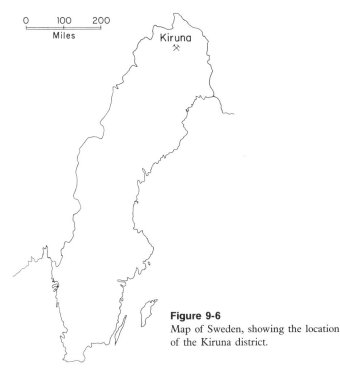

Figure 9-6
Map of Sweden, showing the location of the Kiruna district.

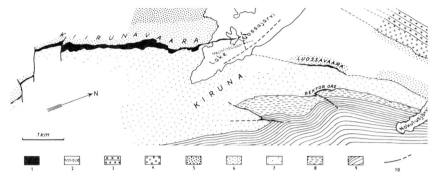

Figure 9-7

Map of the Kiruna district, Sweden. (1) Iron ore; (2) zone containing ore veins (ore breccia); (3) Kiruna greenstones; (4) Kurravarra conglomerate; (5) syenite; (6) syenite porphyry; (7) quartz-bearing porphyry; (8) Lower Hauki complex; (9) Vakko sedimentary series; (10) fault. (From Geijer, 1960.)

The ore consists of fine-grained magnetite or, more rarely, hematite, and contains variable amounts of fluorapatite. The phosphorus content is generally high (greater than 2 percent), due to the apatite; in fact, the term "Kiruna type ore" is universally used to refer to high-phosphorus ores. After apatite, which is common, actinolite and diopside are the most abundant nonmetallic minerals. In places the prisms of apatite exhibit a trachytoidal arrangement, and workers have even observed streaked out or "stratified" alternations of pure apatite and magnetite-apatite laminae. At Kiirunavaara both the apatite and the iron minerals are fine grained, and a well-defined layering or banding is developed. Geijer said the banding was a result of flow.

Contacts between the ore and wall rocks are sharply defined and definitely intrusive in nature. A thin amphibole skarn is found in places, and small veins or apophyses of ore branch out into both the footwall and hanging wall. Separate systems of ore veins are referred to locally as *ore breccia*. Porphyry dikes, apparently intermediate in composition and texture between the footwall and hanging-wall porphyries, intrude the footwall and spread out beneath the hanging wall. The dikes, in turn, are broken up and engulfed by the ore, though one such dike definitely intrudes the magnetite. These relations imply that the ore was emplaced after most of the porphyry dikes were intruded but before the igneous activity had completely stopped. The ore is offset by faults with predominately strike-slip movement. Granophyre dikes apparently unrelated to the ore and older igneous rocks were intruded during the faulting.

Geijer (1931a, 1931b, 1960, 1967) concluded that the Kiirunavaara deposit resulted from the intrusion of highly mobile ore magmas, presumably before the country rocks were tilted. The magnetite-apatite fluid became concentrated during magmatic differentiation as an immiscible fraction within the mother magma. Separation of the two immiscible fractions took place deep within the earth's crust, prior to intrusion. This interpretation is directly supported by Fischer's (1950) experimental confirmation that an immiscible magnetite-apatite liquid can exist.

Many other lines of evidence support Geijer's hypothesis. The orebodies invade the porphyries, and show liquid flowage by their trachytoidal texture. The contact zones underwent slight metasomatism, indicating the presence of volatiles. These volatile constituents were presumably active in separating the two immiscible magmatic fractions within the mother magma. Furthermore, all minerals of the orebody are also in the associated igneous rocks. Other orebodies in the district are clearly related in mineralogy, structure, and chronology to the Kiirunavaara deposit, indicating a closely related genesis; all types of ore, from magmatic to hydrothermal, can be seen in these deposits, and it is clear that some formed at lower temperatures than the Kiirunavaara ores. In the entire district, then, there is evidence of closely related magmatic and hydrothermal ore deposition, wherein the early, high-temperature ores were products of direct magmatic segregation and the later ores were influenced by greater concentrations of volatiles.

PALABORA COPPER-CARBONATITE DEPOSIT

One of the most fascinating ore deposits to be developed in recent years is the one at Palabora, immediately east of the Kruger National Park in northeastern Transvaal, Republic of South Africa (Lombaard *et al.*, 1964; Ed. Staff, *Eng. Mng. Jour.*, 1967; Herbert, 1967). Here a large pipe known as the Palabora Complex, approximately five miles long by two miles wide, vertically invades the old Precambrian granitic basement rocks. Copper, apatite, and vermiculite are, or have been, mined from subsidiary carbonatite pipes and their associated rocks within the much older Palabora Complex. The Complex is mainly pyroxenite, a diopside-phlogopite-apatite rock in which apatite constitutes about 15 percent of the rock.

A mafic pegmatitic rock composed largely of serpentine and hydrated phlogopite (vermiculite) crops out in the northern part of the Complex and is being mined for vermiculite. Figure 9-8 shows the general geology of the Complex.

In the west-central part of the Complex is a nearly vertical pipe of carbonatite and phoscorite, a rock composed of partly serpentinized olivine, magnetite, apatite, and accessory mica. Minor amounts of baddeleyite, ZrO_2, are also present.

Carbonatite is said to have been intruded in two stages. The older is a medium- to coarse-grained rock with conspicuous banding roughly paralleling

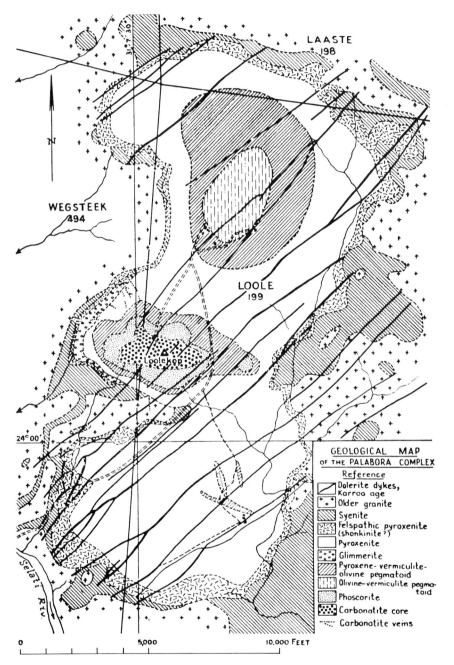

Figure 9-8
Palabora Complex, Transvaal, Republic of South Africa. (From Lombaard *et al.*, 1964.)

the walls of the pipe. This carbonatite, as well as much of the younger material, consists of calcite intergrown with dolomite and containing accessory chondrodite, olivine, phlogopite, and biotite (Lombaard *et al.,* 1964). Banding results from the alignment of magnetite, which makes up as much as 25 percent of the rock, and silicate grains. Contacts between the older carbonatite and phoscorite are commonly gradational; toward the borders of the pipe the two rocks are intimately interlayered. Micaceous pyroxenite, a mica-rich phlogopite-biotite rock, lies close to the outer border of the phoscorite.

The younger carbonatite, generally called transgressive carbonatite because it cross-cuts other rocks, is especially abundant near the center of the pipe and tails out, east and west. Figure 9-9 (a plan of the 400-foot level) shows this relationship very well; part of the younger carbonatite is fine grained and is banded locally like the older rock, thus making it difficult to distinguish the two.

Dolerite dikes, which are apparently not related to the economic mineralization and which range in width from less than a centimeter to more than 50 meters, trend northeastward in a swarm (see Figs. 9-9 and 9-10). The dikes appear to be late in geologic history; much of the pipe shows tight, closely spaced shearing, especially near the center, where the principal elongated pipe-like body of transgressive carbonatite was emplaced. The shearing trends generally to the north and is considered postcarbonatite.

Copper mineralization spreads throughout the carbonatites, mostly in the transgressive carbonatite near the core of the pipe. This is attributed to the localization of the younger carbonatite where the rocks were repeatedly fractured. Both the older carbonatite and the phoscorite contain an earlier carbonatite, copper mineralization. Blebs in the phoscorite seem to have resulted from replacement of interstitial carbonate grains not related to shearing. Much of the older carbonatite is below ore grade, except in conjunction with the younger carbonatite.

The copper orebody, about 4,000 by 2,000 feet in plan view, is said to contain about 315 million tons of ore averaging 0.68 percent copper. Chalcopyrite is the principal sulfide, though bornite, chalcocite, and locally minor amounts of other copper minerals are present. Magnetite is a major constituent of the ore and is of possible value as a byproduct. Titanium is present in much of the magnetite in the carbonatites; the titanium content of the copper ore seems to be below average. Around the periphery of the ores the magnetite contains from 0.6 percent to 2.8 percent TiO_2; this may handicap the use of the magnetite in the steel industry. The transgressive carbonatite in the center of the plug contains only minor amounts of apatite, but elsewhere, especially in the phoscorite, apatite is abundant. It is a valuable byproduct, mined for use in fertilizers. The distribution of copper, magnetite, apatite, and titania, is shown in Figure 9-11.

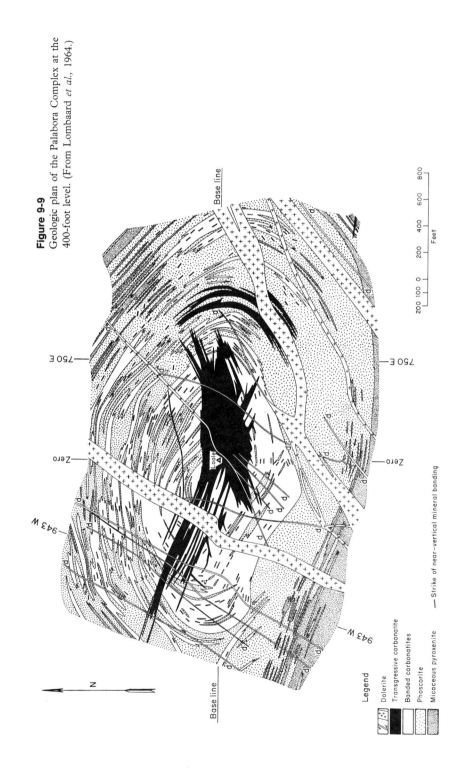

Figure 9-9
Geologic plan of the Palabora Complex at the 400-foot level. (From Lombaard *et al.*, 1964.)

Legend

Dolerite

Transgressive carbonatite

Banded carbonatites

Phoscorite

Micaceous pyroxenite

—— Strike of near-vertical mineral banding

Feet

800 600 400 200 0 100 200

Base line

750 E

Zero

943 W

Base line

N

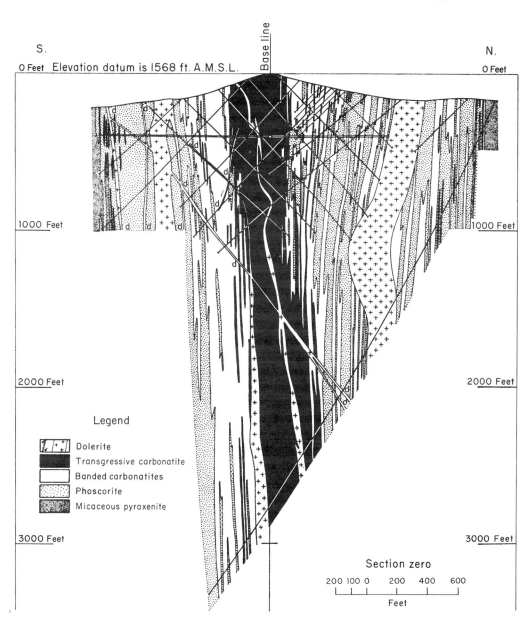

Figure 9-10
Cross section through the Palabora Complex pipes. (From Lombaard *et al.*, 1964.)

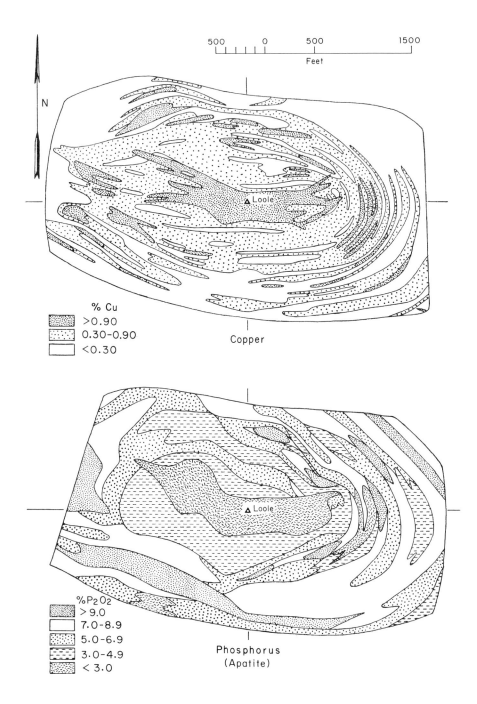

500　　　0　　　500　　　　　1500
Feet

N

% Cu
>0.90
0.30-0.90
<0.30

Copper

Loole

%P₂O₂
>9.0
7.0-8.9
5.0-6.9
3.0-4.9
<3.0

Phosphorus
(Apatite)

Loole

Figure 9-11
Distribution of copper, iron, phosphorus, and titanium at the 400-foot level, Palabora
Complex. (From Lombaard *et al.*, 1964.)

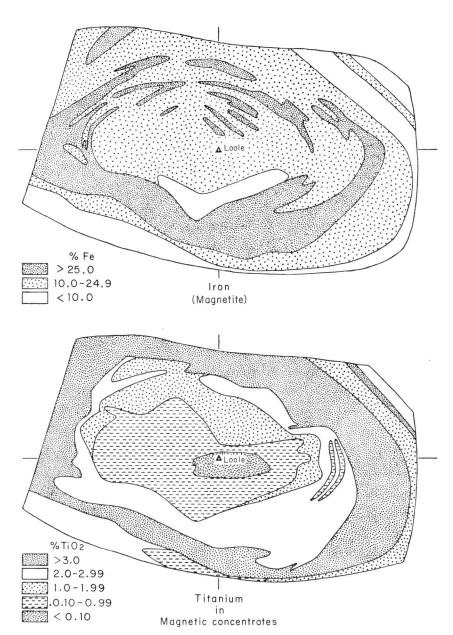

% Fe
> 25.0
10.0-24.9
< 10.0

Iron
(Magnetite)

%TiO2
>3.0
2.0-2.99
1.0-1.99
.0.10-0.99
< 0.10

Titanium
in
Magnetic concentrates

The source of the copper mineralization is unknown, but the copper appears to have been introduced in hydrothermal solutions. This brings the deposits classification into question: as a hydrothermal deposit, should it not be described as a hypothermal or mesothermal deposit rather than as a magmatic segregation? This might well be done, but the close and unique association of the ore with intrusive carbonatites and the fact that copper was introduced in two stages, before and after the transgressive carbonatite was emplaced, argues for close genetic associations. Likewise, the presence of apatite, baddeleyite, and possibly magnetite, also suggest close magmatic affiliations.

SUDBURY NICKEL DEPOSIT

The Sudbury district, Ontario, Canada, is the world's largest producer of nickel. It has yielded about six million tons of this metal, about the same amount of copper, and minor quantities of the platinum metals, gold, silver, cobalt, iron, and sulfur. The geology of the area has been studied and restudied, and the literature is voluminous (Coleman, 1905, 1913; Collins, 1934, 1936; Collins and Kindle, 1935; Dickson, 1904; Hawley, 1962; International Nickel Company Staff, 1946; Yates, 1938, 1948).

The Sudbury basin is a large elliptical depression about 37 miles long by 17 miles wide, with the long axis trending east-northeast (Fig. 9-12). The Whitewater Group occupies the central part of the basin and is separated from the old basement rocks of Archean and Huronian ages by the Sudbury eruptive, an ultramafic mass that contains most of the ore deposits of the district (Phemister, 1937; Walker, 1897).

The Onaping Formation is a complex of breccias about 5,000 feet thick that lies at the base of the Whitewater Group. Fragments in this breccia are said to be somewhat coarser near the base than they are toward the top, though the Formation is generally unsorted. The fragments are composed of various types of country rock cemented by a matrix of dark, devitrified glass (?). The origin of this breccia has been the subject of interesting debate and is still not entirely clear. The explanation that formerly was widely held is that the breccia is a tuff or ignimbrite (Speers, 1957; Hawley, 1962). Williams, (1956) and Thomson and Williams (1959) referred to it as having been deposited from a "glowing avalanche." In the past few years many authors have accepted the thesis, first advanced by Dietz (1964), that the breccia resulted from the impact of a large meteorite (Bray *et al.*, 1966; French, 1968; Guy-Bray and Peredery, 1971; Guy-Bray, 1971). This conclusion is based largely upon the presence in surrounding rocks of shatter cones, the apexes of which are oriented toward the center of the basin and upon other features attributed to impact metamorphism.

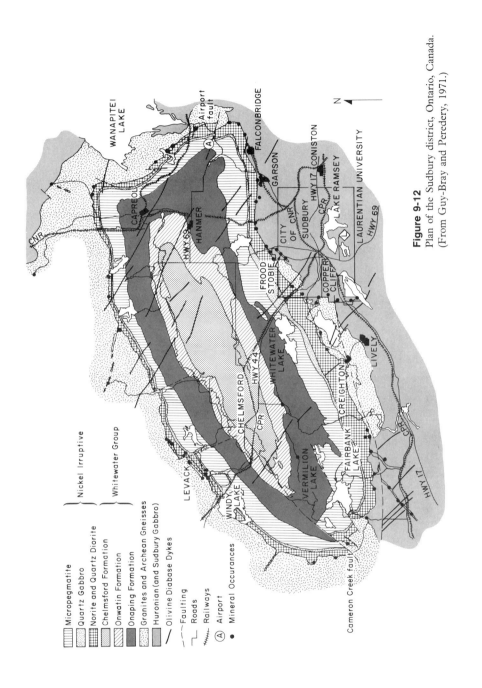

Figure 9-12
Plan of the Sudbury district, Ontario, Canada. (From Guy-Bray and Peredery, 1971.)

Micropegmatite
Quartz Gabbro } Nickel Irruptive
Norite and Quartz Diorite
Chelmsford Formation
Onwatin Formation } Whitewater Group
Onaping Formation
Granites and Archean Gneisses
Huronian (and Sudbury Gabbro)
Olivine Diabase Dykes
Faulting
Roads
Railways
Ⓐ Airport
• Mineral Occurances

WANAPITEI LAKE
Airport fault
FALCONBRIDGE
CAPREOL
CNR
GARSON
CONISTON
HWY 69
HANMER
HWY 17
SUDBURY
LAKE RAMSEY
CITY OF
CNR
CPR
LAURENTIAN UNIVERSITY
HWY 69
FROOD STOBIE
COPPER CLIFF
CHELMSFORD
HWY 44
WHITEWATER LAKE
CPR
LEVACK
VERMILION LAKE
FAIRBANK LAKE
CREIGHTON
LIVELY
WINDY LAKE
CPR
HWY 17
Cameron Creek fault

N

The formations above the Onaping are unbrecciated. Directly above the Onaping is the Onwatin Slate, locally more than 1,000 feet thick, of fine-grained, carbonaceous, pyritic argillite that in places contains beds of limestone and chert. The Onwatin Slate in turn is overlain by the Chelmsford Sandstone or Graywacke, which fills the central part of the basin. Extensive geological mapping throughout the region has failed to correlate the members of the Whitewater Group with strata outside of the Sudbury basin.

North of the Sudbury basin the rocks are mainly granites and gneisses, while to the south lavas, graywackes, and arkoses are widespread. Both of these sequences have been brecciated, and widely distributed shatter cones are well developed.

Intruded as a complex ring-dike mass between the Whitewater Group and the older basement complex is the nickel irruptive. At the top of the intrusive mass is the micropegmatite, a layer consisting of subhedral sodic feldspars in a matrix of granophyre and quartz (Souch *et al.*, 1969; Stevenson, 1963). The micropegmatite reportedly grades into the underlying norite in many places, though age-dating by Fairbairn and his colleagues (1968) indicate an age of about 1,700 million years for the micropegmatite and about 2,000 million years for the underlying norite. Guy-Bray and Peredery (1971) concluded that the micropegmatite is definitely younger than the norites and that it is not part of a simple layered differentiate.

The norite ranges considerably in composition and much of it is highly altered. The feldspars are dusty with sericite, and the pyroxenes are commonly changed to amphiboles. The felsic norite at the top grades abruptly downward to mafic norite, characterized by coarse poikilitic feldspar enclosing pyroxene. It may contain as much as 50 percent hypersthene and minor amounts of augite. At the base of the mafic norite is a discontinuous sublayer of inclusion-rich, sulfide-bearing noritic rock. The ores are most commonly found in this sub-layer and in quartz diorite dikes called *offsets* (Naldrett and Kullerud, 1967, Naldrett *et al.*, 1970; Naldrett *et al*, 1972; Wandke and Hoffman, 1924).

The structure of the basin is far from simple. The northern contact dips about 42° south and the southern contact dips 65° north, as indicated by magnetic studies (Souch *et al.*, 1969). Two systems of faulting are recognized. The first trends nearly east-west and is typified by the Cameron Creek fault on the west and the Airport fault on the east. The Cameron Creek fault dips to the south and seems to have a vertical offset of about three miles, with the south side up relative to the north. The second prominent fault system trends north-northwest; individual faults dip steeply west or are vertical.

Souch and his colleagues (1969) recognized three types of orebodies, exemplified by the North Range, the South Range, and the offsets. The Creighton mine on the south limb of the basin, the Levack mine on the north limb, and the Frood-Stobie offset deposit north of Sudbury, will be used here as examples of the several environments of ore deposition (Symposia, 1948; 1957).

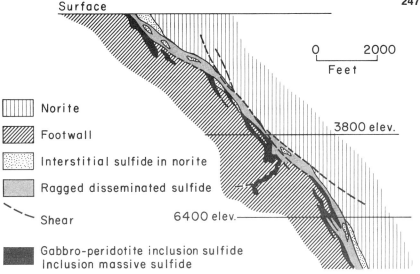

Figure 9-13
Generalized section through the Creighton ore zone, looking west. (From Souch *et al.*, 1969.)

The Creighton ore zone is a series of individual orebodies that have been outlined to a depth of approximately 8,000 feet (Souch *et al.*, 1969; Yates, 1948). Yates reported that an irregular slab of quartz diorite underlies the norite along the east side of a large bay in the norite contact—the Creighton embayment. The contact with the older gabbro, granite, and greenstone on the footwall is sharp, but against the norite the contact is obscure and poorly defined. The quartz diorite is cut by a series of faults that parallel each other and are concentrated close to the quartz diorite contact. Large rolls are recognized in the faults, both along the strike and down the dip. In most places the faults lie close to the footwall and control the position of the ore. Massive sulfides occur along the faults and in the brecciated footwall rock adjacent to the faults. The quartz diorite near the faults contains disseminated sulfides that gradually decrease in quantity away from the faults.

The faults are tenuous rather than simple structures, and in places there are wide zones of shearing that may branch or split and form braided networks of small slips and shears. The most important faulting at the Creighton mine is localized along the contact with the footwall; ore lenses are developed there in an enechelon pattern. Dips of the lenses are steeper than the dip of the contact (see Fig. 9-13).

The southward-dipping Levack mine has an underlying sublayer of sulfides and inclusion-bearing basic norite, sulfide-bearing granitic breccia, and inclusion massive sulfides. The ore is in an embayment that follows the regional trend of the irruptive contact (see Fig. 9-14). Sulfides in the norite are sparsely distributed but rich enough locally to constitute ore. A second type of ore in the granite breccia has sulfides in blebs that coalesce locally and form minable pods. A third type of ore is disseminated in the granite breccia that was intruded along the norite contact and in shattered areas in the underlying granitic gneiss. The controlling structures are somewhat obscure, and faulting similar to that of the Creighton mine is absent.

The Frood-Stobie offset is a mineralized dike-like body of quartz diorite that cuts at low angles across the northeasterly trend of steeply dipping volcanics and sediments. The composition of the quartz diorite varies considerably, but is dominantly a fine- to medium-grained rock containing plagioclase, quartz, hornblende, and biotite, with lesser amounts of sulfides, magnetite, and apatite. Much of the quartz diorite is the peculiar breccia described as the "Frood breccia" (Hawley, 1965; Zurbrigg *et al.*, 1957; Yates, 1938, 1948; Thomson, 1956). The Frood breccia forms a discontinuous envelope around much of the orebody and is considered to be an intrusive expression of the activity that produced the volcanics that lie above the main intrusive mass. Hawley concluded that the offset material had been forced downward into a tension crack.

A cross section of the Frood-Stobie mine is shown in Figure 9-15. Zurbrigg and his colleagues (1957) reported that the orebodies were in and along the

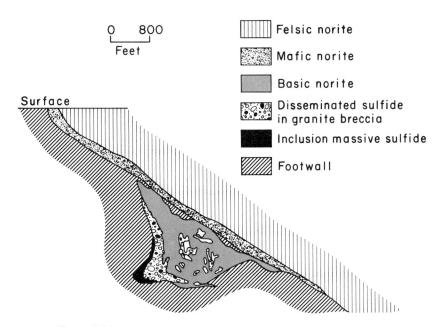

Figure 9-14
Generalized section through a typical Levack mine ore body, looking east. (From Souch *et al.*, 1969.)

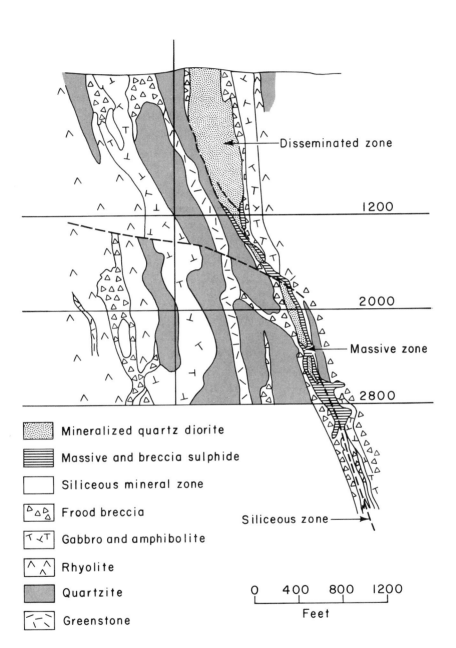

Mineralized quartz diorite

Massive and breccia sulphide

Siliceous mineral zone

Frood breccia

Gabbro and amphibolite

Rhyolite

Quartzite

Greenstone

Figure 9-15
Vertical cross section at Frood mine, showing major zones in orebody.
(From Hawley, 1965.)

underside of a quartz outlier of the nickel irruptive. Two types of ore are distinguished: disseminated and nondisseminated. They include massive sulfides, breccia sulfides, and a breccia containing mineralized fragments in a sulfide matrix, stringer ore, and a low-grade siliceous ore. The mineralogy of the ores is ordinarily simple; the ores consist principally of pyrrhotite, pentlandite, and chalcopyrite, with lesser amounts of pyrite and cubanite. In detail, however, the mineralogy is highly varied; minor amounts of at least 64 species are recognized. Hawley (1962) includes 10 secondary or supergene minerals and 18 nonmetallic and gangue minerals.

The sulfide minerals are dominantly pyrrhotite, pentlandite, and chalcopyrite, with minor amounts of cubanite, though cubanite exceeds chalcopyrite in places in the massive sulfides. Zurbrigg lists accessory minerals such as niccolite, maucherite, galena, sperrylite, hessite, and several minerals of the platinum group. The rather sharp local variations in content of pyrrhotite, pentlandite, and chalcopyrite ratios led Zurbrigg to describe zoning in the Frood-Stobie deposit. Three zones were discussed by Zurbrigg (and later by Hawley): disseminated ore, massive ore, and siliceous ore at depth (Zurbrigg *et al.*, 1957; Hawley, 1965). According to Hawley (1965), this zoning resulted from gravitative settling of the sulfides, which formed the disseminated and massive zones at the base. Because the youngest minerals are concentrated at the base, Hawley described the zoning as "upside down."

REFERENCES TO A SELECTED MAGMATIC SEGREGATION DEPOSIT: THE BUSHVELD IGNEOUS COMPLEX

Beath, C. B., C. A. Cousins, and R. J. Westwood, 1961. The exploitation of the platiniferous ores of the Bushveld igneous complex with particular reference to the Rustenburg platinum mine, *7th Commonw. Congr. Mineral. Metall. Trans.*, Johannesburg, 1:217-243.

Cameron, E. N., 1963. Structure and rock sequences of the critical zone of the eastern Bushveld complex, *Mineral. Soc. Amer. Spec. Pap.* 1, pp. 93-107.

——, 1964. Chromite deposits of the eastern part of the Bushveld complex, in *The Geology of Some Ore Deposits of Southern Africa*, vol. 2, ed. S. H. Haughton. Geological Society of South Africa.

Cameron, E. N., and G. A. Desborough, 1964. Origin of certain magnetite-bearing pegmatites in the eastern part of the Bushveld complex, South Africa, *Econ. Geol.* 59:197-225.

——, 1969. Occurrence and characteristic of chromite deposits—eastern Bushveld complex, *Econ. Geol. Monogr.* 4, pp. 23-40.

Cameron, E. N., and M. E. Emerson, 1959. The origin of certain chromite deposits in the eastern part of the Bushveld complex, *Econ. Geol.* 54:1151–1213.

Coertze, F. J., 1958. Intrusive relationships and ore deposits in the western part of the Bushveld igneous complex, *Geol. Soc. S. Afr. Trans.* 61:387–392.

———, 1962. The Rustenburg fault as a controlling factor of ore deposition, southwest of Pilanesberg, *Geol. Soc. S. Afr. Trans.* 65:253–257.

———, 1963. Structures in the Merensky Reef at the Rustenberg platinum mine, *Ann. Geol. Surv. Pretoria* 2:69–77.

Cousins, C. A., 1959. The structure of the mafic portion of the Bushveld igneous complex, *Geol. Soc. S. Afr. Trans.* 62:179–189.

———, 1962. The stratigraphy, structure and igneous rocks of the Transvaal System at the Western Areas gold mine, *Geol. Soc. S. Afr. Trans.* 65:13–40.

———, 1964a. Additional notes on the chromite deposits of the eastern part of the Bushveld complex, in *The Geology of Some Ore Deposits of Southern Africa,* vol. 2, ed. S. H. Haughton, Geological Society of South Africa.

———, 1964b. The platinum deposits of the Merensky Reef, in *The Geology of Some Ore Deposits of Southern Africa,* vol. 2, ed. S. H. Haughton, Geological Society of South Africa.

———, 1969. The Merensky Reef of the Bushveld igneous complex, *Econ. Geol. Monogr.* 4, pp. 239–251.

Cousins, C. A., and G. Feringa, 1964. The chromite deposits of the western belt of the Bushveld complex, in *The Geology of Some Ore Deposits of Southern Africa,* vol. 2, ed. S. H. Haughton, Geological Society of South Africa.

Ferguson, J., and E. Botha, 1964. Some aspects of igneous layering in the basic zones of the Bushveld complex, *Geol. Soc. S. Afr. Trans.* 66:1–19.

Hall, A. L., 1932. The Bushveld igneous complex of the Central Transvaal, *Geol. Surv. S. Afr. Mem.* 28.

Kupferbürger, W., B. V. Lombaard, B. Wasserstein, and C. M. Schwellnus, 1937. The chromite deposits of the Bushveld igneous complex, Transvaal, *Geol. Surv. S. Afr. Bull.* 10.

Kuschke, G. S. J., 1940. The critical zone of the bushveld igneous complex, Lydenburg District, *Geol. Soc. S. Afr. Trans.* 46:57–81.

Lombaard, B. V., 1934. On the differentiation and relationships of the rocks of the Bushveld igneous complex, *Geol. Soc. S. Afr. Trans.* 37:5–52.

———, 1956. Chromite and dunite of the Bushveld igneous complex, *Geol. Soc. S. Afr. Trans.* 59:59–76.

Schmidt, E. R., 1952. The structure and composition of the Merensky Reef and associated rocks of the Rustenburg platinum mine, *Geol. Soc. S. Afr. Trans.* 55:233–279.

Schweigart, H., 1965. Genesis of the iron ores of the Pretoria Series, South Africa, *Econ. Geol.* 60:269–299.

Schwellnus, C. M., and J. Willemse, 1943. Titanium and vanadium in the magnetic iron ores of the Bushveld complex, *Geol. Soc. S. Afr. Trans.* 46:23–38.

Van der Walt, C. F. J., 1941. Chrome ores of the western Bushveld complex, *Geol. Soc. S. Afr. Trans.* 44:79–112.

Willemse, J., 1964. A brief outline of the geology of the Bushveld igneous complex, in *The Geology of Some Ore Deposits of Southern Africa,* vol. 2, ed. S. H. Haughton, Geological Society of South Africa.

———, 1969a. The vanadiferous magnetite iron ore of the Bushveld igneous complex, *Econ. Geol. Monogr.* 4, pp. 187–208.

———, 1969b. The geology of the Bushveld igneous complex, the largest repository of magmatic ore deposits in the world, *Econ. Geol. Monogr.* 4, pp. 1–22.

REFERENCES CITED

Balsley, J. R., Jr., 1943. Vanadium-titanium-iron ores, Lake Sanford, New York, *U.S. Geol. Surv. Bull.* 940-D.

Bray, J. G., et al., 1966. Shatter cones at Sudbury (Canada), *J. Geol.* 74:243–245.

Buddington, A. F., 1939. Adirondack igenous rocks and their metamorphism, *Geol. Soc. Amer. Mem.* 7, pp. 64–67.

Coleman, A. P., 1905. The Sudbury nickel field, *Ontario Bur. Mines Ann. Rept.* 14, pt. 3.

Collins, W. H., 1934. The life history of the Sudbury irruptive, I: petrogenesis, *Roy. Soc. Can. Trans.* 28(sec. 4):123–177.

———, 1936. The life history of the Sudbury irruptive, III: environment, *Roy. Soc. Can. Trans.* 30(sec. 4):29–53.

Collins, W. H., and E. D. Kindle, 1935. The life history of the Sudbury irruptive, II: intrusion and deformation, *Roy. Soc. Can. Trans.* 29(sec. 4):29–47.

Deardon, E. O., 1958. Lac Tio ilmenite deposit, *Amer. Inst. Mining Eng. Preprint* 5818A4.

Dickson, W. C., 1904. The ore deposits of Sudbury, *Amer. Inst. Mining Eng. Trans.* 34:1–67.

Dietz, R. S., 1964. Sudbury structure as an astrobleme, *J. Geol.* 72:412–434.

Editorial staff, 1967. Palabora, *Eng. Mining J.* (Nov.), pp. 87–111.

Ewers, W. E., and D. R. Hudson, 1972. An interpretive study of a nickel-iron sulfide ore intersection, Lunnon Shoot, Kambalda, Western Australia, *Econ. Geol.* 67:1075–1091.

Fairbairn, H. W., P. M. Hurley, and W. H. Pinson, 1968. Rb-Sr whole-rock age of the Sudbury lopolith and basin sediments, *Can. J. Earth Sci.* 5:707–714.

Fischer, R., 1950. Entmischungen in Schmelzen aus Schwermetalloxyden Silikaten und Phosphaten, *Neues Jahrb. Mineral. Abh.* 81:315–364.

Fischer, R. P., 1946. German iron ores yield vanadium, *Amer. Inst. Mining Eng. Tech. Pub.* 2070.

Fisher, L. W., 1929. Origin of chromite deposits, *Econ. Geol.* 24:691–721.

French, B. M., 1968. Sudbury structure, Ontario (Canada): some petrographic evidence for an origin by meteoric impact, in *Shock Metamorphism of Natural Materials,* ed. B. M. French and M. N. Short. Baltimore: Mono Book Corp.

Frietsch, R., 1973. The origin of the Kiruna iron ores, *Geol. Foren. Stockholm Forh.* 95:375–380.

Geijer, P., 1910. Igneous rocks and iron ores at Kiirunavaara, Lousavaara and Toulluvaara (Sweden), *Econ. Geol.* 5:699–718.

———, 1931*a.* The iron ores of the Kiruna type, *Sveriges Geol. Unders. Arsb. Ser. C Avh. Uppsatser,* no. 367.

———, 1931*b.* Pre-Cambrian geology of the iron-bearing region of Kiruna-Gallivare-Pajala, *Sveriges Geol. Unders. Arsb. Ser. C Avh. Uppsatser,* no. 366, pp. 185–221.

———, 1960. The Kiruna iron ores, *21st Int. Geol. Congr. Guide Excursions A25 & C20,* pp. 3–17.

———, 1967. Internal features of the apatite-bearing magnetite ores, *Sveriges Geol. Unders. Arsb. Ser. C Avh. Uppsatser,* no. 624.

Geijer, P., and N. H. Magnusson, 1952. The iron ores of Sweden, in *Symposium sur les Gisements de Fer du Monde,* vol. 2, 19th International Geological Congress.

Gillson, J. L., 1956. Genesis of titaniferous magnetites and associated rocks of the Lake Sanford district, New York, *Amer. Inst. Mining Eng. Trans.* 205:296–301.

———, 1960. Intriguing examples of geology applied to industrial minerals, *Econ. Geol.* 55:629–644.

Guild, P. W., 1947. Petrology and structure of the Moa district, Oriente province, Cuba, *Amer. Geophys. Union Trans.* 28:218-246.

Guy-Bray, J. V., 1971. Sudbury: the ores, the irruptive and the meteorite impact theory [abstr.], in *Symposium on Archean Rocks*, ed. J. D. Glover, Geological Society of Australia (spec. pap. pub. 3).

Guy-Bray, J. V., and W. V. Peredery, 1971. *Guide Notes, Sudbury Excursion, May 1971*, Geological Association of Canada.

Hammond, P., 1951. Allard Lake ilmenite deposits, *Geol. Soc. Amer. Bull.* 62:1448.

———, 1952. Allard Lake ilmenite deposits, *Econ. Geol.* 47:634-649.

Hawley, J. E., 1962. The Sudbury ores: their mineralogy and origin, *Can. Mineral.* 7(pt. 1).

———, 1965. Upside-down zoning at Frood, Sudbury, Ontario (Canada), *Econ. Geol.* 60:529-575.

Herbert, I. C., 1967. Palabora, *Mining Mag. London* 116:4-25.

Hutton, C. O., 1945. Vanadium in the Taranaki titaniferous iron-ores, *N.Z. J. Sci. Tech.* 27:15-16.

International Nickel Company, 1946. Geology of the Sudbury district, Ontario, in *The Operations and Plants of the International Nickel Company of Canada, Limited*, Canadian Mining Journal.

Lombaard, A. F., N. M. Ward-Able, and R. W. Bruce, 1964. The exploration and main geological features of the copper deposit in carbonatite at Loolekop, Palabora complex, in *The Geology of Some Ore Deposits in Southern Africa*, ed. S. H. Haughton, Geological Society of South Africa.

Naldrett, A. J., J. G. Bray, E. L. Gasparrini, T. Podolsky, and J. C. Rucklidge, 1970. Cryptic variation and the petrology of the Sudbury nickel irruptive, *Econ. Geol.* 65:122-155.

Naldrett, A. J., L. Greenman, and R. H. Hewins, 1972. The main irruptive and the sub-layer at Sudbury, Ontario (Canada), *24th Int. Geol. Congr. Rept. Sec. 4 Mineral Deposits*, pp. 206-214.

Naldrett, A. J., and G. Kullerud, 1967. A study of the Strathcona mine and its bearing on the origin of the nickel-copper ores of the Sudbury district, Ontario (Canada), *J. Petrology* 8:453-531.

Newhouse, W. H., 1936. Opaque oxides and sulphides in common igneous rocks, *Geol. Soc. Amer. Bull.* 47:1-52.

Phemister, T. C., 1937. A review of the problems of the Sudbury irruptive, *J. Geol.* 45:1-47.

Ramdohr, P., 1940. Die Erzmineralien in gewöhnlichen magmatischen Gesteinen, *Preuss. Akad. Wiss. Abh.* no. 2.

Ruiz Fuller, C., 1965. *Geología y Yacimientos Metalíferos de Chile*, Instituto de Investigaciones Geológicas de Chile.

Sampson, E., 1929. May chromite crystallize late? *Econ. Geol.* 24:632-641.

———, 1931. Varieties of chromite deposits, *Econ. Geol.* 26:833-839.

Schneiderhöhn, H., 1958. *Die Erzlagerstätten der Frühkristallisation*, vol. 1, Stuttgart: Gustav Fischer.

Singewald, J. T., Jr., 1917. The role of mineralizers in ore segregations in basic igneous rocks, *Johns Hopkins Univ. Contrib. Geol.* (March), pp. 24-35.

Souch, B. E., et al., 1969. The sulfide ores of Sudbury: their particular relationship to a distinctive inclusion-bearing facies of the nickel irruptive, *Econ. Geol. Monogr.* 4, pp. 252-261.

Speers, E. C., 1957. The age relation and origin of common Sudbury breccia, *J. Geol.* 65:497-514.

Stephenson, R. C., 1948. Titaniferous magnetite deposits of the Lake Sanford area, New York, *Amer. Inst. Mining Eng. Trans.* 178:397–421.

Stevenson, J. S., 1963. The upper contact phase of the Sudbury micropegmatite, *Can. Mineral.* 7:413–419.

Stonehouse, H. B., 1954. An association of trace elements and mineralization at Sudbury, *Amer. Mineral.* 39:452–474.

Stutzer, O., 1907. Geologie und Genesis der lappländischen Eisenerzlagerstätten, *Neues Jahrb. Mineral. Geol. Paläontol.* 24(supp.):548–675.

Symposia, 1948, 1957. In *Structural Geology of Canadian Ore Deposits,* Montreal: Canadian Institute of Mining and Metallurgy, vol. 1, pp. 580–626; vol. 2, pp. 341–376.

Thayer, T. P., 1946. Preliminary chemical correlation of chromite with the containing rocks, *Econ. Geol.* 41:202–217.

———, 1969. Gravity differentiation and magmatic re-emplacement of podiform chromite deposits, *Econ. Geol. Monogr.* 4, pp. 132–146.

Thomson, J. E., 1956. Geology of the Sudbury Basin, *Ontario Dept. Mines Ann. Rept.* 65(pt. 3):1–56.

Thomson, J. E., and H. Williams, 1959. The myth of the Sudbury lopolith, *Can. Mining J.* 80:3–8.

Tolman, C. F., Jr., and A. F. Rogers, 1916. *A Study of the Magmatic Sulfid Ores,* Stanford, Calif.: Stanford Univ.

Uytenbogaardt, W., 1954. On the opaque mineral constituents in a series of amphibolitic rocks from Norra Storfjället, Vasterbotten, Sweden, *Ark. Mineral. Geol.* 1(5–6):527–543.

Vaasjoki, O., 1947. On the microstructure of titaniferous iron ore at Otanmäki, *Comm. Géol. Finlande Bull.* 140 (P. Eskola Mem. Vol.), pp. 104–114.

Vogt, J. H. L., 1894. Beiträge zur genetischen Classification der durch magmatische Differentiationsprocesse und der durch Pneumatolyse entstandenen Erzvorkommen, *Z. Prakt. Geol.* 2:381–399.

Walker, T. L., 1897. Geological and petrographical studies of the Sudbury nickel district of Canada, *Geol. Soc. London Quart. J.* 53:40–66.

Wandke, A., and R. Hoffman, 1924. A study of the Sudbury ore deposits, *Econ. Geol.* 19:169–204.

Williams, D., 1942. Rio Tinto, Spain, in *Ore Deposits as Related to Structural Features,* ed. W. H. Newhouse, Princeton, N.J.: Princeton Univ. Press.

Williams, H., 1956. Glowing avalanche deposits of the Sudbury Basin, *Ontario Dept. Mines Ann. Rept.* 65(pt. 3):57–89.

Woodall, R., and G. A. Travis, 1970. The Kambalda nickel deposit, Western Australia, in *Mining and Petroleum Geology, Commonw. Mining Metal. Cong. Publ. Proc.* 2:517–533.

Yates, A. B., 1938. The Sudbury intrusive, *Roy. Soc. Can. Trans. Ser. 3* 32(sec. 4):151–172.

———, 1948. Properties of International Nickel Company of Canada, in *Structural Geology of Canadian Ore Deposits,* Montreal: Canadian Institute of Mining and Metallurgy.

Zurbrigg, H. F., and Geological Staff, 1957. The Frood-Stobie mine, in *Structural Geology of Canadian Ore Deposits,* vol. 2, Montreal: Canadian Institute of Mining and Metallurgy.

10 / Pegmatites

Pegmatites are unusually coarse-grained igneous or metamorphic rocks. Those of igneous affiliation apparently form from the residual, volatile-rich fractions of magmas, whereas the metamorphic pegmatites represent the more mobile constituents of a rock that are concentrated during metamorphic differentiation. Although pegmatites can be found in any and all shapes, they are typically dikelike or lensoid. Most pegmatites are small, varying from a few feet to tens of feet in the longest dimension. They rarely form extensive and continuous tabular bodies, such as the deep-seated veins, though exceptional deposits over a mile long are known. Igneous pegmatites characteristically solidify late in igneous activity. They tend to be associated with plutonic hypabyssal intrusives from which the volatile fractions could not readily escape. The great majority of pegmatites, whether igneous or metamorphic, developed in deep-seated, high-pressure environments. They are rare in unmetamorphosed sediments or shallow intrusives, lavas, or tuffs. As with the deep-seated veins, pegmatites rarely develop conspicuous alteration halos.

The best-known igneous pegmatites are intermediate to silicic in composition. They are highly colored, conspicuous bodies whose coarse textures and unusual mineral compositions make them favorite mineral collecting grounds. Ferromagnesian pegmatites, associated with gabbroic or more mafic rocks, are fairly common, but they do not stand out as noticeably as the silicic pegmatites and they seldom contain minerals of economic value. Furthermore, mafic plutons are not as abundant as silicic plutons. For these reasons the mafic pegmatites have been described as rare.

Studies in geothermometry indicate that pegmatites form over a wide range of temperatures. Much of the literature reports a temperature of about 575°C, on the assumption that the quartz forms near the alpha-beta inversion point. However, liquid inclusion studies have indicated a temperature of about 150°C for many pegmatites (Ingerson, 1947), and other geothermometers record temperatures up to 700°C. The consensus indicates that most pegmatites are formed between 700° and 250°C (Jahns, 1955).

Pegmatites are classified mineralogically and genetically as *simple* and *complex*. Originally, simple pegmatites were defined as those having undergone no hydrothermal replacement; complex pegmatites were defined as those containing rare minerals introduced by late-stage hydrothermal fluids (Landes, 1933). But these definitions no longer hold. These terms have been redefined: simple pegmatites have simple mineralogy and no well-developed zoning; complex pegmatites have their minerals arranged in a zonal sequence and many contain assemblages of rare minerals. In practice, both sets of definitions are applicable because few pegmatites of truly simple mineralogy have undergone hydrothermal alteration.

SIMPLE PEGMATITES

The vast majority of pegmatites are simple. They consist mostly of coarse-grained quartz and feldspars with subordinate mica, and are ordinarily uniform from wall to wall, both in composition and texture. Except where they can be mined for feldspars or micas, they have no economic value and are of interest principally in helping to unravel geologic history. Simple pegmatites result from metamorphic differentiation or from one comparatively short period of igneous activity. The palingenesis of sediments may produce a melt corresponding chemically to a mixture of quartz, potash feldspars, and muscovite; it is probable that many simple pegmatites are of this origin. Most of the larger simple pegmatites are of igneous origin, whereas the pegmatites formed during metamorphism are typically small and irregular. Lit-par-lit pegmatites intertongued with metamorphic rocks are common; in places they appear to grade into the country rock.

COMPLEX PEGMATITES

Complex pegmatites result from igneous processes rather than from recrystallization or palingenesis associated with metamorphism. Much has been written regarding their origin. They are generally thought to have formed as a result of one long, continuous period of crystallization, during which the first-formed minerals reacted with a progressively changing residual magmatic fluid. Studies indicate that this change involves the development of a gaseous phase in the silicate melt (Jahns and Burnham, 1958; Roering, 1966).

A special feature of some complex pegmatites is the presence of giant crystals within the inner zones. Abnormally large crystals of quartz, feldspars, micas, beryl, apatite, tourmaline, and other pegmatite minerals have been reported. Individual crystals are measured in feet and even tens of feet. Jahns (1953) argues that evidence for a replacement origin of these large crystals is lacking. He contends that they must have crystallized directly from a volatile-rich pegmatitic liquid under delicately balanced thermal and chemical conditions; that is, the crystals must have grown rapidly, and the ions must have been able to diffuse readily through the pegmatitic fluid.

Many complex pegmatites are mined for metallic or rare earth constituents, though not all zoned pegmatites contain these elements in appreciable quantities. Pegmatite bodies are small compared with most other igneous rocks, and the minerals for which they are mined are common only as accessory constituents. Some of these accessory minerals are of great value and are eagerly sought, even in small amounts. In fact, pegmatites are the only economic sources of some metals. They contain tantalum, niobium, beryllium, lithium, cesium, uranium, cerium, lanthanum, thorium, yttrium, and many other rare elements in a host of uncommon and complex minerals, as well as the better-known collectors' items such as topaz, garnet, spodumene, monazite, tourmaline, cassiterite, tantalite, columbite, beryl, and lepidolite. A very long list could easily be offered. Small, uneconomic amounts of sulfides such as chalcopyrite, molybdenite, and sphalerite are widely distributed in pegmatites.

Uses for the rare minerals and metals greatly expanded after World War II, and many of these metals are in short supply. Increased demand encouraged an intensified search that resulted in considerable information about their genesis (Johnston, 1945; Cameron and Shainin, 1947; Cameron et al., 1949; Jahns, 1946, 1951, 1955; Hanley et al., 1950; Page et al., 1953; Brotzen, 1959).

Although the complex pegmatites have highly variable textural, structural, and mineralogical characteristics they also have certain features in common; foremost among these is zoning. Nearly all complex pegmatites show textural and mineralogic variations that define concentric zones or shells (see Fig. 10-1). Contacts between the zones are generally gradational, though in places they are abrupt. Ideally, there are four zones, designated from outside inward as the

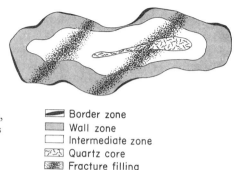

Figure 10-1
Idealized plan of pegmatite body,
showing the distribution of zones
and other superimposed units.

▰ Border zone
▨ Wall zone
▢ Intermediate zone
▨ Quartz core
▨ Fracture filling
    and replacement

border zone, the *wall zone,* the *intermediate zone,* and the *core.* All four zones
are seldom present in any one pegmatite, but in exceptional instances even
more zones have been mapped (additional zones are always described as subdivisions of the intermediate zone). The zones are seldom of uniform thickness
around any one pegmatite—a zone may be thick on one side of a body and thin
or absent elsewhere. Furthermore, individual deposits are irregular and not
arranged in nicely symmetrical layers, though the zones generally reflect the
shape of the pegmatite body, and the outer zones are typically more regular
and continuous than the inner zones (Cameron *et al.,* 1949; Sinkankas, 1968).

The border zone is thin (a few inches or less) in most pegmatites; in some
deposits it cannot be recognized. It is a selvage, or transition, between the
inner, more characteristic pegmatitic materials and the wall rocks, and is typically aplitic in texture. The most common minerals found in border zones are
fine-grained feldspars, quartz, and muscovite; accessory minerals include
garnet, tourmaline, beryl, or even some of the rare species. The valuable metallic constituents are absent from this zone. Some border zones may be the
chilled margins of pegmatitic masses, as suggested by a chemical composition
similar to that of the pegmatite as a whole; other border zones differ distinctly
from the bulk composition of the entire pegmatite (Cameron *et al.,* 1949).

Wall zones are well developed in many pegmatites and absent in others.
In general, the wall zones and border zones contain the same minerals, though
the proportions may differ, and the wall zone is characteristically more coarsely
textured and thicker than the border zone. (the wall zone is, in turn, generally
finer grained than the intermediate zones and the core). The essential minerals
of most wall zones include plagioclase, perthite, quartz, muscovite, and subordinate (but often appreciable amounts of tourmaline, biotite, apatite, beryl,
and garnet. Metallic constituents may be present; in a few deposits they have
been of economic value. Mica and beryl are the principal commercial minerals
recovered from the wall zones.

The intermediate zone includes the greatest concentration of metallic minerals. Although most pegmatites pass directly from the border or wall zones into the core and do not have an intermediate zone, in other pegmatites as many as five or six subdivisions of the intermediate zone can be defined. Different parts of the intermediate zone are generally designated by letter or number; or if only two or three subdivisions are present, they are called the outer, middle, and inner intermediate zones. The intermediate subzones are noted for their varied mineralogy and occasional giant crystals, though the dominant minerals are again feldspars, quartz, and micas. It is in these zones that the minerals of uranium, thorium, lithium, cesium, niobium, tantalum, and rare earth metals are concentrated.

The core in pegmatites is commonly a solid mass of barren white quartz, coarse-grained quartz with feldspar, or quartz with large, euhedral crystals of tourmaline or spodumene. The core is generally near the center of a pegmatitic body and may form discontinuous pods along a central axis. It is ordinarily barren of metallic minerals, though a few exceptions are known.

In places, quartz-filled fractures are superimposed upon the zones or confined to a single zone. The veins may define parallel systems or radial patterns, or they may fill irregular, random fractures. In some deposits the veins are confined to thin joint surfaces, but in others much of the original pegmatite is replaced by minerals introduced along the fractures (see Fig. 10-1). Some of these superimposed vein-filling and replacement materials contain commercial amounts of metals.

Any theory of zoning in the complex pegmatites must take into account certain fundamental relationships, among which are the arrangements of the zones, the gradational to sharp contacts, the transection or replacement of outer zones by inner zones (but not vice versa), the progressive sodium enrichment of plagioclases toward the inner zones, and the paragenetic consistency of zones and mineral assemblages from one pegmatite to another (Cameron *et al.*, 1949).

Of the several explanations that have been advanced, one is that the zonal structure is formed by fractional crystallizations in place under disequilibrium conditions. Thus, the reaction between crystals and rest liquid would be incomplete, creating successive layers of contrasting composition (Brögger, 1890, p. 230). Crystallization in a partially closed system that undergoes repeated pressure releases and consequent resurgent boiling may account for the transgressions of younger zones upon older ones. This hypothesis is supported both in theory and in mineralogical details by Bowen's reaction principle. It is certainly a plausible explanation for those pegmatites that correspond to Bowen's reaction series from the border to the core. The pegmatite from the San Gabriel Mountains (Fig. 10-2) is a good example; it is a mafic pegmatite in coarse-grained norite. The deposit consists of a wall zone with augite-labradorite, an outer intermediate zone of hornblende-labradorite, an inner

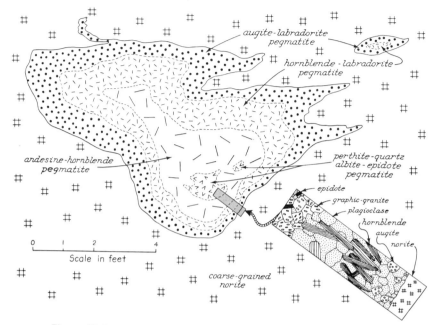

Figure 10-2

Zoning of minerals in a gabbroic pegmatite from the western San Gabriel Mountains, California. (From Jahns, 1954, Fig. 3.)

intermediate zone of andesine-hornblende, and a core of perthite-quartz-albite-epidote. Grain sizes increase gradually from the outer zone through the intermediate zone, and the core is a relatively coarse mixture of graphic granite, albite, and epidote (Jahns, 1954). According to Bowen's series this sequence implies both fractional crystallization and a gradual increase in the volatile content, or at least mobility, of the rest melt.

A second explanation of zoning is that progressively changing solutions deposit materials along the walls of open channels (Hunt, 1871). This hypothesis does not depend upon crystal fractionation and conditions of disequilibrium. The pegmatitic fluids would be expected to vary in composition for any of a number of reasons, such as magmatic differentiation at the source and contamination with wall rocks or other fluids in transit.

A third hypothesis is that complex pegmatites are developed in two stages: (1) formation of a simple pegmatite by the direct crystallization of a pegmatitic fluid, and (2) partial or complete replacement of the pegmatite as hydrothermal solutions pass through it (Hanley et al., 1950; Cotelo Neiva, 1954). The first stage is thought to take place within a relatively closed system; the second, in an open system.

Most modern workers agree with either the first or third hypothesis or with some combination of the two (Schaller, 1933; Landes, 1933; Derry, 1931; Cameron *et al.*, 1949). An argument against the possibility that zones develop along an open channel—as proposed in the second hypothesis—lies in the fact that many interior zones are completely enclosed in outer zones. Since the zones become progressively younger toward the center, it would be difficult for this arrangement to develop in an open system, though crustified zoned veins do form in such an open system. The third hypothesis—the two-stage system—has been objected to on evidence that replacement would have to develop from the outside of a pegmatite toward its core. The zones would represent progressively thinner shells of replacement in fluids of changing composition, in which circumstance the paragenetic sequence would be reversed. In contrast to formation of pegmatites from the walls inward, zoned replacement veins (through-going systems) develop outward from a central fissure. If pegmatite zones do result from replacement reactions in fluids of variable composition, the universal similarity of zonal sequences would seem to be fortuitous rather than a necessary result of the process involved (Cameron *et al.*, 1949).

The source of fracture-filling and replacement material is generally unknown. Secondary minerals that transect all zones probably were derived from outside the immediate pegmatitic system, in which case the pegmatite would fit both the first and third hypotheses; the zoning itself, however, would not have been developed by later replacement processes. Thus, the general concept of a pegmatitic system is one in which the zones develop from the walls inward, within a restricted system. The role of a gaseous phase may be critical in this model. Late in the pegmatitic process hydrothermal fluids that are probably related to the pegmatite travel through fractures and react with the older minerals. This modern concept of the origin of complex pegmatites is precisely what Brögger proposed in 1890.

The vast majority of igneous pegmatites lie near the borders of plutons, either within them or in the nearby country rock. The position and size of these pegmatites have been controlled to a large extent by the structure and fracture pattern in the border areas. In spite of the great amount of work that has been done on pegmatites, very little is actually known about methods of their emplacement.

According to Gevers (1936), the pegmatites of Namaqualand, Republic of South Africa, not only possess internal zoning, but the bodies themselves are distributed in zones in and around an associated granitic batholith. Gevers classed the Namaqualand pegmatites as interior or core pegmatites, marginal or hood pegmatites, and exterior or roof pegmatites. The interior pegmatites are small, scattered, and generally barren of economic minerals. They become increasingly scarce with depth into the batholith. Marginal pegmatites are abundant and large in the upper or outer shell of the batholith. Late-phase hydrothermal alteration effects have developed economic concentrations in many of these pegmatites. The most abundant and most rewarding are the exterior pegmatites, those outside the parent granite body (though some grade into the marginal pegmatites). The exterior pegmatites are distributed over a zone several miles wide (depending upon the size of the parent batholith), and many are highly mineralized.

The relationships pointed out by Gevers fit into a generalization stated by Emmons (1940, p. 22). Emmons noted that the valuable pegmatites are in and near the tops of batholiths—within the intrusives or injected into their roofs. This observation seems to agree with the latest field data. Further, Heinrich (1953) found that simple pegmatites tend to be near or within the batholithic source and the complex pegmatites beyond; complex pegmatites form from the latest pegmatitic fluids, which are more mobile and have a greater rare-element content than the fluids producing simple pegmatites. As with all generalizations in geology, however, there are exceptions. Tôrre de Assunção (1944) pointed out such an exception in Portugal, where the valuable pegmatites are concentrated in the core of a nepheline syenite pluton.

Pegmatites are studied for reasons other than their mineralogy; they are transitions between ordinary intrusive, granitoid masses and the veins and mineral deposits derived from hydrothermal fluids. With the exception of the pegmatites formed by metamorphic differentiation and related processes, most pegmatites are thought to be rest magmas—the residual volatile-rich fractions left after the main magmatic mass has solidified. The pegmatite residues crystallize in place or are squeezed into available openings, where they solidify. Theoretically, it would seem that pegmatites should grade into hydrothermal deposits, especially into quartz veins of hypothermal environments. Many writers state that transitions between pegmatites and quartz veins are known or even common (Schneiderhöhn, 1941; Bateman, 1950). Actual transitions between typical individual pegmatite bodies and typical quartz veins are few; most involve swarms of veins and dikes (Shand, 1943); many are like the Passagem lode in Minas Gerais, Brazil, and to some extent the veins in Southern Piedmont, U.S.A. In these areas the veins are predominately quartz, but they contain minor amounts of scattered garnet, kyanite, tourmaline, apatite, muscovite, phlogopite, biotite, feldspars, and a few other high-temperature minerals. The Passagem lode is considered to be a deep-seated vein, though an early description classified it as a pegmatite because of its mineral composition and coarse texture (Hussak, 1898). Minerals other than quartz are not abundant, however, nor have variations in the vein material been noted either at depth or laterally. Deposits of the Southern Piedmont are similar; most are not typical pegmatites, but at the Old Franklin pit in Clay County, Alabama, a zone of quartz stringers was mined for gold, and one narrow vein grades into what seems to be a characteristic simple pegmatite. The composition of veins in both the Passagem lode and the Southern Piedmont is like that of the enclosing rocks (Ross, 1935), apparently due in part to contamination from materials mobilized metamorphically from within the country rock.

PETACA DISTRICT, NEW MEXICO

Both economically and otherwise, the pegmatites of the Petaca district in northern New Mexico are ideal examples for study. The Petaca district is an area about 15 miles long and four miles wide, trending north-south in Rio Arriba County. The pegmatites there lie within Precambrian rocks south of the San Juan Mountains. Careful studies of the Petaca pegmatites made during World War II were reported in detail by Jahns (1946). His description forms the basis for the following summary.

The pegmatites intrude Precambrian quartzites, quartz-mica schists, granite, and subordinate amphibole schists, andalusite schists, staurolite schists, and metamorphosed rhyolites. Tusas Granite (Just, 1937) invades the metamorphic rocks and appears to be the source of the pegmatitic fluids. Nearly all the commercial pegmatites are in the schists and quartzites, beyond the granite contact. Small irregular pegmatites in the granite are of little or no economic value. The commercial pegmatites reach dimensions of 1400 feet by 275 feet in plan, but most are a few hundred feet long and a few tens of feet wide.

More than fifty pegmatites are exposed in the Petaca district: simple pegmatites in the Tusas Granite and simple sills and dikes as well as irregularly shaped, cross-cutting complex pegmatites in the metamorphics. Only the latter are of economic value. Folds and minor structures controlled the emplacement of many. Associated with the pegmatites are swarms of quartz veins. Both veins and pegmatites apparently have a common origin in the granite, but most of the veins are younger. Some quartz veins cut through the outer zones of pegmatites and merge with their cores; a few veins gradually change along strike from pure quartz to a quartz vein with feldspathic walls.

The pegmatites are primarily of value for their mica, but they also contain significant concentrations of beryllium, columbium, tantalum, bismuth, uranium, thorium, and the rare-earth metals. Mineralogically, they consist of microcline, quartz, albite, and muscovite, with accessory biotite, spessartite ($Mn_3Al_2Si_3O_{12}$, a garnet), fluorite, beryl, columbite-tantalite [$(Fe,Mn)(Nb,Ta)_2O_6$], samarskite [$(Fe,Ca,UO_2)(Ce,Y)(Nb,Ta)_6O_2$], monazite, ilmenite, magnetite, uraninite, lepidolite, tourmaline, and copper sulfides, and rare grains of apatite, native bismuth, bismuthinite, microlite [$(Ca,Na)_2(Ta,Cb)_2(O,OH,F)_7$], cassiterite, fergusonite [$(Y,Er,Ce)(Nb,Ta)O_4$], gadolinite ($Be_2FeY_2Si_2O_{10}$), galena, phenakite (Be_2SiO_4), phlogopite, pyrite, scheelite, topaz, and many secondary alterations of these minerals.

Most commercial pegmatites in the Petaca district are internally zoned and display superimposed replacement minerals. The border zones generally consist of fine- to medium-grained microcline and quartz with subordinate mica and, in places, garnet, fluorite, and beryl. They range from less than an inch to several feet in thickness. Locally, the pegmatitic fluids reacted with the country rocks, developing gradational, mica-rich border zones.

Wall zones of coarse microcline and quartz, with accessory mica, garnet, fluorite, and beryl, are as much as several feet thick. In some pegmatites the wall zones form complete shells, but in others they are only partly developed or were partly removed by reaction with fluids that formed the intermediate zones.

As many as three intermediate zones can be mapped in the Petaca pegmatites, though some have no intermediate zone. The intermediate zones characteristically form hoodlike units along the crests of plunging pegmatite bodies and seldom encompass the whole core, so that these zones vary greatly in thickness. The outer intermediate zone is commonly coarse graphic granite; the middle intermediate zone is coarse blocky microcline; and the inner intermediate zone is generally massive quartz containing sporadic giant microcline crystals.

The cores are near the troughs of plunging pegmatites. They consist of massive quartz (quartz containing large microcline crystals) or, in some cases,

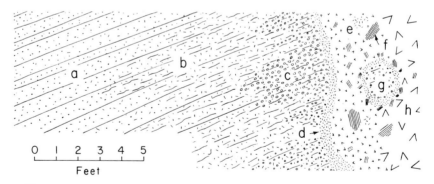

Figure 10-3
Diagrammatic sketch showing typical gradational relations between pegmatite and country rock. Petaca district, New Mexico. (a) Slabby micaceous quartzite; (b) quartzite with muscovite-rich partings and disseminated small flakes of muscovite;
(c) mica-impregnated quartzite with metacrysts of microcline and albite-oligoclase;
(d) fine-grained mica-rich contact zone of pegmatite; (e) medium-grained microcline-quartz-albite-oligoclase-muscovite pegmatite; (f) large block of muscovite;
(g) inclusion of altered quartzite (lithologically similar to that in unit c);
(h) coarse-grained microcline-quartz pegmatite. (From Jahns, 1946, Fig. 5.)

of coarse-grained microcline-quartz pegmatite. Usually they are in the thickest part of the pegmatite and they consist of single ellipsoidal or pipelike masses or series of disconnected pods.

The mineral paragenesis indicates that the zones developed progressively from border to core. Minerals of the outer zones were corroded and replaced by later pegmatitic fluids with which they were no longer in equilibrium; veins of the late fractions cut the peripheral zones. In general, microcline, garnet, and beryl formed early; albite, muscovite, monazite, columbite-tantalite, bismuth, and the sulfides formed later; quartz developed during all stages. The latest minerals to form were samarskite, uraninite, and smoky quartz.

Some of the pegmatites were essentially restricted to fractures, but others permeated the walls and left noticeable alteration zones along gradational contacts. In places the wall rock was replaced by pegmatite, leaving relict, oriented stringers, or ghosts of schist and quartzite. Figure 10-3 demonstrates the typical effect of pegmatitic saturation in the quartzite. The alteration zone averages only a few feet in thickness and projects along permeable parting planes in the quartzite.

Jahns (1946) concluded that the Petaca pegmatites formed in two stages—an early magmatic stage and a later hydrothermal stage. The early stage produced zoned pegmatites of relatively simple mineralogy; the hydrothermal stage, most of the albite, muscovite, and rare minerals. These residual fluids must have been rich in soda, silica, alumina, and significant quantities of columbium, tantalum, beryllium, thorium, uranium, fluorine, bismuth, copper, sulfur, and the rare earths. The hydrothermal fluids traveled along fractures and reacted with the previously formed pegmatitic minerals, leaving behind new minerals of a replacement origin. It was during this hydrothermal stage that most of the metals of economic interest were produced.

REFERENCES TO SELECTED PEGMATITE DEPOSITS

Winnipeg River Area, Manitoba, Canada

Davies, J. F., 1957. Geology of the Winnipeg River area (Shatford Lake-Ryerson Lake) Lac du Bonnet mining division, Manitoba, *Manitoba Dept. Mines Nat. Resources Mines Branch Pub.* 56-1.

Derry, D. R., 1931. The genetic relationships of pegmatites, aplites, and tin veins, *Geol. Mag. Gt. Brit.* 68:454-475.

Springer, G. D., 1950. Mineral deposits of the Cat Lake-Winnipeg River area, Lac du Bonnet division, Manitoba, *Manitoba Dept. Mines Nat. Resources Mines Branch Pub.* 49-7.

Stockwell, C. H., 1933. The genesis of pegmatites of southeast Manitoba, *Roy. Soc. Can. Trans. Ser. 3* 27(sec. 4):37–51.

Wright, J. F., 1930. Tin, lithium, and beryllium deposits of southeast Manitoba, Can. Mining J. 51(22):514–517.

———, 1932. Geology and mineral deposits of a part of southeastern Manitoba, *Can. Geol. Surv. Mem.* 169.

Rajputana, India

Crookshank, H., 1945. Mica, beryl and columbite-tantalite in Rajputana, *Imper. Inst. London* 43(3):228–235.

———, 1948. Minerals of the Rajputana pegmatites, *Mining Geol. Metall. Inst. India Trans.* 42:105–189.

Das, B., 1954. Characters of productive pegmatites, *Indian Mining J.* 2(8):13–15.

Krishnan, M. S., 1946. Beryllium, *Geol. Surv. India Rec.* 76(13).

Sethi, M. L., 1956. Mineral resources of Rajasthan, *Rajasthan Dept. Mines Geol. Bull.* 4.

Wadia, D. N., 1958. Occurrences of beryllium and zirconium in India, *Int. Conf. Peaceful Uses Atomic Energy 2nd Geneva Proc.* 2:107–109.

REFERENCES CITED

Bateman, A. M., 1950. *Economic Mineral Deposits,* New York: Wiley, pp. 54–55.

Brögger, W. C., 1890. Die Mineralien der Syenitpegmatitgänge der Südnorwegischen Augit- und Nephelinsyenite, *Z. Kristall. Mineral.* 16:215–235.

Brotzen, O., 1959. On zoned granitic pegmatites, *Stockholm Contrib. Geol.* 3:71–81.

Cameron, E. N., R. H. Jahns, A. H. McNair, and L. R. Page, 1949. Internal structure of granitic pegmatites, *Econ. Geol. Monogr.* 2.

Cameron, E. N., and V. E. Shainin, 1947. The beryl resources of Connecticut, *Econ. Geol.* 42:353–367.

Cotelo Neiva, J. M., 1954. Pegmatitos com cassiterite e tantalite-columbite da Cabracao (Ponte do Lima–Serra de Arga), *Univ. Coimbra (Portugal) Mem. Not.* 36.

Derry, D. R., 1931. The genetic relationships of pegmatites, aplites, and tin veins, *Geol. Mag. Gt. Brit.* 68:454–475.

Emmons, W. H., 1940. *Principles of Economic Geology,* New York: McGraw-Hill.

Gevers, T. W., 1936. Phases of mineralization in Namaqualand pegmatites, *Geol. Soc. S. Afr. Trans.* 39:331–378.

Hanley, J. B., E. W. Heinrich, and L. R. Page, 1950. Pegmatite investigations in Colorado, Wyoming, and Utah, 1942–44, *U.S. Geol. Surv. Prof. Pap.* 227.

Heinrich, E. W., 1953. Zoning in pegmatite districts, *Amer. Mineral.* 38:68–87.

Hunt, T. S., 1871. Notes on granitic rocks, *Amer. J. Sci.* 101:82–89, 182–191.

Hussak, E., 1898. Der goldführende, kiesige Quarzlagergang von Passagem in Minas Geraes, Brasilien, *Z. Prakt. Geol.* 6:345–357.

Ingerson, E., 1947. Liquid inclusions in geologic thermometry, *Amer. Mineral.* 32:375–388.

Jahns, R. H., 1946. Mica deposits of the Petaca district, Rio Arriba County, New Mexico, *N. Mex. Bur. Mines Mineral Resources Bull.* 25.

———, 1951. Geology, mining, and uses of strategic pegmatites, *Amer. Inst. Mining Eng. Trans.* 190:45–59.

———, 1953. The genesis of pegmatites, *Amer. Mineral.* 38:563–598, 1078–1112.

———, 1954. Pegmatites of Southern California, *Calif. Div. Mines Bull.* 170, pp. 37–50.

———, 1955. The study of pegmatites, *Econ. Geol. (50th Anniv. Vol.),* pp. 1025–1130.

Jahns, R. H., and C. W. Burnham, 1958. Experimental studies of pegmatite genesis: melting and crystallization of granite and pegmatite, *Geol. Soc. Amer. Bull.* 69:1592-1593.

Johnston, W. D., Jr., 1945. Beryl-tantalite pegmatites of northeastern Brazil, *Geol. Soc. Amer. Bull.* 56:1015-1069.

Just, E., 1937. Geology and economic features of the pegmatites of Taos and Rio Arriba Counties, New Mexico, *N. Mex. Bur. Mines Mineral Resources Bull.* 13.

Landes, K. K., 1933. Origin and classification of pegmatites, *Amer. Mineral.* 18:33-56, 95-103.

Page, L. R., et al., 1953. Pegmatite investigations 1942-45, Black Hills, South Dakota, *U.S. Geol. Surv. Prof. Pap.* 247.

Roering, C., 1966. Aspects of the genesis and crystallization sequence of the Karibib pegmatites, South-West Africa, *Econ. Geol.* 61:1064-1089.

Ross, C. S., 1935. Origin of the copper deposits of the Ducktown type in the Southern Appalachian region, *U.S. Geol. Surv. Prof. Pap.* 179, pp. 24-26.

Schaller, W. T., 1933. Pegmatites, in *Ore Deposits of the Western States* (Lindgren Vol.), New York: American Institute of Mining and Metallurgical Engineers.

Schneiderhöhn, H., 1941. *Lehrbuch der Erzlagerstättenkunde*, Jena: Gustav Fischer, pp. 114-120.

Shand, S. J., 1943. *Eruptive Rocks*, New York: Wiley, pp. 8-9.

Sinkankas, J., 1968. Classic mineral occurrences: (1). geology and mineralogy of the Rutherford pegmatites, Amalia, Virginia, *Amer. Mineral.* 53:373-405.

Tôrre de Assunção, C. F., 1944. Algumas observações petrológicas nas Caldas de Monchique, *Lisboa Univ. Mus. Lab. Mineral. Geol. Bol.*, s. 4, no. 11-12, pp. 55-66.

11 / Igneous Metamorphic Deposits

Rocks intruded by igneous masses are commonly recrystallized, altered, and replaced. These changes are caused by heat and by fluids emanating from or activated by the intrusives, and are known as *igneous metamorphism, pyrometamorphism, pyrometasomatism,* or *contact metamorphism.* Although each of these terms means roughly the same thing, "igneous metamorphism" is less restrictive than the others. "Pyrometamorphism" refers only to thermal effects, "pyrometasomatism" refers to replacement reactions, and "contact metamorphism" denotes proximity to the actual igneous contact, whereas many deposits are isolated from any known intrusive mass. "Igneous metamorphism" refers to all forms of alteration associated with the intrusion of igneous rocks, and thus is the preferred term.

Igneous metamorphism is most widely developed around the borders of small- to moderate-sized, discordant, intrusive masses of intermediate composition, such as monzonites and granodiorites, but lesser metamorphic effects are also found around other intrusives, whether silicic or mafic (Edwards and Baker, 1953). Ore deposits of igneous metamorphic origin form under high temperatures and pressures, usually deep within the earth. They are exposed at the surface only after appreciable uplift and erosion.

Characteristics of the igneous metamorphic zone depend upon the nature of both the intruded rock and the emanations activated by the intrusive. Some resistant rocks (such as quartzites) may be unchanged, even at the contact; others (such as carbonates) may be altered up to several miles from the pluton. Two types of alteration are recognized; (1) recrystallization, or rearrangement of the constituents already in the rocks, and (2) addition of materials. Most igneous metamorphic aureoles show both. Considerable discussion has arisen as to the source of metasomatic materials because many small intrusives are altered by the same processes as the bordering rocks. The added materials may have been derived from the local intrusive mass, in which case the resulting ore deposits are likely to be small. Or the ores may have separated at depth in the parent magma chamber, in which case the exposed intrusive will be only a small appendage of a large pluton, that is, a cupola in which the more mobile constituents have accumulated. If the small intrusive body solidified before the metamorphic processes were completed, the igneous rock itself will have been altered or endomorphosed (Lindgren, 1905; Barrell, 1907).

The most striking metamorphic aureoles are developed in the carbonate rocks. Sandstones may be recrystallized, but generally they are unreactive and relatively unaltered except where they are argillaceous or calcareous. In clastic sequences metamorphism is likely to be more widespread where the intrusive mass cuts shale beds, which become baked and hardened or recrystallized into a dense, sugary rock known as granofels and hornfels. Most hornfels are apparently products of simple thermal metamorphism, though in places small additions (especially of silica) have been made. Shaly rocks in igneous metamorphic aureoles are commonly knotty or nodular, the small knots forming around grains of garnet, cordierite, or other porphyroblasts. Likewise, many volcanics are silicified and form hard, dense aureoles.

Skarns or tactites are developed where materials have been added to the country rock. Skarn was defined originally by Swedish miners as the amphibole contact rock with which the magnetite ores of Sweden are associated. The term "tactite" was introduced by Hess (1919) to include all metasomatic products of igneous metamorphism. In present-day usage both terms are commonly used for any silicate rock of complex mineralogy formed in the aureole.

The assemblage of alteration products normally depends upon the character of the invaded rocks, although where large amounts of materials have been added (for example, in ore deposits), the minerals formed may bear little relation to the invaded rock. Generally, the minerals of skarn zones are both diagnostic and conspicuous. In limestones, for example, the skarn is characterized by lime-rich minerals, such as grossularite or andradite, wollastonite, epidote, tremolite, and hedenbergite (or salite). The dolomites develop serpentine, diopside, the humite-chondrodite group, and other minerals comparatively rich in magnesium. Skarns in carbonate hosts are also likely to contain other minerals from the garnet and scapolite families and ilvaite and jeffersonite (a manganese-zinc pyroxene). The carbonate rocks near contacts or skarn zones are generally crystalline or sugary, and many of them have been bleached white.

As would be expected, the thermal metamorphic products in shales are high in alumina, the characteristic minerals including biotite, ottrelite [$H_2(Fe,Mn)Al_2SiO_7$] and other micas, andalusite, sillimanite, hornblende, actinolite, garnets, scapolite, cordierite, and many others. Minerals of skarns and hornfels are commonly distributed in more or less well-defined and regular zones around centers of igneous activity. Their composition and distribution depend not only upon the nature of fluids and rocks, but also upon the temperature and pressure of the reaction. Metamorphic facies or zones may range from sillimanite-rich material near the igneous center to biotite along the outer margin of activities (Hietanen, 1967). Some of the volcanic rocks likewise develop dense, fine-grained skarns that consist mainly of silica and garnet.

Large (1971) published an excellent quantitative discussion of both the endoskarns and the exoskarns of the Bold Head scheelite deposits on King Island, Australia. Here an aureole formed in sediments, mostly in dolomitic limestone members and in volcanics around a stock of adamellite. The temperature of formation is thought to have held fairly steady at 530° to 710°C. At the Christmas mine in Arizona, Perry (1969) described three types of skarns: (1) magnesian skarn in dolomite, with magnetite and forsterite; (2) calcian skarn in limestone with andradite; and (3) endoskarn in porphyry with aluminum-bearing calc-silicates.

Silica is probably the most abundant compound added to igneous metamorphic zones. It may enter into one or more silicate minerals, or it may replace the host completely, producing quartz or chert. The silica sometimes forms small inconspicuous isolated crystals, and elsewhere produces massive deposits of cryptocrystalline quartz called jasperoid. Silicification is common in both carbonate rocks and shales.

Brock (1972) explained the development of banded skarns in the Garnet Hill area of Calaveras County, California, as the result of direct chemical reactions between metasedimentary rocks and marble. This was thought to have produced "diffusion" skarn initially. Although most of the silica, alumina, and magnesia was supplied by the nearby metasedimentary rocks, an aqueous phase from the cooling pluton added some iron, base metals, and silica to form skarn. This skarn, termed "infiltration" skarn, is characterized by iron-rich minerals.

In contrast to the mineralogy of skarns, ore minerals in igneous metamorphic deposits are usually simple sulfides and oxides. The sulfides include sphalerite (which is commonly iron-rich), galena, chalcopyrite, bornite, and in places molybdenite. A few igneous metamorphic zones are comparatively iron-free, but most contain abundant pyrite or even large amounts of the oxides, magnetite and hematite. Scheelite also is found in many igneous metamorphic deposits, and at the Emerald mine in British Columbia the ore occurs in both the skarn and the intrusive rocks.

Igneous metamorphic ore deposits are virtually restricted to the carbonate rock, and are unusual or absent in shales and high-silica sediments. The oxide minerals are commonly within the skarn, forming an integral part of this meta-morphic rock. In places the oxides are concentrated next to the intrusive, but elsewhere they occur along the outer border of the skarn zone. Most of the sulfides are concentrated along the periphery of the skarn, near the contact with the carbonate host. In general, the sulfide minerals were emplaced after the skarn, which they replace. Many of the original sedimentary textures and structures are preserved after the double replacement by skarn and ore, though in some deposits the second replacement obscures the evidence and relict tex-tures can be found only after diligent searching.

The presence of abundant oxide minerals in igneous metamorphic zones presents problems in chemistry. We know, for example, that most, or at least half, of the iron in mafic igneous rocks is in the ferrous state, and hence that the iron concentrated in residual liquids also must be largely reduced. Yet all of the iron in specular hematite and part of that in magnetite is in the ferric state. How is the iron oxidized in the igneous metamorphic environment? Butler (1923, 1927) suggested that the iron is oxidized according to the reaction

$$3FeO + CO_2 \rightleftharpoons Fe_3O_4 + CO$$

by carbon dioxide liberated from the carbonate host rocks during intrusion and replacement. The reaction toward the right is favored by the high temperatures characteristic of igneous metamorphic zones; the reverse reaction is favored by lower-temperature conditions and lower carbon dioxide pressures. The early, near-pluton deposition of magnetite and specular hematite and the later, more distant deposition of ferrous compounds (sulfides, carbonates, and silicates) favors the above reaction. It has been pointed out, however, that the iron is probably not in solution as FeO. Accordingly, Shand (1947) proposed that iron travels as an aqueous colloidal solution of ferrous hydroxide. Upon de-hydration, the ferrous hydroxide is oxidized to magnetite:

$$3Fe(OH)_2 \rightleftharpoons Fe_3O_4 + 2H_2O + H_2.$$

This reaction was proposed for the oxidation process in general and not specifically for igneous metamorphic environments, where the presence of colloids is not very probable, especially since the replacement nature of the ores demands widespread diffusion. Perhaps a compromise between these two hypotheses or a modification of either one is correct. The modern concept of free ions in solution, rather than molecules of compounds, can be applied to Butler's equations. Similarly, it is reasonable that even a true solution of $Fe(OH)_2$, as Fe^{+2} and $2OH^-$, will become oxidized at elevated temperatures. Both reactions have been supported experimentally, but it is possible that neither the compound FeO nor the colloid $Fe(OH)_2$ was the actual substance.

The oxidation state of iron may be controlled by the presence of sulfur. According to Kullerud's experiments, the concentration of sulfur in a silicate system controls the concentration of ferric ion. Thus, at a given temperature, such as 650°C,

$$\underset{\text{hedenbergite}}{4CaFeSi_2O_6} + 2S \rightarrow FeS_2 + Fe_3O_4 + \underset{\text{wollastonite}}{4CaSiO_3} + 4SiO_2$$

when the hedenbergite/sulfur ratio is 4/2. If the ratio is 3/2, then

$$3CaFeSi_2O_6 + 2S \rightarrow FeS_2 + Fe_2O_3 + 3CaSiO_3 + 3SiO_2.$$

In these reactions the iron ore minerals are produced by the reaction of skarn minerals with hydrothermal sulfur. Note that the $CaSiO_3$ (wollastonite) is produced by sulfurization; this is contrary to the classic idea that calcite and quartz react under specific pressure and temperature conditions. To be effective, sulfurization requires large amounts of sulfur; though large amounts are not always present, it is logical to assume that ore minerals replace skarn in many deposits where sulfurization is suspected. Thus, reaction of sulfur with silicates would produce metal sulfides and oxides as well as other skarn minerals. This helps to explain why ore minerals are commonly in skarn rather than in nearby limestone.

A resurgence of economic interest in many skarn deposits is currently making a great deal of theoretical information available. Although many skarn deposits are highly irregular and difficult to mine, others have been highly profitable.

CENTRAL DISTRICT, NEW MEXICO

The Central district in southwestern New Mexico is a highly mineralized area including iron and zinc ores of igneous-metamorphic origin, disseminated copper deposits, replacements of zinc in limestone, and zinc-lead-copper veins. The district has been well studied and is unusually well known (Lindgren et al., 1910; Paige, 1916; Schmitt, 1933, 1942; Spencer and Paige, 1935; Jones et al., 1967; Hernon and Jones, 1968; Lasky and Hoagland, 1950; Horton, 1953).

The Central district contains Precambrian granites, gneisses, schists, and hornfels; Paleozoic limestones and shales underlain by a basal unit of sandstone; and Upper Cretaceous to Tertiary clastics and igneous rocks. The igneous rocks include diorite sills; andesite intrusives and associated pyroclastics; quartz diorite sills and laccoliths; composite stocks of granodiorite and quartz monzonite; dikes of quartz monzonite, granodiorite, and quartz latite; rhyolite pyroclastics; and basalt flows (Lasky and Hoagland, 1950; Hernon *et al.*, 1953). Jones, Hernon, and Moore (1967) list about 25 kinds of igneous rocks that intrude as sills, laccoliths, stocks, and dikes of Late Cretaceous-Early Tertiary age.

The Late Cretaceous or Early Tertiary metallization is related to the granodiorite-quartz monzonite stocks and dikes, though a minor amount of mineralization accompanied Late Tertiary igneous activity. There are three stocks in the central district: the Hanover-Fierro stock, the Santa Rita stock, and the Copper Flat stock (see Fig. 11-1). The regional geology is complex; the district lies on the northeast limb of a broad, shallow syncline modified by minor folds and local domes associated with the stocks and laccoliths. Faulting in the region is also complex; it spans at least six periods, ranging from prestock to postlava flows. Recurrent movement took place along many faults during several or all of these episodes.

The mineralized area is in a horst, bounded on the northeast by the Mimbres fault and on the southwest by the Silver City fault, within which the rocks are highly faulted and mineralized and structural details are complex (see Fig. 11-1). The most conspicuous faults trend northeast; many are filled with dikes and some are bordered by skarns.

The Hanover-Fierro stock is the largest of the plutons. It is somewhat oval, extending about $2\frac{1}{2}$ miles in a north-south direction and averaging $\frac{1}{2}$ mile or more in width. The towns of Hanover and Fierro lie at the south and north ends, respectively. The Santa Rita stock is about one mile southeast of Hanover; the Copper Flat stock, the smallest of the three, is about two miles southwest of Hanover. Each stock is surrounded by an igneous metamorphic aureole, chiefly manifested by replacement skarns in carbonate rocks and shales. Differences in the compositions of intruded rocks caused considerable differences in the alteration products formed in the metamorphic zones. For example, the higher magnesium content in the older Paleozoic rocks along the northern end of the Hanover-Fierro stock and the higher calcium content of the younger rocks near the south end of the stock caused striking differences in the metamorphic products at the two extremities. At Fierro the igneous metamorphic zone contains abundant serpentine, wollastonite, tremolite, and magnetite; at Hanover the mineral assemblage is garnet, epidote, hedenbergite, tremolite, ilvaite, and sphalerite. Some metamorphic minerals are restricted to certain host rocks—notably epidote, which selects aluminous rocks such as shales and argillaceous limestones, and sphalerite, which favors the pure limestones (Schmitt, 1939). Furthermore, the skarn aureole itself is zoned with garnet-pyroxene-ore near the intrusives; this zone is in turn separated by marble from an outer zone of actinolite-tremolite skarn (Lasky and Hoagland, 1950).

The deposition of the igneous metamorphic ores resulted from proximity to the intrusive bodies and from structural control. The ore-bearing fluids

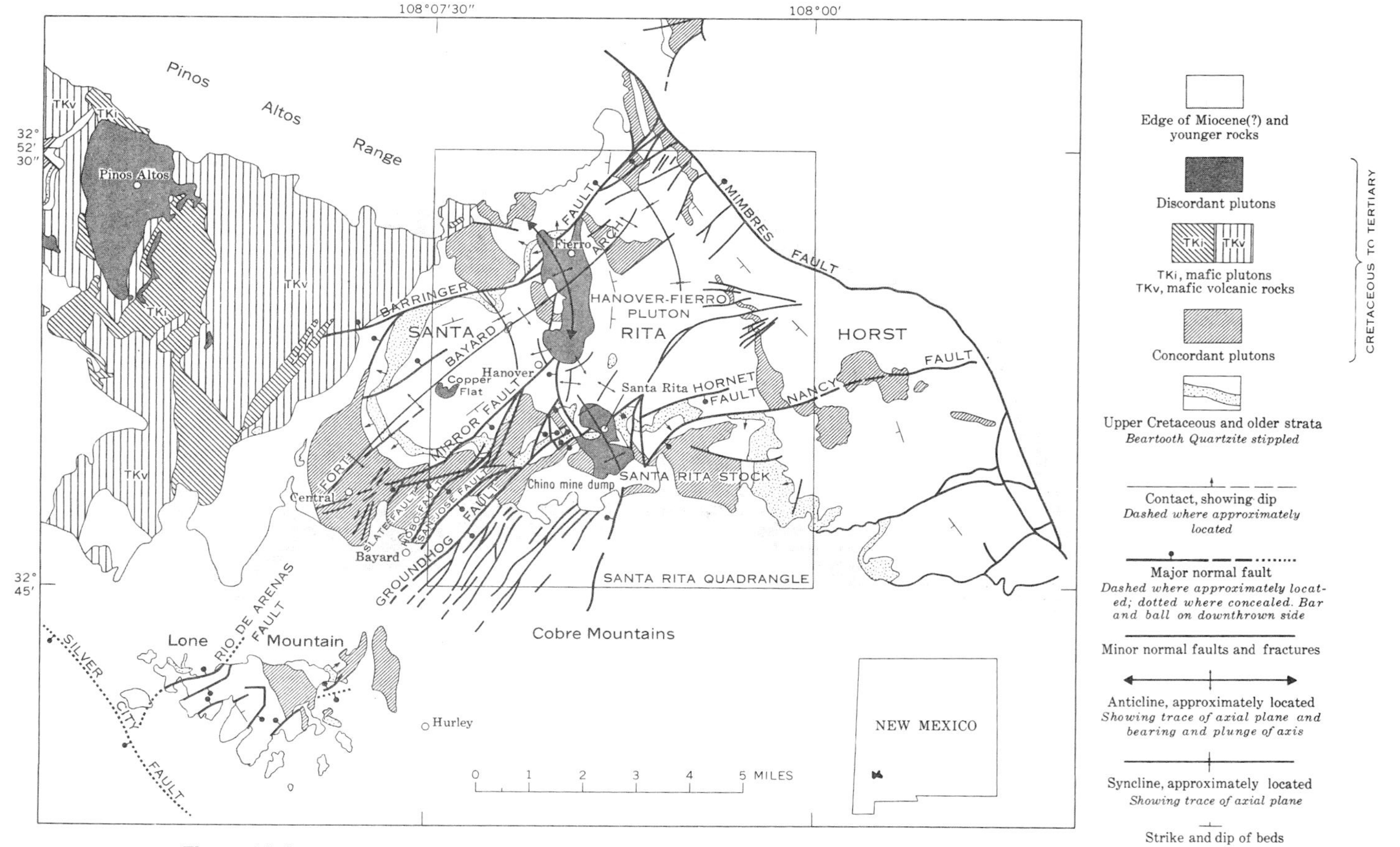
108°07'30"
108°00'
32° 52' 30"
32° 45'
Pinos Altos Range
Pinos Altos
TKv
TKi
BARRINGER
SANTA RITA
BAYARD
Hanover
Copper Flat
MIRROR FAULT
FORT
Central
SLATE FAULT
HOBO FAULT
SAN JOSE FAULT
GROUNDHOG FAULT
Bayard
Fierro
FAULT
ARCH
HANOVER-FIERRO PLUTON
Santa Rita
HORNET FAULT
NANCY
FAULT
HORST
MIMBRES FAULT
SANTA RITA STOCK
Chino mine dump
SANTA RITA QUADRANGLE
Cobre Mountains
Lone Mountain
RIO DE ARENAS FAULT
SILVER CITY FAULT
Hurley
0 1 2 3 4 5 MILES
NEW MEXICO
Edge of Miocene(?) and younger rocks
Discordant plutons
TKi, mafic plutons
TKv, mafic volcanic rocks
Concordant plutons
CRETACEOUS TO TERTIARY
Upper Cretaceous and older strata
Beartooth Quartzite stippled
Contact, showing dip
Dashed where approximately located
Major normal fault
Dashed where approximately located; dotted where concealed. Bar and ball on downthrown side
Minor normal faults and fractures
Anticline, approximately located
Showing trace of axial plane and bearing and plunge of axis
Syncline, approximately located
Showing trace of axial plane
Strike and dip of beds

Figure 11-1
Structural geologic setting of the Santa Rita quadrangle, New Mexico. (From Jones, Hernon, and Moore, 1967.)

ascended most readily along fractures, and the skarn and ore zones are most extensive near them. Most of the metamorphic aureole is within several hundred feet of the intrusives, but apophyses along the faults and dikes extend much farther. The rocks around the southern border of the Hanover-Fierro stock form an overturned anticline that strikes parallel to the contact (see Fig. 11-2). This fold is readily recognized, even in the massive garnet skarn, where bedding is accentuated by differences in texture and size of mineral grains.

The metamorphic zones merge into a single continuous band between Hanover and Santa Rita. About halfway between the two stocks is a small hill known as Bully Hill (not shown on the map, Fig. 11-2) that contains fine-grained silica and considerable amounts of alunite.

Zinc and iron ores form separate deposits of igneous metamorphic origin in the Central district. The largest and richest zinc sulfide orebodies are in the upper crinoidal part of the Lake Valley Limestone (known locally as the Hanover Limestone) of Mississippian age. Lesser amounts of sphalerite replace other limestone, but the Lake Valley Limestone contains an 18-foot shale bed (the Parting Shale), which efficiently dammed ascending ore fluids. Much of the ore at Hanover is concentrated in the limestone immediately below the Parting Shale; locally, however, mineralization extends into the shale and, in places, above it into the limestone.

In effect, the skarn zone was localized beneath the Parting Shale; in turn, the skarn controlled the deposition of sulfide ore minerals. Thus, the igneous metamorphic deposits are generally in the skarn, which they replace. Lasky and Hoagland (1950) observed that the orebodies at Copper Flat lie well within the garnet-pyroxene zone, but at Hanover they are at the outermost edge of this zone, against the marble but still replacing skarn. Consequently, most of the Hanover orebodies have at least one marble wall. Zinc ores also formed in limestone at considerable distances from skarn, beyond the zone of igneous metamorphism; these lower-temperature ores are replacement deposits and are typical of the mesothermal zone.

According to Burt (1968), oxygen fugacity is one form of control in ore-forming processes, particularly in hedenbergite-zinc deposits, such as at Hanover. In the igneous metamorphic zones sphalerite is commonly associated with hedenbergite rather than with the accompanying and nearby garnet. In fact, most of the hedenbergite, or salite, at Hanover contains several percent of zinc. Burt gives the following equation:

$$9CaFeSi_2O_6 + 2O_2 \rightarrow 3Ca_3Fe_2Si_3O_{12} + Fe_3O_4 + 9SiO_2.$$
$$\underset{\text{hedenbergite}}{} \qquad \underset{\text{andradite}}{} \quad \underset{\text{magnetite}}{} \quad \underset{\text{quartz}}{}$$

Burt states that this buffer controls oxygen fugacity at a concentration considerably lower than that of the hematite-magnetite buffer.

Figure 11-2
Major structural features in the Santa Rita quadrangle, New Mexico.
(From Jones, Hernon, and Moore, 1967.)

Landslide deposits

Upper Miocene or younger gravel deposits

Miocene(?) volcanic and detrital rocks

Lower Tertiary
Preore plutons of intermediate composition

Pre-Miocene rocks

Contact
Dashed where approximately located

Normal fault
Bar and ball on downthrown side. Dashed where approximately located. Chain of ellipses marks probable site before obliteration by intrusion or related phenomena

Fractures or minor faults

Area of brecciated rock

Anticline   Overturned syncline

Folds
Showing trace of axial plane and direction of plunge of fold axis. Dashed where approximately located. Arrows indicate direction of dip of limbs

Strong vertical joint set

Mine dump

Some igneous metamorphic sphalerite deposits are relatively large. A typical orebody in the Pewabic mine, near the southeast edge of the Hanover-Fierro stock, was localized along the intersection of a thrust fault and a vertical fracture zone, forming a cigar-shaped mass about 40 feet in diameter and 600 feet long (Schmitt, 1939). An analysis of the early zinc ore from the Hanover area is given in Table 11-1. As in most mining districts, the earliest ore was higher grade than later shipments, though the mineral composition has not varied. In the past the ore brought a premium price because of its extremely low lead content. As the deposits are worked farther and farther from the intrusive contacts, however, the amount of galena increases, with the result that lead is now an important ore constituent.

Igneous metamorphic deposits of iron ore have also been mined in the Central district. In contrast to the sphalerite, which occurs near the outer fringe of the metamorphic aureole, the magnetite is concentrated close to the igneous contacts (Spencer and Paige, 1935). Although magnetite is widespread throughout the whole district, the principal commercial deposit is at Fierro, near the northern end of the Hanover-Fierro stock, especially where the contact of the igneous rock parallels the bedding. The best ores are concentrated in the Lower Ordovician El Paso Limestone; lesser amounts are in the Cambrian Bliss Sandstone. The magnetite orebodies are roughly tabular, conforming to the stratification of the rocks. They alternate with layers of serpentine, wollastonite, and lesser amounts of garnet, hedenbergite, tremolite, and epidote. Small veinlets of pyrite and chalcopyrite, which are in much of the

Table 11-1
Analysis of early ore from Hanover, New Mexico.

Constituent	Percent
Zn	17.49
Pb	0.13
Fe	6.19
Mn	0.94
CaO	8.09
MgO	1.81
CO_2	3.37
S	10.19
Insoluble	49.77
TOTAL	97.98

Source: Spencer and Paige, 1935.

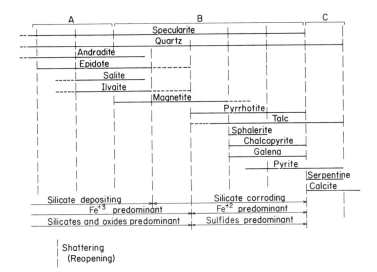

Figure 11-3
Paragenesis of the Pewabic ores, Central district, New Mexico. (From Schmitt, 1939, Fig. 4.)

magnetite, signal a slightly later phase of mineralization. Schmitt (1939) found that the silicates and oxides of the Pewabic ores preceded the sulfides and that ferric iron predominated during the early silicate-oxide phase, whereas ferrous iron predominated during the sulfide phase (see Fig. 11-3). Specular hematite was the first ore mineral deposited contemporaneously with the silicates, but most metallization followed the formation of skarn.

The skarn and ore deposits are not products of the accompanying stocks. The stocks solidified before the ores were deposited and in places were altered to skarn along with the country rocks. During the episode of mineralization, the stocks were intruded by granodiorite dikes, which in turn are partly mineralized. Evidently, the skarn and ore deposits are due largely to hydrothermal fluids that were generated within the same magmatic chamber as the stocks, and each was intruded in sequence into the overlying rocks. The postulated hydrothermal activity is evident throughout the district. It began when the stocks were emplaced, and continued in augmenting proportions through subsequent periods of fracturing, intrusion of dikes, and ore deposition. All of these phenomena represent phases of an extended period of igneous activity (Spencer and Paige, 1935; Jones *et al.*, 1961). The presence of disseminated veinlets of copper ore in the Santa Rita stock indicates that the ore-bearing fluids ascended from depth and were not direct differentiates of the stocks themselves. This origin is also shown by the presence of ore away from the stocks and by the similarity between this ore and that in the skarn. Zoning in the Santa Rita stock has been described by Nielson (1968).

IRON SPRINGS DISTRICT, UTAH

The magnetite deposits of the Iron Springs district, west of Cedar City in southwestern Utah (see Fig. 11-4), furnish another example of igneous meta-morphic deposition (Leith and Harder, 1908; Young, 1947; Mackin, 1947, 1968; Mackin and Ingerson, 1960). This district was mentioned in Chapter 2 in connection with deuteric leaching as a source of metals; here the manner of ore deposition is of interest.

Ore deposition in the Iron Springs district is associated with three intrusive masses: Iron Mountain, Granite Mountain, and Three Peaks. The intrusives are aligned in a northeast direction; each is oval in plan, 3 to 5 miles long, with its long axis parallel to the regional alignment. Besides replacement ores along the borders of the intrusives, vertical veins of magnetite traverse the plutons. Much more ore has been obtained from Iron Mountain and Granite Mountain than from Three Peaks. These deposits differ from the magnetite bodies at Fierro, New Mexico, in that most of the ores at Iron Springs replaced the limestone directly, rather than replacing a previously deposited skarn. In fact, the calc-silicate minerals that are typical of skarns are virtually absent from the ore zone.

The lowest stratigraphic unit exposed in the district is a massive, 200- to 300-foot, blue and gray limestone of Jurassic age, known as the Homestake Formation. It includes a 15- to 25-foot basal siltstone and is capped by 5 to 10 feet of argillaceous limestone that is distinctive for its ripple marks and mudcracks. Overlying the Homestake Formation is the Entrada Formation of Late Jurassic age. The Entrada consists of maroon and gray shales interlensed

Figure 11-4
Map of the Iron Springs
district, Utah.

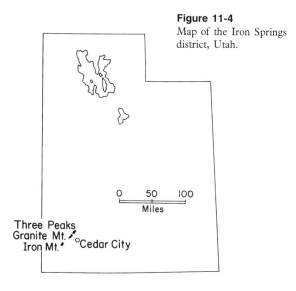

with medium- to coarse-grained arkosic sandstone. It varies in thickness from 50 to 220 feet, the variation being due to erosion. The eroded surface of the Entrada is covered disconformably by a basal conglomerate, which grades upward into a sequence of variegated, fresh-water limestones, shales, sandstones, and conglomerates of the Iron Springs Formation, a Cretaceous unit as much as 5,000-feet thick. The Iron Springs Formation is folded, eroded, and overlain unconformably by the Claron Formation, a Cretaceous to Eocene assemblage of coarse boulder conglomerate, gray and red sandstone, and shale, and white to pink limestone. The pre-Claron folding and erosion left the Iron Springs Formation with a variable thickness of 1,000 to 5,000 feet; rarely is the full 5,000-foot section preserved. At places the Iron Springs and Entrada were completely removed, so that the Claron Formation rests directly on the Homestake (Mackin, 1947).

The three intrusive bodies are all laccolithic; they intruded along the base of the Homestake Formation, from which they arced upward with the overlying sediments. The three laccoliths consist principally of quartz monzonite (Mackin, 1954) or granodiorite porphyry (Mackin and Ingerson, 1960), with fine-grained, chilled borders that definitely resulted from the forceful injection of a viscous melt (Mackin, 1947).

Mackin (1947) delineated three zones in the laccoliths; each zone reflected a different phase of crystallization. Each intrusive is bordered by a shell of fine-grained quartz monzonite (100 to 200-feet thick), an erosion resistant rock that forms ledges above the interior zone. This peripheral rock is finer-grained, less friable, and has a higher magnetite content than the quartz monzonite of the interior region; it also contains fresh biotite and hornblende phenocrysts.

The interior phase of the quartz monzonite is more coarsely grained than the peripheral phase, and the biotite and hornblende have been deuterically altered. Because of this alteration, the rock is readily weathered and generally forms low, crumbly knobs, or flat, barren stretches. Unlike the peripheral quartz monzonite, very little fresh rock crops out in the interior zones.

Both the interior and the peripheral shells of each laccolith are jointed along radial and concentric planes that are spaced from a few feet to a few tens of feet apart. Most radial joints strike at nearly right angles to the concentric joints and show no associated selvage (alteration).

Mackin (1947) described a third zone in the laccoliths, which he called the zone of selvage joints. This zone separates the interior from the peripheral shell and has a rugged topography. The rock on each side of the individual joints is bleached through a zone that ranges from a fraction of an inch to more than six inches in width (see Fig. 11-5). Differential weathering in the zone of selvage joints produces a ribbed physiography because each selvage zone forms a hard surface crust that stands in relief above the surrounding quartz monzonite. This bleached crust grades, within half an inch, into the soft, deuterically altered quartz monzonite characteristic of the interior zone. Mafic minerals in

Figure 11-5
Selvage joint through the Three Peaks quartz monzonite, showing the bleached zone
bordering a magnetite veinlet. ×4. (Specimen collected by A. S. Radtke; photo by
W. J. Crook.)

the selvage zones have been bleached to light green or white. About 20 percent of the rock volume in this zone of selvage joints has been bleached. Differences between the three laccolithic zones seem to be due to the nature and degree of deuteric alteration rather than to any differences in the original rock type.

Selvage joints are radial, concentric, and oblique. The radial set is a continuation of the radial joints in the interior, normal to the intrusive contacts, and with vertical dips. By contrast, the concentric joints parallel the igneous contacts and dip into the intrusives, normal to the contacts. Curved joints that swing from radial to concentric are classified as oblique (see Fig. 11-6). Possibly 30 to 50 percent of the selvage joints contain magnetite with accessory pyroxene, apatite, calcite, and hematite. Comb structures and vugs indicate that the fissures were filled rather than replaced. The veins are unusually abundant; thousands are less than four inches wide, scores of them are several feet wide, and some are over 10 feet wide. Most larger veins occupy radial joints, but some also fill concentric and oblique joints. In general, the radial

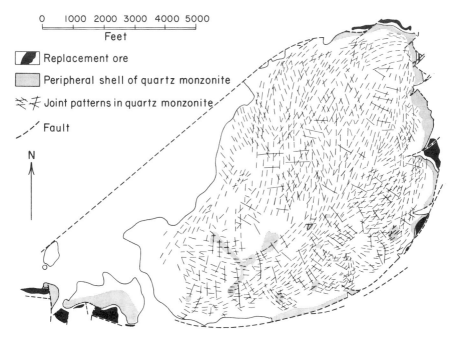

Figure 11-6
Map of quartz monzonite outcrops on Granite Mountain, Utah, showing the joint pattern and areas of peripheral-shell facies. Note the relationship between replacement orebodies and peripheral-shell facies. See Figure 11-7 for the relationship of the quartz-monzonite and ore to sedimentary rocks. (From Mackin, 1954.)

veins are wedge-shaped (thickening toward the margins of the laccoliths) both in plan and in cross section.

Chemical analyses indicate that the iron deposited in fissures and as replacement of limestone was derived from the zone of selvage joints. The deuterically altered interior zone contains about as much iron as the fresh peripheral zone (about three percent), but the bleached selvage zones are all relatively deficient in iron. It is believed that the volume of rock leached could have supplied more than enough iron to account for the deposits in and around the laccoliths.

Detailed mapping shows that the ore deposits are associated with bulges in the laccolith roofs (Mackin, 1947, 1954). Where the intrusive was not arched strongly upward, both the joints and the surrounding limestones are barren of ore. There was apparently a critical period during crystallization of the magmas when fracturing of the roof permitted egress of the volatiles. The Three Peaks laccolith has a relatively flat roof and did not provide joints for escaping mineralizers; consequently, this intrusive has only a thin zone of selvage joints and is relatively unproductive in iron. It is especially noteworthy that roof pendants in the Three Peaks laccolith are barren in contrast to Granite Mountain and Iron Mountain, where they are highly mineralized. Only where late-intrusion tension cracks and breccia zones tapped the selvage zone was ore deposited beyond the laccolithic contact. The iron-rich fluids capable of moving from the zone of selvage joints through the peripheral shell left igneous metamorphic deposits in the Homestake Limestone.

Replacement orebodies in the Homestake Limestone measure as much as 1,000 feet in dip and strike dimensions and are as much as 230 feet thick in places. They are pod-shaped bodies scattered irregularly along the igneous contact. In general, the ore is separated from quartz monzonite by the basal siltstone of the Homestake Formation, and the outer limits of ore conform to the bedding only where the entire thickness of limestone was replaced (Fig. 11-7). Calculations by Mackin (1968) indicate that for every million tons of iron, 40,000 tons of silicon, 20,000 tons of magnesium, and 10,000 tons of aluminum were added to the limestone; this is surprisingly little when compared to the amount of iron introduced. Phosphorus (in apatite) and fluorine (in phlogopite and apatite) are troublesome in the ore because they can cause difficulties in smelting.

Ore deposition was not preceded by the development of skarn, though mica, quartz, garnet, and a few lime silicates are found with the ore. Some preore silicification took place, but it did not control the localization of ore. Apatite in both the vein and replacement ore is concentrated along the footwall. Copper and iron sulfides are minor constituents of the replacement ores, and they are found farthest from the intrusive.

In summary, Mackin (1947, 1954) believes that the Iron Springs ores originated locally within the laccoliths. They were leached from ferromagnesian

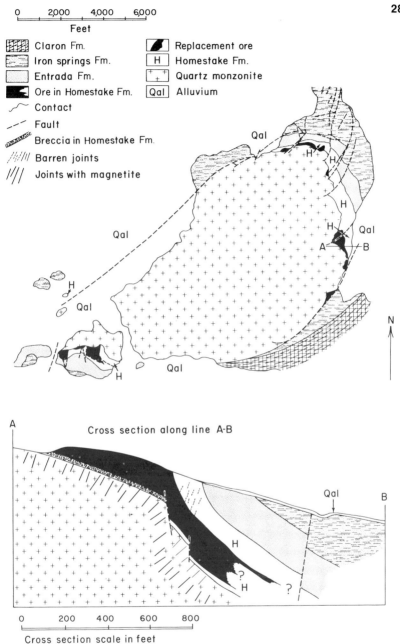

Figure 11-7
Geologic map and cross section of Granite Mountain, Utah. (From Mackin, 1947, Fig. 12.)

minerals during late stages of igneous activity, leaving behind areas of bleaching and alteration. Before the interiors of the laccoliths were completely crystallized, but after the outer zones had solidified enough to support tension cracks, upward surges of the interior crystal mush developed bulges in the laccolith roofs, causing distension and local jointing. The entire history of laccolithic emplacement was a sequence of roof rupturing, rather than a sudden release of the tensile forces across the whole intrusive. The shearing and brecciation that accompanied jointing eased some of the dilative forces. Wherever the peripheral shell was breached by joints or breccia zones, the iron-bearing interior fluids ascended into the overlying limestones, forming replacement deposits. If the ore-bearing fluids could have moved freely out of the laccoliths, a continuous blanket of iron ore probably would have developed. The actual occurrence of ore is sporadic, however, reflecting the fact that only local avenues were available to tap the mineralizers during this critical phase of crystallization.

Remnants of the peripheral shell are found along the eroded tops of laccoliths, suggesting that the original sedimentary cover was not far above the present surface of quartz monzonite. At the time of intrusion and ore deposition the overlying strata were 3,000 to 6,000 feet thick (Mackin, 1954). Thus, the ores were formed at relatively shallow depths for an igneous metamorphic environment. The ore solutions selectively replaced only limestone and bypassed a bed of siltstone, implying a chemical control of ore deposition. If the limestone had not been available, the ore-bearing fluids presumably would have migrated some distance from the laccoliths before precipitating the iron.

Aeromagnetic studies of the Iron Springs district indicate that the three igneous masses of the area are not connected near the surface. The Iron Mountain mass is thought to extend westward a considerable distance, and the Granite Mountain mass extends north, west, and southwest. The Three Peaks body extends east and southeast beneath the alluvium (Blank and Mackin, 1967).

REFERENCES TO SELECTED IGNEOUS METAMORPHIC DEPOSITS

Marysville District, Montana

Barrell, J., 1907. Geology of the Marysville mining district, Montana, *U.S. Geol. Surv. Prof. Pap.* 57.
Knopf, A., 1950. The Marysville granodiorite stock, Montana, *Amer. Mineral.* 35:834-844.
Pardee, J. T., and F. C. Schrader, 1933. Metalliferous deposits of the greater Helena mining region, Montana, *U.S. Geol. Surv. Bull.* 842.

Oslo, Norway

Goldschmidt, V. M., 1911. Die Kontact-metamorphose im Kristianiagebiet (Norway), *Nor. Vidensk. Akad. Oslo Math.-natur-vidensk.-klasse Skrifter* no. 1.

Vogt, J. H. L., 1894. Über die durch pneumatolytische Processe an Granit gebundenen Mineral-Neubildungen, *Z. Prakt. Geol.* 2:458-465.

———, 1895. Beitrage zur genetischen Klassifizierung der durch magmatische Differentiationsprocesse und der durch Pneumatolyse entstandenen Erzvorkommen, *Z. Prakt. Geol.* 3:145-156, 367-370, 444-459, 465-484.

Von Groddeck, A., 1879. *Die Lehre von den Lagerstätten der Erze,* Leipzig.

Suian District, Korea

Higgins, D. F., 1918. Geology and ore deposits of the Collbran contact of the Suian mining concession, Korea, *Econ. Geol.* 13:1-34.

Kato, B., 1910. Journeys through Korea: contrib. 2, the geology and ore deposits of the Hol-gol gold mine, *Imper. Univ. Tokyo Coll. Sci. J.* 27.

Watanabe, T., 1943. Geology and mineralization of the Suian district, Tyosen (Korea), *Hokkaido Imper. Univ. J. Fac. Sci.* (ser. IV) 6(3-4).

REFERENCES CITED

Barrell, J., 1907. Geology of the Marysville mining district, Montana, *U.S. Geol. Surv. Prof. Pap.* 57.

Blank, H. R., Jr., and J. H. Mackin, 1967. Geologic interpretation of an aeromagnetic survey of the Iron Springs district, Utah, *U.S. Geol. Surv. Prof. Pap.* 516-B.

Brock, K. J., 1972. Genesis of Garnet Hill skarn, Calaveras County, California, *Geol. Soc. Amer. Bull.* 83:3391-3404.

Burt, D. M., 1968. Control of oxygen fugacity during ore deposition in some pyrometasomatic zinc deposits, *Geol. Soc. Amer. Abstr. 1968 Mtgs.,* p. 44.

Butler, B. S., 1923. A suggested explanation of the high ferric oxide content of limestone contact zones, *Econ. Geol.* 18:398-404.

———, 1927. Some relations between oxygen minerals and sulphur minerals in ore deposits, *Econ. Geol.* 22:233-245.

Edwards, A. B., and G. Baker, 1953. Scapolitization in the Cloncurry district of northwestern Queensland, *Geol. Soc. Aust. J.* 1:1-33.

Hernon, R. M., and W. R. Jones, 1968. Ore deposits of the central mining district, Grant County, New Mexico, in *Ore Deposits of the United States* (L. C. Graton-R. Sales Mem. Vol.), ed. J. D. Ridge, American Institute of Mining Engineers.

Hernon, R. M., W. R. Jones, and S. L. Moore, 1953. Some geological features of the Santa Rita quadrangle, New Mexico, in *Guidebook of Southwestern New Mexico, 4th Field Conference,* New Mexico Geological Society.

Hess, F. L., 1919. Tactite, the product of contact metamoprhism, *Amer. J. Sci.* 198:377-378.

Hietanen, A., 1967. On the facies series in various types of metamorphism, *J. Geol.* 75:187-214.

Horton, J. S., 1953. Geology of the Hanover mine, *Mining Eng.* 5:1228-1229.

Jones, W. R., R. M. Hernon, and S. L. Moore, 1967. General geology of Santa Rita quadrangle, Grant County, New Mexico, *U.S. Geol. Surv. Prof. Pap.* 555.

Jones, W. R., R. M. Hernon, and W. P. Pratt, 1961. Geologic events culminating in primary metallization in the Central mining district, Grant County, New Mexico, *U.S. Geol. Surv. Prof. Pap.* 424-C, pp. 11–16.

Large, R. R., 1971. Metasomatism and scheelite mineralization at Bold Head, King Island (Australia), *Australas. Inst. Mining Metall. Proc.* 238, pp. 31–45.

Lasky, S. G., and A. D. Hoagland, 1950. Central mining district, New Mexico, *18th Int. Geol. Congr. Rept.*, pt. 7, pp. 97–110.

Leith, C. K., and E. C. Harder, 1908. The iron ores of the Iron Springs district, southern Utah, *U.S. Geol. Surv. Bull.* 338.

Lindgren, W., 1905. The copper deposits of the Clifton-Morenci district, Arizona, *U.S. Geol. Surv. Prof. Pap.* 43.

Lindgren, W., L. C. Graton, and C. H. Gordon, 1910. The ore deposits of New Mexico, *U.S. Geol. Surv. Prof. Pap.* 68, pp. 305–318.

Mackin, J. H., 1947. Some structural features of the intrusions in the Iron Springs district, Utah, in *Guidebook to the Geology of Utah, No. 2,* Utah Geological Society.

———, 1954. Geology and iron ore deposits of the Granite Mountain area, Iron County, Utah, *U.S. Geol. Surv. Mineral Invest. Field Stud. Map* MF 14.

———, 1968. Iron ore deposits of the Iron Springs district, southwestern Utah, in *Ore Deposits of the United States* (L. C. Graton-R. Sales Mem. Vol.), ed. J. D. Ridge, American Institute of Mining Engineers.

Mackin, J. H., and E. Ingerson, 1960. An hypothesis for the origin of ore-forming fluid, *U.S. Geol. Surv. Prof. Pap.* 400-B, pp. 1–2.

Nielson, R. L., 1968. Hypogene texture and mineral zoning in a copper-bearing granodiorite porphyry stock, Santa Rita, New Mexico, *Econ. Geol.* 63:37–50.

Paige, S., 1916. The Silver City folio (New Mexico), *U.S. Geol. Surv. Folio* 199.

Perry, D. V., 1969. Skarn genesis at the Christmas mine, Gila County, Arizona, *Econ. Geol.* 64:255–271.

Schmitt, H. A., 1933. The Central mining district, New Mexico, *Amer. Inst. Mining Eng. Contrib.* 39 (class 1).

———, 1939. The Pewabic mine, *Geol. Soc. Amer. Bull.* 50:777–818.

———, 1942. Certain ore deposits in the Southwest, Central mining district, New Mexico, in *Ore Deposits as Related to Structural Features,* ed. W. H. Newhouse, Princeton, N.J.: Princeton Univ. Press.

Shand, S. J., 1947. The genesis of intrusive magnetite and related ores, *Econ. Geol.* 42:634–636.

Spencer, A. C., and S. Paige, 1935. Geology of the Santa Rita mining area, New Mexico, *U.S. Geol. Surv. Bull.* 859.

Young, W. E., 1947. Iron Springs, Iron County, Utah, *U.S. Bur. Mines Rept. Invest.* 4076.

12 / Hypothermal Deposits

Hypothermal deposits form at high temperatures and high pressures where connection with the surface is impeded. The general temperature range as determined by methods of geothermometry is 300° to 500°C. Textures and structures indicative of replacement are common, whereas vugs and open-fissure fillings, so characteristic of the shallower deposits, are scarce or absent. The rocks may be sheeted or sheared, often so thoroughly that the deposits contain many shadowy fragments of the wall rocks. Most hypothermal ores are coarse-grained, though there are some notable exceptions. For example, the Sullivan mine in British Columbia contains large orebodies of galena and sphalerite with an aphanitic texture. Wall-rock alteration is usually incon-spicuous around hypothermal deposits because the great depth of the environ-ment precludes a significant temperature differential between the ore-bearing fluids and the wall rocks—the veins and their accompanying alteration products grade into the country rock rather than form sharp boundaries.

The most common ore minerals in the hypothermal zone are gold, wolfram-ite, scheelite, pyrrhotite, pentlandite, pyrite, arsenopyrite, löllingite, chalcopy-rite, sphalerite, galena, stannite, cassiterite, bismuthinite, uraninite, and the cobalt and nickel arsenides. Small amounts of fluorite, barite, magnetite, ilmenite, and specularite may be present. Pyrite, probably the most common sulfide in all ore zones, is abundant in hypothermal deposits.

Many minerals of the igneous metamorphic zone continue without interruption into the hypothermal zone. Gangue minerals and products of wall-rock alteration include black tourmaline (schorl), phlogopite, muscovite (commonly stained green with trace amounts of chromium), biotite, zinnwaldite, topaz, apatite, sillimanite, hedenbergite, hornblende, tremolite, actinolite, cummingtonite, the spinels (especially magnetite and gahnite), kyanite, and the feldspars. Many of these minerals are characteristic of igneous metamorphic deposits but are also found in hypothermal veins. Garnets may be present, but more commonly as scattered individual crystals rather than massive bodies as in igneous metamorphic deposits. Similarly, quartz forms in both environments, but it is more abundant and more conspicuous in the vein deposits.

Since hypothermal and mesothermal zones are defined and distinguished by arbitrary criteria, the ore, gangue, and alteration minerals of each show some repetition, especially in the transition environment along the cooler fringes of hypothermal deposits. Gradations between the two have been found in individual mining districts; several ore minerals are common to both, the most notable being chalcopyrite, sphalerite, galena, pyrrhotite, stannite, and uraninite. Alteration minerals commonly grade from high-temperature varieties into typical mesothermal species. In general, the lower-temperature minerals are present in small amounts, forming a zone on the wall-rock side of the hypothermal alteration products. Sericite mica is a good example; it is characteristic of mesothermal alteration zones but is also found along the fringes of many hypothermal deposits. Even chlorite is present in some hypothermal deposits, though it is more common in the epithermal zone. In fact, chlorite is one of the most abundant alteration products at the Homestake mine in South Dakota (Noble, 1950; Noble and Harder, 1948), a deposit whose origin has been well established as hypothermal. Carbonates are widely distributed in many deposits; they are generally ankeritic, but calcite, dolomite, siderite, and rhodochrosite are found locally.

Quartz is probably the most common gangue mineral. In many deposits the quartz is dark gray or bluish gray; in others it may be white or glassy. The color is usually due to myriads of tiny vacuoles and inclusions. At Hog Mountain, Alabama, the quartz contains many fluid inclusions and appears to be under internal stress (Pardee and Park, 1948, pp. 48–49). During mining operations, small razor-sharp chips of this bluish quartz sometimes burst free from the walls and flew violently through the workings. Weathering causes fractures to develop between vacuoles, releasing gases and liquids and creating a sugary texture. Additional weathering removes the sharp edges of individual grains, so that eventually the rock superficially resembles a sandstone (see Fig. 12-1). Within about 50 feet of the surface, the quartz at Hog Mountain contains bluish remnants in a sugary matrix.

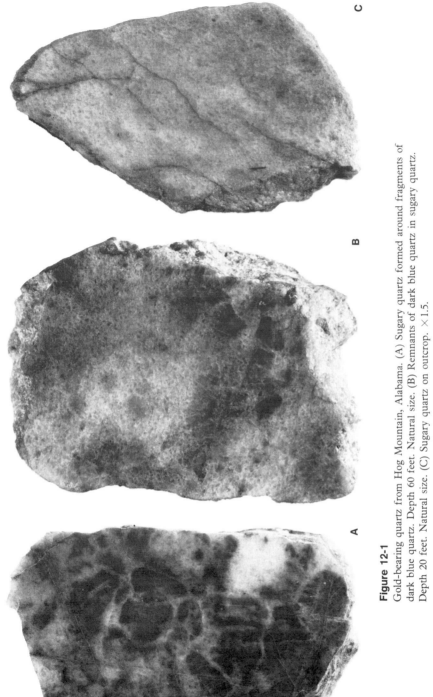

Figure 12-1

Gold-bearing quartz from Hog Mountain, Alabama. (A) Sugary quartz formed around fragments of dark blue quartz. Depth 60 feet. Natural size. (B) Remnants of dark blue quartz in sugary quartz. Depth 20 feet. Natural size. (C) Sugary quartz on outcrop. ×1.5.

As would be expected in deposits formed under high temperatures and pressures, hypothermal orebodies tend to assume irregular shapes. Many deposits, however, are roughly tabular or veinlike. They commonly occupy attenuated crests of folds or shear zones; they also have a tendency to follow drag folds and to replace the country rocks selectively. The development of large, extensive lodes is favored at great depths because here there are no abrupt changes in temperature and pressure. Accordingly, the large persistent veins and replacement masses are characteristically hypothermal. But the reverse is not true; hypothermal deposits may also occur as small pods or stringers. For example, the Kolar gold fields in Mysore, India, have been worked to depths of more than 10,500 feet below the surface, with little or no change in the character of mineralization throughout this range (Bichan, 1947). Deposits such as those of the Southern Piedmont in the United States, however, are small and of limited economic value (see Fig. 4-14, p. 78). They have been described as roots or remnants of more extensive deposits, but the field evidence does not support such a conclusion. The deposits are in schists, which unless thoroughly silicified or otherwise "prepared" should be expected to develop only small, irregular, discontinuous openings under hypothermal conditions. Moreover, the same kinds of deposits are found in the same environment throughout an extensive area.

Ores of the hypothermal zone are deposited at considerable depths and have been brought to the surface only through orogenic processes and erosion. Consequently, these ores are more abundant in metamorphic rocks and rocks of the older geologic periods. They generally are near masses of plutonic igneous rocks, though some large deposits cannot be linked genetically with any particular plutons.

MORRO VELHO, BRAZIL

The Morro Velho mine is an excellent example of a hypothermal gold deposit. It is at the town of Nova Lima, just south of Belo Horizonte, in the state of Minas Gerais, Brazil (Fig. 12-2). Extending 2,454 meters (8,051 feet) below the collar of the shaft, Morro Velho was once the deepest mine in the world (it now is exceeded in depth by several mines in the Witwatersrand district of South Africa and by the Champion Reef in Mysore, India).

The Morro Velho mine has been worked for over two centuries, and has been in continuous operation since 1834. Until recently, however, only a few comprehensive reports of the geology of the district were published. Early reports by Henwood (1884, 1871) and Scott (1903) discuss the history of the mining and the character of the ore. Good short notes on the geology have been published by Derby (1903) and Lindgren (1933), and more complete

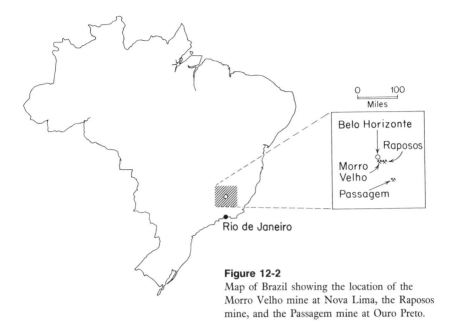

Figure 12-2
Map of Brazil showing the location of the
Morro Velho mine at Nova Lima, the Raposos
mine, and the Passagem mine at Ouro Preto.

articles were published by Oliviera (1933) and Matheson (1956). More recently, Gair (1962) and Tolbert (1964) contributed a great deal to the understanding of the geology. The regional geology has been clarified by the
publication of a map of the Quadrilátero Ferrífero by the Departamento
Nacional da Produçâo Mineral in 1959, and by extensive studies of the surrounding iron ore deposits.

The oldest known rocks at Morro Velho are the Precambrian schists and
phyllites of the Nova Lima group that contain lenses of dolomitic and calcic
white marble, and thin beds of iron formation. Unconformably overlying the
Nova Lima group is the well-known Minas series of younger Precambrian age.
The rocks of the Minas consist of quartzites, schists, phyllites, metaconglom-
erates, and carbonates, and they also include most of the extensive and valuable
iron ore deposits of Minas Gerais (Gair, 1962; Dorr *et al.*, 1957).

Several types of intrusive rock are known in the Morro Velho area. Serpentinite intrudes the Nova Lima rocks and is considered to be older than the
Minas series. Metagabbro is associated with the serpentinite and is also thought
to be of pre-Minas age (Gair, 1962). Metadiabase dikes are common in the
workings of the Espirito Santo and Raposos mines (Fig. 12-3) and elsewhere.
Their precise age is unknown, but Gair says that they are older than post-Minas
series metamorphism. Granitic, gneissic, and pegmatitic rocks are most abundant north of the mine area, though a few exposures have been reported near

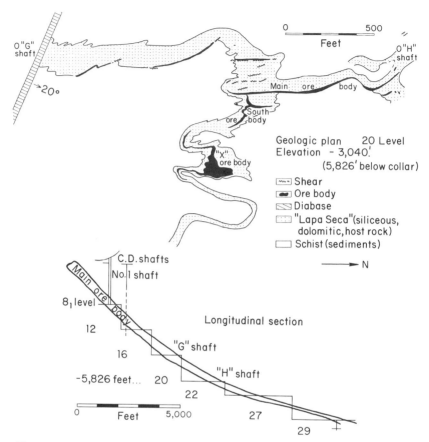

Figure 12-3

Plan and section through the Morro Velho mine, Brazil. (From Matheson, 1956, Fig. 2.)

the mines. Exposures of these rocks are poor, and knowledge of their relationships is slight. The earlier literature reported that the granitic and pegmatitic rocks were older than the Nova Lima Group, but Gair recognized that they intruded into the younger upper-Minas series. The relationship, if any, between the origin of the ore deposits and the granitic-pegmatitic rocks is unknown. The original character of the gneissic rocks is also unknown; their origin may be complex, since both intrusive and sedimentary rocks were present.

The metasediments of the district are tightly folded into eastward plunging structures, which can be readily recognized on maps of the mine levels (see Fig. 12-3). Several inferred faults trend easterly, but have no apparent control or relationship to the mineralization. Regional schistosity trends north-northeast and dips steeply east; near the mines it totally obscures the bedding.

Within the immediate mine area the Nova Lima series consists of meta-morphosed graywackes, shales, sandstones, conglomerates, and discontinuous lenticular beds of quartz-dolomitic or quartz-ankeritic rock that is known locally as Lapa Sêca, or "dry rock." The Lapa Sêca is in schists of similar composition; it grades into the schists both along and across the bedding (Gair, 1962, p. A-23). Lapa Sêca is of special interest because the ores of Morro Velho are restricted to it.

The shape of the main Morro Velho orebody is probably best described as a flattened rod that stands edgewise and plunges eastward with the folded sediments (see Fig. 12-3). Its maximum thickness exceeds 45 feet, and its flat sides average 500 to 600 feet in width. The orebody is continuous from the surface to the deepest workings, its plunge diminishing gradually from about 45° at the top to about 15° at the bottom. This single ribbon of ore has been mined along the plunge for about three miles. Selectively replacing the Lapa Sêca, the lode apparently occupies a sheared or attenuated zone along the axial plane of a tight fold. Its plunge is parallel to the neighboring fold axes and to several smaller orebodies found along the strike of the Lapa Sêca. In many ways the main lode resembles a saddle reef, but not as strikingly so as the associated orebodies, which conform closely to the crests of adjacent folds.

The ore is predominantly quartz containing massive sulfides, of which pyrrhotite is the most abundant. Some of the ore is more than half sulfides. Arsenopyrite and pyrite are conspicuous and widely distributed, and a small amount of chalcopyrite is normally present; the arsenic in arsenopyrite is re-covered as a valuable by-product. Wolframite and schleelite are accessories, and some tetrahedrite, bornite, sphalerite, galena, and stibnite have been re-ported. In addition to abundant quartz, the gangue includes smaller amounts of tourmaline, garnet, kyanite, ankerite, sericite, and albite.

Gold in the Main oreshoot, from which most production has come, averaged about 0.25 to 0.30 ounces per ton; although it is distributed throughout the quartz lode, it is, as in all deposits of this nature, of variable grade. Parts of the lode are comparatively high grade, whereas others are too low to be mined at a profit. Although gold is seldom seen (because it is very finely grained), it is known to be concentrated with the arsenopyrite and pyrrhotite more than with the other sulfides. Although the main orebody is about three miles long, there are no perceptible changes from one end to the other. As expected in ore deposits of hypothermal environment, there is no recognized zoning. Ore from the deepest workings is indistinguishable from that at the surface. Even the arsenic-gold ratio is remarkably constant (Lindgren, 1933). The variations in mineralogy that have been noted can be attributed to differences in the host rocks and not to the distance along the plunge. Studies of the fineness of gold as a function of depth have not been made.

Several additional ore shoots in adjacent folds have been developed (see Fig. 12-3). These shoots, together with branch veins from the Main orebody into the hanging wall, are comparable in grade to the Main lode. They were not mined in the early heyday of the property, so they now furnish ore in the upper levels where they can be mined at a profit.

Since the deposit has been developed along a sequence of vertical winzes and drifts (see Fig. 12-3), the deeper levels have no direct communication with the surface. The lower levels are very dry and dusty and far below the groundwater; water for drilling must be piped down from above. Wall-rock temperatures increase steadily with depth, reaching about 140°F at the bottom of the mine.

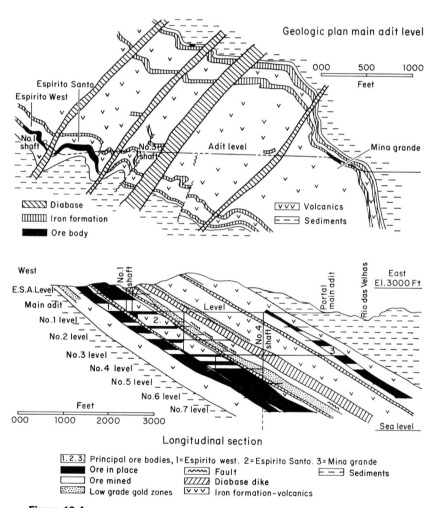

Figure 12-4

Geological plan and section of the Espirito Santo and Raposos mine. (From Matheson, 1956, Figs. 4 and 5.)

Owing to the extremely hot and uncomfortable conditions, the miners must make frequent retreats to the shaft for fresh air and rest. Even at a depth of about 6,600 feet (on the 22nd level) the rock temperature is said to be about 125°F. In order to improve working conditions, 60,000 to 80,000 cubic feet of conditioned air (45°F and 40 percent humidity) was pumped underground every minute. Because of the working difficulties and the high costs of deep mining, operations in recent years have been restricted to the branch veins in the hanging wall of upper levels, above a depth of about 7,000 feet.

About three miles east-northeast of the Morro Velho mine there is another deposit, known as the Espirito Santo and Raposos mine, or simply the Raposos mine. This deposit is similar to Morro Velho, except that the host rock is principally sideritic iron formation instead of sideritic quartzite. Two separate ore shoots are being mined, each resembling a somewhat irregularly dipping, flattened and slightly contorted, rod-shaped body that appears to occupy the nose of a plunging fold (see Figs. 12-4 and 12-5). The workings of the Raposos mine are about 5,000 feet stratigraphically above the workings of the Morro Velho mine, and the orebodies in the two mines are parallel; yet until 1955 no exploration work between them had been attempted. If the orebodies occupy noses of the same fold, or even of nearby folds, the possibility of finding additional ore shoots or saddle reefs between the two known bodies is promising,

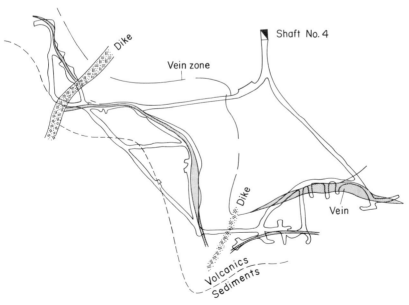

Figure 12-5
Detailed plan of the 1500 (No. 5) level, Espirito Santo mine, Brazil. (Courtesy of A. F. Matheson.)

although diamond drilling since 1955 has failed to disclose ore. Gold values in ores from the Raposos mine are slightly less than those in Morro Velho ores, owing to a smaller sulfide content in the quartz veins. Also, the Raposos ore-bodies are smaller, but they seem to persist in depth. This mine was started in 1934; by 1952 some of the workings had reached 2,500 feet (Matheson, 1956).

Matheson (1956) says that shards in the chlorite schist between the two layers of iron formation indicate a volcanic origin for the schist, but Tolbert (1964) does not mention this possible origin. He calls the rock a chlorite schist and emphasizes the high chromium content. Much of the country rock near the lodes has been changed to chromiferous sericite, which is not found in the altered rocks of the Morro Velho mine. Tolbert (1964, p. 797) states that the wall-rock alteration did not produce mineral assemblages that are characteristic of high temperatures and pressures. Sericite is the main alteration mineral. Yet chromiferous sericite seems to be a common companion of gold in hypothermal and deeper mesothermal deposits, such as the gold deposits of the Southern Piedmont and the Mother Lode of California. Also, the abundant arsenopyrite and pyrrhotite indicate that the deposit should be in the hypothermal zone.

The Morro Velho and Raposos mines have much in common with the Passagem mine near the old town of Ouro Preto, about 45 miles southeast of Belo Horizonte (see Fig. 12-2). This mine is developed on a quartz vein with a mineral suite like that at Morro Velho but containing more tourmaline and other high-temperature minerals. Passagem's quartz vein varies from 5 to 45 feet in thickness, but the wider zones are relatively barren of gold. On the average it is thinner than the Morro Velho lode. Hussak (1898) described the Passagem vein as a pegmatite; Derby (1911) considered it a pegmatite that had been reopened and then invaded by ore-bearing fluids. The deposit is still cited by many geologists as an example of a mineralized pegmatite (Schneiderhöhn, 1941, p. 120). Quartz gangue is predominant within the vein, but isolated crystals and small concentrations of tourmaline, muscovite, oligoclase, and other pegmatite-like silicates are not rare. Since the continuity and structure are more like those of a vein, it seems preferable to classify the deposit as a vein—a vein that possibly represents a gradation into pegmatite.

BROKEN HILL, AUSTRALIA

Another type of hypothermal deposit is illustrated by the Broken Hill district, in westernmost New South Wales (Fig. 12-6). Originally staked (mistakenly as a tin claim) in 1883 it is one of the principal lead-zinc-silver producing areas of the world, exceeding the billion-dollar mark in production over a decade ago. The district produces more than a million tons of ore annually and

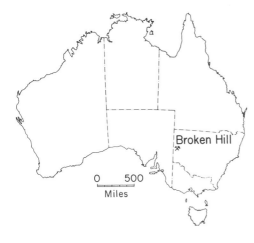

Figure 12-6
Map of Australia showing the
location of the Broken Hill
district, New South Wales.

has maintained a steady, developed ore reserve of 12 or 13 million tons (Andrews, 1948). Average ore yields about 15 percent lead, 12 percent zinc, and five ounces of silver per ton (King and O'Driscoll, 1953). In addition to these metals, the Broken Hill mines annually produce about 60,000 tons of sulfuric acid, about 200 tons of cadmium, and lesser amounts of gold, antimony, copper, and cobalt.

The Broken Hill district has been studied by many able geologists, and excellent reports concerning all phases of the geology are available (Jaquet, 1894; Andrews, 1922; Gustafson *et al.*, 1950; King and Thomson, 1953; King and O'Driscoll, 1953; Stillwell, 1953; Carruthers and Pratten, 1961; Carruthers, 1965; Lewis, Forward and Roberts, 1965). The district is in a semi-arid region of sandy plains and rough, rocky ridges. Highly contorted Precambrian rocks underlie the entire district.

Precambrian rocks are divided into two series, the Willyama and the Torrowangee, which are separated by a widely recognized unconformity. The Willyama Series (probably Lower Precambrian) was originally a thick sequence of sandstones and shales. Only a few stratigraphic members have been sufficiently resistant to metamorphism to serve as marker beds in the mapping. Before (or possibly during) metamorphism, the sediments were intruded by pegmatites and sills of granite and gabbro. Regional metamorphism converted this assemblage to sillimanite-biotite-garnet gneisses, serpentines, itabirites, granulites, quartzites, augen gneisses, amphibolites, and sericite, andalusite, hornblende, and staurolite schists. The pegmatites range from large masses to irregular veinlets and lit-par-lit replacements or injections (Gustafson *et al.*, 1950; King and Thomson, 1953). Andrews (1948) noted that the ore-bearing zone coincides with the centrolineal belt of maximum metamorphic intensity.

Unconformably overlying the Willyama Series is a thick section of Proterozoic sediments that includes claystones, quartzites, shales, limestones, conglomerates, and several glacial deposits and constitute the Torrowangee Series. Locally, near its base, the Torrowangee Series is metamorphosed, but in general post-Willyama metamorphic effects are minor.

Structural interpretation of the region has radically changed in recent years, from the concept of irregular complex basins separated by zones of intense plastic flowage and shearing. The Broken Hill lode is now considered to occupy a small drag fold on the eastern flank of a major syncline that extends throughout the mineral field. This relationship, first proposed by Lewis, is shown in Figure 12-7 with sections from earlier interpretations. The overall plunge of the major syncline is at a low angle to the south. The dominant structural features are thus zones of strong localized isoclinal folding, complicated by intense shearing along bedding planes and faults. Faults, although found in many places, apparently played a minor role during mineralization (Lewis *et al.*, 1965). Deformation was such that the rocks yielded by plastic flow rather than by faulting. The folding was essentially one of shearing adjustments and plastic flow, but without the development of axial cleavage. Folding is also en echelon on both the hand-specimen and regional scales.

The Broken Hill lode strikes north-northeast, parallel to the metamorphosed Willyama sediments. The mineralized zone crops out about three miles along the strike and plunges downward at each end. The Willyama Series consists of tightly folded sillimanite-garnet gneisses that include subordinate thin quartzite beds and numerous folded sills of augen gneiss (granite), amphibolite (gabbro), and pegmatite. Postfolding peridotites, granites, and

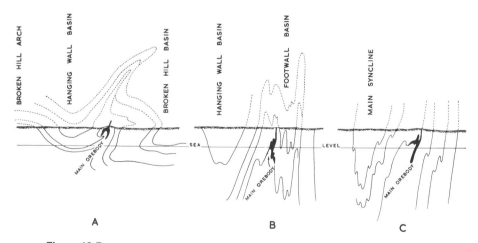

Figure 12-7
Structural interpretation cross sections (looking north), Broken Hill district, New South Wales. (A: From E. C. Andrews, 1922. B: From J. K. Gustafson, 1939. C: From B. R. Lewis, 1957.)

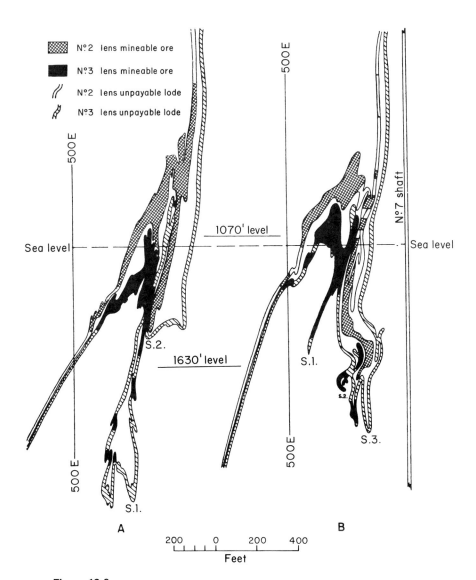

Figure 12-8
Cross sections (looking north) of the Main lode, Broken Hill district, New South
Wales. (A) Cross section at coordinate 2200S; (B) cross section at coordinate 1800S.
(From Lewis *et al.*, 1965.)

pegmatites are also present. Younger pegmatite dikes were intruded after considerable erosion had taken place, and, according to Gustafson and his colleagues (1950), the ore-bearing fluids followed shortly thereafter.

Understanding fold structure is of great value in the mapping of the deposits because the minor folds display marked curves in plunge. As pointed out by Lewis and his associates (1965, p. 324), the peaks of the curves of the folds form at the point of maximum fold development. Cross sections vary markedly in shape, folds develop and fade, and new folds develop in offset positions, but always to the right side of fading members.

Folding in and near the lode is intense and complex (Figs. 12-7 and 12-8); all the folds are deep and isoclinal and have nearly vertical limbs. Gustafson and his colleagues (1950) described second-order folds that crossed main structures, causing abrupt reversals of pitch and divergent plunges in adjacent folds. Vertically plunging buckles and shear zones that cut and offset the major structures were thought to be postfolding and preore fault movements.

The latest version of the lode structure describes two sections that are 400 feet apart; the striking difference between the two is attributed to en echelon folding (Fig. 12-8). Ore shoots in the Main lode occupy complex minor folds; the entire lode is not mineralized, however.

The longitudinal projection of the ore-bearing zone shows an irregular ribbon of ore 2,000 to 3,000-feet high, and of variable thickness and length (Fig. 12-9). Lewis and his colleagues (1965) recognized seven separate en echelon folds on the 1,630-foot level of the Broken Hill South over a strike length of 2,500 feet.

The ore-bearing zone is actually a composite lode of two or more mineralized beds of contorted gneiss, nestled one above the other (Fig. 12-10). Each lode, or lens, is distinguished by details of its gangue mineralogy and

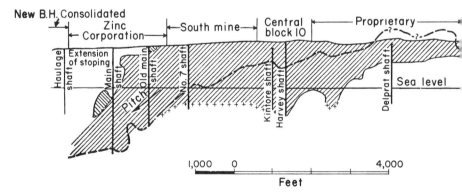

Figure 12-9
Longitudinal projection of Broken Hill lode, New South Wales.
(From Gustafson *et al.*, 1950, Fig. 3.)

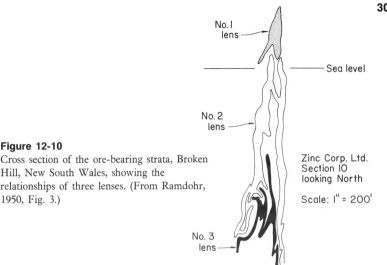

Figure 12-10
Cross section of the ore-bearing strata, Broken Hill, New South Wales, showing the relationships of three lenses. (From Ramdohr, 1950, Fig. 3.)

No. I lens

Sea level

No. 2 lens

No. 3 lens

Zinc Corp. Ltd.
Section 10
looking North

Scale: I" = 200'

metal ratios. Number 2 lens (the upper) is characterized by calcite, bustamite $[(Mn,Ca)SiO_3]$, wollastonite, and both rhodonite and hedenbergite with low refractive indices. By contrast, Number 3 lens (the lower) contains fluorite, garnet, rhodonite, and pyroxmangite $[(Mn,Fe)SiO_3]$, the latter three all distinguished by high refractive indices. Both lodes contain other minerals in common, including garnets with low to intermediate refractive indices, quartz, potash feldspar, apatite, rutile, and species of rare occurrence. Moreover, the lower horizon has relatively high zinc-lead and silver-lead ratios compared to the upper horizon, and rhodonite is much more abundant in the lower lens than in the upper. In the southern part of the district in the mines of the Zinc

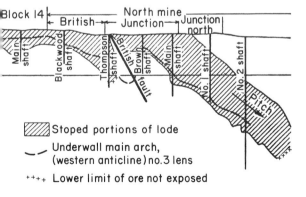

Stoped portions of lode

Underwall main arch,
(western anticline) no.3 lens

++++ Lower limit of ore not exposed

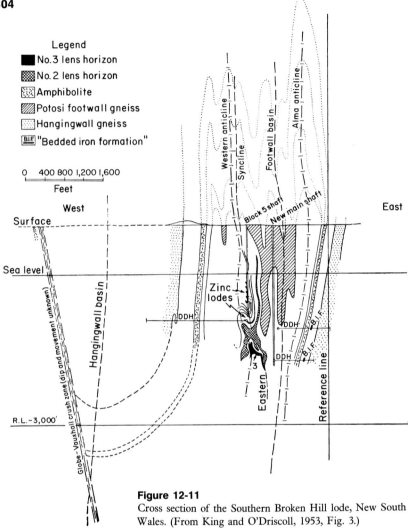

Figure 12-11
Cross section of the Southern Broken Hill lode, New South Wales. (From King and O'Driscoll, 1953, Fig. 3.)

Corporation and the New Broken Hill Consolidated as many as six layers of ore, known as the Zinc Lodes, have been developed (Carruthers, 1965; see Fig. 12-11).

Broken Hill ore is mined principally for sphalerite and galena, but it also contains tetrahedrite, pyrrhotite, marcasite, chalcopyrite, arsenopyrite, löllingite, dyscrasite (Ag_3Sb), pyrargyrite, gudmundite (FeSbS), and cubanite ($CuFe_2S_3$). Stillwell (1953) reported small amounts of many other minerals, including wolframite, scheelite, molybdenite, cobaltite, pyrite, stannite, breithauptite (NiSb), niccolite, bornite, jamesonite, native gold, and native antimony. This mineral assemblage appears to represent formation over a wide temperature range. Arsenopyrite, pyrrhotite, wolframite, and cobaltite are typical hypothermal minerals; tetrahedrite, bornite and jamesonite are usually found in mesothermal deposits; and pyrargyrite and marcasite typify the epithermal environment. An unexplained peculiarity of the mineralization is the presence in white quartz of a few shotlike spheroids of galena ($\frac{1}{8}$ to $\frac{1}{4}$ inch in diameter).

The paragenesis of these minerals is obscure, thus complicating the problem further (King and O'Driscoll, 1953). Petrographic studies indicate that the sulfides followed the principal gangue minerals and that pyrrhotite and arsenopyrite were deposited before zinc, lead, and silver minerals, but Ramdohr (1950, 1953) suggested that the mineral relationships are due to postdepositional recrystallization during the period of metamorphism. If Ramdohr is correct, the ores may represent a metamorphosed epithermal deposit or, as King and Thomson (1953) advocate, a metamorphosed syngenetic deposit. Geothermometry cannot solve this controversy because the ores would indicate high temperatures whether they were hypothermal or metamorphic. If the deposit were hydrothermal, the presence of chalcopyrite-cubanite exsolution textures would indicate a temperature of deposition above 400° to 450°C (Gustafson et al., 1950). What has been interpreted to be partial fusion of antimony would indicate a temperature of about 630°C (Ramdohr, 1950) if the antimony was altered during metamorphism. The low-temperature minerals, such as pyrargyrite and marcasite, may be supergene; if they are, the problem is greatly simplified.

Stanton (1972) discusses the relationship of the ore to the iron formation that seems everywhere to accompany it. He assumes as a working hypothesis that both the ores and the iron formation are of volcanogenic origin.

Although each lens has its own characteristic mineral assemblage, the ore deposits are in general remarkably uniform throughout the entire length of the lode. Differences between the lenses seem to be related to original variations in host rock rather than to zoning effects. A possible hypogene zonation is reflected in slightly greater concentrations of both pyrrhotite and arsenopyrite at the south end of the lode, suggesting that the ore bearing fluids migrated up the pitch from the south (Gustafson et al., 1950).

Wall-rock alteration is not widespread, nor is it uncommon to find ore against relatively unaltered gneiss; however, some of the gneiss around the lodes has been silicified, garnetized, or sericitized. Bleached zones are left in the walls where biotite and sillimanite have been replaced by sericite. The most extensive alteration is replacement by silica, and where silicification has been thorough, the gneisses look like fine-grained quartzites (Gustafson et al, 1950).

The genesis of Broken Hill ore has been debated by two strongly opposed schools, one advocating a hydrothermal origin and the other a syngenetic origin. According to the hydrothermal theory, the ore was emplaced by hot fluids that moved up the dip along the crests of folds, replacing two or more favorable stratigraphic horizons; folding and metamorphism are believed to antedate the ore. The alternative hypothesis is that the ores were deposited as chemical sediments and were subsequently folded and recrystallized during metamorphism. Even the syngenetic ore would show some replacement textures because metamorphism would mobilize or at least force it to flow in plastic fashion from the limbs to the crests of the folds. Both hypotheses are supported by careful geologists who are thoroughly acquainted with the area, and neither hypothesis can answer all the questions.

The syngenetic hypothesis first proposed by King and Thomson (1953) was relatively novel for the Broken Hill lode. Whether advocating igneous or metamorphic processes, sponsors of earlier theories had all argued that the ore is genetically related to the granites and pegmatites. But King and Thomson were impressed by the persistence of Number 2 and Number 3 Lenses, which contain dissimilar mineral assemblages in spite of being stratigraphically close. They suggested that the selective replacement of two favorable strata by two different ore fluids (as reflected in the Zn:Ag:Pb ratios) along $3\frac{1}{2}$ miles of highly contorted rocks would not be so perfect. The transmission of ore-bearing fluids along the lode presents a further problem, in that the mechanism of transport for such large quantities of solutions along a three-mile replacement lode is difficult to envisage. Consequently, King and Thomson suggested that the ores were deposited with the Willyama sediments and were later folded, metamorphosed, and granitized. This interpretation has been supported by radiometric age determinations based upon the lead isotope ratios of the galenas, which indicate that most of the lead was formed 1,400 to 1,600 million years ago and did not experience a complex geochemical history prior to deposition (Russell *et al.,* 1957; Russell and Farquhar, 1960). Some of the ores, however, give anomalous ages, and hence support the hypothesis that at least part of the mineralization is epigenetic.

Most advocates of syngenesis are impressed by the continuity of individual layers and the fact that compositions of these layers do not change. Richards (1966) clearly states his thesis that geometrical and constitutional relationships between the orebodies, garnetiferous and manganiferous quartzites, and itabirites, "suggest that the three rock types are cogenetic." He states further that since two of these types, the garnet- and manganese-bearing quartzites and the itabirites, are metamorphosed sedimentary beds, then the third type, which is the ore bed, is also likely to be of sedimentary origin. Under the syngenetic hypothesis, intense metamorphism and deformation account for many of the features that appear to contradict the sedimentary hypothesis.

Gustafson (1954) pointed out the difficulties inherent in hypothesizing a syngenetic origin, not the least of which is explaining the source of the metals. He argued against a premetamorphism age for the mineralization because the ore and gangue minerals show no effects of differential stress. Petrofabric studies of the quartz gangue support Gustafson's contention that ore deposition took place after metamorphism (den Tex, 1958). Spry (1962) restudied the work of den Tex and pointed out alternative conclusions. He stated that petrofabric studies cannot be used as definite evidence concerning the origin of the ores. It has also been pointed out that the sulfides have replaced some postmetamorphism dikes. Away from the ore the dikes are fresh, cutting clearly across the metamorphosed rocks; but in at least one mine the dikes are uralitized and mineralized (Stillwell and Edwards, 1956).

The advocates of a hydrothermal origin feel that ore deposition was controlled by wall-rock chemistry as well as structure. Attenuation along the crests of folds having left these zones relatively permeable, they acted as conduits for ore-bearing fluids migrating up the plunge. Further localization was caused during the period of ore deposition by repeated fracturing in the favorable strata (Gustafson *et al.*, 1950). Crumpling of the weak beds between layers of stronger rocks produced irregular, saddle-shaped structures that nearly parallel the original bedding. According to the hydrothermal theory, the ores are replacements of the favorable beds along this zone of maximum deformation.

The many factors bearing on the genesis of the ores have been reviewed by Lewis and his colleagues (Lewis *et al.*, 1965). These include the ore mineralogy, the conformity of the ore bodies with the wall rocks, the rock alteration associated with the ores, the stratigraphic distribution, the ore localization on a system of complex folds, the metal ratios of the different beds, the similarity with the nearby cross-cutting Consols Lode, the relations with dolerite dikes, the similarity in age of the country rock and the ore, the presence of gahnite (which is attributed to magmatic and metamorphic origin), the regional metamorphic zoning, the relationships to sedimentary structures, the lack of an obvious igneous rock source, and the distribution as it relates to itabirite. Although Lewis and his associates favor an epigenetic origin for the ores, they state that the origin of the ores is unsettled, and that all of the evidence "dissolves into ambiguity on careful examination." On the other hand, Carruthers (1965), using the same evidence, states that the mineralized layers were originally sedimentary members.

Until a source for the metals can be demonstrated, the genesis of Broken Hill ores will be continuously debated. No satisfactory source for the tremendous quantity of metals in the deposits has been demonstrated. Everyone agrees that the country rocks have been subjected to a high degree of metamorphism, but not everyone will agree that the ores have been through the same metamorphic processes. The criteria for the recognition of metamorphosed ores are deficient. The suggestion that the ores may be of volcanogenic origin must also be considered.

The minerals are predominantly of high-temperature and high-pressure origin, and are of the types usually associated with hypothermal deposits. Whether the ores will ultimately prove to be of igneous, volcanogenic, or sedimentary origin makes no difference in the present mineral suite; the deposit has the characteristics of the hypothermal zone and is so classified.

REFERENCES TO SELECTED HYPOTHERMAL DEPOSITS

Homestake Mine, Lead, South Dakota

Gustafson, J. K., 1933. Metamorphism and hydrothermal alteration of the Homestake gold-bearing formation, *Econ. Geol.* 28:123-162.

McLaughlin, D. H., 1931. The Homestake enterprise: ore genesis and structure, *Eng. Mining J.* 132:324-329.

————, 1949. The Homestake mine, *Can. Mining J.* 70(12):49-53.

Noble, J. A., 1950. Ore mineralization in the Homestake gold mine, Lead, South Dakota, *Geol. Soc. Amer. Bull.* 61:221-251.

Noble, J. A., and J. O. Harder, 1948. Stratigraphy and metamorphism in a part of the northern Black Hills and the Homestake mine, Lead, South Dakota, *Geol. Soc. Amer. Bull.* 59:941-975.

Noble, J. A., and A. L. Slaughter, 1949. Structure of a part of the northern Black Hills and the Homestake mine, Lead, South Dakota, *Geol. Soc. Amer. Bull.* 60:321-352.

Paige, S., 1923. The geology of the Homestake mine, *Econ. Geol.* 18:205-237.

Rye, D. M., and R. O. Rye, 1974. Homestake gold mine, South Dakota, I: stable isotope studies, *Econ. Geol.* 69:293-317.

Slaughter, A. L., 1968. The Homestake mine, in *Ore Deposits of the United States* (L. C. Graton-R. Sales Mem. Vol.), ed. J. D. Ridge, American Institute of Mining Engineers.

Wright, L. B., 1937. Gold deposition in the Black Hills of South Dakota and Wyoming, *Amer. Inst. Mining Eng. Trans.* 126:390-425.

Kolar Gold Field, Mysore, India

Bichan, W. J., 1947. Structural principles controlling the occurrence of ore in the Kolar gold field, *Econ. Geol.* 42:93-136.

Dougherty, E. Y., 1939. Some geological features of Kolar, Porcupine, and Kirkland Lake, *Econ. Geol.* 34:622-653.

Emmons, W. H., 1937. *Gold Deposits of the World*, New York: McGraw-Hill, pp. 378-382.

LeGraye, M., 1942. *Origine et Formation des Gisements d'Or*, Liege, pp. 74-76.

Narayanaswami, S., M. Ziauddin, and A. V. Ramachandra, 1960. Structural control and localization of gold-bearing lodes, Kolar gold field, India, *Econ. Geol.* 55: 1429-1459.

Pryor, T., 1923. The underground geology of the Kolar gold field, *Inst. Mining Metall. London Trans.* 33:95-135.

Smeeth, W. F., and P. S. Iyengar, 1916. Mineral resources of Mysore, *Mysore Dept. Mines Geol. Bull.* 7, pp. 1-55.

Smith, A. M., 1904. The geology of the Kolar gold field, *Inst. Mining Metall. London Trans.* 13:152-180.

Tufty, B., 1957. Kolar gold field mines, rock bursts feature at all levels, plastic flow at great depths, *Can. Mining J.* 78(11):87-91.

REFERENCES CITED

Andrews, E. C., 1922. The geology of the Broken Hill district, *Geol. Surv. N.S.W. Mem. Geol.* no. 8.

————, 1948. Geology of Broken Hill, New South Wales, *18th Int. Geol. Congr. Rept.*, pt. 7, pp. 117-122.

Bichan, W. J., 1947. Structural principles controlling the occurrence of ore in the Kolar gold field, *Econ. Geol.* 42:93–136.

Carruthers, D. S., 1965. An environmental view of Broken Hill ore occurrences, *8th Commonw. Mining Metall. Congr. Trans.* 1:339–351.

Carruthers, D. S., and R. D. Pratten, 1961. The stratigraphic succession and structure in the Zinc Corporation Ltd. and New Broken Hill Consolidated Ltd., Broken Hill, New South Wales, *Econ. Geol.* 56:1088–1102.

Den Tex, E., 1958. Studies in comparative petrofabric analysis: the Broken Hill lode and its immediate wall rock, in *F. L. Stillwell Anniversary Volume,* Melbourne: Australasian Institute of Mining and Metallurgy.

Departamento Nacional da Produçao Mineral, 1959. *Esbôço Geológico do Quadrilátero Ferrífero de Minas Gerais, Brasil* (pub. especial), Departmento Nacional da Produçao Mineral, Brazil.

Derby, O. A., 1903. Notes on Brazilian gold-ores, *Amer. Inst. Mining Eng. Trans.* 33: 282–287.

———, 1911. On the mineralization of the gold-bearing lode of Passagem, Minas Geraes, Brazil, *Amer. J. Sci.* 32:185–190.

Dorr, J. V. N., II, J. E. Gair, J. B. Pomerene, and G. A. Rynearson, 1957. Revision of stratigraphic nomenclature in the quadrilatero ferrifero, Minas Gerais, Brazil, *Minist. Agric. Div. Fom. Prod. Mineral Avulso* 81, pp. 1–31.

Gair, J. E., 1958. Age of gold mineralization at the Morro Velho and Raposos mines, Minas Gerais, *Soc. Bras. Geol. Bol.* 7(2):39–45.

———, 1962. Geology and ore deposits of the Nova Lima and Rio Acima quadrangles, Minas Gerais, Brazil, *U.S. Geol. Surv. Prof. Pap.* 341-A.

Gustafson, J. K., 1954. Discussion: geology of Australian ore deposits, Broken Hill, *Econ. Geol.* 49:783–786.

Gustafson, J. K., H. C. Burrell, and M. D. Garretty, 1950. Geology of the Broken Hill ore deposit, N.S.W., Australia, *Geol. Soc. Amer. Bull.* 61:1369–1437.

Henwood, W. J., 1844. Descriptive notice of the Morro Velho mine, province of Minas Geraes, and on the relations between the structure of the containing rocks and the directions of the shoots of gold in Brazilian mines, *Roy. Geol. Soc. Cornwall Trans.* 6:143–146.

———, 1871. On the gold-mines of Minas Geraes, in Brazil, *Roy. Geol. Soc. Cornwall Trans.* 8(pt. 1):168–370.

Hobbs, B. E., D. M. Ransom, R. H. Vernon, and P. F. Williams, 1968. The Broken Hill orebody, Australia: a review of recent work, *Mineralium Deposita* 3:293–316.

Hussak, E., 1898. Der goldführende, kiesige Quarzlagergang von Passagem in Minas Geraes, Brasilien, *Z. Prakt. Geol.* 6:345–357.

Jaquet, J. B., 1894. Geology of the Broken Hill lode and Barrier Ranges mineral field, *Geol. Surv. N.S.W. Mem.* 5.

King, H. F., and E. S. O'Driscoll, 1953. The Broken Hill lode, in *Geology of Australian ore Deposits,* ed. A. B. Edwards, Melbourne: Australasian Institute of Mining and Metallurgy.

King, H. F., and B. P. Thomson, 1953. The geology of the Broken Hill district, in *Geology of Australian Ore Deposits,* ed. A. B. Edwards, Melbourne: Australasian Institute of Mining and Metallurgy.

Lewis, B. R., P. S. Forward, and J. B. Roberts, 1965. Geology of the Broken Hill lode, reinterpreted, *8th Commonw. Mining Metall. Congr. Trans.* 1:319–332.

Lindgren, W., 1933. *Mineral Deposits,* 4th ed., New York: McGraw-Hill, pp. 676–677.

Matheson, A. F., 1956. The St. John del Rey Mining Company, Limited, Minas Geraes, Brazil, *Can. Mining Metall. Bull.* 77:1–7.

Noble, J. A., 1950. Ore mineralization in the Homestake gold mine, Lead, South Dakota, *Geol. Soc. Amer. Bull.* 61:221-252.

Noble, J. A., and J. O. Harder, 1948. Stratigraphy and metamorphism in a part of the northern Black Hills and the Homestake mine, Lead, South Dakota, *Geol. Soc. Amer. Bull.* 59:941-975.

Oliviera, E. P. de, 1933. Jazida de ouro de Morro Velho, Minas Geraes, Brasil, *Acad. Brasil Sci. Ann.* 5(1):21-40.

Pardee, J. T., and C. F. Park, Jr., 1948. Gold deposits of the Southern Piedmont, *U.S. Geol. Surv. Prof. Pap.* 213.

Ramdohr, P., 1950. Die Lagerstätte von Broken Hill in New South Wales im Lichte der neuen geologischen Erkenntnisse und erzmikroskopischer Untersuchungen, *Hedelberger Beitr. Min. Pet.* 2:291-333.

————, 1953. Über Metamorphose und sekundäre Mobilisierung, *Geol. Rundsch.* 42 (1):11-19.

Richards, S. M., 1966. The banded iron formations at Broken Hill, Australia, and their relationships to the lead-zinc orebodies, *Econ. Geol.* 61:72-96, 257-274.

Russell, R. D., and R. M. Farquhar, 1960. Dating galenas by means of their isotopic constitutions—II, *Geochim. Cosmochim. Acta* 19:41-52.

Russell, R. D., and J. E. Hawley, 1957. Isotopic analyses of leads from Broken Hill, Australia, with spectrographic analyses, *Amer. Geophys. Union Trans.* 38:557-565.

Schneiderhöhn, H., 1941. *Lehrbuch der Erzlagerstättenkunde*, Jena: Gustav Fischer.

Scott, H. K., 1903. The gold-field of the state of Minas Geraes, Brazil, *Amer. Inst. Mining Eng. Trans.* 33:406-444.

Spry, A., 1962. Discussion on studies in comparative petrofabric analysis: the Broken Hill lode and its immediate wallrock, by E. den Tex, *Australas. Inst. Mining Metall. Proc.* 204, pp. 185-196.

Stanton, R. L., 1972. A preliminary account of chemical relationships between sulfide lode and "banded iron formation" at Broken Hill, New South Wales, *Econ. Geol.* 67:1128-1145.

Stillwell, F. L., 1953. Mineralogy of the Broken Hill lode, in *Geology of Australian Ore Deposits*, ed. A. B. Edwards, Melbourne: Australasian Institute of Mining and Metallurgy.

Stillwell, F. L., and A. B. Edwards, 1956. Uralite dolerite dykes in relation to the Broken Hill lode, *Australas. Inst. Mining Metall. Proc.* 178, pp. 213-232.

Tolbert, G. E., 1964. Geology of the Raposos gold mine, Minas Gerais, Brazil, *Econ. Geol.* 59:775-798.

13 / Mesothermal Deposits

Mesothermal deposits are formed at moderate temperatures and pressures. According to Lindgren's classification, the ores are deposited at about 200° to 300°C from solutions that probably have at least a tenuous connection with the surface. In effect, the mesothermal zone has characteristics of both hypothermal and epithermal zones; it is an intermediate rather than a distinctive zone. Probably no single mineral is diagnostic of the mesothermal zone—each can be found elsewhere—but the absence of typically hypothermal or typically epithermal minerals is an important character of mesothermal deposits. Although most mesothermal deposits show abundant replacement phenomena, the textures are not definitive because some vugs and open-cavity fillings are commonly present. The ores are emplaced in many environments and in numerous forms. The host rock may be igneous, metamorphic, or, most commonly, sedimentary. The disseminated, or "porphyry," copper deposits are considered mesothermal. Veins and pipes are common, and mantos and irregular replacement bodies may develop where they cut carbonate rocks. Many deposits have sharp boundaries; others grade into the country rock. Veins commonly develop ribbon structures that layer parallel to the walls and

312

Figure 13-1
Ribbon rock. Hamme tungsten mine, North Carolina. Black is hubnerite; white is quartz.

form by partial replacement of host rock along repeatedly open shears (see Fig. 13-1).

The most abundant products of mesothermal deposits are copper, lead, zinc, molybdenum, silver, and gold. Among the more characteristic ore minerals are chalcopyrite, enargite, bornite, tetrahedrite, tennantite, sphalerite, galena, and chalcocite, as well as many less common minerals. Gangue minerals include quartz, pyrite, and the carbonates; a typical copper-lead-zinc deposit may include all these minerals and others.

Extensive alteration zones surround many mesothermal deposits; alteration products include sericite, quartz, calcite, dolomite, pyrite, orthoclase, chlorite, and clay minerals. Secondary orthoclase and clay minerals have been recognized in many disseminated copper deposits. Some of these minerals, such as chlorite and the clays, are more characteristic of epithermal zones, but they are also common along the outer fringes of mesothermal deposits. Lindgren (1933, p. 530) pointed out that mesothermal deposits do not contain garnet, topaz, pyroxenes, amphiboles, or tourmaline, which are high-temperature minerals, nor do they contain the zeolites, which are stable only at lower temperatures.

Many mesothermal deposits are closely related to igneous rocks, both spatially and genetically; for others, the genetic association is unclear. The copper deposits of Matahambre, Cuba, for example, are $2\frac{1}{2}$ miles from the nearest outcrops of igneous rocks (Pennebaker, 1944). Other mesothermal ores, such as disseminated copper deposits, are in the upper cupolas of quartz monzonite or granodiorite masses.

The wide variety of mesothermal deposits makes it difficult to characterize this zone with a few examples. Nevertheless, most ores are in veins or replace favorable beds, or they permeate the rocks as disseminated veinlets and spots. The Coeur d'Alene district, Idaho, is an example of vein deposits; the Magma mine, Superior, Arizona, is an example of both vein and replacement ore; and Chuquicamata, Chile is an example of a disseminated copper deposit.

COEUR D'ALENE, IDAHO

The Coeur d'Alene district in northern Idaho (Fig. 13-2), one of the largest lead-zinc-silver districts in the world, is an excellent example of a mesothermal deposit. It has produced more than two billion dollars worth of ore since it was discovered in 1879. In the past most of the ore was lead-zinc-silver, but the recent discovery of tetrahedrite deposits in the southern part of the district have added copper and antimony to the metals produced in quantity. Several good descriptions of the deposits are available (Ransome and Calkins, 1908; Umpleby and Jones, 1923; McKinstry and Svendsen, 1942; Sorenson, 1947,

Figure 13-2
Map of Idaho showing the location of the Coeur d'Alene district.

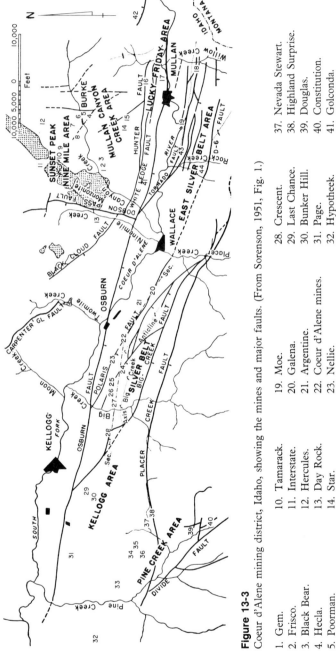

Figure 13-3
Coeur d'Alene mining district, Idaho, showing the mines and major faults. (From Sorenson, 1951, Fig. 1.)

1. Gem.	19. Moe.	37. Nevada Stewart.
2. Frisco.	20. Galena.	38. Highland Surprise.
3. Black Bear.	21. Argentine.	39. Douglas.
4. Hecla.	22. Coeur d'Alene mines.	40. Constitution.
5. Poorman.	23. Nellie.	41. Golconda.
6. Tiger.	24. Silver Summit.	42. Snowstorm.
7. Sherman.	25. Chester.	43. Gem State.
8. Standard Mammoth.	26. Polaris.	44. Rock Creek.
9. Custer.	27. Sunshine.	
10. Tamarack.	28. Crescent.	
11. Interstate.	29. Last Chance.	
12. Hercules.	30. Bunker Hill.	
13. Day Rock.	31. Page.	
14. Star.	32. Hypotheek.	
15. Morning.	33. Liberal King.	
16. Gold Hunter.	34. Denver.	
17. Lucky Friday.	35. Sidney.	
18. Atlas.	36. Pittsburg.	

1951; Shenon, 1948; Fryklund, 1964; Hobbs *et al.*, 1965; Hobbs and Fryklund, 1968).

The Coeur d'Alene mines extend over an area spanning 25 miles from east to west and 10 miles from north to south (Fig. 13-3). At the northern extreme the mines were worked for gold in the early days. Near the center the principal commodities recovered are lead and zinc; the southern edge of the district contains copper-silver ore. Coeur d'Alene is thus an example of district zoning.

Three groups of rocks underlie the district: the Upper Precambrian sediments of the Belt Series, igneous rocks that intrude the Belt Series, and unconsolidated Tertiary and Quaternary clastics. The Belt Series consists principally of quartzites, argillites, and calcareous rocks, reaching a maximum thickness of 20,000 feet or more. Abundant mud cracks, ripple marks, cross-bedding, and stromatolites indicate that most of this sequence is of shallow-water origin. The Belt Series has been divided into six lithologic units. They are, in chronologic order: the Prichard Formation, 12,000 feet of black argillite and argillaceous quartzite; the Burke Formation, an 1,800 to 2,400-foot sequence of gray to white quartzite, some of which is argillaceous; the Revett Formation, 2,100 to 3,400 feet of massive gray to white quartzite with some interbedded argillaceous quartzite; the St. Regis Formation, consisting of thin-bedded purple to purplish-gray argillite and quartzite; the Wallace Formation, 4,500 to 6,000 feet of gray-black calcareous argillite and white to gray calcareous quartzite; and the Striped Peak Formation, which comprises 1,500 feet or more of purple, pink, and green quartzite and calcareous argillite, and apparently contains no ore (Sorenson, 1951).

During the Laramide orogeny the Coeur d'Alene district was strongly folded and faulted. Monzonite stocks were intruded after folding, accompanied by diabase and lamprophyre dikes and sills (McKinstry and Svendsen, 1942). The stocks, which vary from diorite and quartz monzonite to syenite, appear to be satellitic to the Idaho batholith (Umpleby and Jones, 1923). At least one diabase sill predates the monzonite, and even the folding; but the lamprophyre dikes, and probably most of the diabase dikes, are younger than the monzonite.

In spite of its age and the degree of folding, the Belt Series has undergone surprisingly little metamorphism. The argillites have a slaty cleavage, and the arenites are mostly quartzitic, but in general the metamorphic effects are minor. Nevertheless, the sediments around the borders of some monzonite stocks show igneous metamorphic alteration in the form of bleaching, sericitization, and even granitization (Anderson, 1949).

As shown in Figure 13-3, the district is cut by several west-northwesterly faults, the most pronounced being the great Osburn fault. This structure can be traced well beyond the Coeur d'Alene district. It dips 55° to 65° south and has 15,000 feet of normal, dip-slip offset (Shenon, 1948), with possibly 16 miles of right lateral strike-slip displacement (Hobbs and Fryklund, 1968).

The actual fault zone is 100 to 200 feet wide and consists of sheared, brecciated, and powdered rock. South of the Osburn fault, there are many parallel and oblique faults, which appear to be second-order structures caused by the Osburn faulting. All the second-order faults dip south. Tensional stresses set up by movements along these major structures developed local openings through which the mineralizing fluids ascended (Sorenson, 1951). The major structural elements differ on opposite sides of the Osburn fault. To the south, the fold axes trend east-west, or parallel to the faults; but north of the Osburn fault, both the folds and the faults strike north to northwest, parallel to a broad anticlinal arch heading toward the Idaho batholith.

Most of the ore deposits lie along fractures and shear zones in the Belt Series. Minor porous fractures that contain only small amounts of gouge are more favorable for mineralization than are the faults with larger displacements. Minor mineralization in the monzonite stocks and diabase dikes proves their preore age, but the lamprophyre dikes were emplaced after the ore. Concentrations of ore lie along areas of strong structural deformation and in the brittle, brecciated rock units (Sorenson, 1951; McKinstry and Svendsen, 1942). Where the rocks were brittle and the bedding planes were oblique to the directions of shear, open-fissure zones were produced. Thus, the quartzites are more favorable hosts than the argillites; for example, in the Polaris mine many fissures are mineralized in the St. Regis Formation but not along their continuations in the Wallace Formation (Fig. 13-4).

The ore minerals include galena, sphalerite, tetrahedrite, chalcopyrite, pyrrhotite, magnetite, arsenopyrite, and minor amounts of bornite, chalcocite, stibnite, boulangerite, bournonite, gersdorffite ($NiAsS$), scheelite, and uraninite. Arsenopyrite is considered an indicator of economically favorable areas because it forms envelopes around ore shoots (Mitcham, 1952a). Quartz, siderite, other carbonates, pyrite, and, locally, barite are the principal gangue minerals. These ore and gangue minerals form a typical mesothermal assemblage. Geothermometric studies have been somewhat inconclusive, but they indicate the general mesothermal temperature range (Fryklund and Fletcher, 1956). Hobbs and Fryklund (1968) state that depth-temperature classification has no meaning in this district, probably because the mineralization ranges from the epithermal zone through the hypothermal zone. But this is true of nearly all major mesothermal mining districts. Since most of the deposits seem to fall within the mesothermal range, the district is regarded as mesothermal.

The lead-zinc ore averages 10 percent combined lead and zinc and about one ounce of silver per unit of lead (Shenon, 1948). This does not mean, however, that the silver is in the galena; actually, it is in the tetrahedrite, which in turn tends to be associated with galena (Warren, 1934).

The mineral paragenesis is difficult to establish for the district as a whole.

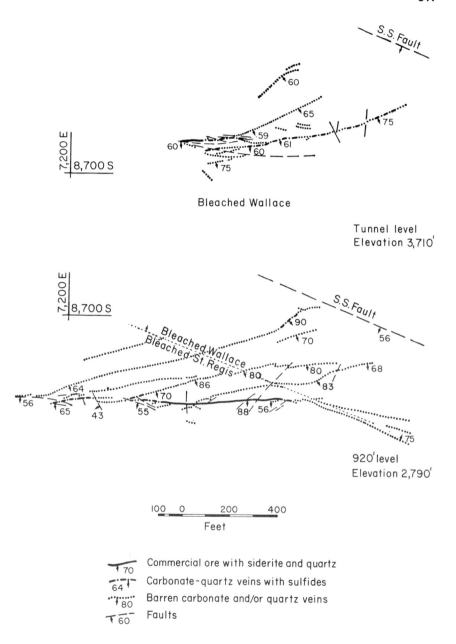

Figure 13-4
Plan of the 920-foot and tunnel levels of the Polaris mine, Coeur d'Alene district, Idaho.
(From Sorenson, 1951, Fig. 3.)

In general, the gangue minerals were deposited early, and most of them continued to be deposited throughout the sulfide phase. Magnetite, pyrrhotite, and arsenopyrite were among the earliest ore minerals; tetrahedrite, galena, and sphalerite were intermediate; and chalcopyrite was one of the last to be deposited (Hosterman, 1956; Anderson, 1940; Ransome and Calkins, 1908). The accessory minerals have not been included in this paragenetic sequence because their order of deposition is not clear; indeed, the paragenesis of even the principal minerals is questionable. A few of the accessory minerals apparently formed as reaction products between two other compounds. For example, bournonite seems to represent an intermediate step in the replacement of tetrahedrite by galena (Anderson, 1940). Much of the mineralization lies in the altered (bleached) country rock that trends parallel to anticlinal axes and faults, and has generally been a good guide to ore.

Hobbs and Fryklund (1968, p. 1,431) attribute the bleaching to presulfide solutions, and state that some orebodies are independent of bleaching. Geologists in the past attributed the bleaching to potash metasomatism and the alteration of argillaceous materials to sericite, but more recent studies have shown that the unaltered argillites contain as much sericite as the bleached rock (Anderson, 1949; Sorenson, 1951; Mitcham, 1952a, 1952b). Besides the controversial sericite, the bleached zones contain disseminated pyrite, quartz, carbonates, and in places chlorite.

The Coeur d'Alene ores have long been considered Laramide, but lead-isotope studies indicate either that the principal metallization took place during Precambrian times or that the galena was derived from a lead source of Precambrian age (Long *et al.*, 1960; Doe, 1973). According to proponents of the Laramide origin, monzonite stocks invaded the previously folded Belt Series, and subsequent hydrothermal fluids moved along the major faults that were repeatedly opened as the deeper parts of the source batholith crystallized. Within this deeper source, the late magmatic fractions contained concentrations of potassium, lead, zinc, copper, silver, gold, iron, sulfur, arsenic, and other elements. The earliest phase of mineralization took place within the pluton itself, where the outer shell underwent deuteric alteration by potassium-rich solutions. The original rock was probably dioritic in composition and was deuterically altered to monzonite and syenite (Anderson, 1949). Perhaps some sericitization of the country rocks accompanied this, but probably not to the degree assumed in the earliest studies. Sericite was one of the few persistently stable minerals in the alteration halo, reflecting the mesothermal conditions of the environment. The metallization took place late in the hydrothermal activity, and the intrusion of lamprophyre dikes marks the close of the igneous period.

Six separate periods of mineralization, ranging from Precambrian to Tertiary, have been described by Fryklund (1964). Geologic mapping has demonstrated that the principal lead mineralization is later than the monzonite stocks of known Cretaceous age. The isotopic composition of the lead is "ancient," and could not have been derived from an igneous source in Cretaceous time. Possibly the lead and the other metals came from a deep sulfide-rich, and uranium- and thorium-poor level of the Earth's mantle in Cretaceous time. Similarly, older leads cutting younger rocks have been found at Butte, Montana, at many places in the Colorado Plateau, and at Broken Hill, Australia.

MAGMA MINE, ARIZONA

The Magma mine (Fig. 13-5) at Superior, Arizona, has both vein and replacement mesothermal ores. Copper, silver, gold, zinc, and some lead, have been produced from the mine (Ransome, 1912; Short and Ettlinger, 1927; Short *et al.*, 1943; Hammer and Peterson, 1968). Early operations were restricted to a pair of rich veins, but around 1950 an extensive deposit of replacement ore was discovered.

Well-exposed rocks of Precambrian and Paleozoic age, overlain to the east with angular discordance by Tertiary dacitic volcanics, make up the geologic column in the Superior district (Fig. 13-6). The Precambrian rocks include the metamorphic Pinal Schist and a younger sequence of conglomerates, shales, quartzites, and some limestones. The latest Precambrian is represented by the Troy Quartzite, 0 to 730 feet thick, which uncomformably overlies the older Precambrian sediments. Deposition of the Troy Quartzite was followed by deformation and erosion that produced a widespread unconformity, although

Figure 13-5
Map of Arizona showing the location of the Magma mine at Superior.

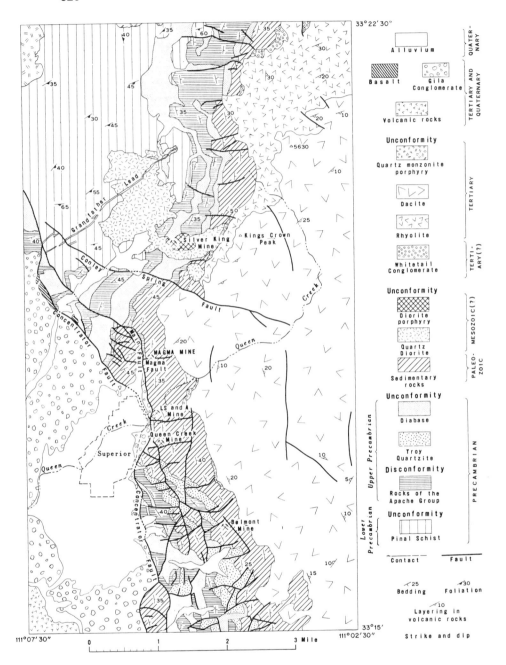

Figure 13-6
Geologic map of the vicinity of the Magma mine, Superior, Arizona.

at the Magma mine there is no angular discordance at the top of the Pre-cambrian section. A pair of diabase sills, 3,100 feet thick, invaded the Upper Precambrian rocks. The diabase is probably Precambrian (Peterson, 1962), although not all geologists agree on this (Sell, 1960).

The Cambrian is represented by a quartzite that is lithologically similar to the Troy, and with which it has been confused. This Cambrian quartzite is 0 to 360 feet thick in the mine area (Hammer and Peterson, 1968) and is prob-ably equivalent to the Bolsa Quartzite, or the Abrigo Formation of southern Arizona (Krieger, 1961; D. W. Peterson, personal communication, 1963). Other Paleozoic units include 350 to 450 feet of Devonian Martin Limestone, 500 feet of Mississippian Escabrosa Limestone, and up to 1,400 feet of Penn-sylvanian Naco Limestone (Hammer and Peterson, 1968). Post-Carboniferous dikes of quartz monzonite porphyry and younger (Late Cenozoic) basalt dikes are also found in the mine area. The quartz monzonite is mineralized wherever it is cut by the principal vein. Basalt was extruded much later than the ores and is unmineralized.

The sediments were faulted and tilted during the Laramide orogeny or possibly earlier. East-west faults, which formed normal to regional compressive forces, were utilized by quartz monzonite magmas and shortly thereafter acted as passageways for ascending ore fluids. Apparently, late faulting continued into the period of ore deposition. Owing to these movements, the quartz mon-zonite porphyry was brecciated, and repeated opening of the fault zone ad-mitted the ore-bearing solutions. The quartz monzonite and ore may have originated in a buried plutonic mass of batholithic dimensions (Ettlinger, 1928). A quartz diorite stock near the Silver King stockwork crops out less than two miles north of the Magma mine (Short *et al.*, 1943) and is believed to be part of this batholith. Postore faulting, associated with the period of dacitic igneous activity, produced structures trending predominantly north to northwest. Tilting of the strata to the east may have resulted from this de-formation.

Two parallel east-west faults contain the vein deposits at Magma. The bulk of ore has been recovered from the Main, or Magma, vein, a fault zone with about 500 feet of vertical displacement and even more strike-slip offset (Short and Wilson, 1938). At the surface the fault strikes east-west and dips about 65°N; but below the 900-foot level it dips 78°S and finally changes in strike to N80°E (Short *et al.*, 1943). Most of the ore is in a large continuous shoot that plunges steeply to the west along the Magma vein, but there are smaller, iso-lated zones to the east and west (Fig. 13-7). Mineralization is found along 10,000 feet of strike and to a depth of 4,900 feet. On individual levels the main shoot extends more than 2,000 feet along the fault. The ore filled open spaces along the fault, which ranges from less than a foot to over 50 feet in width (Wilson, 1950), but replacement of the breccia and sheared material accounts for the bulk of mineralization (Wilson, 1950; Short *et al.*, 1943). Another

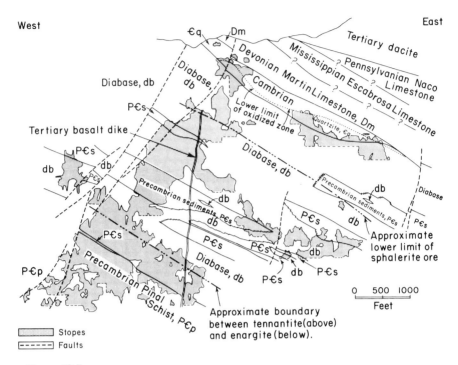

West   East

Figure 13-7

Longitudinal section along the Magma vein, Superior, Arizona, showing the main ore shoot and smaller shoots. (Courtesy of Magma Copper Company.)

mineralized fault, known as the Koerner vein, is about 1,100 feet south of the Magma vein. Although parallel in structure and identical in ore mineralogy to the Magma vein, the Koerner deposit is much smaller and is no longer being mined.

The width of ore along the veins is partly dependent upon the brittleness and reactivity of the wall rocks. Wide rich shoots are found in diabase, which became permeable where brecciated by faulting, but the less competent Pinal Schist caused branching along the faults. Moreover, certain carbonate beds also favored the deposition of massive replacement ores (Wilson, 1950).

One section of Martin Limestone was especially favorable for ore deposition—a 30- to 50-foot thickness beginning about 20 feet stratigraphically above the Bolsa Quartzite. Not only was this section replaced for as much as 30 feet on both sides of the Magma vein, but it was the host for an extensive manto of ore located east of the vein, between the Magma and Koerner faults (Webster, 1958)—see Figures 13-8 and 13-9.

The manto deposit is localized along an east-west vein that may be a subsidiary southern branch of the Magma fault. Replacement was confined to a single stratigraphic section in the Martin Limestone, forming a manto deposit that averages about 20 feet thick and dips 30°E with the bedding. The bed is continuously replaced by sulfides for about 750 feet south of the vein and as much as 200 feet north of the vein. The replacement along the dip exceeds 5,000 feet. Development indicates that the manto fingers out updip at about the 2,000-foot level (Webster, personal communication, 1962). A north-south fault of postore age divides the manto into two segments, with an apparent strike-slip component of about 400 feet of left-lateral displacement (see Fig. 13-8). An incompletely developed manto was recently uncovered in the Escabrosa limestone in the eastern part of the mine (Hammer and Peterson, 1968, p. 1,295).

Both the veins and the replacement mantos consist of massive sulfide ore—bornite, chalcopyrite, pyrite, sphalerite, enargite, tennantite, galena, chalcocite, digenite, and stromeyerite. But these minerals are not distributed evenly throughout the deposits; in places the veins are richest in copper where they pass through diorite. By contrast, the upper eastern part of the Magma vein is deficient in copper and enriched in sphalerite. Marked differences in ore mineralogy define zones parallel to the bedding; as a result, the zonal distribution of ore forms a pattern that is inclined to the east (see Fig. 13-6). However, the mineralogical zones do not conform to individual lithologic units. The upper zone is characterized by zinc and silver mineralization; the middle zone, by chalcocite-bornite-chalcopyrite-tennantite ore; and the lowest zone, by enargite, bornite, pyrite, and chalcopyrite (Short and Wilson, 1938; Webster, 1958). In general, the manto contains specularite, pyrite, chalcopyrite, bornite, chalcocite, and minor amounts of quartz, magnetite, and barite. Silver, gold, zinc, and lead minerals characterize the middle zone (Webster, 1958).

Paragenetic studies indicate that pyrite was the earliest sulfide deposited, and it clearly replaces the hematite (Fig. 13-10). Sphalerite and enargite were next. Galena and stromeyerite were deposited later, probably at about the same time as the principal copper minerals—tennantite, bornite, chalcopyrite, and digenite (Short *et al.*, 1943).

Wall-rock alteration—sericitization and silicification—accompanied the deposition of ores. It was most intensely developed in the diabase sills, but it also affected the schists and clastic sediments (Short *et al.*, 1943). Some of the limestone beneath the manto is bleached (Webster, 1958), but most carbonates are relatively unaltered.

Although there is some correlation between the ore mineralogy and type of wall rock, the overall zoning seems independent of the stratigraphy. If the zones were controlled by wall-rock chemistry, they would correspond to single lithologic units. In the Magma vein, however, zone boundaries fall within stratigraphic units—in the diabase, for example—rather than along contacts.

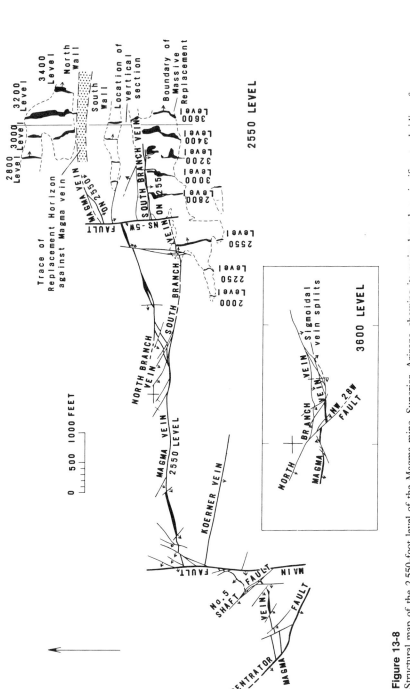

Figure 13-8

Structural map of the 2,550-foot level of the Magma mine, Superior, Arizona, showing its major branches, significant subsidiary fractures, offsets by major faults of the north-striking set, and the Koerner vein. The outline of the east-dipping limestone-replacement orebodies is projected to the diagram. The inset shows a structural map of part of the 3,600-foot level. (From Hammer and Peterson, 1968.)

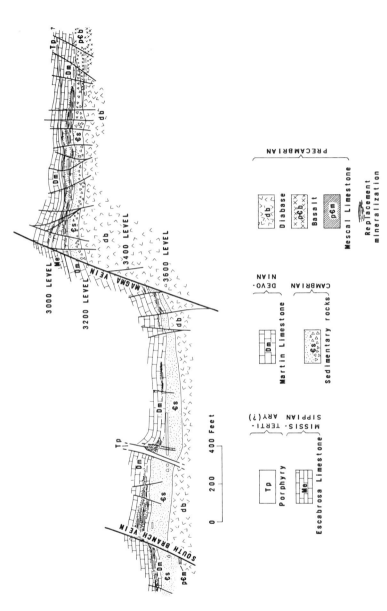

Figure 13-9

Vertical north-trending section (looking west) through the limestone-replacement deposit, Magma mine, Arizona. Some ore shoots show direct connection with veins, but others are isolated. (From Hammer and Peterson, 1968.)

Figure 13-10
Pyrite pseudomorphous after specular hematite, manto ore. Magma mine, Superior, Arizona. White is pyrite; black and gray are gangue; light gray is hematite. ×240. (Courtesy of Ray Robison.)

In general, the mineral assemblages that constitute the deeper ores formed at higher temperatures than those of the shallow ores, thus suggesting that the zoning was produced by a temperature gradient. Ascending isotherms should approach horizontality at any appreciable distance from the heat source; accordingly, the presence of ore zoning parallel to bedding indicates that the ore was probably deposited before the beds were tilted.

Near the surface, the ores were oxidized, and where pyrite was available to produce ferric sulfate and sulfuric acid, meteoric waters leached some of the copper, carrying it down to a supergene enrichment zone. The oxidized zone dips eastward, roughly parallel to the sediments. This is further evidence that tilting is postore (Short and Ettlinger, 1927; Wilson, 1950).

In summary, the Magma ores were deposited by ascending fluids that traveled along the east-west faults shortly after quartz monzonite dikes had been introduced along these same structures. The thickness and width of any portion along the ore shoot depended on the degree of earlier brecciation and shearing, which in turn was a function of the physical and chemical properties of wall rocks. The ores were distributed zonally because of either a thermal gradient or the wall-rock chemistry, or both. Where the ore fluids encountered a thin, reactive section of the lower Martin Limestone, they formed highly selective massive replacement deposits. Essentially all of the ore mineralization, from the earliest sphalerite to the latest copper minerals, took place under mesothermal conditions.

CHUQUICAMATA, CHILE

Chuquicamata, in the Atacama Desert of northern Chile (Fig. 13-11) is one of the largest copper deposits known (López, 1939, 1942). The mine is 130 miles northeast of Antofagasta and about 90 miles from the coast, at an elevation of 9,500 feet in the foothills of the main Andes Mountains. Because it is an extremely arid region, large amounts of water-soluble sulfate minerals have been preserved in the zone of oxidation. Chuquicamata is an excellent example of a disseminated or "porphyry" copper deposit, and, along with other deposits of this type, it is classed as mesothermal. Copper ornaments found in the graves of local Indians indicate that the deposits were worked in a small way before the Spanish conquistadors arrived (Taylor, 1935). Small amounts of shaped turquoise also are found in places along the outcrops. The earliest mining on an appreciable scale—confined principally to narrow but rich veins— was done chiefly by English and Chilean companies between 1879 and 1912.

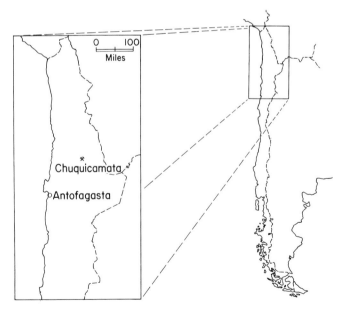

Figure 13-11
Map of Chile showing the location of Chuquicamata.

The mineralization is apparently related to a period of faulting, folding, and igneous intrusion during the late Mesozoic or Early Tertiary. Upper Jurassic sediments and older metamorphic rocks were first gently folded and covered with a series of volcanic flows, and then strongly deformed as the Andean batholith was being intruded. After considerable erosion had exposed the plutonic rocks, the area was again uplifted, and more volcanics were deposited. Slight warping and volcanism have continued into Recent time (Taylor, 1935; Jarrell, 1944).

Near Chuquicamata the Andean batholithic complex is represented by three nested stocks, or possibly three facies of a single batholith. The plutonic rocks invaded the crest and eastern limb of a broad anticline, forming an elongate mass 20 miles long by 4 miles wide, with a northerly or northeasterly trend. Similar localized outcrops of granitic rocks along a narrow belt 95 miles long suggest cupolas of an underlying batholith that erosion has just begun to expose (Perry, 1952). The three facies of granitic rocks at Chuquicamata include two apparently different granodiorites, known as the Fortuna Granodiorite and the Elena Granodiorite, and a highly altered quartz monzonite, called the Chuquicamata Porphyry. In the mine area (Fig. 13-12), the Chuquicamata Porphyry is an elongate, tabular mass, separating the two different kinds of granodiorite. All contacts between quartz monzonite and granodiorite are either faults or gradational and altered zones; consequently, the genetic relationships between facies cannot be determined. Fortuna Granodiorite is unaltered and is separated from altered Chuquicamata Porphyry by a strong fault zone. East of this fault, known as the West Fissure, the Chuquicamata Porphyry and part of the Elena Granodiorite are mineralized. In general, the large open-pit operation is confined to a highly sheared zone in the Chuquicamata Porphyry between the two granodiorites.

The Chuquicamata Porphyry is made up of several varieties of altered rock. Some parts of it are characterized by plagioclase or orthoclase phenocrysts; others are almost aplitic in texture. The rock is laced with stringers of orthoclase and quartz. This altered porphyry is strikingly similar to the hosts for other great disseminated copper deposits (Perry, 1952).

Two strongly sheared zones define the western and eastern limits of the mineralized zone. These shears vary in strike from N10°E to as much as N76°E, forming a violin-shaped ore body with the neck pointed south. Between the eastern and western shear zones there are several intermediate shear zones and a complex system of tension fractures, most of which form a horsetail pattern branching northeastward from the shears. López (1939, 1942) suggested that the shears and tension fractures are all related to a simple shearing couple acting horizontally in a north-south direction. Given this, the maximum shear stresses would have been oriented northeast-southwest, parallel to the horsetail tension fractures. Another prominent set of tension fractures strikes northwest, roughly normal to the horsetail fissures. During shearing, secondary

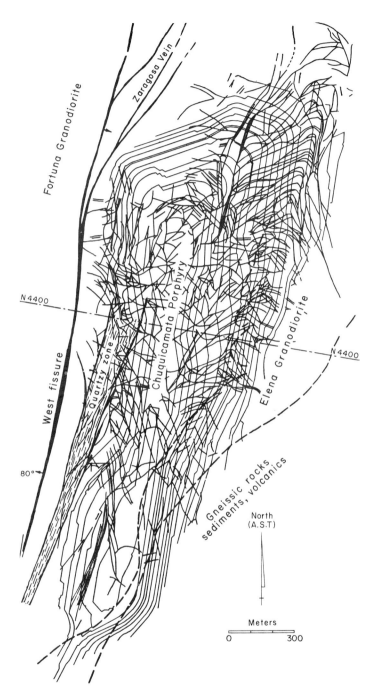

Figure 13-12
Geologic plan of Chuquicamata, Chile. (From Perry, 1952.)

tensional stresses were set up on opposite sides of the shear plane. The opposing tension fractures at Chuquicamata look suspiciously like products of such secondary forces, especially where they turn into the shear zones.

Several intergradational types of alteration have been described at Chuquicamata (Taylor, 1935; López, 1939; Perry, 1952). They include silicification, sericitization, chloritization, albitization, and epidotization (Fig. 13-13). Taylor (1935) classified the altered rocks in five categories: (1) siliceous rock (2) sericitic rock (3) normal rock (4) flooded rock, and (5) transition rock. The siliceous rock consists almost entirely of quartz and sericite, and the original texture is completely obliterated; repeated phases of silicification are recorded by interstitial quartz and cross-cutting veinlets. A wide belt of siliceous rock strikes nearly parallel to the West Fissure along the west side of Chuquicamata Porphyry. The sericitic rock is marginal to the silicified zone; it consists of Chuquicamata Porphyry containing sericitized feldspars and veinlets of quartz. Normal rock retains the original porphyritic texture, though some of the feldspars are albitized or sericitized. The large orthoclase crystals and quartz "eyes" that characterize normal rock are thought to be relict phenocrysts rather than metacrysts. Flooded rock is similar to normal rock except that it has been permeated with hydrothermal quartz. Between the fresh Elena Granodiorite on the east and the normal facies of Chuquicamata Porphyry, there is a narrow zone of transition rock that consists of Elena Granodiorite containing chloritized ferromagnesian minerals, slightly sericitized and albitized feldspars, a little epidote, and veinlets of specular hematite. In general, the alteration is

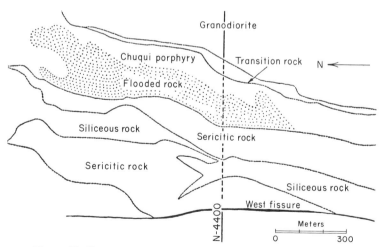

Figure 13-13
Alteration facies at Chuquicamata, Chile. (From Lopez, 1939, Fig. 2.)

most intense along the western border of the Chuquicamata Porphyry and dies out gradually toward the east (Taylor, 1935; López, 1939, 1952).

The open-pit orebody is about two miles long and has a maximum width of 3,600 feet. Copper minerals were introduced after the rock had been fractured, altered, and fractured again. They fill the fracture systems and permeate veinlets throughout the rock mass. Most of the ore that has been mined is oxidized, but supergene-enriched sulfide zones and primary hypogene ore are also recovered. Listed in paragenetic order, the minerals of the hypogene veinlets are quartz, pyrite, enargite, chalcopyrite, bornite, and tennantite-tetrahedrite (López, 1939). The bulk of hypogene copper is contained in enargite. A zonal distribution of the primary minerals has been established, with molybdenum along the west, enargite in the sericitized rock, and tennantite-tetrahedrite along the east (López, 1939). The pattern seems to reflect deposition from west to east. Supergene enrichment formed chalcocite and covellite at the expense of all other sulfides. The oxidized zone contains many rare minerals, many of which are water soluble but are preserved in the extreme aridity of the Atacama Desert. Sulfates, oxides, and carbonates characterize the oxidized zone. The copper minerals include cuprite, melaconite (CuO), native copper, brochantite [$Cu_4SO_4(OH)_6$], antlerite [$Cu_3SO_4(OH)_4$], kröhnkite [$Na_2Cu(SO_4)_2 \cdot 2H_2O$], chalcanthite ($CuSO_4 \cdot 5H_2O$), natrochalcite [$NaCu_2$ $(SO_4)_2OH \cdot H_2O$], cuprocopiapite [$CuFe_4(SO_4)_6(OH)_2 \cdot nH_2O$], salesite ($CuIO_3$ OH), turquoise, leightonite [$K_2Ca_2Cu(SO_4)_4 \cdot 2H_2O$], lindgrenite [$Cu_3$ $(MoO_4)_2(OH)_2$], chenevixite [$Cu_2Fe_2(AsO_4)_2(OH)_4 \cdot H_2O$], bellingerite [$3Cu(IO_3)_2 \cdot 2H_2O$], atacamite [$Cu_2Cl(OH)_3$], and numerous other species (Jarrell, 1944; Bandy, 1938; Palache and Jarrell, 1939; Berman and Wolfe, 1940). Sulfates of iron are also abundant in the oxidized zone. The presence of chlorine and iodine in some of the compounds has been attributed both to wind transport of these halides from the ocean (Jarrell, 1944) and to recent volcanism (Bandy, 1938).

Chuquicamata has experienced a complex history of oxidation, leaching, and supergene enrichment. During the Late Tertiary uplift of the area, the climate changed from humid to arid, and the water table fluctuated radically. Copper leached from the oxidation zone was carried down to the water table, where it was reprecipitated as a sulfide enrichment of the primary ore. As the land surface was eroded, the water table and its associated supergene ore moved downward through the hypogene ores. The sudden change to an arid climate quickly lowered the water table and left the supergene zone high and dry above the newly established water table. Subsequent leaching has, in general, been slow, but where the ore contained sufficient pyrite to react with the water and form sulfuric acid, and where there was very little feldspathic material to neutralize this acid, the copper was leached. Consequently, the belt of silicified rock along the western side of the Chuquicamata Porphyry is barren of ore.

Similarly, where the porphyry is thoroughly sericitized, it is relatively impermeable and has shielded patches of hypogene ore from the leaching action of descending waters (Jarrell, 1944). The sulfide-enriched blanket that was developed before the period of final uplift contains less pyrite than the hypogene ore, because the supergene copper was deposited at the expense of iron in pyrite. As a result, when this sulfide zone was stranded above the water table, it could not generate enough sulfuric acid to dissolve the copper minerals and carry them down to the new zone of enrichment. Instead, the copper was merely oxidized in place to sulfates, forming the present oxidized orebody. The hypogene ore that had previously underlain the supergene blanket contained sufficient pyrite to leach the copper away; as a result, the rich zone of oxide ore is separated in places from the present zone of supergene enrichment by a layer of iron-oxide waste. Furthermore, the old supergene enriched zone was not uplifted uniformly; it was tilted, leaving its south end lower than its north end. The present-day water table truncates this surface, isolating some sulfides in the north and submerging part of the oxidized zone in the south.

The Chuquicamata ores owe their concentration to several factors. First, there must have been a special supply of copper in the source batholith. Second, the West Fissure must have tapped this supply and conducted the ore-bearing fluids up into the cupola. Third, alternating periods of silicification and fracturing prepared the rock for ore deposition; the ore ascended along the West Fissure, spreading eastward when it reached the altered and fractured zone. Finally, the copper was leached and redeposited by meteoric waters, concentrating what would otherwise have been a more dispersed zone of mineralization.

Recent developments at Chuquicamata include a drainage tunnel driven 1,300 feet below the original surface; this tunnel lowered the water table, allowing the zone of supergene enrichment to be mined. As a result, mining is now in the sulfide zone below the oxidized ores, and Chuquicamata promises to be a major producer of copper for many years.

REFERENCES TO SELECTED MESOTHERMAL DEPOSITS

Butte, Montana

Hart, L. H., 1935. The Butte district, Montana, in *Copper Resources of the World*, vol. 1, 16th International Geological Congress.

Lindgren, W., 1927. Paragenesis of minerals in the Butte veins, *Econ. Geol.* 22:304–307.

Locke, A., D. A. Hall, and M. N. Short, 1924. Rôle of secondary enrichment in genesis of Butte chalcocite, *Amer. Inst. Mining Eng. Trans.* 70:933–963.

Meyer, C., E. P. Shea, C. C. Goddard, Jr., and staff, 1968. Ore deposits at Butte, Montana, in *Ore Deposits of the United States* (L. C. Graton–R. Sales Mem. Vol.), vol. 2, ed. J. D. Ridge, American Institute of Mining Engineers.

Ray, J. C., 1914. Paragenesis of the ore minerals in the Butte district, Montana, *Econ. Geol.* 9:463–482.

Sales, R. H., 1910. Superficial alteration of the Butte veins, *Econ. Geol.* 5:15–21.

——, 1914. Ore deposits at Butte, Montana, *Amer. Inst. Mining Eng. Trans.* 46:3–109.

Sales, R. H., and C. Meyer, 1948. Wall rock alteration at Butte, Montana, *Amer. Inst. Mining Eng. Trans.* 178:9–35.

——, 1949. Results from preliminary studies of vein formation at Butte, Montana, *Econ. Geol.* 44:465–484.

——, 1951. Effect of post-ore dike intrusion on Butte ore minerals, *Econ. Geol.* 46: 813–820.

Weed, W. H., 1912. Geology and ore deposits of the Butte district, Montana, *U.S. Geol. Surv. Prof. Pap.* 74.

Mount Isa, Queensland, Australia

Bennett, E. M., 1965. Lead-zinc-silver and copper deposits of Mount Isa, *8th Commonw. Mining Metall. Congr. Trans.* 1:233–246.

Blanchard, R., 1942. Mount Isa ore geology, in *Ore Deposits as Related to Structural Features*, ed. W. H. Newhouse, Princeton, N.J.: Princeton Univ. Press.

——, 1948. The alleged mineral zoning at Mount Isa, *Amer. Inst. Mining Eng. Trans.* 178:107–133.

Blanchard, R., and G. Hall, 1937. Mount Isa ore deposition, *Econ. Geol.* 32:1042–1057.

——, 1942. Rock deformation and mineralization at Mount Isa, *Australas. Inst. Mining Metall. Proc.* 125, pp. 1–60.

Carter, E. K., and J. H. Brooks, 1965. Geology and mineralization of northwestern Queensland, *8th Commonw. Mining Metall. Congr. Trans.* 1:221–232.

Carter, E. K., J. H. Brooks, and K. R. Walker, 1961. The Precambrian mineral belt of north-western Queensland, *Aust. Bur. Mineral Resources Geol. Geophys. Bull.* 51.

Carter, S. R., 1948. Mount Isa geology, paragenesis and ore reserves, *18th Int. Geol. Congr. Rept.*, pt. 7, pp. 195–205.

——, 1953. Mount Isa mines, in *Geology of Australian Ore Deposits*, ed. A. B. Edwards, Melbourne: Australasian Institute of Mining and Metallurgy.

——, 1958. Notes on recent Mount Isa ore discoveries, in *F. L. Stillwell Anniversary Volume*, Melbourne: Australasian Institute of Mining and Metallurgy.

Crawford, N. J. W., 1964–65. Origin and significance of volcanic potash-rich rocks from Mount Isa, *Inst. Mining Metall. London Trans.* 74 (pt. 2):33–43.

Elliston, J., 1960. Ore localization by preconsolidation structures, *Australas. Inst. Mining Metall. Proc.* 196, pp. 29–49.

Grondijs, H. F., and C. Schouten, 1937. A study of the Mount Isa ores, *Econ. Geol.* 32:407–450.

Hall, G., 1939. Geology as applied to the mining of silver-lead-zinc ores at Mount Isa, *Australas. Inst. Mining Metall. Proc.* 115, pp. 215–234.

Knight, C. L., 1953. Regional geology of Mount Isa, in *Geology of Australian Ore Deposits*, ed. A. B. Edwards, Melbourne: Australasian Institute of Mining and Metallurgy.

Love, L. G., and D. O. Zimmerman, 1961. Bedded pyrite and micro-organisms from the Mount Isa shale, *Econ. Geol.* 56:873–896.

Murray, W. J., 1961. Notes on Mount Isa geology, *Australas. Inst. Mining Metall. Proc.* 197, pp. 105–136.

Soloman, P. J., 1965. Investigations into sulfide mineralization at Mt. Isa, Queensland, *Econ. Geol.* 60:737-765.

Stillwell, F. L., and A. B. Edwards, 1945. The mineral composition of the Black Star copper ore body, Mount Isa, Queensland, *Australas. Inst. Mining Metall. Proc.* 139, pp. 149-159.

Ajo, Arizona

Bryan, K., 1925. The Papago country, Arizona, *U.S. Geol. Surv. Water-Supply Pap.* 499, pp. 208-210.

Dixon, D. W., 1966. Geology of the New Cornelia mine, Ajo, Arizona, in *Geology of the Porphyry Copper Deposits,* ed. S. R. Titley and C. L. Hicks, Tucson: Univ. Arizona Press.

Gilluly, J., 1935. The Ajo district, Arizona, in *Copper Resources of the World,* vol. 1, 16th International Geological Congress.

————, 1937. Geology and ore deposits of the Ajo quadrangle, Arizona, *Ariz. Bur. Mines Geol. Ser. 9 Bull.* 141.

————, 1942. The mineralization of the Ajo copper district, Arizona, *Econ. Geol.* 37: 247-309.

————, 1946. The Ajo mining district, Arizona, *U.S. Geol. Surv. Prof. Pap.* 209.

Joralemon, I. B., 1914. The Ajo copper-mining district, *Amer. Inst. Mining Eng. Trans.* 49:593-609.

Schwartz, G. M., 1947. Hydrothermal alteration in the "porphyry copper" deposits, *Econ. Geol.* 42:319-352.

REFERENCES CITED

Anderson, A. L., 1949. Monzonite intrusion and mineralization in the Coeur d'Alene district, Idaho, *Econ. Geol.* 44:169-185.

Anderson, R. J., 1940. Microscopic features of ore from the Sunshine mine, *Econ. Geol.* 35:659-667.

Arnold, R. G., R. G. Coleman, and V. C. Fryklund, 1962. Temperature of crystallization of pyrrhotite and sphalerite from the Highland-Surprise mine, Coeur d'Alene district Idaho, *Econ. Geol.* 57:1163-1174.

Bandy, M. C., 1938. Mineralogy of three sulphate deposits of northern Chile, *Amer. Mineral.* 23:669-760.

Berman, H., and C. W. Wolfe, 1940. Bellingerite, a new mineral from Chuquicamata, Chile, *Amer. Mineral.* 25:505-512.

Doe, B. R., 1973. Lead isotopes, ore genesis, and ore prospect evaluation [abstr.], *Econ. Geol.* 68:1206.

Ettlinger, I. A., 1928. Ore deposits support hypothesis of a central Arizona batholith, *Amer. Inst. Mining Eng. Tech. Pub.* 63.

Fryklund, V. C., Jr., 1964. Ore deposits of the Coeur d'Alene district, Shoshone County, Idaho, *U.S. Geol. Surv. Prof. Pap.* 445.

Fryklund, V. C., Jr., and J. D. Fletcher, 1956. Geochemistry of sphalerite from the Star mine, Coeur d'Alene district, Idaho, *Econ. Geol.* 51:223-247.

Hammer, D. F., and D. W. Peterson, 1968. Geology of the Magma mine area, Arizona, in *Ore Deposits of the United States* (L. C. Graton-R. Sales Mem. Vol.), vol. 2, American Institute of Mining Engineers.

Hobbs, S. W., and V. C. Fryklund, Jr., 1968. The Coeur d'Alene district, Idaho, in *Ore Deposits of the United States* (L. C. Graton-R. Sales Mem. Vol.), vol. 2, American Institute of Mining Engineers.

Hobbs, S. W., A. B. Griggs, R. E. Wallace, and A. B. Campbell, 1965. Geology of the Coeur d'Alene district, Shoshone County, Idaho, *U.S. Geol. Surv. Prof. Pap.* 478.

Hosterman, J. W., 1956. Geology of the Murray area, Shoshone County, Idaho, *U.S. Geol. Surv. Bull.* 1027-P.

Jarrell, O. W., 1944. Oxidation at Chuquicamata, Chile, *Econ. Geol.* 39:215-286.

Krieger, M. H., 1961. Troy Quartzite (younger Precambrian) and Bolsa and Abrigo Formations (Cambrian), northern Galiuro Mountains, southeastern Arizona, *U.S. Geol. Surv. Prof. Pap.* 424-C, pp. 160-164.

Lindgren, W., 1933. *Mineral Deposits*, 4th ed., New York: McGraw-Hill.

Long, A., A. J. Silverman, and J. L. Kulp, 1960. Isotopic composition of lead and Precambrian mineralization of the Coeur d'Alene district, Idaho, *Econ. Geol.* 55:645-658.

López, V. M., 1939. The primary mineralization at Chuquicamata, Chile, *Econ. Geol.* 34:674-711.

———, 1942. Chuquicamata, Chile, in *Ore Deposits as Related to Structural Features*, ed. W. H. Newhouse, Princeton, N.J.: Princeton Univ. Press.

McKinstry, H. E., and R. H. Svendsen, 1942. Control of ore by rock structure in a Coeur d'Alene mine, *Econ. Geol.* 37:215-230.

Mitcham, T. W., 1952a. Indicator minerals, Coeur d'Alene silver belt, *Econ. Geol.* 47:414-450.

———, 1952b. Significant spatial distribution patterns of minerals in the Coeur d'Alene district, Idaho, *Science* 115:11.

Palache, C., and O. W. Jarrell, 1939. Salesite, a new mineral from Chuquicamata, Chile, *Amer. Mineral.* 24:388-392.

Pennebaker, E. N., 1944. Structural relations of the copper deposits at Matahambre, Cuba, *Econ. Geol.* 39:101.

Perry, V. D., 1952. Geology of the Chuquicamata orebody, *Mining Eng.* 4:1166-1168.

Peterson, D. W., 1962. Preliminary geologic map of the western part of the Superior quadrangle, Pinal County, Arizona, *U.S. Geol. Surv. Mineral Invest. Field Studies. Map* MF 253.

Ransome, F. L., 1912. Copper deposits near Superior, Arizona, *U.S. Geol. Surv. Bull.* 540(pt. 1), pp. 139-158.

Ransome, F. L., and F. C. Calkins, 1908. The geology and ore deposits of the Coeur d'Alene district, Idaho, *U.S. Geol. Surv. Prof. Pap.* 62.

Sell, J. D., 1960. Diabase at the Magma mine, Superior, Arizona, *Ariz. Geol. Soc. Digest* 3:93-97.

Shenon, P. J., 1948. Lead and zinc deposits of the Coeur d'Alene district, Idaho, *18th Int. Geol. Congr. Rept.*, pt. 7, pp. 88-91.

Short, M. N., and I. A. Ettlinger, 1927. Ore deposition and enrichment at the Magma mine, Superior, Arizona, *Amer. Inst. Mining Eng. Trans.* 74:174-222.

Short, M. N., F. W. Galbraith, E. N. Harshman, T. H. Kuhn, and E. D. Wilson, 1943. Geology and ore deposits of the Superior mining area, Arizona, *Ariz. Bur. Mines Geol. Ser. 16 Bull.* 151.

Short, M. N., and E. D. Wilson, 1938. Magma mine area, Superior, *Ariz. Bur. Mines Geol. Ser. 12 Bull.* 145, pp. 90-98.

Sorenson, R. E., 1947. Deep discoveries intensify Coeur d'Alene activities, *Eng. Mining J.* 148(10):70-78.

———, 1951. Shallow expressions of Silver Belt ore shoots, Coeur d'Alene district, Idaho, *Amer. Inst. Mining Eng. Trans.* 190:605-611.

Taylor, A. V., Jr., 1935. Ore deposits at Chuquicamata, Chile, in *Copper Resources of the World,* vol. 2, 16th International Geological Congress.

Umpleby, J. B., and E. L. Jones, Jr., 1923. Geology and ore deposits of Shoshone County, Idaho, *U.S. Geol. Surv. Bull.* 732.

Warren, H. V., 1934. Silver-tetrahedrite relationship in the Coeur d'Alene district, Idaho, *Econ. Geol.* 29:691-296.

Webster, R. N., 1958. Exploration extends Magma's future, *Mining Eng.* 10:1062-1065.

Wilson, E. D., 1950. Superior area, *Ariz. Bur. Mines Geol. Ser. 18 Bull.* 156, pp. 84-98.

14 / Epithermal Deposits

Epithermal deposits are products of hydrothermal origin formed at shallow depths and low temperatures. Deposition normally takes place within 3,000 feet of the surface, in the temperature range 50° to 200°C. Most are in the form of vein fillings, irregular branching fissures, stockworks, or breccia pipes. Replacement is recognized in many of the ores, but open-space fillings are common and in some deposits are the dominant form of emplacement. Drusy cavities, comb structures, crustifications, and symmetrical banding are generally conspicuous. The fissures have a direct connection with the surface, allowing the ore-bearing fluids to flow with comparative ease; in fact, some hot springs and steam vents are probably surface expressions of underlying epithermal systems. Colloform textures are also characteristic of the epithermal zone, reflecting the moderate temperatures and free circulation.

A few epithermal deposits can be related directly to deep-seated intrusive bodies, but this relationship is demonstrable only under special conditions of erosion. Many epithermal deposits have no observable association with intrusive rocks. Most ores are in or near areas of Tertiary volcanism, especially near volcanic necks and other structures that tap underlying source materials. Because these deposits are formed near the surface they are most abundant in young rocks, otherwise they would ordinarily have been removed by erosion. The volcanic environment engenders hot waters in some mines; for example, hot waters were encountered at depth in the Comstock Lode of Nevada and in several of the mercury mines of California.

The country rocks near epithermal veins commonly are altered extensively, even though the vein walls may be sharply defined. Relatively high porosity allows the associated fluids to permeate the wall rocks for great distances, and a favorable temperature differential promotes reactions between the host and invading solutions. As a result, wall-rock alteration is typically both widespread and conspicuous. Among the principal alteration products are chlorite, sericite, alunite, zeolites, clays, adularia, silica, and pyrite. Chlorite is probably the most common alteration mineral in this zone. In intermediate to mafic volcanics propylitization is the dominant process, propylite being an aggregate of secondary chlorite, pyrite, epidote, sericite, carbonates, and albite. The silica and pyrite of epithermal alteration halos are generally fine grained. Carbonate minerals, especially calcite, dolomite, and rhodochrosite, are also found as alteration products. Furthermore, the clay minerals are abundant and conspicuous (Sudo, 1954), forming zones of different colors parallel to the walls of veins. Although sericite is more characteristic of mesothermal deposits, it is not uncommon in epithermal ores; where present, the sericite is normally subordinate to chlorite, forming a narrow zone of relatively high-temperature alteration adjacent to the veins.

The gangue minerals in epithermal veins include quartz (in places, amethystine), chalcedony, adularia, calcite, dolomite, rhodochrosite, barite, and fluorite. Typical hypothermal minerals, such as tourmaline, topaz, and garnet, are absent.

Ore minerals characteristic of epithermal deposits include the sulfantimonides and sulfarsenides of silver (polybasite, stephanite, pearceite, pyrargyrite, proustite, and others), the gold and silver tellurides (petzite $[(Ag,Au)_2Te]$, sylvanite $[(Au,Ag)Te_2]$, krennerite $[(Au,Ag)Te_2]$, calaverite $(AuTe_2)$, hessite (Ag_2Te), and so on), stibnite, acanthite, cinnabar, and native mercury. Some of the world's richest concentrations of native gold and electrum (the natural gold-silver alloy), were deposited under epithermal conditions; these are the famous bonanza deposits, such as Goldfield, Nevada, and Hauraki, New Zealand. Other epithermal bonanza deposits contain gold tellurides and silver sulfides, sulfosalts, and selenides. The famous native copper deposits of the Keweenaw Peninsula in northwestern Michigan, deposited in propylitized basaltic lavas and interbedded conglomerates, are also classified as epithermal. Galena, sphalerite, chalcopyrite, and other sulfides commonly found in the mesothermal zone extend into the epithermal zone, but rarely in large concentrations. By definition, the high-temperature minerals characteristic of hypothermal veins and igneous metamorphic deposits are not present in epithermal ores.

It is not uncommon to find large, highly colored gossans (iron oxide cap-pings) covering epithermal ores. During weathering, the widespread pyrite in the altered wall rock is oxidized to limonite and hematite, forming a conspic-uous guide to ore deposits.

CHINESE ANTIMONY DEPOSITS

The world's principal resources of antimony lie in southern and southwestern China, in the provinces of Hunan, Kweichow, Kwangsi, Kwangtung, and Yunnan (Fig. 14-1). The deposits lie in three crudely defined east-west belts associated with Late Mesozoic to Early Tertiary plutonic rocks, principally quartz diorites. The northern belt, extending from northeastern Kweichow through northern Hunan, is the richest; the southern belt, which reaches from southern Yunnan to central Kwangsi, is the least productive; the central belt follows a mountain range along the Kweichow-Kwangsi and Hunan-Kwang-tung borders. Hunan, the chief source, contains the richest deposits and about 88 percent of the reserves (Collins, 1918; di Villa, 1919; Tegengren, 1921; Juan, 1946).

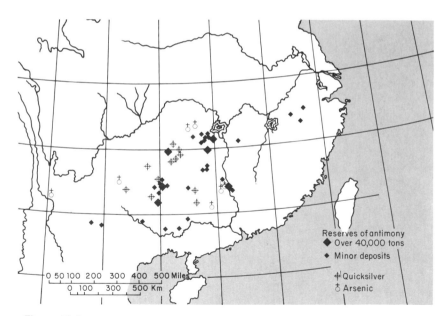

Figure 14-1
Map of China showing the location of antimony, quicksilver, and arsenic deposits. (From Juan, 1946, Plate 7.)

All the deposits are in sedimentary rocks. There are veins in shales, slates, sandstones, and quartzites, and there are replacement lodes in limestones and dolomites. Most of the ores are in Paleozoic rocks, but the Hunan ores are in Devonian clastics and Carboniferous carbonates. The Kweichow ores are in Precambrian slates, and some small deposits in Yunnan are in Triassic rocks.

There are two principal kinds of antimony deposits in this extensive region: stibnite-quartz veins in clastic host rocks, and stibnite-galena-arsenopyrite replacement deposits in carbonate rocks. Although some of the most productive mines are in replacement ores, the dominant ore is the stibnite-quartz vein variety.

The veins are restricted to zones of shearing or brecciation. In many places individual fissures open into stockworks or irregularly distributed veinlets. Apparently, the slates favor continuous veins and the quartzites favor stockwork deposits. Few of the veins are uniform in their content of stibnite, which occurs sporadically in pockets or bunches. Quartz, which forms crystal-lined vugs in many places, is generally the only gangue mineral. Cinnabar and pyrite are found in some of the stibnite veins, but only in subordinate amounts. Because of differences in host rocks, stockworks predominate in Hunan and veins carry the ore in Kweichow. The average vein is about one foot wide; widths of more than three feet are exceptional. Stibnite-quartz ore ranges from 6 to 25 percent stibnite; it is hand-sorted to much higher grades. Because no effort is known to have been made to mine or explore beyond a few hundred feet in depth, the vertical extent of vein ore is unknown.

Replacement deposits of stibnite in carbonate country rocks form irregular pods, with associated galena, arsenopyrite, and pyrite. Relatively continuous and uniform veins carrying the same minerals as the replacement deposits are also found in limestones. The stibnite content of limestone deposits is generally higher than that of stibnite-quartz veins; it ranges from 20 to 57 percent (Juan, 1946).

In mountainous country along the Tzu River in central Hunan there are several vein deposits known as the Panshih (Pan Hsi) mines (di Villa, 1919)—see Figure 14-2. The tilted sediments (clay, slates, shales, and quartzites) have been invaded by granites, which caused widespread jointing but only slight igneous metamorphism. Vertical veins cut obliquely through the eastward striking sediments. The veins are very narrow, seldom exceeding 15 inches in width, and consist of stibnite in a gangue of quartz or interlaminated schist and quartz. Economic quantities of antimony are restricted to short shoots within narrow footwall seams, which, in turn, are accompanied by a strong clay selvage. The deepest of the mines extend only about 600 feet. The lower levels show that the ore shoots decrease in length, even though well-defined fissures continue to greater depths.

Not far from the Panshih (Pan Hsi) mines, in the rugged country between the Tzu and Yuan rivers, stibnite deposits are associated with gold-bearing

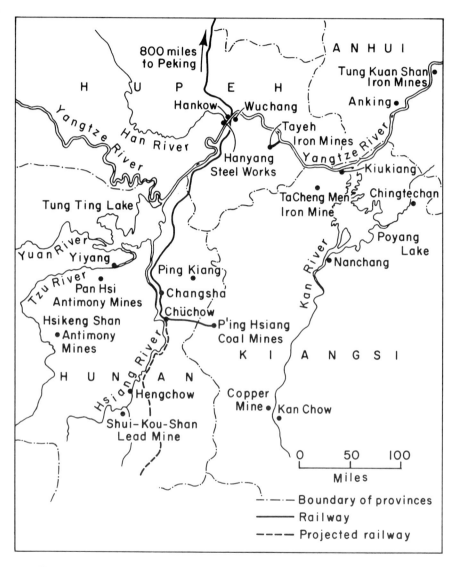

Figure 14-2
Map of the chief mining centers of the Lower Yangtze Valley. (From Collins, 1918.)

quartz (di Villa, 1919). These are the Wushih mines. The deposits are veins in folded slates, shales, and a few interbedded quartzites. Originally, the mines were worked for gold, and the stibnite was discarded as waste (the quartz gangue contains 0.20 to 0.25 ounces of gold per ton). Quartz veins, having sharply defined, regular footwalls and indistinct hangingwalls, carry the stibnite, which is generally concentrated along the footwall side. The principal veins are persistent in both strike and dip, the deepest workings being at the water table, 540 feet from the surface. In places, the veins are as much as five feet wide, and hand-sorted ore from these wide sections carries 20 to 30 percent antimony. At depth, the shoots contract sharply, and stibnite gives way to pyrite.

The Hsi-K'uang-Shan (Hsikeng) district in central Hunan (Fig. 14-2) contains the largest and richest structure in the entire region (Tegengren, 1921; di Villa, 1919). Gently folded dolomitic limestone and a few beds of shale, sandstone, and low-grade bituminous coal overlie a quartzitic sandstone that contains the ore deposits. The orebodies are concentrated along the crest of a gentle anticline, the folding of which caused jointing and brecciation in the brittle quartzites sandwiched between shales. Seams and large pockets of stibnite ore extend for about a mile. The most favorable bed for ore is a quartzite along the upthrown side of a prominent fault. This bed, more than 150 feet thick, contains several favorable strata, so that the deposits seem scattered throughout it. Most stibnite is in veins, irregular veinlets, or lenticular bodies; veins of pure stibnite more than a foot wide are not uncommon. Open-space filling was apparently dominant (Tegengren, 1921). Three miles south of Hsi-K'uang-Shan a smaller deposit follows the same fault. The ore consists almost entirely of long, radiating or columnar crystals of stibnite, oxidized in places to stibiconite ($SbSb_2O_6OH$), which contain only a fraction of one percent of combined arsenic, lead, and copper.

The Chinese antimony deposits, especially those of Hunan Province, are said to have been mined since the sixteenth century, yet most of the workings are confined to the zone above the water table. Large-scale, systematic mining was started in 1927, but much hand sorting and hand labor are still used. The available resources are said to be adequate for many years.

ALMADÉN, SPAIN

The Almadén mine, about 130 miles southwest of Madrid in the western part of Ciudad Real Province, is another famous epithermal deposit (see Fig. 14-3). It is the world's richest mercury mine and has been so for as long as mercury or cinnabar has been sought. The deposits were well known to the ancients, who used the cinnabar as a paint pigment. According to Bennett (1948), the cinnabar was used as far back as the fourth century B.C., but the tenth-century Moors

Figure 14-3
Map of Spain showing the
location of the Almadén district.

were the first to commercially distill the mercury. In recent years the deposits have been a fruitful source of income to the Spanish government, which owns and operates the mines (De Kalb, 1921; Hernandez Sampelayo, 1926; Menendez y Puget, 1949; Schuette, 1931; Saupé, 1973).

The rocks near Almadén consist of Lower Paleozoic clastics intruded by Tertiary(?) diabases and silicic porphyries (Almela Samper, 1959; Almela Samper *et al.*, 1962). Slates of unknown age underlie a thick, cross-bedded quartzite of early Ordovician age. A prominent ridge trending east-west to west-northwest just south of Almadén is formed by the quartzite, which dips vertically and is about 900 feet thick. Upper Ordovician sandstones, slates, and quartzites overlie the ridge-forming quartzite unit, and Silurian graptolitic shales, sandstone, and quartzite follow in sequence. The ore-bearing zone lies near the top of the Silurian section in a unit consisting of quartzite and slates with basalt sills and lava flows. Overlying the mineralized strata are Devonian quartzite, sandstone, and a few beds of fossiliferous limestone. Just north of Almadén shallow intrusives of diabase and various types of porphyries project into the Devonian rocks; however one small quartz porphyry mass near the ore deposits seems to intrude the Silurian rocks conformably.

All the strata are either vertical or dip steeply to the north (see Fig. 14-4). The stratigraphy has baffled most workers, thus leaving in doubt the true structural configuration, but a few isoclinal folds have been worked out. The intrusives, although poorly exposed, seem to parallel the sediments, which strike generally east-west. East of Almadén are two prominent strike-slip faults. The first is followed by a Tertiary (?) dike along part of its length and has about $2\frac{1}{2}$ miles of right-lateral displacement (Almela Samper, 1959).

The nature of the igneous activity associated with ore deposition at Almadén is controversial. Some geologists have concluded that there are no igneous rocks near the ore deposits; others insist that there exists an intimate association between the cinnabar and altered dikes (DeKalb, 1921; Ransome, 1921; Van der Veen, 1924). Much of the controversy results from inadequate field study; the intrusive silicic porphyries and diabases that crop out just north of Almadén

344

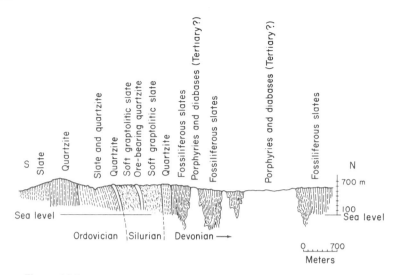

Figure 14-4
Cross section, approximately north-south, through the Almadén district, Spain.
(From Almela Samper, 1959.)

have been overlooked in most of the arguments. Almela Samper (1959) may
have solved the problem by mapping the area around rather than within the
mine. The large intrusive masses are strong evidence that there has been igneous activity in the area; they are younger than the ore-bearing strata and therefore represent a logical source of the ore.

The problem is further confused, however, by the presence of a strange
breccia near the mercury deposits. Locally, this rock is known as *piedra frailesca*
(friarlike rock) because of its textural resemblance to the robes of Franciscan
monks. This rock has been at the center of the debate over the presence or
absence of igneous activity. The piedra frailesca is a fine-grained granulated
rock containing fragments of quartzite, slate, porphyry, and limestone (Almela
Samper, 1959). It seems to be restricted to a single stratigraphic position
(Almela Samper and Febrel, 1960). Near the principal workings at Almadén
the piedra frailesca forms a cone or wedge-shaped structure some 800-feet
wide at the surface but absent or inconspicuous in the deepest workings.
Petrographically, the matrix of piedra frailesca is similar to a crushed quartzite
(Ransome, 1921); thus, it has long been considered a tectonic breccia. Almela
Samper and Febrel (1960) carefully studied it and concluded that it is a basaltic
tuff formed contemporaneously with associated lava flows. This interpretation
precludes the possibility that the piedra frailesca and the mercury deposits are
cogenetic because the period of metallization followed an episode of postbasalt
volcanism. An alternative hypothesis might be that the piedra frailesca represents material formed by explosive gaseous penetrations—that is, products of

fluidization (compare the description of piedra frailesca with, for example, the pseudoaplite of Tsumeb, South-West Africa, or the pipe at Black Peak, Arizona). This interpretation would seem to be suggested by the presence of both igneous and sedimentary fragments in the material; the pipe, or cone, shapes; and the association of mineral deposits. Nevertheless, it has not been advanced by any of the geologists who have examined the rocks in the field.

The ore at Almadén consists predominantly of cinnabar and smaller quantities of native mercury and pyrite (Van der Veen, 1924; Raynaud, 1941). Gangue minerals include quartz, calcite, dolomite, barite, and natrolite (or a related zeolite). Three closely spaced quartzite strata contain the orebodies (Fig. 14-5). These beds are separated by 10 to 20 feet of slate near the surface, but at depth they appear to merge. They dip from vertical to 70°N (see Fig. 14-6). The lodes are named, in stratigraphic order (from south to north), San Pedro-San Diego, San Francisco, and San Nicolás; they are, respectively, 17 to 36 feet wide, 8 feet wide, and 10 feet wide on the lower levels, where all of them are about 1,000 feet long (De Kalb, 1921). The vein material is brecciated quartzite cemented with quartz, a small amount of sericite, and zeolite. Pyrite is widely distributed, even in the adjacent slates. Weathering of the pyritized slate produces conspicuous outcrops that distinguish this slate from nonpyritiferous slate.

Rich zones, one to three meters wide, are restricted to the central plane of the lodes, and the values decrease gradually toward the walls. Until 1928 mining was confined to the area between two cross faults that cut the orebodies on the east and west sides, but underground exploration beyond these faults has proven the extension of ore. The true configuration of the orebodies is not yet known. Early reports that the lengths were less at depth than at the surface

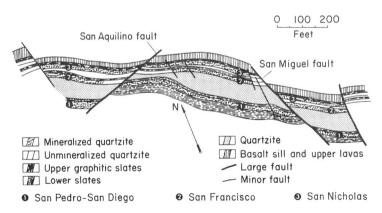

Figure 14-5
Plan of the 14th level, Almadén mine, Spain. (Courtesy of A. Almela Samper, 1962.)

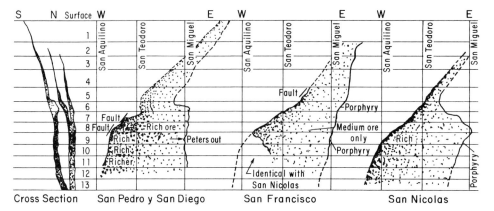

Figure 14-6
Outline of orebodies at the Almadén mine, Spain. (From Schuette, 1931.)

are no longer valid (Menendez y Puget, 1949); however, the three mineralized beds apparently converge and become thinner at depth (Bennett, 1948).

Petrographic studies have demonstrated clearly that the cinnabar replaces quartzite and is not merely an interstitial filling around grain borders (Beck, 1909; Ransome, 1921; Raynaud, 1941). This relationship was well illustrated by Ransome (1921) (see Fig. 14-7). Silicification, pyritization, and sericitization preceded the ore deposition. Subsequent fracturing of the rocks permitted the mercurial fluids to penetrate the quartzite beds, where cinnabar was deposited at the expense of sericite, quartz, and pyrite. The zeolite was deposited at the same time as the cinnabar.

The Almadén mine is unique among the quicksilver deposits of the world. Most mercury deposits contain irregular pockets of ore, unevenly distributed throughout brecciated and highly altered surficial materials. Almadén is the deepest and most continuous mine of its type. But even here, the evidence of near-surface deposition abounds. Although replacement of the quartzite has accounted for most of the ore, the cinnabar was introduced through innumerable small fissures, leaving vugs, druses, and banded open-space fillings. Nearly all mercury mines are epithermal; further evidence for this classification is the presence of a contemporaneous zeolite. The sericite, which might represent a slightly higher temperature, was deposited before the cinnabar. The age of metallization cannot be accurately determined, but is thought to be late and related to a period of thermal spring activity in the area (Menendez y Puget, 1949). If so, the ore must have been deposited very close to its present depth. In all respects, then, the Almadén mercury deposits conform to their classification as epithermal.

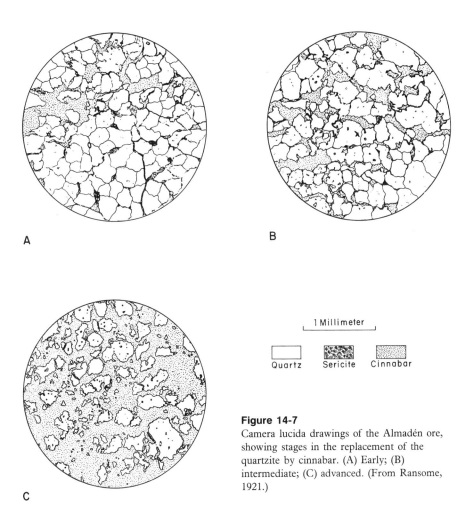

1 Millimeter

Quartz   Sericite   Cinnabar

Figure 14-7
Camera lucida drawings of the Almadén ore,
showing stages in the replacement of the
quartzite by cinnabar. (A) Early; (B)
intermediate; (C) advanced. (From Ransome,
1921.)

Although mineralization is traceable on the surface for more than 12 miles, with numerous scattered showings of cinnabar, most of the mining is confined to the immediate area of Almadén. The ore averages about 6 percent mercury, a grade that is three or four percent above their richest competitor. Since 1499, the date of the earliest records, the total production from Almadén has been about 240,000 metric tons of mercury. The reserves are adequate for at least another century of continuous production.

PACHUCA – REAL DEL MONTE DISTRICT, MEXICO

The famous bonanza silver districts of Pachuca and Real del Monte are about 60 miles north-northeast of Mexico City (Fig. 14-8). They would represent a single district were they not effectively separated by the crest of the Sierra de Pachuca. Pachuca lies along the west flank of the mountain range; Real del Monte is on the east flank, only about $3\frac{1}{2}$ miles away. Because Real del Monte receives about twice as much rainfall as Pachuca, their geographies differ considerably. But geologically they represent two ends of a single mineralized area (Ordoñez, 1902; Wisser, 1937, 1942; Winchell, 1922; Geyne, 1956; Geyne *et al.*, 1963).

The Sierra de Pachuca is made up of Tertiary volcanic rocks overlying Cretaceous sediments. Thick andesite flows and associated tuffs and breccias constitute the bulk of these, which are overlain by rhyolites, dacites, and local basalt flows. The ores were emplaced after the rhyolite and before the basalt. The volcanic rocks total about 6,500 feet in thickness. Premineral dikes of andesite, dacite, and rhyolite strike generally N75°W across the district. The most abundant dikes are rhyolitic, representing the last phase of igneous activity just before the ore-bearing fluids were introduced (Geyne, 1956).

Fossil plants in the volcanic rocks make it possible to date the periods of igneous activity fairly closely. The andesites are Oligocene to Miocene; the rhyolites, Late Miocene or Early Pliocene; and the younger lavas, Pliocene. The age of the ore, then, is probably Early Pliocene, slightly younger than the rhyolites.

The structure has been studied in great detail, because the ore deposits are localized along faults. Regionally, the volcanic rocks describe a broad, shallow syncline, but in the immediate mining district the bedding varies in dip from horizontal to 40° to 50° in all directions. East-west and northwest-southeast faults dominate the structural pattern, which is emphasized further by the northwest-southeast dike system (Fig. 14-9). In addition to the east-west faults, the eastern part of the district, in Real del Monte, has a well-developed set of mineralized north-south faults. The parallel fracture zones apparently represent regional faulting. The dikes ascended along the earliest fractures, but faulting continued throughout the period of mineralization. Nearly all of the east-west

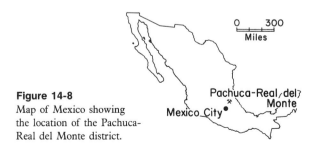

Figure 14-8
Map of Mexico showing the location of the Pachuca-Real del Monte district.

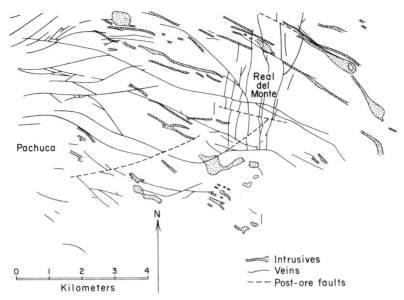

Figure 14-9
Veins, faults, and intrusives in the Pachuca–Real del Monte district, Mexico. The paired
lines and spotted areas are dikes and other intrusives, mostly rhyolite but including some
dacite; the solid lines are veins along faults; and the broken lines are postore faults.
Intrusives are as mapped on the surface; veins and faults are projected to the common
level. (From Geyne, 1956, Figs. 3 and 4.)

faults have steep dips and normal displacements. The north-south faults
formed near the end of the period of east-west faulting, just at the time of ore
deposition. These late faults are small, but recurrent movements and the timing
made them especially favorable as avenues for ascending ore fluids (Wisser,
1942; Geyne, 1956).

The ore deposits are in quartz veins along the faults, forming shoots where
local conditions favored development of breccias or wide zones. Paragenetically, the hypogene sulfide sequence is pyrite, sphalerite, galena, chalcopyrite,
acanthite, polybasite, and stephanite (Bastin, 1948; Geyne, 1956). Chalcocite,
covellite, and bornite occur locally, but they are believed to represent secondary
enrichment. By far the most abundant compound of silver is acanthite (Ag_2S),
a product of both hypogene and supergene processes. Other secondary minerals
include native silver, silver chlorides and bromides, malachite, azurite, anglesite, and oxides of manganese and iron. Besides quartz, the veins contain calcite, dolomite, rhodochrosite, rhodonite, bustamite, and barite.

Propylitization is the most widespread wall-rock alteration. It affects extensive areas of the volcanics. Minor amounts of sericitized, kaolinized, pyritized, and silicified wall rocks serve as valuable guides to ore (Thornburg, 1952;
Geyne, 1956).

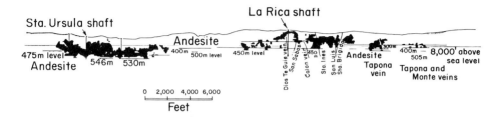

Figure 14-10
Longitudinal section (looking northerly) along the Vizcaina-Tapona vein, Pachuca-Real del Monte district, Mexico. (From Thornburg, 1952, Fig. 6.)

The minable ore shoots reach more than 3,000 feet in length, but the vertical dimensions are generally shorter (see Fig. 14-10). As a result of leaching by meteoric waters, most orebodies do not reach the surface. Except where exposed in Recent deep canyons, the tops of the ore deposits occur anywhere from 60 feet beneath the surface (in the east-west fault system) to more than 800 feet beneath the surface (in the north-south faults), and the ore shoots continue down 1,000 to 2,000 feet lower. The reasons for localization of ore along the veins cannot always be ascertained. Some shoots occupy fault intersections; others are in open zones where the faults changed attitude or formed branches, loop structures, and feather joints. In some places ore was deposited along the contacts between veins and rhyolite dikes (Wisser, 1942; Geyne, 1956; Thornburg, 1945). Veins average three to five feet in width, but the silver and gold values typically extend into the wall rocks as disseminations or as stringers that trend parallel to the veins. Consequently, the average stopes are eight to ten feet wide (Thornburg, 1952).

The evidence for shallow, low-temperature deposition of ore minerals is abundant. Repeated movements along the faults caused brecciation of the early quartz, allowing later quartz and ore minerals to enter. Banded veins, vugs, and replacement textures are all common. Some of the quartz is even chalcedonic, indicating possible deposition from a colloidal system. At depth the veins branch, the quartz content diminishes, and the ore values decrease abruptly. Geyne and his colleagues (1963) estimated that the zone of ore deposition was between 300 and 1,000 meters below the original ground surface. Table 14-1 shows the paragenesis of a typical vein, the Vizcaina-Tapona-Monte. This paragenesis is characteristic of many epithermal vein deposits.

Oxidation and supergene leaching have removed the ore minerals from the upper parts of veins. Much of the silver that was dissolved from the zone of oxidation was reprecipitated beneath the water table at the expense of iron, copper, and sulfur. This supergene-enriched zone accounts for some of the most valuable ore in the district.

Metallization is related to an underlying magmatic source, expressed at the

Table 14-1
Paragenetic sequence of the minerals in the Vizcaina-Tapona-Monte vein, Mexico.

	Primary	Secondary
ORE		
Pyrite	——	
Sphalerite	——	
Galena	——	
Polybasite	——	
Argentite-acanthite	——————	
Chalcopyrite	——	
Covellite		——
Chalcocite		——
Sternbergite	——	
Goethite		—
Silver	——	
Manganese oxides		——
Bornite	——	
Hematite		——
Neodigenite	——	
GANGUE		
Quartz	————	
Calcite	————	
Selenite		——
Kaolin	——	——
Chlorite	——	
Albite	——	
Prehnite	——	
Epidote	——	

Source: Geyne *et al.*, 1963.

surface as rhyolite dikes and small stocks or plugs. Some of the tectonic features are related to these stocks; the contemporaneity of fault movement and ore deposition implies a cogenetic source for the magmas and mineralizing fluids.

Since about 1526, when mining began in the Pachuca-Real del Monte area, the district has produced more than 1.2 billion Troy ounces of fine silver and 6.2 million Troy ounces of fine gold, the combined output being worth over 1.3 billion dollars at 1960 prices (Geyne *et al.*, 1963). This amounts to about six percent of the world's silver production for this period, most of which has been achieved during the present century. As evidence of this productivity, the district is honeycombed with 1,200 to 1,300 miles of underground workings.

REFERENCES TO SELECTED EPITHERMAL DEPOSITS

Comstock Lode, Nevada

Bastin, E. S., 1923. Bonanza ores of the Comstock Lode, Virginia City, Nevada, *U.S. Geol. Surv. Bull.* 735-C.

Becker, G. F., 1882. Geology of the Comstock Lode and the Washoe district, *U.S. Geol. Surv. Monogr.* 3.

Calkins, F. C., and T. P. Thayer, 1939. Preliminary map and mimeographed text of the Comstock Lode district, Nevada, U.S. Geological Survey.

Coats, R., 1940. Propylitization and relation types of alteration of the Comstock Lode, *Econ. Geol.* 35:1-16.

Gianella, V. P., 1934. New features of the geology of the Comstock Lode, *Mining & Metall.* 15(331):298-300.

———, 1936. Geology of the Silver City district and the southern portion of the Comstock Lode, Nevada, *Univ. Nev. Bull.* 30(9).

Thompson, G. A., 1956. Geology of the Virginia City quadrangle, Nevada, *U.S. Geol. Surv. Bull.* 1042-C.

Lake Superior Copper Deposits

Broderick, T. M., 1929. Zoning in Michigan copper deposits and its significance, *Econ. Geol.* 24:149-162, 311-326.

———, 1931. Fissure vein and lode relationships in Michigan copper deposits, *Econ. Geol.* 26:840-856.

Broderick, T. M., and C. D. Hohl, 1935a. Differentiation in traps and ore deposition, *Econ. Geol.* 30:301-312.

———, 1935b. The Michigan copper district, in *Copper Resources of the World,* vol. 1, 16th International Geological Congress.

Broderick, T. M., and H. N. Eidemiller, 1946. Recent contributions to the geology of the Michigan copper district, *Econ. Geol.* 41:675-725.

Butler, B. S., W. S. Burbank, et al., 1929. The copper deposits of Michigan, *U.S. Geol. Surv. Prof. Pap.* 144.

Cornwall, H. R., 1951. Differentiation in lavas of the Keweenawan Series and the origin of the copper deposits of Michigan, *Geol. Soc. Amer. Bull.* 62:159-202.

Lane, A. C., 1909. The Keweenawan Series of Michigan, *Mich. Geol. Biol. Surv. Pub.* 6, ser. 4, vols. 1-2.

Stoiber, R. E., and E. S. Davidson, 1959. Amygdule mineral zoning in the Portage Lake lava series, Michigan copper district, *Econ. Geol.* 54:1250-1277, 1444-1460.

Hauraki Peninsula, New Zealand

Bell, J. M., and C. Fraser, 1912. The geology of the Waihi-Tairua subdivision, Hauraki division, *N.Z. Geol. Surv. Bull.* 15.

Finlayson, A. M., 1909. Problems in the geology of the Hauraki gold fields, New Zealand, *Econ. Geol.* 4:632-645.

Jarman, A., 1916. The geology of the Waihi Grand Junction mine, *Inst. Mining Metall. London Trans.* 25:3-40.

Morgan, P. G., 1924. The geology and mines of the Waihi district, Hauraki gold field, New Zealand, *N.Z. Geol. Surv. Bull.* 26.

Park, J., 1897. The geology and veins of the Hauraki goldfields, New Zealand, *N.Z. Inst. Mining Eng. Trans.* 1:1-105.
Von Bernewitz, M. W., 1934. The Waihi: one of the world's greatest gold-silver mines (New Zealand), *Chem. Eng. Mining Rev.* 26:197-201, 233-236.

Cripple Creek, Colorado

Cross, W., and R. A. F. Penrose, Jr., 1895. The geology and mining industries of the Cripple Creek district, Colorado, *U.S. Geol. Surv. Ann. Rept.* 16(pt. 2):1-209.
Koschmann, A. H., 1947. The Cripple Creek district, Teller County, in *Mineral Resources of Colorado*, ed. J. W. Vanderbilt, State of Colorado Mineral Resources Board.
————, 1949. Structural control of the gold deposits of the Cripple Creek district, Teller County, Colorado, *U.S. Geol. Surv. Bull.* 955-B.
Lindgren, W., and F. L. Ransome, 1906. Geology and gold deposits of the Cripple Creek district, Colorado, *U.S. Geol. Surv. Prof. Pap.* 54.
Loughlin, G. F., 1927. Ore at deep levels in the Cripple Creek district, Colorado, *Amer. Inst. Mining Eng. Tech. Pub.* 13; also *Amer. Inst. Mining Eng. Trans.* 75:42-73.
Loughlin, G. F., and A. H. Koschmann, 1935. Geology and ore deposits of the Cripple Creek district, Colorado, *Colo. Sci. Soc. Proc.* 13(6):217-435.
Lovering, T. S., and E. N. Goddard, 1950. Cripple Creek district, *U.S. Geol. Surv. Prof. Pap.* 223, pp. 289-312.

REFERENCES CITED

Almela Samper, A., 1959. Esquema geológico de la zona de Almadén (Ciudad Real), *Inst. Geol. Minero España Bol.* 70:315-330.
Almela Samper, A., M. Alvarado, J. Coma, C. Felgueroso, and I. Quintero, 1962. Estudio geológico de la región de Almadén, *Inst. Geol. Minero España Bol.* 73:197-327.
Almela Samper, A., and T. Febrel, 1960. La roca frailesca de Almadén: un episodio tobáceo en una formación basáltica del Siluriano superior, *Inst. Geol. Minero España Not. Comun.* 59, pp. 41-72.
Bastin, E. S., 1948. Mineral relationships in the ores of Pachuca and Real del Monte, Hidalgo, Mexico, *Econ. Geol.* 43:53-65.
Beck, R., 1909. *Lehre von der Erzlagerstätten*, 3rd ed., vol. 1, Berlin, p. 521.
Bennett, E., 1948. Almadén, world's greatest mercury mine, *Mining & Metall.* 29:6-9.
Collins, W. P., 1918. *Mineral Enterprise in China*, London: Wm. Heinemann.
De Kalb, C., 1921. The Almadén quicksilver mine, *Econ. Geol.* 16:301-312.
Geyne, A. R., 1956. Las rocas volcánicas y los yacimientos argentíferos del distrito minero de Pachuca-Real del Monte, Estado de Hidalgo, *20th Int. Geol. Congr. Excursions A-3 & C-1*, pp. 47-57.
Geyne, A. R., C. Fries, Jr., K. Segerstrom, R. F. Black, I. F. Wilson, and A. Probert, 1963. Geology and mineral deposits of the Pachuca-Real del Monte district, state of Hidalgo, Mexico, *Mex. D.F. Consejo Recursos Naturales no Renovables Pub.* 5E.
Hernandez Sampelayo, P., 1926. Minas de Almadén, *14th Int. Geol. Congr. Guidebook Excursion B-1*.
Juan, V. C., 1946. Mineral resources of China, *Econ. Geol.* 41:399-474.
Menendez y Puget, L., 1949. The riches of Almadén, *Mining World* 11(7):34-36, 11(8):38-41, 11(9):35-37.

Ordoñez, E., 1902. The mining district of Pachuca, Mexico, *Amer. Inst. Mining Eng. Trans.* 32:224–241.

Ransome, F. L., 1921. The ore of the Almadén mine, *Econ. Geol.* 16:313–321.

Raynaud, J., 1941. Le minerai de la mine d'Almaden (Espagne), *Soc. Géol. Belg. Bull.* 64:226–237.

Saupé, F., 1973. La géologie du gisement de mercure d'Almadén, *Sci. Terre Mem.* 29.

Schuette, C. N., 1931. Occurrence of quicksilver orebodies, *Amer. Inst. Mining Eng. Trans.*, gen. vol., pp. 403–488.

Sudo, T., 1954. Types of clay minerals closely associated with metalliferous ores of the epithermal type, *Tokyo Kyoiku Daigaku Sci. Rept.* 3(23):173–197.

Tegengren, F. R., 1921. The Hsi-K'uang-Shan antimony mining fields, Hsin-Hua district, Hunan, *Geol. Surv. China Bull.* 3, pp. 1–26.

Thornburg, C. L., 1945. Some applications of structural geology to mining in the Pachuca–Real del Monte area, Pachuca silver district, Mexico, *Econ. Geol.* 40: 283–297.

——, 1952. The surface expression of veins in the Pachuca silver district of Mexico, *Amer. Inst. Mining Eng. Trans.* 193:594–600.

Van der Veen, R. W., 1924. The Almaden mercury ores and their connection with igneous rocks, *Econ. Geol.* 19:146–156.

Villa, E. M. di, 1919. The examination of mines in China, *North China Daily Mail* (Tientsin), pp. 71–73.

Winchell, H. V., 1922. Geology of Pachuca and El Oro, Mexico, *Amer. Inst. Mining Eng. Trans.* 66:27–41.

Wisser, E., 1937. Formation of the north-south fractures of the Real del Monte area, Pachuca silver district, Mexico, *Amer. Inst. Mining Eng. Trans.* 126:442–487.

——, 1942. The Pachuca silver district, Mexico, in *Ore Deposits as Related to Structural Features*, ed. W. H. Newhouse, Princeton, N.J.: Princeton Univ. Press.

15 / Telethermal Deposits

Some ore deposits are formed by hydrothermal fluids that have migrated so far from their source they have lost most of their heat and most of their potential to react chemically with the surrounding rocks. These terminal phases of the hydrothermal plumbing system are called *telethermal fluids* (Greek, *tele:* far). The telethermal zone is a shallow environment where temperatures and pressures are low and where the general characteristics of minerals are similar whether precipitated from descending meteoric waters or from ascending hydrothermal fluids diluted by cooler ground waters.

The mineralogy of telethermal ores is simple and nondiagnostic. It includes such minerals as iron-poor sphalerite and silver-poor galena, chalcopyrite, pyrite, marcasite, chalcocite, and very minor amounts of other sulfides. Native copper is deposited in the telethermal zone, and the oxide minerals are practically unlimited in variety. Many deposits of fluorite and barite are classified here. Gangue minerals include calcite, dolomite, quartz, fluorite, and barite, the latter two reaching economic proportions at places. In some districts telethermal galena, sphalerite, barite, and fluorite are found together in minable quantities.

One characteristic of telethermal deposits is the paucity of associated wall-rock alteration effects. In contrast to the widespread wall-rock alteration of the

Table 15-1
Factors for and against syngenetic, diagenetic, and epigenetic deposits in teletermal zones (Snyder, 1967).

SYNGENETIC ORE

For:

   Uniform mineralization at a given stratigraphic position or within a restricted range
   Close relationship between mineralization and particular sedimentary lithologies, facies, and features

Against:

   Transgressive relationships unrelated to facies changes
   Marked changes in height, width, and tenor of ore
   Mineralization of post-depositional structures
   Mineralization in a variety of lithologies
   Isotopes

DIAGENETIC ORE

For:

   Mineralization within a restricted stratigraphic interval
   Close relationship between mineralization and sedimentary lithologies and facies
   Mineralization related to diagenetic features and structures

Against:

   Extensive transgressive features unrelated to depositional or diagenetic structures
   Mineralization of post-lithification structures
   Mineralization in a variety of lithologies
   Extensive open-space filling
   Isotopes

EPIGENETIC ORE

For:

   Mineralization of post-lithification structures
   Marked changes in height, width, and tenor of ore that cannot be related to sedimentary or diagenetic features or environment
   Extensive open-space filling, vein, breccia, or bedding
   Distribution of mineralization relative to tectonic structures; fractured sediments not normally mineralized carry ore
   District-wide lack of close control of mineralization by specific sedimentary environments
   Isotopes

epithermal zone, the teletermal zone generally exhibits no alteration or only minor silicification, pyritization, and carbonatization. Cryptocrystalline silica may replace limestone near the ores, and scattered pyrite crystals may define a zone of reduced iron in clastic sediments, but the alteration halo is rarely conspicuous enough to be a noteworthy guide to ore. However, argillization of porous sediments may effectively bleach wide zones near teletermal deposits, although this alteration typically predates ore deposition, thus reducing the degree of correlation between bleached zones and orebodies.

Textures and structures are also nondiagnostic because the ores are deposited by replacement as well as by open-space filling, and the minerals may vary from aphanitic to very coarse-grained. The evidence for open-cavity filling is an abundance of crystal-lined vugs, comb structures, and rhythmically banded ores. Replacement is dominant in carbonate rocks, although many permeable limestones and dolomites contain ores that fill pore spaces and other cavities.

Telethermal deposits are likely to be structurally simple. They were formed a long distance from possible magmatic centers and commonly a long way from areas of strong tectonic activities. Circulation of fluids and deposition of ores were controlled by all types of permeability. Most of the ores are in flat-lying beds and show little or no evidence of deposition from ascending fluids. Their general appearance and character are such that they could readily (and have often been) interpreted as products of meteoric or sedimentary processes.

Many large deposits are classified as telethermal. These include such districts as Kennecott copper, Alaska; Aachen, Germany; Sardinia, Italy; Mississippi Valley and Tri State lead-zinc and fluorite; Colorado Plateau and Wyoming uranium-vanadium; Mascot-Jefferson City zinc, Tennessee; Upper Silesian lead-zinc, and many others in Europe, North Africa, and elsewhere. The recurrence of certain ore types is remarkable; the list includes copper, lead, zinc, and fluorite, or various combinations of these.

Most telethermal deposits are stratiform. Their lack of diagnostic characteristics has resulted in disagreement over their origin. Three genetic types are possible: syngenetic, diagenetic, and epigenetic. In an attempt to clarify the discussion concerning genesis, Snyder (1967) listed the criteria for and against each type. These are given in Table 15-1.

Ohle (1970) summarized the methods by which the telethermal (Mississippi Valley) type lead and zinc deposits might have formed. He pointed out six methods for which substantial arguments have been advanced.

1. Original syngenetic deposition.

2. Original dispersed and low-grade syngenetic deposition, with later concentration by regional metamorphism.

3. Original dispersed syngenetic deposition, with later concentration by groundwater moving upward in artesian flow.

4. Original dispersed syngenetic deposition, with later concentration by downward moving ground water.

5. Deposition from fluids of igneous derivation by hydrothermal or gaseous transport, either with volatile aid or simply as metallic vapor.

6. Deposition from connate basinal water that moved updip by compaction or other loading.

Figure 15-1
Map of Poland showing the location of
the Upper Silesian zinc-lead deposits.

The zinc-lead deposits of Upper Silesia, Poland, are the classic example of telethermal ore in carbonate rocks. Deposits of this type differ in many ways from the sandstone deposits. The principal mines are 10 to 25 kilometers northwest of Katowice (see Fig. 15-1). These ores have been exploited for more than six centuries, accounting for a major part of European production (Stappenbeck, 1928; Wernicke, 1931; Schneiderhöhn, 1941; Beyschlag et al., 1916; Zwierzycki, 1950; Sachs, 1914).

The Upper Silesian ores are in the Muschelkalk Limestone, a Triassic, shallow-water marine formation that includes dolomites and shales as well as limestones. No evidence of igneous activity is known in the region. Concentrations of ore are in the lower 350 feet of the Muschelkalk strata, which were faulted and folded into broad gentle warps during an Early Jurassic orogeny. The ore is irregularly distributed along the deeper parts of these structures, especially where cross-cutting faults have provided avenues for ore-forming solutions. Lower Muschelkalk beds are preserved only in the synclines. Four of these synclines contain most of the ore: the Bytom, Wilkoszyn-Trzebinia, Chrzanow, and Tarnowskie-Góry. A small amount of ore is present in the Krzeszowice-Siewierz monocline, an extensive structure that causes the Muschelkalk Limestone to plunge to the northeast under younger sediments (Fig. 15-2). The cross-cutting faults extend upward from the underlying Paleozoic rocks and apparently resulted from renewed movements along older faults. Displacements in the Muschelkalk Limestone are generally less than 100 feet.

Deposition of the ore was preceded by extensive dolomitization, which is most thoroughly developed along a single limestone bed. The dolomitized zones are widely fractured and brecciated, particularly where they are folded. They thus form a permeable host for the circulating ore fluids.

A somewhat unusual feature of the district is the presence of a clay layer at the base of the dolomitized stratum. Known locally as the "vitriolic clay," it is a slightly permeable bed, 20 to 40 centimeters thick, and marks the lower limit of mineralization. The clay bed was apparently formed by removal of argillaceous impurities from the limestone during dolomitization and concomitant recrystallization. Concentrated at the base of the dolomite, the clay bed impeded the circulation of ore-bearing fluids. All dolomite and all ore are above the clay layer. The deposition of ores above an impermeable clay and near the base of a permeable dolomite is more readily explained as a result of groundwater circulation than as a product of ascending hydrothermal fluids. Nevertheless, the ore-bearing fluids joined the groundwater system via faults, as evidenced by traces of mineralization along the faults at depth (Zwierzycki, 1950).

The ore deposits closely follow the base of the dolomite, where they form layers 5 to 13 feet thick. Two similar layers are stratigraphically above and within 100 feet of the principal bed in the Bytom syncline, and in several other parts of the region a third layer is recognized. All of the ore lies within 300 feet of the surface; none of the ore is continuous, even in the principal bed; orebodies are largest and richest near faults and fissures. Shattering of the dolomite was more extensive along troughs of synclines than along crests of anticlines, so it is doubtful that extensive ore deposits were removed by erosion of the anticlinal folds. The upper bed, which is connected to the lower bed by veins, forms a horizontal body, irrespective of the structure, but the other orebodies conform to the bedding.

The primary minerals include sphalerite, wurtzite, galena, pyrite, marcasite, and small amounts of the arsenic and antimony sulfides, jordanite ($4PbS \cdot As_2S_3$), gratonite ($9PbS \cdot 2As_2S_3$), and meneghinite ($4PbS \cdot Sb_2S_3$). A regional zoning is shown by the lead-zinc ratios; in the rich central areas zinc greatly predominates over lead, but marginal ores contain more lead than zinc. The mines have produced an estimated five to ten times more zinc than lead. A typical analysis of ore from the rich Bytom syncline showed 1.7 percent lead and 10.7 percent zinc. The galena contains very small amounts of silver, some of the sphalerite contains cadmium, and some of the pyrite contains thallium (Zwierzycki, 1950).

Ascending solutions entered the dolomite along faults, and ore deposition took place at low temperatures after the fluids had mixed and migrated with groundwaters. The ore minerals were deposited in crevices, solution cavities, and joints in the dolomite, just above the clay stratum, or in overlying zones where brecciation was especially prevalent. The presence of encrustations and stalactites of ore shows that much of the deposition took place in open spaces, but, in addition, sulfides replaced the dolomite.

The mineral textures are typically colloform, both superficially and in detail (Kutina, 1952; Krusch, 1929). Sphalerite was deposited in rhythmically

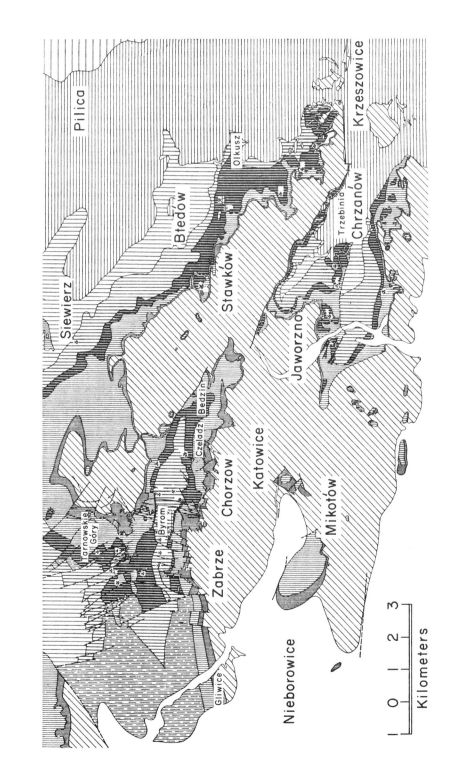

Pilica

Siewierz

Błedów

Olkusz

Stawków

Trzebinia

Chrzanów

Krzeszowice

Jaworzno

Czeladź

Będzin

Katowice

Tarnowskie
Góry

Byrom

Chorzow

Zabrze

Mikołów

Gliwice

Nieborowice

0   1   2   3
Kilometers

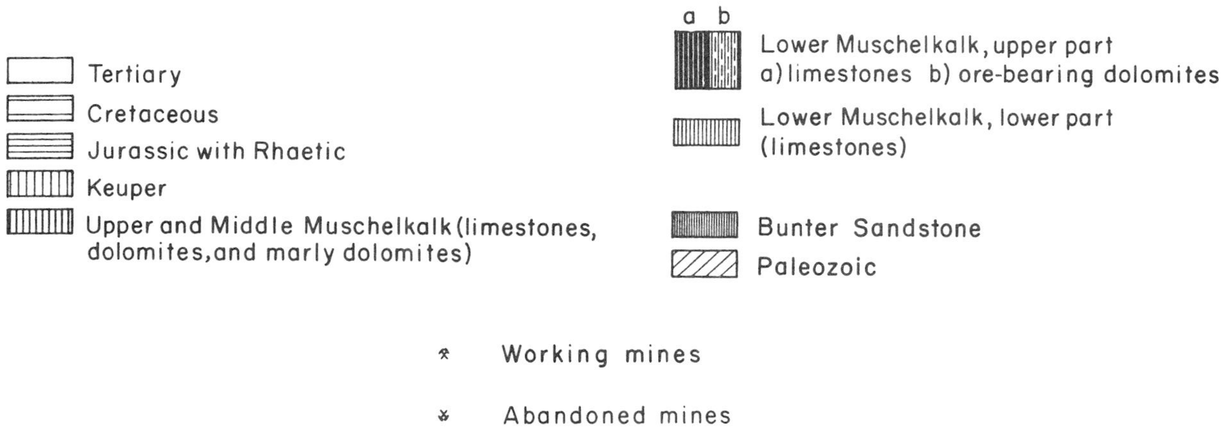

Tertiary

Cretaceous

Jurassic with Rhaetic

Keuper

Upper and Middle Muschelkalk (limestones, dolomites, and marly dolomites)

a b
Lower Muschelkalk, upper part
a) limestones b) ore-bearing dolomites

Lower Muschelkalk, lower part
(limestones)

Bunter Sandstone

Paleozoic

✷ Working mines

✷ Abandoned mines

Figure 15-2

Geologic map of the lead-zinc field of Upper Silesia, Poland. (From Zwierzycki, 1950, Fig. 82.)

banded, botryoidal masses that show numerous shrinkage cracks, which trend both normal and parallel to orbicular surfaces. Liesegang rings are also present, and spectrographic analyses show slight differences in iron content between the light and dark bands. The excellent structural and mineralogical papers published by Haranczyk emphasize the colloform nature of most of the ore. Haranczyk also considers that the ores are of hydrothermal origin (Haranczyk, 1958, 1959, 1961, 1963a, 1963b). Figure 15-3 shows the oölitic form of fine-grained sphalerite. According to Krusch (1929), the zinc sulfide was originally deposited as a colloidal gel that gradually crystallized to wurtzite and finally to sphalerite. Marcasite was deposited contemporaneously with the zinc sulfide; it too exhibits colloform textures, such as spherical shapes and desiccation cracks. Galena was probably deposited from a true solution; it forms alternating layers with sphalerite and marcasite, indicating that both chemical precipitation and flocculation of colloids took place in the same hydrothermal system. Younger generations of galena were deposited after the zinc-iron sulfide phase was completed. Haranczyk's description of the ore in the Boleslaw mine near Olkusz and other nearby properties points out that ZnS is in both isometric and hexagonal forms. ZnS may either be in oölites or form crusts of cryptocrystalline ZnS. Inclusions of galena, some being measured in the hundredths of a

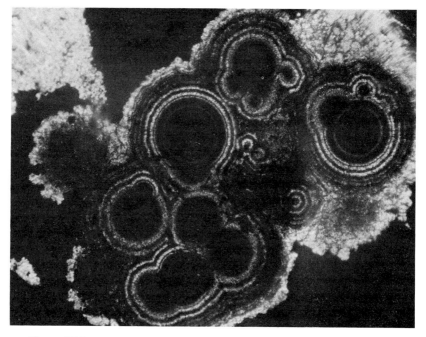

Figure 15-3
Oölitic sphalerite or "brunckite" ore. Thin section, crossed nicols. Boleslaw mine, Poland. ×70. (From C. Haranczyk, 1959.)

millimeter, form droplets and streaks in the oölitic zinc sulfide. Sometimes the galena droplets run together and form partly developed skeleton crystals (Haranczyk, 1958, Figs. 1 and 2). Massive galena is present at many places; where it is etched it also shows peculiar colloform structures (see Fig. 15-4).

Along the shallow edges of the syncline and on the Krzeszowice-Siewierz monocline, the ore beds are deeply oxidized to a product known locally as galman. In places the zone of galman is as much as 60 to 100 feet thick. The higher solubility of zinc caused the zinc and iron to separate during weathering. Relatively pure iron oxides are concentrated along the outcrops, and zinc enrichment zones lie beneath. Enrichment of cadmium also accompanied oxidation. The most abundant oxidation products are smithsonite, hemimorphite, cerussite, and limonite, but minor amounts of other minerals, such as anglesite, goslarite ($ZnSo_4 \cdot 7H_2O$), tarnowitzite [$(Ca,Pb)CO_3$], and phosgenite ($PbCl_2 \cdot PbCO_3$) are also found (Zwierzycki, 1950).

The origin of the Upper Silesian ores has been widely discussed, but few workers agree completely. Certainly the ores were deposited at low temperatures, and the presence of mineralized fissures continuing from the Muschelkalk Limestone downward into the Paleozoic rocks is strong evidence that ascending solutions deposited the ore. But Sachs (1914) argued that the ores were

Figure 15-4
Massive galena etched with HNO_3. Oölites may be seen on the sides. Boleslaw mine, Poland. ×70. (From C. Haranczyk, 1959.)

deposited from descending solutions that were dammed against the vitriolic clay and precipitated by organic matter in the clay. Stappenbach (1928) envisioned a process of lateral secretion whereby artesian waters leached syngenetic metals from the limestones and reconcentrated them as ore deposits. Krusch (1929) and Kutina (1952) emphasized the colloform structures and textures and attributed great significance to colloidal transportation. At some mines the evidence for colloidal deposition is so striking that this origin remains virtually unchallenged. Other deposits in the district indicate that part, but not all, of the zinc sulfides and marcasite were precipitated as colloids.

Haranczyk decided, as a result of careful study, that the ores are of hydrothermal origin. Galkiewicz (1967) summarized the evidence concerning the origin of the deposits and concluded that it is hydrothermal. Arguments in support of this theory are: (1) the vertical interval of ore mineralization that extends into the Devonian, Carboniferous, Permian, Triassic, and Jurassic beds; (2) the position of the larger deposits along W-NW fractures; (3) the great variability in form and content of the orebodies; (4) the characteristic ore textures; (5) the vertical zonality of mineralization; (6) the big haloes around the orebodies; (7) different physicochemical conditions in the formation of wall rocks and orebodies, the differences being confirmed by the distribution of marcasite, wurtzite, and sulfosalts in the carbonate rocks; (8) temperature of formation of sphalerite, 120°C; and (9) the isotopic composition of lead, which indicates Late Tertiary age for the ore mineralization.

On the other hand, Gruszczyk (1967) favors a sedimentary origin for the deposits and cites the following evidence: (1) a lack of intrusive or extrusive rocks; (2) the regional extent of the mineralization not only in Poland but throughout much of Central Europe; (3) interdependence of the deposits and facies of the Triassic system, regional zoning, subordination of deposits to stratigraphic factors, deposits in the form of beds and dislocations being postore; (4) simple mineral composition throughout, and negligible minerals typical of hydrothermal deposits; and (5) the uniform textures and the explanation of veined and brecciated ores as having been formed by diagenetic processes. Kautzsch also concluded that the Upper Silesian-Cracovian ores, as well as many from other European districts, are of primary sedimentary origin (Kautzsch, 1967).

Summarizing the many conflicting opinions and statements concerning the origin of the deposits is difficult. Certainly the ores were formed near the surface at comparatively low temperatures and pressures. Comparison with ores of the Mississippi Valley, especially those of Southeast Missouri, shows a striking similarity. Recent developments in the Mississippi Valley favor a hydrothermal origin for the ores, and this, combined with the facts listed by Haranczyk (1963) and Galkiewicz (1967) appear to favor this origin for the Upper Silesian-Cracovian deposits. Ridge and Smolarska (1972) point out that the isotope ratios in the Upper Silesian ores are remarkably uniform.

REFERENCES TO SELECTED TELETHERMAL DEPOSITS

Tri-State District, Mississippi Valley

Bastin, E. S., ed., 1939. Contributions to a knowledge of the lead and zinc deposits of the Mississippi Valley region, *Geol. Soc. Amer. Spec. Pap.* 24.

Behre, C. H., Jr., A. V. Heyl, Jr., and E. T. McKnight, 1950. Zinc and lead deposits of the Mississippi Valley, *18th Int. Geol. Congr. Rept.*, pt. 7, pp. 51-69.

Fowler, G. M., 1942. Ore deposits in the Tri-State zinc and lead district, in *Ore Deposits as Related to Structural Features*, ed. W. H. Newhouse, Princeton, N.J.: Princeton Univ. Press.

Fowler, G. M., and J. P. Lyden, 1932. The ore deposits of the Tri-State district, *Amer. Inst. Mining Eng. Trans.* 102:206-251.

————, 1935. The ore deposits of the Tri-State district—discussion, *Econ. Geol.* 30: 565-575.

Newhouse, W. H., 1933. The temperature of formation of the Mississippi Valley lead-zinc deposits, *Econ. Geol.* 28:744-750.

Ohle, E. L., 1958. Some considerations in determining the origin of ore deposits of the Mississippi Valley type, *Amer. Inst. Mining Eng. Preprint* 5817A6.

————, 1970. Mississippi Valley-type ore deposits—a general review, *Wash. Div. Mines Geol. Bull.* 61, pp. 5-15.

Ridge, J. D., 1936. The genesis of Tri-State zinc and lead ores, *Econ. Geol.* 31:298-313.

Snyder, F. G., 1966. Criteria for origin of stratiform orebodies, with application to Southeast Missouri, *Econ. Geol. Monogr.* 3, pp. 1-13.

Snyder, F. G., and P. E. Gerdemann, 1968. Geology of the southeast Missouri lead district, in *Ore Deposits of the United States* (L. C. Graton-R. Sales Mem. Vol.), vol. 1, American Institute of Mining Engineers.

Stoiber, R. E., 1946. Movement of mineralizing solutions in the Picher field, Oklahoma-Kansas, *Econ. Geol.* 41:800-812.

Morocco

Aubert de la Rue, E., 1928. *Observations sur Quelques Gisements Marocains de Plomb et de Zinc*, Paris: Mines et Carrières.

Blondel, F., 1935. Le plomb, le zinc, et l'argent, in *Resources Minerales France Outre-mer*, Paris, pp. 1-82.

Bouladon, J., 1952. Plomb et zinc, in *Géologie des Gites Minéraux Marocains*, 19th International Geological Congress (Monogr. Regional., 3rd ser., no. 1).

Claveau, J., J. Paulhac, and J. Pellerin, 1952. The lead and zinc deposits of the Bou Beker-Touissit area, eastern French Morocco, *Econ. Geol.* 47:481-493.

Jouravsky, G., F. Permingeat, J. Bouladon, and J. Agard, 1950. Deux types de gisements de plomb au Maroc Français, *18th Int. Geol. Congr. Rept.*, pt. 7, pp. 222-233.

REFERENCES CITED

Beyschlag, F., J. H. L. Vogt, and P. Krusch, 1916. *The Deposits of the Useful Minerals and Rocks*, vol. 2, tr. S. J. Truscott, London: Macmillan, pp. 723-730.

Galkiewicz, T., 1967. Genesis of Silesian-Cracovian zinc-lead deposits, *Econ. Geol. Monogr.* 3, pp. 156-168.

Gruszczyk, H., 1967. The genesis of the Silesian-Cracow deposits of lead-zinc ores, *Econ. Geol. Monogr.* 3, pp. 169–177.

Haranczyk, C., 1958. Skeletal and colloform textures of galena from Silesian-Cracovian lead-zinc deposits, *Acad. Sci. Poland Ser. Chem. Geol. Geogr.* 7(1):55–58.

————, 1959. Brunckite from the Silesian-Cracow zinc and lead deposits, *Acad. Sci. Poland Ser. Chem. Geol. Geogr.* 7(5):359–366.

————, 1961. The PbS gel-boleslavite, *Acad. Sci. Poland Ser. Chem. Geol. Geogr.* 9(2): 85–89.

————, 1963a. Vertical ore-zoning in the zone of faulting observed in Klucze near Olkusz (Silesian-Cracovian zinc and lead deposits), in *Symposium on Problems of Postmagmatic Ore Deposition*, vol. 1, Prague: Czechoslovak Academy of Sciences.

————, 1963b. Silesian-Cracovian type of Zn-Pb ore deposits and their comagmatic relation to alkaline igneous rocks [article in Polish], *Rudy Metale Niezelazne* 10(3): 132–139, 187–193.

Hawkes, H. E., and J. S. Webb, 1962. *Geochemistry in Mineral Exploration*, New York: Harper and Row.

Kautzsch, E., 1967. Genesis of stratiform lead-zinc deposits in central Europe, *Econ. Geol. Monogr.* 3, pp. 133–137.

Krusch, P., 1929. Über kolloidal Vorgänge bei der Entstehung der oberschlesischen Zink-Bleierzlagerstätten, *Z. Oberschlesischen Berg. Huttenmännischen Vereins Katowice*, nos. 6–7. Abstr. in *Z. Deut. Geol. Ges.* 81:169–170.

Kutina, J., 1952. Mikroskopischer und spektrographischer Beitrag zur Frage der Entstehung einiger Kolloidalstrukturen von Zinkblende und Wurtzit, *Geologie* 1:436–452.

Ohle, E. L., 1970. Mississippi Valley type ore deposits, a general review, *Washington Div. Mines Bull.* 61:5–15.

Ridge, J. D., and I. Smolarska, 1972. Factors bearing on the genesis of the Silesian-Cracovian lead-zinc deposits in southern Poland, *24th Int. Geol. Congr. Rept.*, sec. 6, pp. 216–229.

Sachs, A., 1914. Die Bildung schlesischer Erzlagerstätten, *Centralbl. Mineral. Geol. Paläontol. Jahrb.*, pp. 12–19, 186–190.

Schneiderhöhn, H., 1941. *Lehrbuch der Erzlagerstättenkunde*, Jena: Gustav Fischer, pp. 573–579.

Snyder, F. G., 1967. Criteria for origin of stratiform orebodies, *Econ. Geol. Monogr.* 3, pp. 1–13.

Stappenbeck, R., 1928. Ausbildung und Ursprung der oberschlesischen Bleizinkerzlagerstätten, *Arch. Lagerstättenforsch.* 41.

Wernicke, F., 1931. Die primären Erzmineralien der Deutsch-Bleischarley-Grube bei Beuthen O.S., *Arch. Lagerstättenforsch.* 53.

Zwierzycki, J., 1950. Lead and zinc ores in Poland, *18th Int. Geol. Congr. Rept.*, pt. 7, pp. 314–324.

16 / Xenothermal Deposits

Plutons intruded to shallow depths expel high-temperature fluids into low-pressure environments. Under these conditions the temperature and pressure gradients are exceptionally steep, causing ore fluids to undergo rapid cooling and sudden losses of pressure during their ascent. As a result, the ore minerals are deposited over only a short distance and in a confused paragenesis. The earliest minerals to form are high-temperature varieties, but rapid cooling to near-surface temperatures requires deposition of typical low-temperature minerals during the waning stages of hydrothermal activity. Furthermore, most of the early, high-temperature minerals are not in equilibrium with the cooler phases, and so are etched and altered during later mineralization. Thus, the pressure and temperature indications may be complex and confused. Mixed high- to low-temperature ores deposited near the surface are known as *xeno-thermal* deposits (Greek, *xeno:* strange, abnormal). This category was introduced by Buddington (1935) as a necessary addition to Lindgren's classification.

Depending upon how fast the temperature and pressure decrease as the ore-bearing fluids ascend toward the surface, the high-temperature and low-temperature minerals may be either "dumped" together or spread along a restricted course. "Dumping" results when minerals that ordinarily are not found together are precipitated practically simultaneously. In some xenothermal deposits the shallower minerals closely follow the higher-temperature, deeper minerals, but the sequence of deposition is recognizable and similar to the normal paragenesis of vein minerals. This type of deposit is said to be "telescoped" because each zone overlaps the next. Telescoping and dumping characterize xenothermal deposits.

Most xenothermal deposits are associated with volcanic and tuffaceous rocks of comparatively recent age, but they also are found at shallow depths in rocks of all types and ages. The deposits generally form composite veins, developed by the periodic reopening of fissures and the deposition of progressively lower temperature materials. Telescoped veins show a simple gradation in space, from high-, through intermediate-, to low-temperature minerals. Open-fissure textures tend to predominate over replacement textures because the systems are throughgoing and there is not enough time for most replacement reactions to take place. The host rocks are typically fractured, crackled, or sheared, and the hydrothermal minerals are generally fine-grained.

The mineralogy of xenothermal deposits is usually complex due to the extreme range of temperatures involved. Typical high-temperature minerals, such as cassiterite, wolframite, magnetite, specularite, scheelite, and molybdenite, occur with minerals characteristic of low-temperature environments, such as the silver sulfosalts. Furthermore, any of the minerals common to mesothermal deposits may be present in xenothermal assemblages. The gangue minerals include such diverse associations as orthoclase, tourmaline, topaz, augite, diopside, phlogopite, chalcedony, apatite, and alunite; but beryl, alkali tourmalines, spodumene, and other high-pressure minerals are not formed in the xenothermal environment. Wall-rock alteration ranges from tourmalinization to kaolinization and alunitization, depending upon the temperatures or the phase of hydrothermal activity.

IKUNO-AKENOBE DISTRICT, JAPAN

The Late Tertiary rocks of Japan contain several excellent examples of xenothermal deposits (Kato, 1928). One of the best is the Ikuno and Akenobe district in Hyogo Prefecture, southwestern Honshu (Fig. 16-1). This district has produced gold, silver, copper, lead, zinc, tin, tungsten, bismuth, and arsenic from Tertiary volcanic rocks and underlying sediments (Kato, 1920, 1927, 1928; Geological Survey of Japan, 1960; Nakano, 1931; Kondo and Kawasaki, 1943; Yamaguchi, 1939; Sato and Kaneko, 1952; Maruyama, 1957; Abe, 1963; Imai, 1966; Imai, Katayama, and Fukuoka, 1970).

The mines are in Cenozoic, Mesozoic, and Paleozoic rocks ranging from slates, phyllites, and quartzites at the base, to Mesozoic and Early Tertiary shales, conglomerates, and sandstones, assorted Tertiary volcanics, and Quaternary basalt. Mineralization accompanied a late phase of igneous activity, but

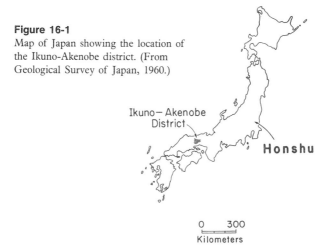

Figure 16-1
Map of Japan showing the location of
the Ikuno-Akenobe district. (From
Geological Survey of Japan, 1960.)

Ikuno–Akenobe
District

Honshu

0   300
Kilometers

the basalts postdate the ore. The igneous activity began with the ejection of a
rhyolitic tuff breccia, followed by thick rhyolite flows. Overlying the rhyolite
is a sequence of pyroxene andesite flows, which are extensively propylitized
in places. Numerous dikes of andesite, quartz porphyry, and basalt cut the
sedimentary rocks and flows. The ore-bearing veins are younger than some
dikes but older than others. Late Mesozoic gabbros and diorites intrude the
sediments, but these plutonic rocks are not related to the period of ore
deposition.

Ikuno and Akenobe are the principal mines in a regional swarm of veins.
The veins are fissure fillings; some lie alongside brecciated structures and
others occupy repeatedly opened fractures. Banding and ribbon structures
abound. Near the Ikuno mine the veins are in Tertiary volcanics, but in the
Akenobe province they are in the older metamorphic and plutonic rocks.
Nevertheless, the veins in these provinces are similar in nearly all aspects; they
are all composite, telescoped fissure fillings that record a sequence of high-
to low-temperature surges of mineralization. Each vein may have a slightly
different history than the others, but the general succession is the same everywhere—that is, tin and tungsten are followed by copper, then zinc-lead, and
finally by silver-gold-quartz.

The principal vein in the Ikuno mine is the Kanagase lode. It strikes roughly
north-south and dips 60° to 80°E. The vein has been followed for about two
miles along strike and passes from rhyolite at the surface into the underlying
Tertiary sediments. At places the Kanagase vein is as much as 25 feet wide,
but it pinches and swells irregularly. It is a typical composite vein, formed by
successive depositions along a repeatedly opened fracture (Kato, 1927). Near
the surface the vein is principally a copper producer, but within 600 feet of
depth it also carries appreciable amounts of tin and tungsten.

Four successive stages of mineralization are readily distinguished in the Kanagase vein (Kato, 1927, 1928). The first stage is represented by cassiterite and quartz and minor amounts of pyrrhotite, chalcopyrite, and pyrite that surround and replace the cassiterite and quartz. Small amounts of scheelite fill the interstices between quartz crystals.

Sharply defined veins of wolframite-cassiterite-quartz constitute the second stage. They cut through the earlier cassiterite-quartz veins. Again, there are minor quantities of sulfide minerals and a late-phase addition of scheelite. Tiny crystals of topaz are enclosed in the granular quartz. The sulfides are late in the paragenesis; they fill intergranular spaces and transect even the scheelite.

The third stage consists of massive chalcopyrite in veins and irregular impregnations, with associated quartz, sphalerite, galena, tetrahedrite, and other sulfides. This was the main stage of sulfide deposition. Chalcopyrite and quartz were the first minerals deposited during this stage, and they in turn are cut and replaced by the other sulfides.

The fourth, and last, stage of mineralization is represented by gold-silver-quartz veins that transect all previous stages of mineralization. Some of the veins consist of massive quartz, but wide zones and sporadic stretches contain dark patches caused by minute grains of tetrahedrite, chalcopyrite, pyrite, sphalerite, galena, and other sulfides. The sulfide minerals fill the interstices between quartz grains and associated sericite. Gold and silver occur primarily as trace elements in the sulfides, but native gold has also been found.

Five stages of mineralization have been described in the Akenobe district (Kato, 1927). In chronologic order they are (1) chalcedonic quartz with cassiterite and minor amounts of sulfides; (2) wolframite-cassiterite-quartz with accessory bismuth, fluorite, and topaz; (3) massive chalcopyrite veins; (4) veinlets of sphalerite; and (5) barren quartz. This sequence is almost identical to that in the Kanagase vein.

Imai described five zones of mineralization in the Akenobe mine. The zones are: (1) tin-tungsten; (2) tin-copper; (3) copper-zinc; (4) zinc-lead, and (5) gold-silver. The first four zones are shown in Figure 16-2.

The ore throughout the district is in composite veins that carry chalcopyrite, bornite, cassiterite, stannite, wolframite, scheelite, native bismuth, bismuthinite, native gold, argentite, pyrite, sphalerite, galena, and arsenopyrite in a gangue of quartz, calcite, chlorite, fluorite, siderite, apatite, sericite, barite, and adularia (Geological Survey of Japan, 1960; Kato, 1927). Wall-rock alteration effects are appreciable and seem to depend upon the host-rock composition. The rhyolites were silicified, the andesites propylitized, and the slates chloritized. Kato (1927, 1928) interpreted many of the textures of both the ore and gangue minerals as products of colloidal deposition. Even the earliest cassiterite is in globular aggregates of radial crystals that appear to be spherulites of recrystallized colloidal gel. Much of the quartz forms botryoidal structures and also has a radial texture. If these minerals were deposited as colloids,

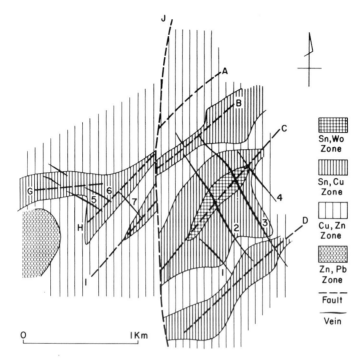

Figure 16-2
Zonal distribution of the ores in the adits of the Main level of the
Akenobe mine, Japan. (From Imai, 1966.)

they represent an unusually high-temperature example of this type of deposition. Perhaps they reflect the rapid aggregation of solid particles in a system that did not provide sufficient time for equilibrium reactions.

All the veins in this district are related genetically to the Late Tertiary volcanic activity, represented by large volumes of extrusive rhyolite and minor andesitic and felsitic dikes. The hydrothermal fluids apparently represent the latest fraction of an underlying magma that had previously supplied the volcanic materials.

Zoning is evident in many of the veins of the area. Imai gives a longitudinal section along the Senju vein in the Ikuno mine; this is reproduced in Figure 16-3.

Takahashi and his associates (1955) studied about 70 samples of ore collected from a vertical range of 500 feet in the Akenobe mine. Using the decrepitation method of geothermometry, they discovered that the temperatures of formation on the 500-foot level averaged about 300°C, whereas those from the 1,000-foot level averaged 335°C. However, the fact that the veins are composite probably means that the temperature at any one level varied widely. Thus,

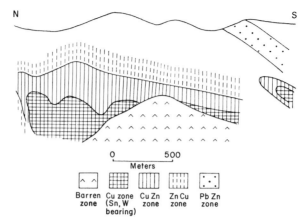

Figure 16-3
Longitudinal section of the Senju vein in the Ikuno mine, Japan, showing the zonal distribution of vein materials. (From Imai, 1966.)

the temperature *range* is probably more significant than the average temperature gradient. The maximum temperature was 350°C, and the minimum temperature was 160°C; this range apparently represents a decrease in temperature with time. Imai used the decrepitation method to study the temperatures of formation of the ores and zones of the Akenobe mine (Imai, 1966). His results, based mainly upon ores of the lower zones, agree in general with those of Takahashi.

The sequence of mineral deposition in the veins agrees well with the standard zonal sequence. The paragenesis further shows that the earliest minerals formed at high temperatures and that the latest minerals formed at moderate temperatures. Since all stages of this sequence are found either together or in telescoped zones, the deposit is logically classified as xenothermal, and most workers interpret the ores as the near-surface expression of an underlying shallow intrusive mass.

POTOSÍ DISTRICT, BOLIVIA

Several tin and tin-silver deposits in central and southern Bolivia are classic examples of xenothermal ore deposition. The region as a whole contains many ore districts, which range over a wide belt along the Eastern Andes of central Bolivia (Fig. 16-4). This belt extends across southwestern Bolivia along a curve from north-south to northwest-southeast (Ahlfeld, 1946). It is one of the

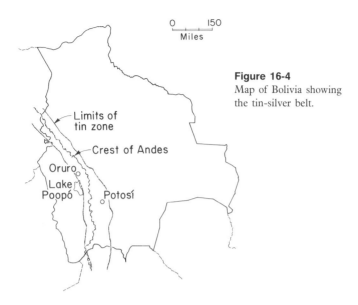

0   150
Miles

Figure 16-4
Map of Bolivia showing
the tin-silver belt.

Limits of
tin zone

Crest of Andes

Oruro

Lake
Poopó   Potosí

world's most productive metallogenetic provinces, reaching beyond the borders of Bolivia to attain a total length of about 580 miles. The most intense mineralization lies within a more restricted area, about 150 miles long, between Oruro on the northwest and Potosí on the southeast, coinciding with a major bend in the Andean Cordillera (Ahlfeld, 1941). In northern Bolivia the deposits are hypothermal and mesothermal (Turneaure and Welker, 1947), but in southern and central Bolivia they are characteristically xenothermal. Tin and silver are the principal metals recovered from the xenothermal deposits, but some also produce tungsten, lead, zinc, antimony, and gold. Potosí is said to have produced more silver than any other district in the world.

The entire belt of mineralization is related to Middle(?) Tertiary intrusives—granitic batholiths, stocks, and shallow intrusives that apparently represent the uppermost cupolas and apophyses of an underlying batholithic complex. Nearly all deposits were emplaced under near-surface conditions over a wide range of temperatures and at low pressures. Composite veins, sheeted zones and stockworks carry the ores, and only a minor amount of erosion followed their emplacement. Lindgren and Creveling (1928) estimated that the ores of the xenothermal zone were deposited at a depth of 3,000 feet or less.

The ore deposits at Potosí are restricted to an area of about one-half square mile. As shown in Figure 16-5, Potosí lies in a region of Paleozoic (Ordovician) shales unconformably overlain by Tertiary volcanic rocks and their clastic derivatives. The principal host rock is a quartz porphyry stock, the Potosí intrusive, which forms a pyramidal hill called Cerro Rico. It is funnel shaped,

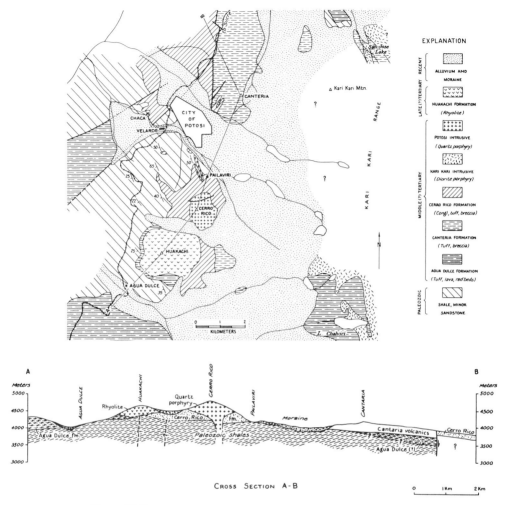

Figure 16-5
Geologic map and cross section of the Potosí district, Bolivia. (From Turneaure, 1960, Figs. 2 and 3.)

decreasing in cross section from about 5,600 × 4,000 feet at the surface to about 325 × 325 feet at a depth of 1,650 feet (Turneaure, 1960). Although the original composition of the Potosí intrusive was masked by strong propylitization, sericitization, pyritization, and silicification, petrographic studies have indicated that it was either dacite or quartz latite (Jaskolski, 1933; Ahlfeld, 1941; Lindgren and Creveling, 1928).

Most veins at Potosi fill shear fractures and normal faults of only slight displacement. In general, they strike northeast and dip steeply. The shallow veins are concentrated in the quartz porphyry, but the deeper veins are equally abundant in the shales and overlying sedimentary and pyroclastic rocks (Fig. 16-6). A remarkably productive sheeted zone, made up of small, parallel veinlets, dips easterly through the crest of Cerro Rico, reaching a maximum width of over 550 feet and a maximum length of 1,150 feet (Turneaure, 1960).

The ores are in composite, anastomosing, veins or lodes as well as in parallel fissures within the sheeted zone. The veins range from three or more feet to only a few inches in width. Several overlapping episodes of ore deposition have been recognized, though for descriptive purposes they can be classified as early and late (Turneaure, 1960). The early mineralization produced cassiterite, pyrite, wolframite, bismuthinite, arsenopyrite, quartz, and minor pyrrhotite. Subsequently, stannite, tetrahedrite, sphalerite, ruby silver, and minor amounts of chalcopyrite, andorite ($AgPbSb_3S_6$), matildite ($AgBiS_2$), jamesonite, boulangerite, and galena were deposited. Late-stage alunite veinlets cut the sulfide veins. The characteristic alteration products are chlorite, sericite, quartz, kaolinite, alunite, and possibly tourmaline. Deeper parts of the stock are sericitized; the shallower part is intensely silicified (Turneaure, 1960).

Paragenetic studies have shown that quartz, pyrite, and cassiterite were deposited early, followed by stannite and chalcopyrite, then tetrahedrite and andorite, and finally lead sulfosalts, galena, ruby silver, and the late-stage gangue minerals. This sequence represents a general drop in temperature during the history of mineralization. The early minerals were thought to be deposited at 400° to 500°C, and the late minerals at 100° to 150° (Turneaure, 1960).

Ahlfeld (1941) argued that the ores were deposited as colloids, emphasizing that "wood tin" is abundant; but the existence of high-temperature colloids is not probable, nor do the sulfide textures support this hypothesis. Davy (1920) observed, however, that the colloidal tin is only within oxidized layers, which suggests that it was deposited as a secondary mineral at the expense of stannite.

The general zonal arrangement conforms to the paragenesis. A deep, central zone, characterized by bismuthinite and wolframite, is surrounded by a zone

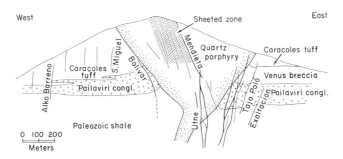

Figure 16-6
Cross section of Cerro Rico showing the vein system of the Potosí district, Bolivia. (From Turneaure, 1960, Fig. 15.)

containing silver sulfosalts and, beyond this, by a zone containing lead and zinc minerals (Ahlfeld, 1941).

Ore deposition at Potosí was probably controlled by changes in pressure and temperature rather than by wall-rock chemistry. There is no consistent difference between lodes in altered quartz porphyry, shale, and bedded volcanics. Apparently, metallization occurred wherever the ore-bearing fluids were flushed into a fissure or fracture zone, so that any and all competent formations were favorable hosts. The high-grade ore shoots, complex textures, and unusual mineral assemblages all reflect rapid changes in temperature and pressure.

The accepted view of ore genesis as outlined by Turneaure (1960), is as follows. A centrolineal belt of magma, differentiating at depth, supplied volcanics and shallow intrusives to the long belt of the Bolivian Andes. Early hydrothermal fluids ascended fracture systems developed by stresses along the zone of igneous activity. These fluids altered the host rocks and intrusives, causing extensive chloritization, sericitization, and silicification. Further fracturing allowed high-temperature tin-bearing fluids to enter the altered rocks, forming cassiterite deposits. With waning temperatures, and probably after further differentiation at the source, the vein systems received deposits of sulfides and sulfosalts. The latest ores and gangue minerals were deposited at relatively low temperatures. This picture of ore deposition at shallow depths with temperatures ranging from those characteristic of hypothermal veins to those typical of the epithermal zone is practically a classic example of the xenothermal environment.

REFERENCES TO SELECTED XENOTHERMAL DEPOSITS

Oruro, Bolivia

Ahlfeld, F., 1954. *Los Yacimientos Minerales de Bolivia*, Bilbao: El Banco Minero de Bolivia y la Corporación Minera de Bolivia.

Campbell, D. F., 1942. The Oruro silver-tin district, Bolivia, *Econ. Geol.* 37:87-115.

Chace, F. M., 1948. Tin-silver veins of Oruro, Bolivia, *Econ. Geol.* 43:333-383, 435-470.

Davy, W. M., 1920. Ore deposition in the Bolivian tin-silver deposits, *Econ. Geol.* 15:463-496.

Lindgren, W., and A. C. Abbott, 1931. The silver-tin deposits of Oruro, Bolivia, *Econ. Geol.* 26:453-479.

Singewald, J. T., Jr., 1929. The problem of supergene cassiterite in Bolivian tin veins, *Econ. Geol.* 24:343-364.

Turneaure, F. S., 1960. A comparative study of major ore deposits of central Bolivia, *Econ. Geol.* 55:217-254, 574-606.

Mount Bischoff–Renison Bell District, Tasmania

Fisher, N. H., 1953. The Renison Bell tinfield, in *Geology of Australian Ore Deposits,* ed. A. B. Edwards, Melbourne: Australasian Institute of Mining and Metallurgy.

Knight, C. L., 1953. Mount Bischoff tin mine, in *Geology of Australian Ore Deposits,* ed. A. B. Edwards, Melbourne: Australasian Institute of Mining and Metallurgy.

Stillwell, F. L., and A. B. Edwards, 1943. The mineral composition of the tin-ores of Renison Bell, Tasmania, *Australas. Inst. Mining Metall. Proc.* 131-132, pp. 173-186.

Weston-Dunn, J. G., 1922. The economic geology of the Mount Bischoff tin deposits, Tasmania, *Econ. Geol.* 17:153-193.

REFERENCES CITED

Abe, M., 1963. Zonal distribution of ore deposits at the Akenobe mine [article in Japanese], *Mining Geol. Jap. J.* 13:101-114.

Ahlfeld, F., 1941. *Los Yacimientos Minerales de Bolivia,* La Paz: Ministerio de Economía Nacional, Dirección General de Minas y Petróleo.

———, 1946. *Geología de Bolivia,* La Plata, Argentina: Ministerio de Economía Nacional, Dirección General de Minas y Petróleo.

Buddington, A. F., 1935. High-temperature mineral associations at shallow to moderate depths, *Econ. Geol.* 30:205-222.

Davy, W. M., 1920. Ore deposition in the Bolivian tin-silver deposits, *Econ. Geol.* 15:463-496.

Geological Survey of Japan, 1960. *Geology and Mineral Resources of Japan,* 2nd ed., Hisamotocho, Kawasaki-shi: Geol. Surv. Japan.

Imai, H., 1966. Formation of fissures and their mineralization in the vein-type deposits of Japan, *Univ. Tokyo Fac. Eng. J.* 28(3):255-302.

Imai, H., B. Katayama, and I. Fukuoka, 1970. *Geology and Mineral Deposits of the Akenobe Mine and the Ningyo-Toge Uranium Deposits,* International Association for the Genesis of Ore Deposits (Tokyo-Kyoto mtg., Guidebook 8, Excursion B4).

Jaskolski, S., 1933. Les gîsements argento-stannifères de Potosí en Bolivie, *Arch. Minér. Soc. Sci. Varsovie* 9:47-92.

Kato, T., 1920. A contribution to the knowledge of the cassiterite veins of pneumo-hydatogenetic or hydrothermal origin, *Imper. Univ. Tokyo Coll. Sci. J.* 43, art. 5.

———, 1927. The Ikuno-Akenobe metallogenetic province, *Jap. J. Geol. Geogr.* 5(3): 121-133.

———, 1928. Some characteristic features of the ore deposits of Japan, related genetically to the Late Tertiary volcanic activity, *Jap. J. Geol. Geogr.* 6(1-2):31-48.

Kondo, N., and S. Kawasaki, 1943. Survey of the Ikuno mine, *Geol. Surv. Jap. Spec. Rept.* 3.

Lindgren, W., and J. G. Creveling, 1928. The ores of Potosí, Bolivia, *Econ. Geol.* 23:233-262.

Maruyama, S., 1957. The relation between ore veins and igneous intrusives at the Ikuno mine [article in Japanese], *Mining Geol. Jap. J.* 7:281-284.

Nakano, O., 1931. Some microscopic structures of copper ore from the Akenobe mine, *Jap. Ass. Mining Petrol. Econ. Geol. J.* no. 5, pp. 217-222; no. 6, pp. 21-24.

Sato, K., and J. Kaneko, 1952. Geophysical exploration at Ikuno mine, Hyogo Prefecture, *Geol. Surv. Jap. Bull.* 3(10):27-34.

Takahashi, T., S. Takeuchi, S. Nishio, and H. Imai, 1955. Temperature of mineral formation in some types of deposits in Japan, as measured by the decrepitation method (II), *Mining Geol. Jap. J.* 5(15):9-17.

Turneaure, F. S., 1960. A comparative study of major ore deposits of central Bolivia, *Econ. Geol.* 55:217-254, 574-606.

Turneaure, F. S., and K. K. Welker, 1947. The ore deposits of the eastern Andes of Bolivia: the Cordillera Real, *Econ. Geol.* 42:595-625.

Yamaguchi, K., 1939. Ore deposits of the Ikuno mine and their zonal arrangement, *Jap. Ass. Mining Petrol. Econ. Geol. J.* 21:257-275, 22:25-37.

17 / Volcanogenic Deposits

The term *volcanogenic*, or *volcanogenic-exhalative*, refers to mineral deposits, commonly stratabound, that have been formed by volcanic processes and activities of thermal springs beneath bodies of water (Oftedahl, 1958). Although many types of deposits are of this origin, recent authors have tended, incorrectly, to restrict the term "volcanogenic" to the massive sulfide bodies deposited from thermal springs during periods of volcanic activity. Mineralizing fluids were either poured out on the sea floor or permeated and replaced shallow layers of volcanic tuffs and associated sediments. The assumption is made that if the ore minerals had been deposited on the surface of the land, they would have been quickly destroyed; only under exceptional circumstances would they have been preserved.

The idea that an ore deposit may form by deposition from volcanic and thermal springs pouring out on the sea floor is not a new one, but has been held for many years; particularly, it has been used to describe layered manganese and iron oxide deposits (Van Hise and Leith, 1911; Park, 1942; Hewett, 1966; Gilmour, 1965). Only in recent years, however, has the concept been widely applied to the accumulations of massive sulfides associated with thick volcanic piles. Many massive sulfide deposits are now thought to be stratabound and closely related to the processes of volcanism, especially to the formation of rhyolites and rhyolite domes (Kinkel, 1962, 1966; Hodge, 1967, Griffitts *et al.*, 1972; Goodwin, 1965; Clark, 1971; de Rosen-Spence, 1969; Sinclair, 1971; Anderson and Nash, 1972).

The theory of volcanogenesis in massive sulfides was given considerable support by the discovery of the metal-bearing deposits now being formed by hot brines in the Red Sea. Oceanographic studies in this area, off the coast of Sudan, were first reported by Miller and later by other scientists (Miller, 1964; Hunt, Hays, Degens, and Ross, 1967; Degens and Ross, 1969). Hot concentrated saline waters or metals are found here in as many as 18 separate deeps. In one of the largest of the basins, Atlantis II deep, the temperature is about 56° and the salinity is about 255 parts per thousand. The basin is said to resemble a caldera. The deeps are aligned along what appears to be a rift zone and volcanic activity is clearly involved in the generation of the highly heated and saline waters. Iron, copper, zinc, and silver are present in appreciable amounts. Further observations of the Red Sea deposits should prove to be of great interest in the study of ore genesis.

KUROKO ORES

The Kuroko (Kuromono) ores of Japan are the classic example of volcanogenic massive sulfide deposits (Kinoshita, 1924, 1929, 1931; Griggs, 1947; Collins, 1950; Hayashi, 1961; Hashimoto, Kamono, and Hayashi, 1962; Kamono and Ishikawa, 1965; Horikoshi, 1969; Matsukuma and Horikoshi, 1970; Horikoshi and Sato, 1970; Suga and Takahashi, 1970; Ohmoto, 1972; Ogura, 1972). The Kuroko ores have been the subject of interesting debate for many years. Among geologists who have studied them, they are now widely accepted as unmetamorphosed (or only weakly metamorphosed in places by thermal processes), stratabound, polymetallic deposits genetically related to submarine volcanic activities during the Tertiary (Neogene) period.

Although the discussion of Kuroko ores is generally limited to the deposits in Japan, deposits occur in the Philippine Islands (Bryner, 1967), in Fiji (Frenzel and Ottemann, 1967), and possibly in Turkey (Griffitts, Albers, and Oner, 1972). As the emplacement of volcanogenic massive sulfides becomes more widely understood, it is entirely reasonable that other deposits will be considered volcanogenic.

"Kuroko" means "black ore." The term "Kuroko-type" is commonly applied to three categories of ore, based upon mineral compositions. The siliceous (Keiko) ores contain sulfides, particularly chalcopyrite, disseminated through highly silicified rock. Yellow (Oko) ores are primarily pyrite with minor amounts of chalcopyrite and quartz. The black (Kuroko) ores are intimate mixtures of sphalerite, galena, barite, and minor quantities of pyrite and chalcopyrite; wurtzite, enargite, tetrahedrite, marcasite, and numerous other minerals are found locally in small amounts. Veins and large masses of gypsum are in related but separate bodies, and large discrete bodies of barite are frequently present.

Kuroko-type orebodies, especially common in northeastern Japan, occupy definite stratigraphic horizons characterized by the accumulation of sandy and muddy sediments that contain abundant molluscan fossils of warm water species. Deposition is thought to have taken place in relatively quiet, shallow, and isolated basins rather than on the open sea floor. Volcanic activity began with the accumulation of pyroclastic debris that was reworked by turbidity currents during later formation of lava domes and flows. The volcanic materials with which the ores are usually associated are silicic rhyolites and dacites, commonly brecciated (Matsukuma and Horikoshi, 1970).

Solutions carrying sulfur and base metals moved upwards through and around the rhyolite domes, forming a network of veins in the relatively brittle volcanic rocks. Where the solutions reached the unconsolidated, salt-water saturated tuffs and breccias over and around the domes, without reaching the sea floor, they reacted quickly. Massive sulfides, together with silica, gypsum, and barite, extensively replaced the brecciated materials. Changes in volume and density caused a slumping and the incorporation of blocks of different kinds of replaced materials in the sandy tuffs or in other types of ore. Various kinds of solutions were involved; locally, massive bodies of gypsum, barite, or mixed materials of several types were produced. After the main period of mineralization the deposits were covered with pyroclastics and flows of dacitic composition. Waning fluid movements from the subvolcanic source caused mild alteration of the higher volcanic beds (Jenks, 1966, 1971).

Watanabe (1970) states that many of the Kuroko-type ores are sedimentary and localized at specific horizons; prospecting has been successful where this criterion has been used. Both replacement at shallow depths, probably within a few hundred feet of the surface, and sedimentation were active processes; Kuroko-type ores may be either bedded or irregular and massive replacement bodies. Below many of the ores are bodies of stringer ores consisting of chalcopyrite veinlets disseminated through silicified volcanics and sediments. The best evidence for sedimentary origin includes the stratabound character of the deposits, extremely fine but well-formed laminations, false bedding, and graded bedding. Direct precipitation on the sea floor seems to be the most reasonable explanation for the genesis of much of the ore.

The Kuroko-type ores are extremely fine grained. They commonly show framboidal textures, concentric banding, nodules, and colloform structures. In fact, these textures are so abundant that Kinoshita (1924, 1929) thought the deposits precipitated from colloidal solutions. Most certainly the ores are chemical precipitates. The presence of veins of silica jel were noted in the upper parts of the massive sulfide deposits of Turkey by Griffitts, Albers, and Öner (1972). This jel was considered to be of supergene origin.

Four zones of alteration are widely recognized near the Kuroko-type ores of Japan: (1) strong silicification in the footwalls of the orebodies, commonly accompanied by small amounts of sericite and chlorite; (2) sericite, chlorite, and quartz intimately associated with the ores; (3) sericite, chlorite, and pyrite above the ores; and (4) montmorillonite and zeolite mineralization that grades outward and upward into unaltered rock (Matsukuma and Horikoshi, 1970). Sudo (1954) reported montmorillonite, iron-montmorillonite, sericite, and chlorite as alteration products.

Kuroko orebodies range in size from small nodules to irregular masses as much as 2,400-feet long by 1,000-feet wide by 300-feet deep. Many are flat lenticular bodies.

The Hanoaka-Matsumine district in northern Honshu, Japan, is typical of Kuroko deposits. According to Ogura (1972), the ores from this district are zoned with siliceous rock at the base, gypsum above, followed by siliceous iron sulfides (Keiko) ores and then by yellow (Oko) ores; the black (Kuroko) materials are on top. Colloform and spherulitic textures are abundantly developed, especially in the massive ores.

The Hanoaka-Matsumine orebodies are considered to have formed as submarine hydrothermal sediments in the upper part of the pyroclastic rock sequence and are interlayered with or overlain by mudstones (see Fig. 17-1). Sedimentary structures, such as stratification and fine laminations, are widely recognized. The introduction of the ore followed intrusion of a rhyolite dome and the solutions permeated through and around the dome. The ores were deposited in bordering sediments or, where the thermal springs reached the sea, were spread out on the sea floor. Ogura (1972) determined the temperature of formation of the ore at about 200°C.

The Kuroko-type ores are unmetamorphosed, or at most have been slightly affected by thermal metamorphism. Regional dynamic metamorphism is absent. In older volcanic provinces in other parts of the world, where dynamic metamorphism has been active, it is not uncommon to find massive sulfide bodies that have many features resembling the Kuroko ores. Such deposits are abundant in the Tasman geosyncline, along the eastern parts of Australia. Deposits in this geosyncline are somewhat changed by dynamic metamorphism, but to a far less extent than the massive sulfide ores of eastern Canada.

READ-ROSEBERY DEPOSITS

The Rosebery mine in western Tasmania, about 70 miles south of Burnie on the western slope of Mount Black, is one massive sulfide deposit resembling the Kuroko ores (Hall et al., 1953, 1965; Brathwaite, 1971, 1974; Solomon et al.,

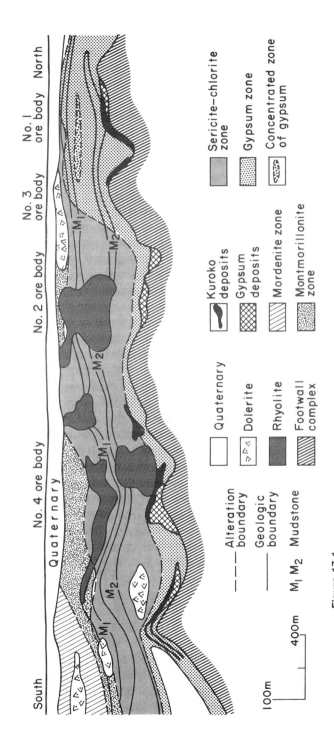

Figure 17-1
Geologic section of the Shakanai mine, Hanoaka-Matsumine district, Akita Prefecture, Japan, showing zonal arrangement of alteration. (From Matsukuma and Horikoshi, 1970.)

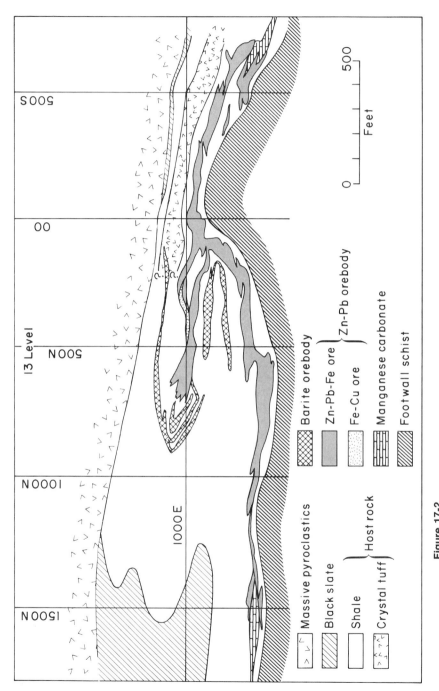

Figure 17-2

Geologic plan of the 13th level of the Rosebery mine, Tasmania. (From Brathwaite, 1974.)

1972). The orebodies at Rosebery are massive galena and sphalerite restricted to a fine-grained tuffaceous shale member of a sequence of pyroclastic rocks. The age of the formation is said to be Lower Paleozoic.

Minerals in the orebody are principally sphalerite, galena, and pyrite, with lesser amounts of chalcopyrite, bournonite, arsenopyrite, pyrrhotite, tetrahedrite, pyrargyrite, and gold. Alteration products, abundantly developed in the shale layers near the ore, consist of sericite, chlorite, kaolinite, and feldspathic blebs.

The geologic structure of the area is complex, and the rocks are both folded and faulted. The shales are sheared so much that bedding is largely destroyed and locally the shale is missing between the underlying tuffs and the overlying pyroclastics. Faulting may be responsible for the lack of shale, though Hall and his colleagues (1965) think an unconformity may explain the absence. The most prominent structure in the area is a large overturned drag fold and this, combined with faulting, resulted in intense shearing and development of cleavage, particularly in the rocks below the massive pyroclastics. Figure 17-2 shows a geologic plan of the 13th level of the Rosebery mine.

The massive sulfide ore of the Tasman geosyncline so closely resembles the Kuroko deposits of Japan that it is considered to be a type intermediate between the Japanese deposits and the more highly metamorphosed massive sulfide ores of Canada (Sangster, 1972).

CANADIAN MASSIVE SULFIDE DEPOSITS

Sangster (1972) considers that the first and lowermost step in the process of deposition of the Canadian massive sulfide deposits was the accumulation of a complex, thick sequence of flows of pillow and vesicular basalts. Above these basement rocks are flows, flow breccias, and tuffs, mainly of andesitic composition. The uppermost materials of the volcanic section consist of abundant dacitic to rhyolitic, massive, and almost textureless flows of pyroclastics. The rhyolite cycle is usually followed by more basalt. When volcanic activity stops, the igneous flows are commonly covered by graywacke type sediments. It is also characteristic of the ore-containing volcanic piles that they are intruded by dikes and irregular masses as diverse as the extrusives.

Many of the massive sulfide ores in Canada are closely associated with coarse fragmental pyroclastics of rhyolitic composition. The second most common host for ore deposits is the volcanic sediments, graywackes and shales. A relationship not reported in either the Japanese deposits or the Tasman geosyncline is the deposition of the chemical sediments chert and iron formation above or in the ores. Mineralization appears to be most abundant near the centers of sedimentary basins, but faults and fissures are numerous and may play dominant roles in localization of the ores.

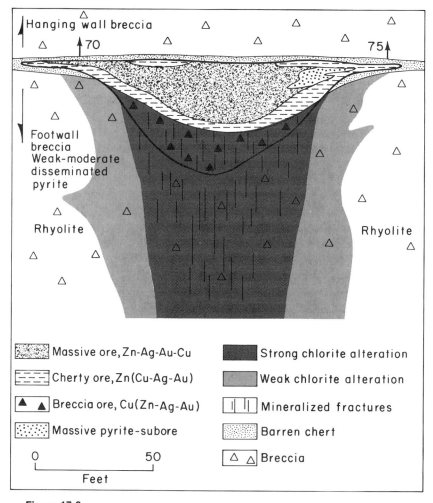

Figure 17-3
Plan of a typical ore shoot, Lens B, 850-foot level, Delbridge mine, Noranda district, Canada. (From Boldy, 1968.)

The mineralogy of the Canadian massive sulfide deposits is rather simple. They generally contain more than 50 percent total volume of sulfides. Pyrite and pyrrhotite are the most abundant constituents. Sphalerite, chalcopyrite, and galena are the principal sulfides that are recovered. Minor quantities of many other minerals are widely distributed. Colloform and framboidal structures, similar to those in the Kuroko ores, are abundant; they are especially common where the carbon content of the rock is high (Sangster, 1972).

Gilmour (1965) says that alteration pipes are well developed in the rocks below undeformed massive sulfide bodies. Alteration in the Canadian deposits is confined to the footwall rocks, whereas in the Kuroko deposits alteration products occur both above and below the massive sulfides. The cores of the alteration pipes in Canada are generally chlorite that contains disseminated sulfides and magnetite. The chlorite content decreases outward in the pipes and the borders are gradational and poorly defined. Sericite is widely distributed and is most easily recognized close to the borders of the chlorite. The chlorite-sericite rock grades outward into massive siliceous rock, particularly into rhyolite. Disseminated sulfides are present in ore shoots in the altered materials. The footwalls of the altered areas are commonly diffuse and vague, but the hanging walls are sharp. Figure 17-3 clearly shows the footwall alteration pipe at the 850-foot level in the Delbridge mine, Noranda district, Canada.

According to Sangster (1972) most volcanogenic massive sulfide deposits are zoned in consistent patterns; the zoning is of four kinds. (1) *Morphologic zoning* exists where both massive ores and stringer ores are known; the stringer ores are everywhere below the massive sulfides. (2) *Mineralogic zoning* in an ideal ore would consist of pyrite-sphalerite (chalcopyrite-galena) underlain by stringer ore of pyrrhotite-chalcopyrite. In the massive ore galena-sphalerite are more abundant in the upper parts of the orebodies. Chalcopyrite increases toward the footwall and grades into stringer ore. (3) *Textural zoning* is defined by the fact that the sphalerite-rich parts are banded, expressed as monomineralic layers of pyrite and sphalerite. The chalcopyrite-rich parts of the orebodies seldom show banding. (4) *Compositional zoning* parallels the distribution of the three major sulfides, sphalerite, chalcopyrite, and galena, but is usually more quantitative and can be detected on routine assay charts. Sphalerite, because of its wide variation in color, is difficult to recognize but shows up clearly on assay plots.

Sangster (1972) lists points of similarity and dissimilarity between the metamorphosed Canadian massive sulfide deposits and the Kuroko ores of Japan. The points of similarity are:

1. Both are with calc-alkaline, submarine, volcanic rocks.

2. Both tend to be in clusters or districts related to centers of volcanic activities.

3. Both show strong spatial correlation with acidic, explosive phases of volcanism.

4. Both consist of two main types of ore, massive sulfides and stringer ore. The massive ore in both areas is essentially conformable with surrounding rocks, but the stringer ore clearly cuts the stratigraphy. The massive ores are banded.

5. Both are commonly capped by a layer of ferruginous chert—hematite in Japan and magnetite in Canada.

6. Both show compositional zoning relative to stratigraphy, with lead-zinc decreasing downward and copper increasing downward.

7. Both are underlain by alteration materials that enclose stringer ores.

The points of dissimilarity are:

1. Footwall alteration in the Kuroko deposits is mainly silicification, whereas in the Canadian deposits silica is removed during alteration and magnesium is increased. Scott (1948) described siliceous alteration in rhyolites beneath the Quemont orebodies in the Noranda district, Canada.

2. Postore alteration, which can be directly related to mineralization of hanging-wall rocks is common in the Kuroko ores but is not recognized in Canada.

3. The anhydrite-gypsum ores of Japan have no known equivalent in Canada.

4. Differences in mineralogy exist. In the Kuroko ores bornite and tetrahedrite-tennantite are major constituents but in Canada are seldom more than minor accessories. Galena is abundant in Japanese ores but seldom encountered in Canada.

Sangster (1972) believes that if volcanogenic sulfide deposits formed from submarine exhalations as proposed, then the chemical differences between Precambrian and Kuroko Tertiary ores may be manifestations of the chemical evolution of the earth's lithosphere, hydrosphere, and atmosphere. Perhaps the Precambrian oceans were considerably different from those of Tertiary times.

As in many areas of volcanic activities, the manifestations tend to be aligned along faults or other structural controls. The volcanogenic ore deposits form clusters, the distribution of which is also controlled by structural features. Because the massive sulfide deposits of Noranda, Canada, are distributed along structural features, many geologists favor volcanogenesis for them (Boldy, 1968; de Rosen-Spence, 1969; Dresser and Denis, 1949; Dugas, 1966; Gilmour, 1965; Goodwin, 1965; Scott, 1948; Spence, 1966).

REFERENCES CITED

Anderson, C. A., and J. T. Nash, 1972. Geology of the massive sulfide deposits at Jerome, Arizona—a reinterpretation, *Econ. Geol.* 67:845-863.

Boldy, J., 1968. Geological observations on the Delbridge massive sulfide deposit, *Can. Inst. Mining Metall. Trans.* 71:247-256.

Brathwaite, R. L., 1971. Structure of the Rosebery ore deposit, *Australas. Inst. Mining Metall. Proc.* 241, pp. 1-13.

———, 1974. The geology and origin of the Rosebery ore deposit, Tasmania, *Econ. Geol.* 69:1086-1101.

Bryner, L., 1967. Geology of the Barlo mine and vicinity, Dasol, Pangasinan Province, Luzon, Philippine Islands, *P.I. Bur. Mines Rept. Invest.* 60.

Clark, L. A., 1971. Volcanogenic ores: comparison of cupriferous pyrite deposits of Cyprus and Japanese kuroko deposits, *Soc. Mining Geol. Jap. Spec. Issue* 3, pp. 206-215.

Collins, J. J., 1950. Summary of Kinoshita's kuroko deposits of Japan, *Econ. Geol.* 45: 363-376.

Degens, E. T., and D. A. Ross, eds., 1969. *Hot Brines and Recent Heavy Metal Deposits in the Red Sea*, New York: Springer-Verlag.

de Rosen-Spence, A., 1969. Genése des roches á cordiéret-anthophyllite des gisements cuprozincíféres de la région de Rouyn-Noranda, Quebec, Canada, *Can. J. Earth Sci.* 6:1339-1345.

Dresser, J. A., and T. C. Denis, 1949. Geology of Quebec, vol. 3: economic geology, *Quebec Dept. Mines Geol. Rept.* 20, pp. 371-383.

Dugas, J., 1966. The relationship of mineralization to Precambrian stratigraphy in the Rouyn-Noranda area, Quebec (Canada), *Geol. Ass. Can. Spec. Pap.* 3, pp. 43-56.

Frenzel, G., and J. Ottemann, 1967. Eine Sulfid-Paragenese mit Kupferhaltigem Zonarpyrite von Nukundamu/Fiji, *Mineralium Deposita* 1:307-316.

Gilmour, P. C., 1965. The origin of the massive sulfide mineralization in the Noranda district, northwestern Quebec (Canada), *Geol. Ass. Can. Proc.* 16:63-81.

Goodwin, A. M., 1965. Mineralized volcanic complexes in the Porcupine-Kirkland Lake-Noranda region, Canada, *Econ. Geol.* 60:955-971.

Griffitts, W. R., J. P. Albers, and Ö. Öner, 1972. Massive sulfide copper deposits of the Ergani-Maden area, southeastern Turkey, *Econ. Geol.* 67:701-716.

Griggs, A. B., 1947. Zinc-lead resources of Japan, *Supreme Command Allied Powers Nat. Resources Sec. Rept.* 65, pp. 23-26.

Hall. G., V. M. Cottle, P. B. Rosenhain, and R. R. McGhie, 1953. The lead-zinc deposits of Read-Rosebery and Mount Farrell, in *Geology of Australian Ore Deposits*, ed. A. B. Edwards, Melbourne: Australasian Institute of Mining and Metallurgy.

Hall, G., V. M. Cottle, P. B. Rosenhain, R. R. McGhie, and J. G. Druett, 1965. Lead-zinc ore deposits of Read-Rosebery (Tasmania), in *Geology of Australian Ore Deposits*, vol. 1, 8th Commonwealth Mining and Metallurgical Congress.

Hashimoto, K., H. Kamona, and S. Hayashi, 1962. On the Uchinotai kuroko (black ore) deposits, Kosaka mine (Japan), *Mining Geol. Jap. J.* 12:129-142.

Hayashi, S., 1961. On the mode of occurrence of kuroko (black ore) in the Motoyame ore deposit of the Kosakak mine, Akita Prefecture (Japan), *Mining Geol. Jap. J.* 11:433-442.

Hewett, D. F., 1966. Stratified deposits of the oxides and carbonates of manganese, *Econ. Geol.* 61:431-461.

Hodge, H. J., 1967. Horne mine, Noranda Mines, Ltd., in *Centennial Field Excursion Guidebook*, Canadian Institute of Mining and Metallurgy.

Horikoshi, E., 1969. Volcanic activity related to the formation of the kuroko-type deposits in the Kosaka district, Japan, *Mineralium Deposita* 4:321-345.

Horikoshi, E. and T. Sato, 1970. Volcanic activity and ore deposition in the Kosaka mine. In *Volcanism and Ore Genesis*, Tokyo, Tokyo Univ. Press, pp. 181-195.

Hunt, J. M., E. E. Hays, E. T. Degens, and D. A. Ross, 1967. Red Sea: detailed survey of hot brine areas, *Science* 156:514-516.

Jenks, W. F., 1966. Some relations between Cenozoic volcanism and ore deposition in northern Japan, *N.Y. Acad. Sci. Trans.* 28:463-474.

——, 1971. Tectonic transport of massive sulfide deposits in submarine volcanic and sedimentary host rocks, *Econ. Geol.* 66:1215-1224.

Kamona, H., and Y. Ishikawa, 1965. On the Ushinotai kuroko deposits, Kosaka mine, Akita Prefecture (Japan), *Mining Geol. Jap. J.* 15:130-142.

Kinkel, A. R., Jr., 1962. Observations on the pyrite deposits of the Huelva district, Spain, and their relation to volcanism, *Econ. Geol.* 57:1071-1080.

————, 1966. Massive sulfide deposits related to volcanism and possible methods of emplacement, *Econ. Geol.* 61:673-694.

Kinoshita, K., 1924. colloidal solutions as the mineralizing solutions of the "kuromono" deposits, *Tohoku Imper. Univ. Sci. Rept.,* 3rd ser., 2:23-30.

————, 1929. On the genesis of the "kuromono" deposits, *15th Int. Geol. Congr. Compt. Rend.* 2:454-474.

————, 1931. On the "kuroko" (black ore) deposit, *Jap. J. Geol. Geogr.* 8:281-352.

Matsukuma, T., and E. Horikoshi, 1970. Kuroko deposits in Japan: a review, in *Volcanism and Ore Genesis* (Watanabe Comm. Vol.), ed. T. Tatsumi, Univ. Tokyo Press.

Miller, A. R., 1964. High salinity in sea water, *Nature* 203:590.

Oftedahl, C., 1958. A theory of exhalative-sedimentary ores, *Geol. Foren. Stockholm Forh.* 80:1-19.

Ogura, H., 1972. Geology and "kuroko" ore deposits of the Hanaoka-Matsumine mine, northern Japan, *24th Int. Geol. Congr. Sec. 4 Mineral Deposits,* pp. 318-325.

Ohmoto, H., 1972. Origin of hydrothermal fluids responsible for the kuroko deposits in Japan [abstr.], *Mining Eng.* 24(12):74.

Park, C. F., Jr., 1942. The manganese deposits of Cuba, *U.S. Geol. Surv. Bull.* 935-B, pp. 75-77.

Sangster, D. F., 1972. Precambrian volcanogenic massive sulfide deposits in Canada: a review, *Geol. Surv. Can. Pap.* 72-22.

Scott, J. S., 1948. Quemont mine, in *Structural Geology of Canadian Ore Deposits,* Montreal: Canadian Institute of Mining and Metallurgy.

Sinclair, W. D., 1971. A volcanic origin for the no. 5 zone of the Horne mine, Noranda, Quebec (Canada), *Econ. Geol.* 66:1225-1231.

Solomon, M., D. I. Groves, and J. Klominsky, 1972. Metallogenic provinces and districts in the Tasman orogenic zone of eastern Australia, *Australas. Inst. Mining Metall. Proc.* 242, pp. 9-24.

Spence, C. D., 1966. Volcanogenic settings of the Vauze base metal deposit, Noranda district, Quebec (Canada); paper presented at Canadian Institute of Mining and Metallurgy Annual Meeting, April 1966.

Sudo, T., 1954. Types of clay minerals closely associated with metalliferous ores of the epithermal types, *Tokyo Kyoiku Daigaku Sci. Rept.,* ser. C, no. 23, pp. 186-192.

Suga, K., and T. Takahashi, 1970. On the mode of acidic volcanism in the Miocene in the region of Hanaoka-kuroko belt, *Mining Geol. Jap. J.* 20:328-344.

Van Hise, C. R., and C. K. Leith, 1911. The geology of the Lake Superior iron region, *U.S. Geol. Survey Monograph* 52.

Watanabe, T., 1970. Volcanism and ore genesis, in *Volcanism and Ore Genesis* (Watanabe Comm. Vol.), ed. T. Tatsumi, Univ. Tokyo Press.

18 / Sedimentary Deposits

Mechanical and chemical weathering supply ore materials to basins of deposition just as they supply quartz, clays, and dissolved solids for the production of clastic and nonclastic sediments. Under favorable conditions of transportation, sorting, and deposition, some of the ore materials become sufficiently concentrated to constitute economic deposits. These sedimentary ores are generally classified, first, as either chemical precipitates or mechanical accumulations and, second, according to their chemical or mineralogical composition. Whether chemically or mechanically derived, sedimentary ores are *syngenetic* deposits.

Many large and valuable sedimentary deposits have been exploited in recent years, and additional discoveries are to be expected as our knowledge of sedimentary processes grows. Unfortunately, most students of ore deposits are poorly prepared in the study of these processes. The mining industry, however, has begun to appreciate the value of studies both of syngenetic deposits and of epigenetic ores whose emplacement was controlled by primary textures and structures. The sedimentary features are especially clear in such deposits as the Blind River uranium ores in Ontario, Canada, the Zambian copper belt, and the Witwatersrand gold ores in South Africa.

CHEMICAL PRECIPITATES

Certain deposits of the metals have been precipitated as primary sediments from surface waters by chemical and biochemical processes (McKelvey, 1950; Mason, 1958). Although most elements are amenable to chemical precipitation in one form or another, only a few have formed large deposits. The principal ores of this type include the oxides, silicates, and carbonates of iron and manganese, such as the Lake Superior banded "iron formation" and the oölitic manganese sediments at Chiaturi in the Russian Urals. Base metal deposits, such as the Kupferschiefer copper-zinc-lead strata at Mansfeld, Germany, are sediments that have accumulated under unusual conditions. The origin of other bedded sulfides is considered by many geologists to be either syngenetic or volcanogenic (Dunham, 1964; Sangster, 1972). Low-grade uranium, vanadium, and rare-element deposits, particularly those associated with marine black shales and phosphorites, also form chemical sediments.

The chemical precipitation of sediments is controlled by many factors, chief among which are the availability of the ions in question and the pH and Eh of the environment. Oxidation-reduction potentials are related to the oxygen content of the water, which is ordinarily a function of depth and, possibly, nearness to the shoreline. Most dissolved solids are supplied to the depositional basins by streams, so mineral deposits are likely to be thickest and best developed along the shoreline. Reeflike deposits are thus more common than broad, equidimensional sheets.

The role of biochemical processes in the precipitation of ore deposits has been a subject of long-standing debate and remains unsolved (Hem, 1960; Temple, 1964; Hariya and Kikuchi, 1964; Trudinger, Lambert, and Skyring, 1972). Certain bacteria and algae can cause the precipitation of oxide compounds by acting as catalysts for oxidation reactions. Furthermore, anaerobic bacteria are able to reduce sulfates, producing H_2S, which in turn may cause the precipitation of ore minerals. Sulfide deposits of the base metals form where the metal ions encounter H_2S, and native sulfur may form through the oxidation of H_2S by inorganic processes or by aerobic bacteria as the gas ascends into an aerated zone (Dessau *et al.*, 1962). Some oxide deposits may even result from the reducing action of H_2S; for example, Jensen (1958) suggested that H_2S produced by anaerobic bacteria caused the reduction of soluble U^{+6} ions to insoluble U^{+4} ions, forming the uraninite ores of sandstone uranium deposits. Iron- and manganese-fixing bacteria have been recognized since the nineteenth century, but their actual role in the formation of ore deposits is unknown. Since most of the reactions they cause may also take place without their aid, although at a slower rate, the physical and chemical environment may be the true controlling factor. The presence of sulfides in the carbonaceous shale facies of basin sediments supports the biogenic hypothesis, but oxide, carbonate, and silicate facies generally lack evidence of organic activities. The problem is clouded by the similarity between the conditions under which microorganisms cause iron and manganese to precipitate and those under which these metals will precipitate inorganically (Beerstecher, 1954).

A distinction between sedimentary ores formed as chemical precipitates and those formed by hydrothermal processes may be difficult or nearly impossible to make. The problem is generally a matter of distinguishing pseudomorphous bedding from original bedding or interstitial hydrothermal deposits from diagenetic cement. Fine-grained, banded sulfide ores similar to those of the Rammelsberg district, Germany (Fig. 17-12), are especially difficult to interpret.

Sedimentary Iron Ore Deposits

Several types of iron ore deposits form directly as chemical precipitates. Probably the most widespread is *iron formation,* also known as *taconite* in the Lake Superior district, as *itabirite* in Brazil, as *jaspilite* in Australia, and as *banded ironstone* elsewhere in the British Commonwealth. Other sedimentary iron ores include oölites (such as the Clinton ores of the eastern United States and the minettes of Alsace-Lorraine), bog iron ores, and iron carbonate beds (otherwise known as "black band" ores).

Iron Formation

The term "iron formation," or "banded iron formation," is widely used by geologists, though not always consistently. Deposits of iron formation generally have features in common, but each deposit has some distinctive characteristics (Gundersen, 1960). Because of these differences, considerable confusion and misunderstanding have arisen over the definition of iron formation. James (1954) attempted to clarify the matter by generalizing the definition. He defined iron formation as "a chemical sediment, typically thin-bedded or laminated, containing 15 percent or more iron of sedimentary origin, commonly but not necessarily containing layers of chert." Since this definition is broad enough to embrace all so-called iron formation, it is used in this book. However, since most iron formation contains between 25 and 35 percent iron, some geologists consider the lower limit of 15 percent too low. Furthermore, the presence of thin layers and nodules of chert is considered by many to be an essential characteristic of iron formation (see Figs. 18-1 and 18-2). The iron contained in iron formation is present in several compounds, including magnetite, hematite, limonite, siderite, chlorite, greenalite [$(Fe^{+2},Fe^{+3})^6Si_4O_{10}(OH)_8$], minnesotaite [$Fe^{+2},Mg,H_2)_3(Si,Al,Fe^{+3})_4O_{10}(OH)_2$], specularite, soft red hematite, stilpnomelane [$K(Fe^{+2},Fe^{+3},Al)_{10}Si_{12}O_{30}(O,OH)_{12}$], grunerite [$(Fe,Mg)_7Si_8O_{22}(OH)_2$], fayalite [$Fe_2SiO_4$], and pyrite.

394

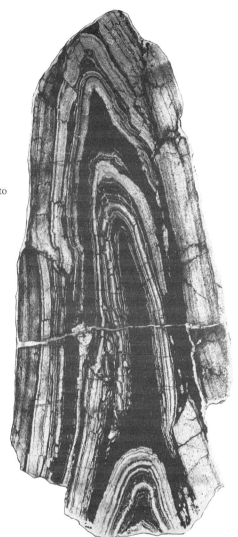

Figure 18-1
Typical folded iron formation
from Gabon, Africa. ×1.5. (Photo
by Ruperto Laniz.)

Figure 18-2
Folded iron formation from outcrop,
Andrade mine, Minas Gerais, Brazil.
Layers of sandy weathered silica fell
out when the specimen was moved.
Natural size. (Photo by Ruperto
Laniz.)

Iron formations are widely distributed in the world, but surprisingly the largest and most abundant deposits are around the Atlantic and Indian Oceans. Deposits of iron formation are few and of little consequence in the Pacific Basin. Major districts of iron formation are present in Brazil, Venezuela, central United States, eastern Canada, along the west coast of Africa (particularly in Gabon, Liberia, and Mauritania), in several districts in Russia, India, Manchuria, and in western Australia, where large deposits were discovered recently.

Most iron formation contains no fossils other than a few algae or nondiagnostic and questionable organisms. It has long been claimed that all iron formation is of Precambrian age, but several Paleozoic deposits have been reported (O'Rourke, 1961). Many hypotheses on the origin of iron formation have been based upon the assumption that these ores formed only during the Precambrian. Whether Precambrian or Paleozoic, the deposits are all very old. Such ancient rocks have usually experienced a complex tectonic history, so that the easily deformed, thin-bedded iron formation is likely to be intricately distorted or even highly metamorphosed (see Fig. 18-3). The apparent lack of similar materials formed during the later geologic periods has led investigators to explain the tremendous accumulations of iron minerals as the result of special Precambrian conditions that have not recurred. For example, Lepp and Goldich (1959) suggested that the Precambrian atmosphere was relatively deficient in oxygen, and that as a result more iron was taken into solution in the ferrous state. Geologists who disagree with the thesis that the Precambrian atmosphere was oxygen deficient claim that longer periods of crustal stability and deeper peneplanation resulted in a higher degree of chemical weathering during Precambrian and early Paleozoic time. River waters carry large amounts of iron, furnishing a ready and adequate supply for chemical deposition. Since iron is more abundant than calcium in the earth's crust, sedimentary iron should, under the proper conditions, be as extensive as limestone.

The origin of banded iron formation has been discussed at length, and the literature on the subject is voluminous. Recent articles dealing with the genesis

Figure 18-3
Sketch of a split drill core of iron formation. Bomi Hills, Liberia. Three-fourths natural size.

are by Goldich (1973), Eugster and I-Ming (1973), Holland, 1973, and Garrels, Perry, and Mackenzie (1973). Six of the better known theories are given below:

1. Silica and iron accompanied volcanism and were probably poured out on the sea floor from magmatic springs (Van Hise and Leith, 1911; Trendall, 1968).

2. Iron and silica were carried in solution from a nearby land mass and were rhythmically deposited as sediments in water, probably in response to seasonal variations in composition of the water. Various explanations have been offered for the mode of deposition, most of which involve direct precipitation of silica and iron or one of several biochemical processes.

3. Beds were originally deposited as fine-grained ferruginous tuffs and other iron-rich sediments; these sediments were quickly oxidized and silicified, more or less contemporaneously with their deposition, under the action of solutions that were partly magmatic in origin. The silicification gave rise to the banded cherts and jaspers that alternate with more iron rich layers (Dunn, 1935, 1941).

4. Deposits of iron formation in Rhodesia apparently resulted from the supergene leaching, silicification, and ferrugination of carbonatized felsite (Zealley, 1918).

5. Iron formation is due to accumulation in a partly enclosed basin, the precipitation and character of the sediments being controlled by the oxidation-reduction conditions of the depositional environment (James, 1951, 1954).

6. Jolliffe envisions a primitive acidic sea, with a pH of 6 or less, an Eh of about 0, and the water in equilibrium with an atmosphere rich in CO_2. Under these conditions the iron released by erosion and by volcanism would remain as ferrous iron in the waters of the sea. As time progressed, the CO_2 of the atmosphere would be gradually depleted and an increase in the pH of the sea would result. A point of saturation was ultimately reached and $FeCO_3$ started to precipitate. The gradual buildup of oxygen and the depletion of CO_2 in the atmosphere eventually led to the wholesale precipitation of the iron in the seawater (Jolliffe, 1966).

The first theory—that the iron formation is a direct result of volcanic activity—has received much support and is still favored by many geologists (Van Hise and Leith, 1911; Royce, 1942; Goodwin, 1956). Advocates of this theory point to the common association of volcanic activity and iron formation. The hot waters associated with the igneous activity might logically furnish a ready supply of iron in solution. However, this relationship between iron formation and volcanic activity is not found everywhere; Morro do Urucum, in Western Brazil, is an example of a large district in which volcanism seems to have been absent.

As a result of his work in India, Dunn (1935, 1941) conceived the idea that the silica layers in the iron formation are not sedimentary but are products of secondary silicification of materials now represented by ferruginous, chloritic, or carbonaceous shales and phyllites, many of which originated as tuffs. The silicification is thought to have been in part contemporaneous with the deposition of the beds and to have resulted from thermal effects accompanying the volcanism. A low temperature is suggested by the fine-grained textures of the silica. The Rhodesian iron ores, however, have been attributed to the leaching of volcanics by circulating meteoric waters.

The issue seems to involve two problems: (1) What is the source of the iron? (2) How did the banded deposit form? Of the six theories, three attribute the iron directly to a magmatic source—either thermal volcanic waters or penecontemporaneous leaching of volcanic rocks. The banded deposits are attributed to rhythmic pulsations of volcanic waters, to silicification parallel to the water table, and to seasonal variations in the environment of deposition. Perhaps several of these theories are valid for individual deposits, but none of the theories relating the ores to volcanism can be applied universally.

The work of James and others in the Lake Superior region throws doubt on the volcanic source of iron and silica in this area (James, 1951, 1954; White, 1954). These geologists relate both the iron-rich rocks and the volcanism to geosynclinal development during Huronian time. The major environmental requirement for deposition of iron formation is thought to be the closed or restricted basin. This condition coincides in time with a normal stage in evolution of the geosyncline—that is, with the structural development of offshore buckles or swells that subsequently developed into the island arcs characterized by volcanism.

Lepp and Goldich (1964) argued against James' "facies" concept because they do not accept his premise that the Precambrian atmosphere was essentially the same as today's. Rather, they prefer to assume an oxygen-free or oxygen-deficient atmosphere until late Precambrian times. They say that, regardless of facies, iron formation contains an average of 26.7 percent of iron and is remarkably uniform. This is attributed to a fundamental control in the source of materials. Calcium and iron should have been equally abundant at the weathering source and the principal minerals deposited were siderite, calcite, or an intermediate carbonate. Silica was carried in solution and was precipitated as chert, replacing the carbonates. Iron silicate minerals, such as greenalite and stilpnomelane, were formed during diagenesis, as were magnetite and hematite. Laterites that were rich in aluminum and titanium were left at the source.

Arguments advanced against the theory of Lepp and Goldich include:

1. Stratigraphers apply the facies concept with excellent results, and iron facies seem to fit paleoecological facies.

2. The origin of the banding is doubtful. Replacement of carbonate mud by silica seems to favor nodules rather than delicate laminae.

3. What happened to bauxites left at the source? Aluminum-rich sediments of the same age as the iron formations are not found (Gross, 1965; Trendall, 1965).

Owens (1965) argued that the iron content is not uniform in iron formation, as Lepp and Goldich state, but really varies greatly. The variations are significantly related to stratigraphy, even within a single formation.

The iron formation (taconite) of the Lake Superior region is divided into four facies—sulfides, oxides, carbonates, and silicates—on the basis of the dominant iron-bearing mineral. Deposition of each facies was largely controlled by the Eh and pH conditions of the environment, especially the oxidation-reduction potential (see Fig. 18-4). Low values of pH and Eh favor the deposition of pyrite, whereas high values promote the oxide minerals; carbonates and silicates form under intermediate Eh-pH conditions (Krumbein and Garrels, 1952; Huber, 1958). The deposition of taconite took place in restricted basins that were separated from the ocean by thresholds that inhibited free circulation and caused the development of abnormal oxidation-reduction environments.

Govett (1966) argued that iron formation cannot come into being in a marine environment; he proposed a lacustrine or closed basin as the earliest iron-formation environment. He said that hematite was the only primary iron mineral in sedimentary rocks and that other iron minerals were formed by diagenesis. According to Govett, organic matter played a significant role in both weathering and diagenetic processes. This suggestion may be valid for some of the smaller iron formations, but it certainly cannot be valid for the extensive materials in Australia, Brazil, and West Africa.

Hough (1958) suggested that the following conditions provided the proper environment for deposition of iron formation: large, deep, freshwater lakes; a subtropical or warm-temperate annual weather cycle; and a watershed in a state of mature geomorphic development. The lakes would have to be deep enough to permit density stratification of the water. During the summer the lower water zone would be isolated from the atmosphere, making it acidic and slightly reducing and retaining iron in solution. During the winter convective overturn of the water would bring about oxidizing, alkaline conditions, thereby causing precipitation of the iron. Biochemical precipitation may account for enrichments of silica during the summer, but direct chemical precipitation also may have been seasonally controlled—especially during the Precambrian, when surface waters were probably saturated with silica (Siever, 1957; Hough, 1958). Furthermore, differential weathering in the source area may have increased the supply of iron during the winters and the supply of silica during the summers (Alexandrov, 1955).

Lake Superior's sulfide facies is represented by black slates containing as much as 40 percent pyrite. The free carbon content of these slates typically

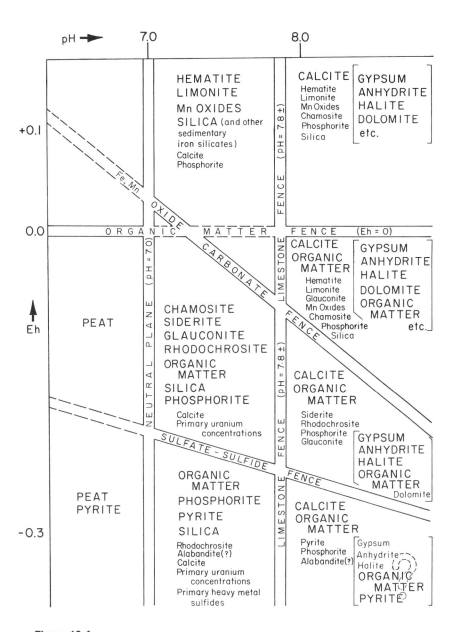

Figure 18-4

Fence diagram showing Eh-ph fields in which chemical end-members of nonclastic sediments are formed under normal sea-water conditions. Associations in brackets are for hypersaline conditions (salinity > 200 ‰). (From Krumbein and Garrels, 1952.)

ranges from 5 to 15 percent, indicating that ultrastagnation conditions prevailed during deposition. Individual crystals of pyrite are microscopic, even in the slate, and the iron content is not conspicuous in hand specimens. When uncontaminated, the carbonate facies consists of interbedded siderite, or iron-rich ankerite, and chert. It is a product of an environment in which the oxygen concentration was high enough to destroy most organic material but not high enough to permit ferric compounds to form. There are two subfacies of the oxide zone, one characterized by hematite and the other by magnetite; both are primary sediments. The magnetite facies, very common in the Lake Superior region, consists of magnetite interlayered with silica, carbonates, iron silicates, or some combination of these minerals; its mineralogy and associations suggest weakly oxidizing to moderately reducing conditions. The hematite rock consists of finely crystalline hematite interlayered with chert or jasper; oölitic structures are common. Evidently the hematite facies accumulated in a strongly oxidizing, near-shore environment like that in which younger iron-rich rocks, such as those of the Clinton formation of the eastern United States, were deposited. The silicate facies contains abundant hydrous ferrous silicates—greenalite, minnesotaite, stilpnomelane, or chlorite, and is most commonly associated with either carbonate- or magnetite-bearing rocks, suggesting that the optimum conditions for deposition ranged from slightly oxidizing to slightly reducing. Figure 18-5 illustrated the depositional zones proposed by James (1954).

Figure 18-5
Depositional zones in a hypothetical basin in which iron compounds are being precipitated. (From James, 1954, Fig. 3.)

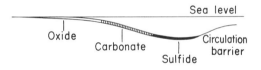

Excellent evidence in support of James's theory was advanced by Gastil and Knowles (1960) as a result of their work in the Wabush Lake area of eastern Canada. They found similar facies in a sequence controlled by oxygen availability as determined by water depth or proximity to a shoreline. Figure 18-6 shows this relationship clearly, and the sequence may be explained by a shoreline to the northwest.

Many geologists believe that ordinary river waters can dissolve and transport enough iron to account for the immense amounts in iron formations; this hypothesis is supported by field and laboratory evidence. Moore and Maynard (1929) made an important discovery when they found that abundant organic matter allows cold water to extract and transport enough iron and silica to build up large deposits of iron formation. They said that the iron and silica traveled and precipitated as colloids protected by organic matter; their experi-

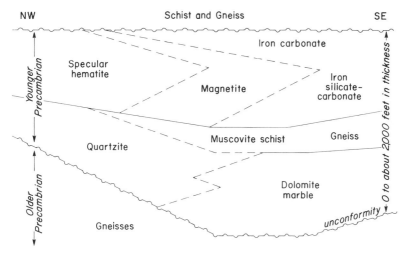

NW Schist and Gneiss SE

Iron carbonate

Specular hematite

Magnetite

Iron silicate-carbonate

Muscovite schist Gneiss

Quartzite

Dolomite marble

Gneisses

unconformity

Younger Precambrian

Older Precambrian

0 to about 2000 feet in thickness

Figure 18-6
Diagrammatic section across the Wabush Lake area, Canada, showing the distribution of sedimentary facies. (From Gastil and Knowles, 1960, Fig. 2.)

ments showed that a mixture of these sols will form a banded sediment, with an iron-rich layer at the base and nearly pure silica at the top. However, Krauskopf (1956a) has cast doubt on the colloidal nature of the silica.

It has been suggested that iron formation is an epicontinental sediment formed as a chemical precipitate from river waters that entered saline lakes or other closed basins; the absence of clastic diluents generally means a land surface that had been reduced to the limit of peneplanation. The combination of closed basins and peneplanation is rare, and the precipitation of iron formation is restricted to periods of exceptionally perfect base-leveling (Woolnough, 1941). Studies in Manchuria by Sakamoto (1950) led to the conclusion that shallow lakes in an oscillating basin, coupled with a monsoonlike climate, were responsible for the banded iron deposits. Periodic fluctuations in the pH of the lake water resulted in cyclic deposition of iron oxides and silica in alternating bands. Sakamoto attributed the iron and silica to mature weathering in neighboring areas and noted that the minerals must have been transported in ways that varied according to the seasons, such as the development of distinct soil horizons in wet and dry climates. During the wet seasons, when the waters were acidic, the iron was carried to lakes; arid seasons would have caused precipitation of the iron and concomitant transportation of silica because of a sharp rise in pH. Accordingly, a uniform climate could not have produced the banded iron formation.

The layers and nodules of silica in iron formation consist of chert, though in many places—especially where it is metamorphosed and later weathered—the silica resembles quartzite or sandstone, and is frequently so called. Examination of many specimens has led to the conclusion that the material is not a mechanical sediment. In some districts the silica has a mosaic texture (Tyler, 1949), and iron formation in general is notably lacking in the heavy accessory minerals expected in a clastic sediment. As in any sequence of sedimentary rocks, local accumulations of mechanical deposits occasionally grade into chemical precipitates. Nevertheless, by far most silica in iron formation is chemically precipitated chert.

Bacteria may be important in the deposition of iron formation, but it is not possible to determine the importance since the role of bacteria is not fully understood. Iron-secreting bacteria exist now and were apparently active in Precambrian times as well (Harder, 1919; Gruner, 1922; Beerstecher, 1954). Under favorable conditions these microorganisms may or may not be the major factor in the deposition of iron. The bacteria may merely speed up an inorganic reaction that would produce the same results if given enough time. Both processes—the organic and the inorganic—are undoubtedly active; the question is which process produced the most iron in a given deposit.

The observation that "the waste of one generation is the ore of the next" is well exemplified by the story of iron formation. Until about the end of World War II, iron formation was not ore; the iron content was considered too low and the silica content too high. Iron formation was mined as ore only where the silica had been leached by weathering or other processes. With the development of fusion or jet-piercing drills and economic methods of concentrating and pelletizing the iron, low-grade taconites became ore almost overnight; low-grade direct-shipping ores (50 ± percentage of iron) now find it increasingly difficult to compete. However, most iron formation is not ore unless it is situated so that it can be cheaply and easily handled—that is, it must be shallow, near a market, and readily amenable to concentration.

Canga is defined as a ferruginous breccia or conglomerate composed of fragments of hematite and itabirite cemented by limonite or hematite, and sometimes by other lateritic constituents (Holmes, 1920). The term originated in Brazil and is widely used for iron rich laterites that usually accumulate at the base of slopes. Silica is leached from many cangas, and where the silica is entirely removed and the iron oxides recrystallized and compacted, hard blue hematite masses may result. Many districts underlain by iron formation contain pods, lenses, and layers of high-grade hard hematitic ore. These concentrations were formed either by leaching of silica or by the addition of iron as a replacement product, resulting from metamorphism, hydrothermal fluid activities, or weathering. The hard blue hematite in Gabon may be directly related to the transformation of canga by meteoric waters (Park, 1959). In many places where the underlying rock is siliceous itabirite or taconite, only the high-grade bodies may be mined profitably.

Iron ores of unquestioned chemical sedimentary origin, but distinct from iron formation, are mined in many places throughout the world. Of these ores, the oölitic deposits are probably the most valuable. Oölitic ores vary considerably; in some the oöids abound and make up most of the rock, but in others they are scattered throughout a clay or limestone matrix. The oöids also vary considerably in composition; they may consist of hematite, limonite, siderite, or chamosite.

Oölitic Iron Ores

One of the world's most extensive deposits of oölitic iron ore is found in the Clinton Formation, which crops out sporadically from upper New York State southward into Alabama, where it is covered by the Coastal Plain sediments. The Clinton Formation, of Silurian age, is composed of thin-bedded, iron-stained sandstone, shale, and oölitic hematite. Locally the beds are calcareous and grade into impure limestones (Smyth, 1892; Burchard and Butts, 1910; Alling, 1947). Fossils are abundant, and many have been replaced entirely by hematite. The hematitic beds attain their greatest development in the Birmingham district of Alabama, where they are as much as 20 feet thick.

Three types of ore have been mined in the Birmingham district: oölitic, in which the hematite forms oöids generally in a hematite, calcite, or somewhat siliceous matrix; "flaxseed," in which the hematite forms small flat grains, or flattened oöids; and fossil ore, in which the hematite replaces numerous fossils, principally molluscoids and bryozoa. As a result of the supergene leaching of calcite matrix, much of the ore in the upper parts of the mines is soft; at depth the ores are harder and more siliceous. The hard ores are relatively low grade, the main impurity being calcite. But since calcite is necessary as a fluxing agent, and since coking coals are mined in the nearby Warrior coalfields, the Birmingham deposits were able to compete successfully with higher grade deposits found elsewhere. With depth the Birmingham ores become increasingly siliceous and more refractory to smelt; the mines are now closed.

The Clinton Formation is of shallow-water, marine origin (Smyth, 1892; Alling, 1947). It exhibits cross-bedding, mud cracks, animal tracks, oölitic structures, and other shallow-water features, and in places it contains lenses of conglomerate. The hematite is probably of both primary and diagenetic origin, as evidenced by replaced fossils and concentric layers in calcareous oöids (Alling, 1947). The iron was carried into shallow marine basins and was slowly oxidized and precipitated at the same time the other sediments were accumulating. Castaño and Garrels (1950) demonstrated experimentally that aerated river waters with a pH of 7 or lower are able to carry significant quantities of ferrous iron in solution. If such a solution enters a marine environment where solid calcium carbonate is at equilibrium with the sea water, the iron will be precipitated as ferric oxide, both in the water and as a direct replacement of the calcium carbonate. This mechanism is the most popular explanation of the Clinton ores. Hence, these ores are considered to be early diagenetic replacements of calcareous ooze, fossil fragments, and oöids, to which additional contributions were made by direct chemical precipitation.

The oölitic limonite ores of Alsace-Lorraine and Luxembourg are in Middle Jurassic shales, sandstones, and marls. The ores average between 30 and 35 percent iron in a gangue of calcium carbonate and silica. These ores furnish the bulk of iron used in the Western European iron and steel industry (Bubenicek, 1961). The oöids consist dominantly of limonite, though siderite, chlorite, and hematite are also present. According to Cayeux (1909), the iron minerals replaced the original calcitic oöids.

A third example of oölitic ore is found at Wabana, along the coast of Newfoundland. These deposits consist of beds that are a few inches to 30 feet thick, and lie within the upper 400 feet of a sequence of Ordovician sandstones and shales. Only the three thickest beds are of economic importance. Shallow-water to subaerial depositional conditions are evidenced by raindrop imprints, oöids, cross-bedding, mud cracks, and the presence of shallow-water marine fossils, such as brachiopods and trilobites. The oöids consist of concentric shells of hematite and chamosite, in places embedded in a siderite matrix. Both the oöids and the matrix are penetrated by algal borings, thereby giving evidence that the ores were approximately in their present condition before they were buried. The deposition of iron oxide apparently terminated abruptly, as evidenced by overlying graptolitic shale, which contains oölitic pyrite (Hayes, 1915, 1928, 1929, 1931). These deposits are exposed along the northwest shore of Bell Island and dip about 9°NNW out under Conception Bay. Mining operations have been carried on for a distance of more than $2\frac{1}{2}$ miles along the dip, placing the deepest workings well out under the bay. The ore averages 51.5 percent iron, 11.8 percent silica, 0.9 percent phosphorus, and 1.5 percent moisture. An area of 50 square miles or more is available for mining, of which hardly more than one-tenth has been developed (Lyons, 1957).

Siderite Deposits

Sedimentary beds containing siderite, commonly known as black-band ores, are widely distributed throughout the world. Efforts to mine these deposits have generally been unsuccessful because of the very low grade, but sedimentary siderite has been successfully worked in Germany and the British Isles. During the colonial period in the United States, small amounts of iron carbonates were extracted from marine sediments in what is now Ohio, Pennsylvania, and a few other areas.

Bog Ores and Spring Deposits

Bog iron ores occur in small, low-grade deposits with much phosphorus, water, clay, and other impurities. They are of interest mainly because some are good examples of biochemical precipitation of iron minerals. The iron content of bog waters is higher than that of most other surface waters, because the iron is stabilized by humic complexes (Rankama and Sahama, 1950, p. 665). Bacterial action causes the precipitation of ferric oxides and hydroxides from the breakdown of humic iron complexes and ferrous bicarbonate. Supplies of iron are transported to the bog waters by streams and springs. These reactions can be observed in some northern glacial lakes. At present, bog ores are of very minor economic significance.

Special environmental conditions may account for local concentrations of both calcium carbonate and iron. An analogous example is offered by spring deposits; travertine forms where the spring waters are charged with calcium carbonate, and iron oxide forms from iron-rich waters. For example, at the Kuchan mine in Hokkaido, Japan, about 5 million tons of limonite were mined where a cold spring flowed from a hill side. The ferrous iron in solution was oxidized as soon as the water was exposed to the air, and the ensuing deposit formed a terracelike embankment.

Lake Superior Region

The Lake Superior region of the north-central United States and southern Canada (Fig. 18-7) is the most productive iron ore province in the world. From the first production in 1884 through 1956, the mines yielded more than 3 billion tons of ore (Wade and Alm, 1957). Although the most accessible and highest quality ores have already been mined, the region will continue to produce good ore for many years, though in decreasing amounts. The future of the region is assured by the current development of new economical methods of handling the hard magnetic taconites from the eastern part of the Mesabi range and the "jasper" ores of the Marquette range.

There are six principal districts—or ranges, as they are called—within the United States sector of the Lake Superior region. These are the Mesabi, the Vermilion, and the Cuyuna in Minnesota; the Penokee-Gogebic across the Wisconsin-Michigan border; and the Marquette and Menominee ranges in Michigan. The Menominee range is further divided into the Iron River, Crystal Falls, and Iron Mountain districts. Of the six ranges, the Mesabi is by far the most productive. Many subordinate deposits, such as the Gunflint range, the Republic trough, the Amasa district, the Baraboo range, and the Florence district, have contributed ore, but in much smaller quantities than the six principal ranges. In Canada the two most productive deposits are the Michipicoten range and the Steep Rock Lake district.

The literature concerning the Lake Superior deposits is voluminous; many excellent discussions of both the region and individual districts have been published (Van Hise and Leith, 1911; Gruner, 1930, 1946; Royce, 1938; Tyler, 1949; Dutton, 1952; James, 1954; Grout and Wolff, 1955; Marsden, 1968; Sims and Morey, 1972).

Extensive tracts between the various ranges are covered with glacial drift and vegetation, handicapping geological studies, so that correlation between the ranges is difficult. Nevertheless, the possibility of finding new ore beneath the drift and of extending the known ranges into covered territory is attractive—

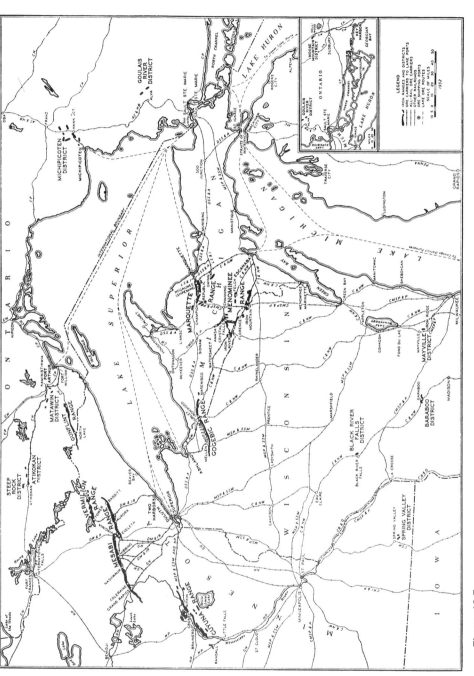

Figure 18-7

Map of the Lake Superior region showing the location of major iron ore districts. (From Lake Superior Iron Ore Association, 1952.)

much exploration has been done and some has been successful. In addition to geological studies, much geophysical work has been done. Magnetic methods using the dip needle and the magnetometer are very popular, and most of the region has now been mapped on a reconnaissance scale with airborne magnetic equipment. Magnetic techniques have helped in the mapping of bedrock structure beneath the drift. Because most high-grade ores are related to structures—usually folds—magnetic methods have contributed directly to the discovery of considerable amounts of ore.

The Precambrian stratigraphy of the Lake Superior region varies considerably from one district to another. Probably the most comprehensive correlations were made by Leith and his co-workers, in 1935; a modified copy of their correlation chart is reproduced in Table 18-1. The great abundance of slates associated with the iron formation is a conspicuous feature. Many of these slates are lithologically similar to one another and are not easy to distinguish in the field. They are also intricately distorted in some places. Most of the slates are uniformly gray or black, but a few show gradational bedding that resembles varves. Other beds contain a few slate fragments of slightly different color, and locally sandy, or even conglomeratic, beds are recognized. The slates contain appreciable amounts of finely divided carbon and pyrite. In the Menominee range many slates ignite spontaneously when exposed to the air, and for this reason are avoided in the mines wherever possible.

Volcanic materials are widely distributed, particularly in the Huronian and Keweenawan series. Greenstones are the most abundant igneous rocks, though intrusive bodies are also common. Many mafic dikes are present, and in the Gogebic range the best ores are concentrated along these intrusives. The Duluth Gabbro borders the western shore of Lake Superior and cuts off the eastern extension of the Mesabi range. Granitic rocks (Giants Range Granite) are also widespread north of the Mesabi range.

Throughout most of the region the iron formation contains between 25 and 35 percent iron. It is a hard, tough rock, difficult to break and expensive to beneficiate. The iron is present as hematite, magnetite, siderite, pyrite, or one of the iron silicates. The direct-shipping ore of the Lake Superior region comes from those areas where natural processes have leached the silica, leaving an enriched iron-bearing product. Only in recent years have methods been developed for mining and concentrating the unenriched taconite. In the eastern Mesabi range the iron formation contains magnetite; this hard rock is crushed, concentrated magnetically, and roasted into high-grade pellets that are much in demand in industry. On the Marquette range a similar process has been developed for hard jasper-hematites, but here the iron is concentrated by flotation. The use of unleached taconites is likely to expand in the future.

A major controversy has developed concerning the ways in which silica was removed from iron formation, leaving the high-grade ore. The issue becomes more and more critical as mining depths increase. If the silica was leached by

Table 18-1

Comparative stratigraphic columns from the principal districts of the Lake Superior region.

Series		Gogebic	Marquette	Menominee
Post-Keweenawan rocks				
		Sandstone	Sandstone	Sandstone
		Unconformity ~		
Keweenawan		Presque Isle Granite Mafic intrusives Sandstones, shales, and conglomerates Silicic flows Mafic flows Quartzite and conglomerate		
		Unconformity ~		
Huronian	Upper	Granite Tyler Slate	Silicic intrusives Mafic intrusives Michigamme Slate Upper slates Bijiki iron formation member Lower slates Clarksburg volcanics Greenwood iron formation Goodrich Quartzite	Silicic intrusives Mafic intrusives Michigamme Slate (including iron formation) South belt of Quinnesec greenstone (position uncertain)
	Middle	Unconformity ~ Mafic intrusives and extrusives Ironwood iron formation Palms Quartzite	Negaunee iron formation Siamo Slate Ajibik Quartzite	Vulcan iron formation Greenstones (Lake Antoine region)
	Lower	Unconformity ~ Bad River Dolomite Sunday Quartzite	Wewe Slate Kona Dolomite Mesnard Quartzite	Randville Dolomite Sturgeon Quartzite
Algoman granite		Granite		Granite
Knife Lake (may be Lower Huronian)				
Laurentian granite		Unconformity ~ Granite and granitoid gneiss	Granite, syenite, peridotite Palmer Gneiss	Granites and gneisses
Keewatin		Greenstones and green schists	Kitchi Schist and Mona Schist	

Source: Leith et al., 1935.

Cuyuna	Mesabi	Vermilion
Cretaceous conglomerate	Cretaceous conglomerate and shale	
～～～～～～～～～	～～～～～～～～～	～～～～～～～～～
Mafic intrusives Silicic intrusives	Duluth Gabbro	Duluth Gabbro
Virginia Slate 　Upper slates 　Deerwood iron 　　formation 　　member 　Lower slates	Virginia Slate	Rove Slate
	～～～～ ? ～～～～	～～～～ ? ～～～～
	Biwabik iron formation Pokegama Quartzite	Gunflint iron formation
Dolomite	Unconformity ～～～～ Giants Range Granite	Vermilion Granite
	Knife Lake Slate	Knife Lake Slate 　Slate 　Agawa iron formation 　　member 　Ogishke conglomerate 　　member
		Granite
	Greenstones, schists, and 　porphyries	Soudan iron formation Ely Greenstone

cold waters, then the ore should be concentrated either near the present surface or along unconformities where weathering took place during earlier geologic times. If the leaching was caused by hydrothermal waters—possibly associated with volcanic activity—there is a favorable chance that more ore will be encountered at depth. This problem has not been solved, and possibly it cannot be in view of our present knowledge. Some of the evidence advanced is contradictory, indicating that many of the basic field relationships have not been solved. The current arguments for cold- versus hot-water leaching are listed below.

EVIDENCE AND ARGUMENTS FOR COLD-WATER LEACHING OF SILICA

1. The ores are near surface, either near the present surface or near an old surface that existed during Precambrian time. Advocates of the cold-water leaching hypothesis emphasize the presence of many unconformities in the stratigraphic column, which they feel are genetically related to the ores.

2. Most of the ore is concentrated in synclines. This is interpreted to signify leaching by descending waters.

3. The orebodies grade in depth into unleached iron formation, similar to other deposits known to have been formed by ordinary weathering processes. That supergene leaching can form high-grade deposits in iron formation is demonstrated in other areas. For example, extensive itabirites in the Republic of Gabon, Africa, were leached by surface agencies, and in places the silica has been removed to depths of more than 300 feet.

4. The mineralogy is simple and is what would be expected from meteoric conditions.

5. Comparison has been made with the formation of bauxite-boehmite deposits, where surface waters have leached tremendous quantities of silica.

6. In the Penokee-Gogebic range ore is found both above and below comparatively impermeable dikes. Here the leaching is attributed to meteoric water moving along an old erosion surface. The ore above the dikes seems difficult to explain by ascending waters.

EVIDENCE AND ARGUMENTS FOR HOT-WATER LEACHING OF SILICA

1. There are numerous dikes in this region, and the ores are commonly found along or near these igneous rocks.

2. The depth of leached iron formation, especially in the Penokee-Gogebic range, is between 3,000 and 4,000 feet. This depth of leaching seems too great for waters related to the present surface, and no conclusive evidence has been found that the leaching is related to old erosion surfaces.

3. Where dikes are inclined at low angles, the ores are likely to be concentrated on the footwall side—a relationship that is common in the Penokee-Gogebic range. (This evidence seems to directly conflict with argument six of the cold-water hypothesis.)

4. In a few places ore is recovered along the crests of anticlines (which contradicts the argument that descending fluids selectively leached silica from the synclines).

5. Much smaller quantities of warm water than cold water would be required for the leaching.

6. The presence of manganese ore pipes containing hausmannite (Mn_3O_4), a member of the spinel family, indicates that hydrothermal fluids have been active in the region.

The Mesabi range (Fig. 18-8) is by far the leading producer in the region. Most of the ore is a mass of soft, reddish-brown hematite and limonite from which all or nearly all of the silica and other minerals have been leached. Structurally the Mesabi range is simple. Except for a Z-shaped bend near the center of the outcrop area, the orebody strikes east-northeast and dips 4° to 7°SE in the western part of the area and 6° to 12°SE in the eastern part. At a depth of a few hundred feet the ore grades into unleached and unoxidized iron formation, which is partly iron carbonate. The eastern end of the range consists

Figure 18-8
Aerial view of the Mississippi Group, Mesabi range, Minnesota. (Courtesy of Hanna Mining Co.)

of hard magnetic taconite, metamorphosed perhaps by the Duluth gabbro to the east (French, 1968). The western end consists of soft weathered taconite in which the silica is still retained. Between these extremes the iron is direct-shipping, high-grade material. The soft western taconite is also mined; it is beneficiated to an economic grade by removing the silica by washing.

In contrast to the simple structure of the Mesabi range, most of the other districts are structurally complex. The iron formation is intricately folded and faulted, and as a result, detailed knowledge of both the stratigraphy and structure are essential to exploration (Schmidt and Dutton, 1957; Pettijohn, 1946; Dutton *et al.*, 1945).

In the Steep Rock Lake deposit of Ontario the ore is massive and not clearly bedded. Boulders of ore found along the lakeshore led to the discovery of this deposit, which, between 1938 and 1942 was outlined by exploratory drilling through the ice of Steep Rock Lake. Before the ore could be developed, however, the river had to be diverted and the lake drained. The ores are localized along the contact of limestones and volcanic rocks, which they seem to replace (Roberts and Bartley, 1943a, 1943b; Quirke, 1943). These relationships led early workers to conclude that the iron was introduced by hydrothermal agencies. The hard ores of Steep Rock Lake are similar to the hard ores of the Vermilion and Marquette ranges, suggesting that all three deposits were formed by a similar process. Jolliffe (1955) remapped the Steep Rock Lake deposit after considerable mining had been done; he concluded that it was not of replacement origin. He interpreted the deposit as a sedimentary bed of limonite that was modified by hydrothermal fluids. This theory is supported by the stratigraphic position of the ore zone, which lies on an erosion surface; by the presence of gradational facies that are undoubtedly sedimentary in origin; by the mineralogy; and by the paragenesis. Jolliffe (1955, 1966) emphasized that the dominant mineral, goethite, is early in the paragenetic sequence—an anomalous position for hydrothermal ore but normal for a sediment that was later altered by hydrothermal activity. Jolliffe (1966) enlarged upon his early ideas of the origin of the Steep Rock Lake ores. He envisaged an iron-rich residuum of a landmass of low relief being transported to a shallow, water-filled basin. Here the colloidal oxide and hydroxide gels were deposited like modern bog iron deposits. The Steep Rock beds were later tilted and deformed, but metamorphism was surprisingly slight; the composition of the beds remains much today as it was originally.

In general, the mineralogy of the Lake Superior ores is simple. Hematite is the principal mineral, either the soft, reddish variety of the Mesabi range, or the hard, massive variety of the Marquette and Vermilion ranges. Magnetite is common and is the dominant mineral in a few areas. On the Menominee and Cuyuna ranges the ore is composed of soft, yellow to brown limonite with some soft hematite. Several beds in the Cuyuna range carry as much as eight percent manganese, which brings them a higher price than the straight iron ore.

Sedimentary Manganese Deposits

Manganese behaves chemically much as iron behaves—the two elements accumulate in similar environments and under similar conditions. Under oxidizing conditions, pyrolusite or some other form of MnO_2 would be expected to form; at intermediate values of Eh and pH, hausmannite or the manganese carbonates or silicates should be deposited; and in extremely reducing environments, alabandite (MnS) or manganosite (MnO) should form (Krauskopf, 1957). The extremely low Eh-pH conditions necessary for alabandite and manganosite are not likely to be attained in sedimentary environments, but the other minerals are common and seem to be deposited according to their thermodynamic restrictions. Figure 18-9 shows the stability fields for many of the manganese minerals. Although the compounds having complex or variable compositions must be generalized by a simple formula, the figure still gives a fairly accurate picture of manganese environments and the facies to be expected under various conditions. For example, the diagram indicates that pyrolusite should be formed in the same environments as manganite and braunite but not under the same conditions as rhodochrosite, alabandite, or hausmannite; the natural mineral assemblages corroborate this expectation.

Both oxides and carbonates of manganese are widely distributed throughout the world. The carbonates are generally complex, containing variable amounts of calcium, magnesium, and iron along with the manganese. Many manganese oxide deposits are nearly pure, but others contain minor amounts of cobalt,

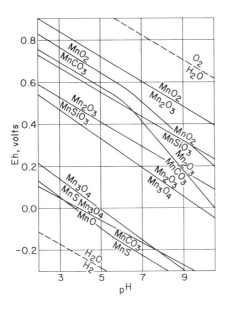

Figure 18-9
Eh-pH diagram for manganese compounds, showing the stability fields for common sedimentary minerals. (From Krauskopf, 1957, Figs. 1 and 3.)

nickel, tungsten, copper, and barium, or such extraneous materials as clay, limestone, chert, and tuff. Manganese silicates ordinarily do not accumulate as sedimentary deposits, although at the San Francisco mine in Jalisco, Mexico, braunite ($Mn^{+2}Mn_6^{+3}SiO_{12}$) is among the most abundant of the sedimentary minerals.

Sedimentary manganese deposits may be classified into three categories:

1. Deposits associated with tuff and clastic sediments of volcanic materials.
2. Deposits independent of volcanic activities.
3. Deposits associated with iron formation.

These three classes are intergradational and also grade into deposits of hydrothermal and metamorphic origin. In places it is difficult or impossible to distinguish among the different types of deposits. For example, where hot springs transport manganese to a lake or ocean floor, part of the manganese precipitates out of the water and becomes an integral part of the ordinary sediments; but ore that has formed near or within the spring orifice may be classified as a hydrothermal deposit. Notwithstanding its limitations, the above classification serves as a convenient outline for studies or descriptions of the genesis of manganese deposits.

According to Mikhalev (1946), all of the principal manganese deposits of the world are either of sedimentary origin or are metamorphosed sediments. He concluded that the deposition of manganese ores takes place in near-shore basins, especially along the peripheries of continents. Many manganese deposits that were formerly described as residual products of weathered metamorphic rocks are now considered to be primary sediments. Other deposits have resulted from the weathering of metamorphosed sediments (Park, 1956b; Nagell, 1962). Whether residual or primary, the manganese is a sedimentary product, perhaps modified by metamorphism or enriched by weathering.

Many valuable manganese ores, usually consisting of oxide minerals, are interlayered with (and genetically related to) highly altered reddish or greenish andesitic tuffs and clastic volcanic sediments. These beds exhibit sedimentary features like those of the overlying and underlying tuffaceous beds. They are generally classified as sedimentary deposits, although where they are localized along faults they have been interpreted as hydrothermal (Wilson and Veytia, 1949). The deposits are thin-bedded, with individual layers four inches thick and less. In a few places the beds resemble ordinary banded iron formation because the manganiferous layers are separated by thin beds and nodules of chert. More commonly, however, the interlayered materials are clays, altered tuffs, or clastic volcanic debris. Locally, the layers between manganese strata pinch out, and the manganese becomes massive.

The origin of the manganese oxides associated with volcanic processes is of considerable theoretical and economic interest. Hot volcanic materials ejected under water tend to become finely and thoroughly fragmented. The fragments are agitated, and because of their textures are ideally prepared for leaching by volcanically heated lake or sea water or by hydrothermal waters contributed during volcanism. Under these conditions, the ferromagnesian minerals, which contain the manganese, are susceptible to alteration. Volcanic emanations that percolate through the tuff contribute toward decomposition of the ferromagnesian minerals, and might conceivably transport manganese. Manganese in the solutions may also come from remote magmatic sources or from land waste. The dissolved manganese migrates upward to the floor of the lake or sea and is deposited near the top of the tuff bed. These ores may be regarded as exhalative or volcanogenic deposits.

If deposition should take place where oxygen is limited or excluded, manganese oxide nodules would form until the available oxygen was exhausted, and the remaining manganese would be deposited as a carbonate or would stay in solution until it migrated to a more oxygen-rich region. The abundance of oxygen is related to water depth and consequently is indirectly related to the shoreline. Manganese deposits thus form under much the same conditions as iron formation, and like many other sediments, they resemble reefs. An example of this type is in the Elqui River valley in central Chile, where a ribbon-shaped bed of manganese oxides averaging about eight inches in thickness is enclosed in highly altered volcanic debris. It has been followed in the valley for about 20 miles along the strike, but as far as can be determined it is only one mile wide across the strike. Field evidence indicates that it is a reeflike deposit that parallels an ancient shoreline. In support of this contention, impure limestones are almost everywhere associated with the manganese and tuffs.

Deposits similar to the Chilean manganese are known in Cuba, at Charco Redondo, El Cristo, Ponupo (Fig. 18-10), and Taratana; in Mexico, at Paridero and San Francisco in the state of Jalisco (Fig. 18-11); and in the United States, near Lake Mead in Nevada and Arizona (Park, 1942, 1956a; Hewett and Webber, 1931; McKelvey et al., 1949).

Silica is a common gangue mineral in sedimentary beds associated with altered tuffs and other volcanic debris. In Cuba and Haiti the ores are clearly associated with fine-grained, reddish-brown silica called "bayate." Bayate resembles the jasperoid that is so common in hydrothermal ore deposits of the western United States. It is probably cogenetic with the manganese minerals; field evidence indicates that it was deposited at the orifices or along the channels of hot springs. In Jalisco, Mexico, a similar kind of silica has been found along fractures that cross the bed of ore and at places along the edges of the orebody.

Sedimentary manganese deposits that are independent of volcanic processes include both carbonate and oxide minerals. A reasonable explanation for the origin of these beds is that the manganese was (1) leached from surrounding rocks during normal weathering; (2) transported in stream waters to nearly

Figure 18-10
Bed of manganese oxides and tuff overlain by massive white limestone. Ponupo mine, Oriente Province, Cuba.

Figure 18-11
Workings along the outcrop of a bed of manganese oxides. San Francisco mine, Autlán, Mexico.

closed and protected basins; and (3) precipitated by electrolytic coagulation or some other chemical process. The deposition of manganese in a lake or sea would be chemically similar to the deposition of manganese contributed by volcanic activities. The ores are commonly concentrated in ribbon-shaped deposits parallel to a shoreline.

Nodules generally containing less than 20 percent of manganese oxides around cores of silica grains or other materials are widely distributed in the deeper parts of ocean basins; the nodules also contain small percentages of copper, nickel, and cobalt. A great deal of research is being done in efforts to develop economic recovery methods of the nodules, which are found in great abundance locally (Bender, Ku, and Broecker, 1966; Mero, 1964). Many geologists consider the nodules to be related to submarine volcanic processes (Bonatti and Nayuyu, 1965), though their presence in parts of the Great Lakes would indicate that volcanism may not be necessary.

Some of the world's largest manganese deposits are sediments that are apparently independent of volcanism. The famous Russian areas of Nikopol on the Dnieper River and the Chiaturi field on the south slope of the Caucasus Mountains are of this type. At Chiaturi the ore beds average four to eight feet thick and cover about 50 square miles. Manganese was derived from the weathering of neighboring granites, syenites, and possibly some andesitic volcanics. The ore consists of oölitic and nodular pyrolusite in a matrix of manganese oxides, and the associated sediments are shales, marls, and sandstones (de la Sauce, 1926). Less extensive deposits of this type in the United States are the oxides of Artillery Peak, Arizona, and the carbonate beds of Arkansas, Virginia, and Maine. In general, carbonate beds are uneconomic unless they are weathered and concentrated as oxides. An exception to this is the extensive carbonate bed near Molango, Hidalgo, Mexico, where the beds, made up of kutnahorite, manganoan calcite, and calcian rhodochrosite, are as much as 60 meters thick and extend along the strike for about 50 kilometers. Some of the beds are 25 percent or more manganese. The ore is crushed, roasted, and then pelletized. The final product is a high lime-manganese oxide pellet which makes an excellent furnace feed for steel mills.

The third type of sedimentary manganese ore is associated with iron formation—an association found in many parts of the world. The manganiferous beds commonly contain oxides concentrated in layers separate from the iron-rich beds, as at Morro do Urucum, Brazil; in other deposits, such as the Cuyuna range, Minnesota, and Lagoa Grande, Minas Gerais, Brazil, the manganese is distributed throughout many of the iron-oxide layers.

The voluminous literature dealing with iron formation seldom discusses the origin of the accompanying manganese. Why manganese is separated from iron in some deposits but distributed throughout the iron-rich strata in other deposits has long puzzled geologists. Certainly the association of manganese oxides with iron formation is common enough that an explanation of the origin of the iron formation must also explain the origin of the manganese oxides. Ljunggren (1955) pointed out that manganese and iron separate during precipitation in bogs and during movement of mineral-bearing waters through soils and subsoils. In both Sweden and Finland iron and manganese have been found as separate precipitates within single lakes or inlets. The degree of separation seems to be directly dependent upon the pH of the waters: acidic waters retain dissolved manganese longer than iron, and alkaline or less acidic waters precipitate the two oxides together (Ljunggren, 1955). The solubilities of both manganese and iron are functions of their oxidation states; in the reduced state they are both readily soluble, but each forms a very insoluble oxide. As Krauskopf (1957) pointed out, iron compounds in nature are uniformly less soluble than the corresponding manganese compounds, and the ferrous ion is more easily oxidized than the manganous ion under any naturally occurring Eh-pH conditions. Consequently, iron should be precipitated before manganese from any solution containing both metals, unless the Mn/Fe ratio is very high. Similarly, manganese should be more readily distributed than iron by weathering or meteoric processes. The problem of iron and manganese separation, both within laminae of iron formation and as pure deposits of each metal, has not been solved. Differential oxidation and solubility seem to be the most logical explanation, except perhaps locally.

Manganese silicates are not found in most unmetamorphosed iron formations, in spite of the abundance of silica in these rocks. Possibly, the development of silicates is directly related to warm or hot waters, which are essentially absent during the deposition of iron formation.

Bacteria are known to secrete manganese as well as iron, though the extent to which they are operative is debatable. Zapffe (1931) studied the manganese in ground waters and concluded that bacteria may play an active part in the precipitation. In support of this hypothesis, the sedimentary ores of Tennengebirge, Austria, have been attributed to bacterial action (Cornelius and Plöchinger, 1952). Algae are also known to precipitate manganese (Twenhofel, 1950). It has also been suggested that the separation of iron and manganese may be due in part to the presence of bacteria that use one of the metals but not the other (Krauskopf, 1957). As with iron, although the effects of bacteria may be considerable, the available evidence is inconclusive, and other processes appear to explain the larger accumulations of manganese satisfactorily.

Stratabound Uranium Deposits

Uranium ore deposits are of many types and origins, ranging from hydrothermal through sedimentary. Some of the largest and most productive are of sedimentary origin; these include the well-known conglomerates of the Witwatersrand, South Africa, and Blind River, Canada. Others are stratabound and their origin is less clear. In general, the stratabound types have been enriched by deposition of uranium minerals from circulating groundwaters.*

The numerous uranium-vanadium deposits in the Colorado Plateau region of the southwestern United States are a good example of stratabound uranium deposits enriched by circulating groundwaters. Although the deposits have been known since 1899, only after the expanded development of fissionable materials was there much interest in the region. In recent years the deposits have been of great economic value. The deposits are widely distributed throughout the watershed of the Colorado River in western Colorado, eastern Utah, northeastern Arizona, and northwestern New Mexico (see Fig. 18-12). As a result of the intense search for uranium during the past few decades, many articles have been published concerning the deposits. Most of the studies have been detailed descriptions of individual deposits; only a few treat the region as a whole (Hess, 1914; Coffin, 1921; Fischer, 1942, 1950, 1968, 1970; Kerr, 1958; Kelley, 1963).

Exploration techniques for uranium ores are varied because of the special properties of uranium minerals and because radioactivity can be detected instrumentally from a distance. The most widely used techniques employ scintillometers and geiger counters, but prospecting has also been based upon fluorescent minerals, groundwater and stream analyses, electrical geophysical methods, stratigraphic characteristics, biogeochemistry, and geobotany. Regarding the last, a few species of *Astragalus* (also known as "poisonvetch" or "locoweed") grow only where selenium is available in the soil, and on the Colorado Plateau selenium is characteristically associated with uranium-vanadium ores (Cannon, 1957; Hawkes and Webb, 1962).

The known uranium-vanadium deposits of economic value are most abundant in the Shinarump Conglomerate, Chinle Formation, Entrada Sandstone, Todilto Limestone, and Morrison Formation of the Triassic and Jurassic periods, but many other Mesozoic formations, as well as some Paleozoic and Tertiary rocks, also contain ore deposits. Most of these formations are similar; they consist of continental sandstones, siltstones, conglomerates, and impure limestones (Isachsen *et al.,* 1955; Fischer, 1956).

The ore-bearing formations are generally horizontal, or nearly so, but they are disturbed in places by moderately strong monoclinal and anticlinal folds and by high-angle faults. Major structural basins, such as the San Juan Basin, the Uinta Basin, and the Piceance Basin, delimit relatively barren areas between which there are clusters of ore deposits. A regional northwesterly structural grain apparently reflects deep-seated igneous and tectonic movements. It is defined by parallel sets of joints and small faults, as well as alignments of intrusive masses. Yet there seems to be no direct relationship between the

---

*Fischer (1974) discusses controls of uranium deposition and the most likely methods of prospecting for uranium ores of the sedimentary or stratabound types.

Figure 18-12
Map of the Colorado Plateau showing the location of the major uranium deposits. (From Kerr, 1958, Fig. 2; and Kelley, 1956, Fig. 30.)

distribution of uranium and the regional structure, except insofar as the tectonic history has controlled sedimentation, geomorphology, and igneous activity, and insofar as the ore deposits lie within the uplifted regions between structural basins (Kelley, 1955; Kerr, 1958). Salt domes and various types of igneous intrusives are widely distributed in the Colorado Plateau. The intrusive bodies include laccoliths, sills, dikes, diatremes, volcanic necks, and small stocks. Lava flows and pyroclastics are also present. Nearly every major type of igneous rock is represented on the Colorado Plateau, but most of the larger masses are silicic to intermediate in composition. Except for some Precambrian batholiths, the intrusives all seem to be Miocene or younger. No general spatial distribution between igneous activity and ore deposits can be demonstrated, except where uranium mineralization is associated with diatremes and volcanic collapse structures (Shoemaker, 1956).

Most of the orebodies are similar in type, regardless of their mineralogy and geographic or stratigraphic position. Some deposits are small irregular pods sporadically distributed within a favorable rock unit, but the larger deposits form mantos several thousand feet long, several hundred feet wide, and more than 10 feet thick. Elongated orebodies tend to follow buried stream courses or lenses of conglomeratic material. Uranium and vanadium minerals commonly fill pore-spaces in sandstones and conglomerates. Local fragments of lignite or bone material acted as centers of reduction and were replaced by rich concentrations of uranium. In addition, disseminations of replacement ore are found in carbonate rocks, and incrustations of uranium minerals occupy small fractures and bedding planes in shales and limestones (Fischer, 1956).

The principal ore minerals recovered on the Colorado Plateau are carnotite $(K_2O \cdot 2U_2O_3 \cdot V_2O_5 \cdot 3H_2O)$ and tyuyamunite—oxidation products of primary uraninite, coffinite $[U(SiO_4)_{1-x}(OH)_{4x}]$, roscoelite $[(Al,V)_2(AlSi_3)(K,Na)O_{10}(OH,F)_2]$, and montroseite $[VO(OH)]$. The associated primary and secondary vanadium-uranium minerals are too numerous to list. Minor amounts of pyrite, marcasite, galena, sphalerite, chalcopyrite, bornite, chalcocite, covellite, and many other sulfide minerals also are found. Native selenium and native arsenic are characteristically associated with the uranium-vanadium ores, and the molybdenum oxide, ilsemannite $(Mo_3O_8 \cdot nH_2O?)$ is also common (Weeks, 1956; Weeks and Thompson, 1954; Kerr, 1958). In general, the ores are low grade and rarely contain as much as one percent U_3O_8 and five percent V_2O_5. The grade of ore mined ranges from 0.16 percent to 0.25 percent or three to five pounds of U_3O_8 per ton.

Local controls of ore deposition are commonly related to permeability and groundwater movements. Lithologic controls are the most noticeable. Thus, the ores may be found along valleys in an old erosion surface, within channel sands, or in sandstones interfingered with shales and mudstones (Wright, 1955; Miller, 1955). Sands containing abundant carbonaceous matter may be selectively enriched in uranium and vanadium. The influence of fracture systems can be seen in places; for example, at Gallup, New Mexico, deposits formed along the basal side of carbonaceous shales where they are intersected by fractures in the underlying sandstone (Kerr, 1958).

Several geologists have emphasized the relationship between uranium ores and minor crenulations, or *rolls*. Shawe and Granger defined a *roll* as a curved ore layer not concordant with the enclosing strata, but one that has discordant C-shaped or S-shaped cross sections (Shawe and Granger, 1965). Rolls vary vertically from a few feet to more than 30 feet. Shawe and Granger distinguish two types of rolls, the Colorado Plateau type and the Wyoming Basin type.

The first, which is surrounded by a wide halo of mildly altered rock, was probably formed in favorable strata before or during their maximum burial prior to major structural deformation. They are not related to uplift and erosion. The second type, found mainly in the Tertiary Basins of Wyoming, is bounded on one side by relatively oxidized altered rock and on the other by relatively reduced rock. These rolls probably formed near the surface and after major uplift, deformation, and erosion. The ore-bearing solutions that gave rise to the Colorado Plateau-type flowed parallel to the axis of the roll, whereas in the Wyoming Basin-type the solutions are thought to have flowed through the roll front. A knowledge of the various spatial relationships that exist between roll-type deposits and altered-rocks deposits is an advantage for those who explore for uranium deposits.

Two types of deposits are widely recognized, based upon form and position. One is enclosed in a peneconcordant manner in sedimentary beds and lacks known cross-cutting hydrothermal feeders. Most of the deposits of Ambrosia Lake district, New Mexico, are of this type. They form rolls and elongated bodies in sandy beds, and show no evidence of any igneous relationship. The second type follows nearly vertical fractures, as in the Kane Creek district, Utah (Davidson and Kerr, 1966, 1968). Other cross-cutting veins of almost certain hydrothermal affiliation have been considered.

The role of oxidation-reduction processes in uranium transportation and deposition has been recognized by most Colorado Plateau geologists. Uranium and vanadium are relatively soluble in the oxidized forms, as hexavalent uranyl dicarbonate and tricarbonate complexes and as tetravalent vanadium complexes. Geochemical studies indicate that the metals would be most stable in solution within a mildly reducing, neutral to alkaline environment that contains abundant CO_2 (Hostetler and Garrels, 1962). Precipitation takes place under more strongly reducing conditions, such as those produced by the action of carbonaceous material or H_2S.

The origin and processes of concentration of the ores have been the subjects of much discussion and disagreement. Four principal ideas have been advanced. It has been argued that (1) the ores are syngenetic and were deposited with the sediments or as an early diagenetic cement; (2) the uranium and vanadium were leached from overlying or interlayered tuffs and volcanic rocks; (3) groundwaters leached the metals from the enclosing sands or from a nearby granitic source, forming deposits where the proper Eh-pH conditions prevailed or where groundwater circulation was impeded; and (4) the ores are products of ascending hydrothermal fluids derived from underlying magmas (McKelvey et al., 1955).

The syngenetic theory is no longer held because radioactive dating methods have shown that the uranium is younger than the host rocks. Early studies, however, were strongly influenced by the apparent lack of structural control, the obvious association between lithologic features and orebodies, and the absence of typical hydrothermal minerals (an erroneous conclusion based on the fact that all the early mines were in oxidized ores). Consequently, the syngenetic hypothesis was one of the first proposed (Coffin, 1921; Fischer, 1950), and was widely accepted until recently (Wright, 1955; Fischer, 1956).

The presence of devitrified volcanic ash near most of these uranium deposits has supported the hypothesis that the uranium is a product of ash-leaching by meteoric waters (Waters and Granger, 1953). Not only on the Colorado Plateau, but in other areas of similar uranium deposits, a striking correlation has been noted between volcanic ash and the ore. Uranium deposits in the Dakotas are in lignites that underlie bentonites, the lignites acting as precipitants for the uranium leached by descending meteoric waters (Denson and Gill, 1956). But sandstone hosts allow greater dispersion by circulating groundwaters, hence a further mechanism of concentration is required for the Colorado Plateau ores. It is puzzling to consider why the uranium and vanadium were not taken up by the montmorillonite formed by devitrification of the volcanic ash, rather than being leached from the bentonitic beds. Hostetler and Garrels (1962) proposed that the ore-bearing fluids were squeezed out of tuff beds by load compaction; that is, connate waters carried the soluble uranium and vanadium salts out of the tuffs and into more permeable sandstone strata. Thus, the Eh-pH conditions need not have changed until the migrating solutions encountered a reducing environment in the permeable sands.

The problems of mineral distribution and precipitation are more directly considered by advocates of the theory that the ores were deposited by circulating groundwaters. In its broadest sense, this theory is independent of the ultimate origin of uranium and vanadium. But in a strict sense, it implies that the ore metals were leached from nearby rocks by the same circulating groundwaters. Gruner (1956) proposed that bicarbonated groundwaters would leach uranium from Precambrian plutonic rocks and carry it into the interior drainage basins, where it would be deposited under local reducing conditions. The arid or semiarid climates in the Plateau province would favor retention of the ore minerals in favorable strata and would account for a gradual increase in the ore grade. Thus, Gruner envisaged a multiple migration-accretion process whereby repeated oxidation and leaching at the source would be accompanied by migration to a depositional site and successive increases in the uranium content of the deposit. It has also been proposed that precipitation results from the reducing action of hydrogen sulfide formed in the sediments by anaerobic bacteria (Jensen, 1958); this precipitating agent would be equally effective for ordinary groundwater and for ascending hydrothermal fluids. A similar reducing environment may be produced locally where fluids containing organic compounds diffuse away from centers of carbonaceous debris; the merging of this reducing medium with oxidized uraniferous waters would cause precipitation of uranium minerals and a consequent diffusion of more uranium ions to the reaction front (Huff and Lesure, 1962).

As a result of their studies of uranium-lead isotopes, Miller and Kulp (1963) concluded that the most probable origin of many of the Colorado Plateau uranium ores was circulating groundwaters in H_2S-rich environments. Deposition generally took place shortly after that of the enclosing sediments, but may also have occurred much later.

Deep exploration in the veins and cross-cutting structures has disclosed a typically hydrothermal suite of minerals, modified near the surface by groundwaters. The minerals at depth include galena, alunite, hard pitchblende, gersdorffite (NiAsS), sphalerite, chalcopyrite, and several others (Benson *et al.*, 1952). The mineral associations and the presence of cracked hydrocarbons may indicate abnormally high temperatures, in the general range 100° to 350°C. Wall-rock alteration effects suggest hydrothermal activity. In places dolomite, clays, and silica—in a zonal sequence—have been recognized as alteration products associated with ore deposits. Additional evidence of hydrothermal alteration is furnished by the presence of a chrome-mica clay, recrystallized sedimentary clays, and alunitization. The widespread presence of molybdenum is also a possible indication of hydrothermal activity because most other molybdenum deposits are clearly related to igneous sources. A strong argument for a hydrothermal origin for at least some of the deposits is based on their spatial and temporal relationship to igneous features (such as diatremes, volcanic vents, and breccia pipes). Tentative radiometric age determinations further support an epigenetic origin and imply that the ore deposits and igneous activity are contemporaneous (Kerr, 1958).

Davidson and Kerr (1966, 1968) give convincing evidence for hydrothermal origin of the uranium ores at Kane Creek, Utah. Here the ore is in veins that cut the Permian Cutler formation, which is about 225 million years old. The cross-cutting, when considered in light of associated host-rock bleaching, weak argillic alteration, base-metal sulfide mineralization, and a 60-million-year isotopic Pb/U age, indicates ore deposition in a hydrothermal environment that is consistent with Larimide mineralization.

The evidence published in favor of the various hypotheses is contradictory and inconclusive. Most workers recognize that groundwaters and supergene processes have strongly influenced the present ore distributions and mineral associations. At present, much mining and geological work is being done, and new facts are being uncovered constantly. It seems likely that these efforts will eventually provide conclusive evidence concerning the origin of the ores, thereby profoundly influencing decisions to explore at depth and illuminating the controversial origin of sandstone deposits of such minerals as copper, lead, and silver.

Sedimentary Deposits of the Base Metals

Wherever there is a source of metal ions and favorable redox conditions for precipitation, an ore deposit has a chance to form. Such conditions may be met in certain basins of deposition where decaying organic debris or bacterial acṭ ṳn generates an exceptional reducing environment and where the accumulation of clastics is practically nil. Several large deposits in the world seem to belong in this category, but in each case there is doubt whether the ore minerals are actually syngenetic. Perhaps the principal objection to this mechanism of ore deposition is the mysterious source of metals; normal seawater does not carry an appreciable supply. Given the base-metal ions, however, there is no chemical objection to precipitation in a sedimentary reducing environment.

Geologists have long advocated a syngenetic origin for the base-metal deposits of the Kupferschiefer near Mansfeld, Germany (Pompeckj, 1920; Schneiderhöhn, 1923: Trask, 1925). Even such ardent magmatists as Lindgren (1933) have been impressed with the strong evidence for a syngenetic origin.

According to Dunham (1964), the Kupferschiefer of Northern Europe is a bituminous-calcareous shale that lies near the base of the Zechstein formation of Middle Permian age. The Kupferschiefer is only about two feet thick, but it is spread from Northern England eastward through the Netherlands, across Germany, and into Poland. It appears to follow a long shallow arm of the Permian sea.

The Kupferschiefer overlies a thin conglomerate and is overlain by limestones. It contains marine animal and plant remains, but they do not represent an ecologic assemblage. Apparently, the environment was similar to that of the present Black Sea—shallow, stagnant waters that were not amenable to a flourishing fauna. The ore deposit possibly represents an adsorption of metallic elements from sea water during the slow accumulation of sediments.

The ore minerals of the Kupferschiefer are sulfides: principally bornite, chalcocite, chalcopyrite, galena, sphalerite, tetrahedrite, and pyrite. Minor amounts of silver, nickel, cobalt, selenium, vanadium, and molybdenum have also been recovered. Copper is the most important product; over wide areas it averages as much as three percent of the deposit. The unusual nature of the deposit is apparent in the east end of the Harz Mountains of Germany, where the shale, which is only 22 centimeters thick, is worked over an area of about 140 sq km. According to Dunham (1964) at least 2,200 square miles is underlain by shale that contains more than one percent zinc; another, almost equal area has similar concentrations of lead. The total area underlain by mineralized shale is at least 8,000 square miles.

Much discussion has focused on the source of the metals. Dunham (1964) favors a sedimentary syngenetic origin with sulfur coming from a biogenic source. The alternative to this hypothesis is an exhalative sedimentary origin; the metals were contributed by submarine volcanic springs, and precipitation took place where stagnant bottom conditions prevailed.

Some geologists maintain that many fine-grained, banded sulfide ores similar to those of the Rammelsberg district, Germany (Fig. 18-13), are sediments (Schneiderhöhn, 1953). The sedimentary hypothesis for the origin of these ores has been supported mainly by European geologists; Americans have been somewhat slower to accept these ideas. The principal problem stems from the

Figure 18-13
Banded sulfide ore from
the Rammelsberg district,
Germany. Natural size.

source of supply of the very large amounts of the metals that must be supplied
to the basins of deposition. With the discovery of submarine deposition of the
sulfides in the deeps of the Red Sea, a large part of this objection is removed.
A volcanogenic-sedimentary origin for many deposits is reasonable. An unexplained factor is the complex history of replacement that is revealed on
polished surfaces of many of the ores. Nevertheless the syngeneticists are sup-

ported by field evidence, especially the great extent of stratabound deposits that are best explained as resulting from sedimentation.

An interesting prospect of sulfur isotope geochemistry is that it may be used to distinguish between igneous and sedimentary processes in the formation of stratiform ores. On the basis of early bacteriological studies, it appeared that sedimentary, presumably biogenic, sulfides tended to be variable, whereas replacement, nonbiogenic, magmatic hydrothermal sulfides would approximate average crustal $^{32}S/^{34}S$ and show a comparatively narrow spread. Unfortunately this has not yet been demonstrated (Stanton and Rafter, 1966).

Schneiderhöhn (1937), Garlick (1953, 1961), Mendelsohn (1961), and Davis (1954) advocated a syngenetic origin for the Zambian copper ores. These deposits, averaging several percent copper (as disseminated grains of chalcopyrite, bornite, and chalcocite), are in Precambrian sediments—principally carbonaceous shale, but also nonargillaceous rocks. Early studies concluded that the ores are hydrothermal, but the postulated igneous source has been shown to be older than the host rock (Garlick and Brummer, 1951). The distribution of mineralization—restricted to narrow stratigraphic limits over an area of several hundred square miles—is difficult to explain as a product of hypogene processes, but it seems reasonable for a syngenetic deposit. Extensive ore-bearing strata lie in a sequence of sandstones, arkoses, shales, conglomerates, and dolomites that underwent regional low-rank metamorphism (greenschist facies), grading into the epidote-amphibolite facies in the southern part of the district (Mendelsohn, 1961a). The sediments and ore minerals exhibit a type of zoning that has been interpreted to be parallel to an old shoreline (Garlick, 1953). Basinward from the shore, the facies change from barren quartzite and dolomite to shale and, within the shale, from chalcocite to bornite, then to chalcopyrite, and finally to pyrite. Thus, the copper-iron ratio decreases outward from the shoreline. These facies have no relationship to folds or other structures (Mendelsohn, 1961b).

In support of the zonal interpretation, Davis (1954) pointed out the apparent transgressive nature of the sedimentary strata and the ore minerals. Nevertheless, there is strong evidence that at least some of the ores are epigenetic. The magmatists' arguments emphasize the presence of replacement textures and small veinlets of ore; the syngeneticists argue that some of the ores were mobilized during metamorphism and point out the remarkable stratigraphic persistence of the deposits. Jensen and Dechow (1962) interpreted the sulfur isotope studies as indicative of a biogenic origin for the sulfur, which they thought favored the syngenetic hypothesis. Davidson (1962) pointed out that the cobalt in the Zambian ores consistently exceeded nickel; this preponderance is not found in either modern or ancient sediments, where nickel is the more abundant element. On the other hand, cobalt is more abundant in many hydrothermal ores that are associated with granitic intrusions and their skarns. From this Davidson concluded that the Zambian ores were both epigenetic and hydrothermal. Although field evidence may be advanced to justify either a syngenetic or an epigenetic origin for the Zambian copper ores, the evidence for both is unconvincing.

Davidson theorized that the interstitial brines derived from the diagenesis and lixiviation of salt deposits circulated at depth and then leached copper from primary sulfide deposits and disseminations of magmatic origin in regions of high geothermal gradients. Where the solutions ascended to shallower and cooler environments, the metal was deposited under favorable structural and chemical conditions (Davidson, 1965).

A further example of stratiform sulfides is furnished by the fahlbands in the Precambrian of southern Norway. Fahlbands are sparse sulfide disseminations that follow the strike of gneiss and schist. They have been traced along the strike for several miles and their width ranges from a few inches to several hundred feet. These layered beds, containing mostly small amounts of pyrite, chalcopyrite, sphalerite, and galena, have been described as being (1) of hydrothermal origin; (2) impregnations related to the injection of basic dikes; or (3) a sedimentary syngenetic deposit.

Gammon favors a syngenetic origin. He envisions their deposition in a nearshore, deltaic environment where conditions were such that organic materials were deposited and covered before decomposition was completed. Possibly, sulfate-reducing bacteria liberated hydrogen sulfide and fixed the iron and copper, leading to the production of pyrite and chalcopyrite. During later metamorphism the sulfides were recrystallized and lost all trace of their original organic origin and textures. Gammon based his conclusions primarily upon the textures that he ascribed to metamorphism. Sulfides are thought to be corroded by rock-forming silicates, and lenses of sulfides do not cut the foliation (Gammon, 1966).

With few exceptions (such as the Kupferschiefer), the status of syngenetic base-metal sulfide deposits remains an unsolved problem. The syngenetic hypothesis may help to explain the stratiform character better than does the hydrothermal theory, but additional problems are created. From where do the tremendous amounts of sulfur and metal come? They are certainly not the usual products of erosion and sedimentation. Although the exhalative-volcanic-sedimentary origin helps to explain some accumulations, until additional information is obtained, the status of many deposits will remain open to debate (Sales, 1960).

Other Chemical Precipitates

Carbonaceous shales, phosphatic shales, and many similar marine sediments contain minor amounts of uranium, vanadium, silver, arsenic, gold, molybdenum, and other metallic elements (Krauskopf, 1955, 1956b). Our knowledge of the metal content of sediments is fragmentary, largely because the amounts of metals are so small that most deposits are not commercial and have not received much attention. Shales appear to contain slightly higher percentages of metals than the other sedimentary rocks, but sampling and distribution studies have not been adequate. It has been observed, however, that some metals are enriched in black, carbonaceous marine shales. Krauskopf (1955) found that several elements are enriched more than a thousandfold in selected organic sediments. These elements are concentrated by chemical precipitation in the reducing environment created by organic sediments; by adsorption on clay particles, colloidal gels, and organic debris; and by organic processes such as bacterial action.

Krauskopf (1956b) studied the concentration of zinc, copper, lead, bismuth, cadmium, nickel, cobalt, mercury, silver, chromium, molybdenum, tungsten, and vanadium in sea water and concluded that the seas are greatly undersaturated in each of these metals. Accordingly, direct chemical precipitation of these elements cannot be responsible for the observed concentrations in sediments. Local precipitation of sulfides may remove some of the metals from sea water. This is not, however, the chief control; the concentrations in solution are unrelated to the sulfide solubilities, and some of the metals do not form stable sulfides. Again, it was concluded that adsorption and organic precipitation must be significant factors in removing many of the dissolved metals present in sea water (Krauskopf, 1956b).

Marine sediments have recently attracted increased attention because certain types contain appreciable, though as yet noncommercial, amounts of uranium and vanadium (McKelvey and Nelson, 1950; McKelvey et al., 1955). The uranium-bearing black shales are typically rich in organic matter and sulfides. Many black shales and phosphorites contain 0.01 to 0.02 percent uranium, and the diagenetic nodules of the alum shale of Sweden carry as much as 0.5 percent uranium. The character of the uranium-bearing mineral in the black shales is not known, but these shales are characteristically more phosphatic than others, so the uranium may be a phosphate or adsorbed ions on organic matter (McKelvey et al., 1955). In the phosphorites the uranium content varies roughly with the phosphate content, thereby giving evidence that the uranium is contained in a phosphatic mineral. The depositional environment of uranium-bearing formations—both the black shales and the phosphorites—is characteristic of low-lying, stable areas, where the influx of clastic material is small. Many of these deposits are associated with diastems or minor unconformities; as far as is known, the nonmarine black shales are not uraniferous. The precipitation of uranium in carbonaceous black shales may be brought about by chemical adsorption on apatite, on living or dead plankton (McKelvey et al., 1955), and by reduction of U^{+6} ions to the less soluble U^{+4} ions through the action of biochemically generated H_2S (Goldschmidt, 1954; Jensen, 1958). Quadrivalent uranium ions and divalent calcium ions are approximately the same size, which permits uranium to substitute for calcium in the apatite structure of phosphorites. Consequently, uranium ions compete for positions with calcium ions, and only the calcite-poor varieties of shale or phosphorite are able to adsorb appreciable quantities of uranium.

Vanadium is also concentrated in shales and organic sediments (Goldschmidt, 1954). It is typically associated with sedimentary uranium in reducing environments, but unlike uranium, it forms stable sulfides. Both solid and liquid hydrocarbons have been known to contain highly abnormal concentrations of vanadium, perhaps a biochemical precipitate or a product of reduction in sapropelic muds. The patronite (VS_4) of Peru was probably formed by the secondary enrichment of vanadiferous hydrocarbons through natural fractionation in an oil seep. Phosphatic rocks, such as the Phosphoria Formation of the northwestern United States, have yielded concentrations of vanadium that are apparently related to organic matter in the sediments rather than to phosphate minerals (Jacob *et al.*, 1933). The presence of both vanadium and uranium in these carbonaceous sediments is thought to be due to reduction from higher, more soluble stages of valency rather than from direct concentration in plant and animal tissues.

MECHANICAL ACCUMULATIONS

Minerals that are chemically stable at the earth's surface are not decomposed by weathering; as the surrounding rocks are dissolved and disintegrated, they either remain in the soil or are carried away by rain, streams, waves, or wind. The lighter particles are readily moved and become dispersed; others break easily along cleavage or fracture planes and become so fine grained that they, too, disperse. But the heavy, stable minerals are left as residual particles in the soil or are transported into sands and gravels of streams and beaches. Further agitation in the stream or at the beach causes the heavy particles to settle to the bottom and concomitantly enrich the deposit by removing the lighter and more brittle gangue. The result is a concentration of the heavy, tough, and chemically resistant minerals. These minerals may accumulate near the outcrops as residual concentrations; they may be washed into streams and accumulate in sand bars or in riffles and irregularities along the channel floors; or they may reach bodies of water where they are reworked by wave action and deposited in beach sands. All such concentrations of clastic minerals are called placers. Where the slopes below the outcrops are steep, or where other conditions encourage movement, the resistant particles slide or creep gradually down the slope until they reach a stream bed. Here the material is moved more rapidly by running waters, which also sort them according to their specific gravities, their shapes, the velocity and gradient of the stream, and other factors. The minerals that have properties favorable for deposition are concentrated at the expense of the lighter, brittle particles, which are broken, scattered, and transported into the deeper basins of deposition.

The most abundant placer minerals include the native metals, especially gold and the platinum group, and many of the heavy, inert oxides and silicates, such as monazite, zircon, cassiterite, chromite, wolframite, rutile, magnetite, ilmenite, and many gemstones. Since sulfides readily break up and decompose, they seldom accumulate in placers. In a few exceptional instances, however, small amounts of relatively insoluble sulfides—for example, the cinnabar at New Almaden, California—have been recovered from placers formed near the lode deposit. Magnetite and ilmenite are among the most abundant minerals in placers, but concentrations of these are rarely sufficiently rich to be of economic interest. Some of the world's greatest tin and diamond deposits are also placers. But the most valuable placer commodity is gold. Placer gold varies widely in composition, depending on the character of the original mineral and its distance from the source lode. Native gold is generally alloyed with silver, and less commonly with copper and other metals. Because both silver and copper are more soluble than gold, they are selectively leached from the alloy. Consequently, gold far removed from its source tends to become purer than the original material. The constant pounding and abrasion that particles of placer gold receive as they travel downstream also result in a gradual reduction in grain size away from the lode.

Placer deposits have formed throughout geologic time, but most are of Cenozoic age. Most placer deposits are small and form at the earth's surface, usually above the local base level; hence, most are eventually removed by erosion before burial. Moreover, the older deposits are likely to be tilted and lithified, so that unless they contain unusually valuable minerals or exceptional concentrations, they cannot compete with the unconsolidated surface materials whose chief virtue may be ease of recovery.

Fossil placers, or those buried under younger rocks, are mined in many areas. In the Sierra Nevada of California and in the Victoria field of Australia, for example, placer gold deposits have been deeply buried beneath other stream materials and lava flows. Lavas that flowed down stream valleys and covered placer deposits, derived from the Mother Lode, have subsequently been left as residual ridges above the eroded interstream terrane. As a result, the old stream valleys are easy to locate, but the sporadic gold concentrations are difficult to find beneath the flows. If placer deposits were evenly distributed along stream courses, the problem of underground mining would be greatly simplified—but this is not the case. Aside from the concentrations along riffles and other irregularities on the stream bottom, placer minerals tend to accumulate on the insides of curves or meanders. Where a meander migrates downstream, a rich streak of ore may be formed (Fig. 18-14). Finding these pay streaks is enough of a challenge in surface deposits, without extrapolating the techniques to buried placers. Consequently, much of the placer gold covered by the lava flows in the Sierra Nevada remains to be discovered.

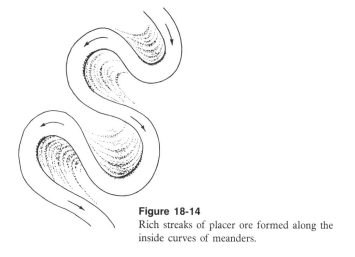

Figure 18-14
Rich streaks of placer ore formed along the inside curves of meanders.

Florida Beach Sands

The world contains many commercial beach placers, but few have received more attention that the titanium deposits of eastern Florida (Martens, 1928, 1935; Creitz and McVay, 1949; Lynd, 1960; Pirkle and Yoho, 1970). Some placers have been deposited along existing beaches; others have been left along elevated shorelines. Two principal deposits are elevated sand bars, one east of Jacksonville, and the other, known as Trail Ridge, east of Starke. The average heavy-mineral content in both deposits is very low, and only the combination of low-cost dredging and beneficiation by the efficient Humphrey spiral concentrator has made the operation possible.

Trail Ridge sand concentrates contain staurolite, zircon, sillimanite, tourmaline, kyanite, rutile, andalusite, pyroxene, and small amounts of corundum, gahnite, garnet, and monazite (Creitz and McVay, 1949; Carpenter *et al.*, 1953). Epidote, hornblende, sphene, and other minerals have been identified elsewhere, but not from Trail Ridge. Most of the titanium at Trail Ridge is in leucoxene, which is an alteration product of ilmenite, sphene, or other titaniferous minerals and consists for the most part of microcrystalline rutile, anatase, or brookite. During the alteration, some of the titania (TiO_2) apparently goes through an amorphous phase, thus some leucoxene is structureless. A long-standing controversy has centered around the true nature of this titanium mineral. Early studies considered it to be ilmenite (Martens, 1928, 1935). Subseqeuntly it was discovered that the iron-titanium ratio does not match the stoichiometry of ilmenite, and it was suggested that it is a nearly amorphous variety of arizonite (Miller, 1945), which is a ferric iron analogue of ilmenite with the formula $Fe_2Ti_3O_9$. Further work with samples of so-called arizonite has essentially discredited this mineral as a valid species; instead of being a new mineral, the material is apparently a cryptocrystalline mixture of hematite, ilmenite, anatase, and rutile (Overholt *et al.,* 1950). Possibly, the alteration of ilmenite takes two courses: the oxidation of iron to produce the material called arizonite, and the leaching of iron to form leucoxene. But whatever the true mineralogy of the Florida titanium ores, the economically important fact is that the material—call it "leucoxene" or "altered ilmenite"—is more amenable than ilmenite to metallurgical extraction of titanium.

Concentrations of titaniferous sands are widely distributed in Florida, but most are not of commercial grade. Studies of the modern beaches, which are similar in mineralogy and structure to the elevated bars (such as Trail Ridge), have revealed the mode of concentration and source of heavy minerals. The deposits form strips or ribbons parallel to the shoreline. Sand dunes built by wind and storm action form a ridge along the back side of the beach. The major concentrations of heavy minerals are at the foot of the dunes on the ocean side, where storm waves have reworked the dune sands (see Fig. 18-15). The titanium bearing beach sands are somewhat finer grained than the normal beach sands; they seem to accumulate where the older dunes are being reworked. Heavy minerals are present throughout most of the sands, but they occur in commercial amounts only where they have been selectively enriched.

Pirkle and Yoho (1970) describe Trail Ridge as a broad sand ridge with a north-south stretch of about 125 miles. The most widespread concentrations of heavy minerals are along a 17-mile stretch at the southern end of the ridge. Evidence indicates that this part of Trail Ridge was formed along an old shore line bordering the northern part of the Lake Wales Ridge. Trail Ridge is localized by currents impinging on the northern end of Lake Wales remnant. The Lake Wales Ridge was crucial to the environment in which the Trail Ridge orebody formed. Younger ridges, referred to as the Atlantic coast type,

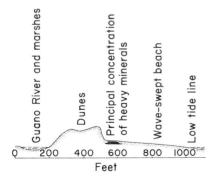

Figure 18-15
Generalized section of beach south of Mineral City, St. John's County, Florida. Vertical scale is exaggerated. (From Martens, 1928, Fig. 28.)

parallel Trail Ridge in the latitude of the orebody. Some of these younger ridges have been confused with Trail Ridge.

All the heavy minerals present in the Florida Beach sands are also found in the old metamorphic and intrusive rocks of the Southern Piedmont. Geologists agree that the most likely source of the heavy minerals lies in the Piedmont province, especially in parts of Georgia and the Carolinas. Evidently, the sands were transported to the coast by streams and reworked along the coast by southward-moving, longshore currents. The heavy sand deposits are found for long distances north of Florida, but in general they become less concentrated (McKelvey and Balsley, 1948). The content of titanium also decreases northward as the magnetite content increases.

Martens (1935) emphasized the fine-grained character of the heavy sands and the presence of coarser fragments of shells and glaucophane from local formations. He suggested that much detrital material in the beach sands did not come directly from the older rocks of the Piedmont, but had rested for varying lengths of time in some of the Coastal Plain formations. Modifications of the texture and composition of sands can be explained as admixtures from local sources, supplied by the direct wave erosion of the land during high tides and storms. One feature that has caused discussion is that the sand in southern Florida is generally coarser than the sand to the north; thus, the grain size seems to increase away from the source of supply. Martens (1935) suggested that the southern beaches winnow out the finer materials more efficiently because the sea floor is steeper in the south than in the north. But the change from fine to coarse grains is fairly abrupt; most of the beaches to the north of Cape Canaveral are fine-grained, and most of those to the south are coarse-grained. This relationship suggests that the southward-moving longshore currents are deflected oceanward at Cape Canaveral; if this is so, the southern beaches must receive most of their sand from direct erosion of the Florida landmass (much of which was derived from the Piedmont region during an earlier cycle of erosion).

Witwatersrand Gold Field

The Witwatersrand district (Fig. 18-16) is in South Africa, near the cities of Johannesburg and Pretoria. The Rand, as it commonly known, has been by far the most productive gold district in the world, yielding over 12 billion dollars in gold and containing large reserves. Although the ore averages only 0.2 to 0.3 ounce of gold per ton, as much as 20 million ounces of gold have been recovered in a single year (Douglas and Moire, 1961). Since the discovery that appreciable amounts of uranium can be recovered as a by-product of the gold ore, the district has become one of the world's principal sources of fissionable materials. Serious mining in the Witwatersrand district was begun in 1886,

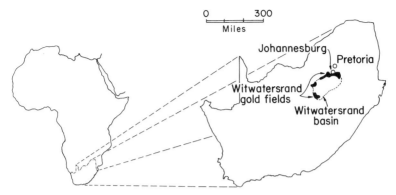

Figure 18-16
Map of Union of South Africa showing the location of the Witwatersrand deposits.

when a gold rush started the activity. Since that time, many mines have been developed along the 250 miles of strike on the Witwatersrand conglomerate beds. Several mines are operating at depths of more than 10,000 feet. The East Rand Proprietary mine, the deepest in the world, operates at a depth of more than 11,200 feet. The distance mined down the dip of the ore-bearing strata exceeds $3\frac{1}{2}$ miles. Refrigeration is needed to cool some of the mines because the rock temperatures at these great depths are as high as 125°F (Barcza and Lambrechts, 1961). According to the *South African Mining and Engineering Journal* (Aug. 23, 1968, p. 427), about nine square miles of deep lying reef are excavated each year in the gold mines. The cost of keeping these workings open amounts to about one percent of the total revenue derived from gold mining.

Mine waters constitute a further problem in the Rand; about 50 million gallons of water are pumped from the mines each day (Dolan, 1961). During 1968 the West Dreifontein mine, one of the most productive in the Rand, encountered a flow of water estimated to be about 80 million gallons a day. Mining was stopped until pump capacity could be obtained and normal water-flow restored.

The literature on the Witwatersrand is voluminous (Hargraves, 1960). Every aspect of the deposits has been studied by capable geologists, but some fundamental issues are still debated. Of special interest to mining geologists is the controversy concerning ore genesis. The Rand deposits are discussed here under placer deposits because this theory of origin is favored among South African geologists; others would list it with the hydrothermal ores.

Four theories have been advanced to explain the origin of the gold. One theory, which is adhered to by most geologists who have worked with these deposits for long periods, argues that the ores are placers. Most supporters of this theory agree that the gold is no longer in the form of simple placer grains; they argue that it was mobilized and recrystallized during the metamorphism of the enclosing rocks (Mellor, 1916, 1931; Reinecke, 1927, 1930; Young, 1931; Macadam, 1936; de Kock, 1940; du Toit, 1940; Frankel, 1940; Liebenberg, 1955; Ramdohr, 1958). Recently, Köppel and Saager (1974) studied the lead isotopes of sulfides in the gold deposits of the Precambrian greenstones of the eastern Transvaal, South Africa, and compared them with the lead isotopes of pyrites from gold deposits of the Witwatersrand basin. They concluded that the isotopic evidence supported the contention that the greenstone belts of the Swaziland Sequence and their numerous gold deposits contributed detritus to the Witwatersrand basin.

A second, much smaller group of geologists says the ores are syngenetic, but advocates chemical precipitation of the gold rather than mechanical deposition of placer grains (Penning, 1888; de Launay, 1896; Garlick, 1953). A modification of this process has also been proposed wherein the gold is considered to be a diagenetic addition to chemically precipitated, syngenetic uranium-carbon deposits (Miholic, 1954). This theory also argues that the ores have been reorganized during metamorphism.

The third theory states that the ores were deposited by hydrothermal fluids that arose directly from an underlying magmatic source (Maclaren, 1908; Hatch and Corstorphine, 1909; Horwood, 1917; Graton, 1930; Davidson, 1953, 1957). This concept has had flashes of popularity, but the Rand geologists have remained opposed to it.

The fourth theory, advanced by Davidson (1964-65a), states that uranium and gold ores were derived from overlying volcanics by highly saline, heated waters. These waters sank to the lowest permeable horizons, principally along unconformities and relatively open conglomerate channels. They deposited their metallic loads toward the cooler outer margins of the basins. Davidson supported this view by a reinterpretation of available data and by isotopic determinations.

Following the publication of Davidson's hypothesis, a flood of interesting and convincing criticism was produced by proponents of the placer theory (Bowie, 1964-65; Dowie, 1964-65; Rice, 1964-65; Way, 1964-65; Greenberg, 1964-65; Collender, 1964-65; Bishopp, 1964-65; and Taylor, 1964-65). The criticisms were manifold, but mainly showed that, at least in the East Rand, pay streaks of gold were not deflected by crosscutting dikes or feeders to the overlying volcanics from which the ores were thought to have been leached (Collender, 1964-65). Gold, uranium, and pyrite are present in boulders in the conglomerates at the base of the Ventersdorp volcanics and in the Black Reef (Way, 1964-65; Taylor, 1964-65; Dowie, 1964-65). This was said to indicate that gold could not have been derived from the overlying volcanics. Bishopp (1964-65) stated that the volcanics solidified rapidly and showed no evidence of extensive leaching. There has been no evidence of saline beds ever having been present; none are present, and collapse structures, where such beds may have existed previously, have not been encountered (Dowie, 1964–65).

To one unfamiliar with the Rand geology, the vast literature and many conflicting statements make the task of reaching an impartial appraisal nearly impossible. The published evidence is contradictory, and may be taken to support either the syngenetic placer theory or the epigenetic hydrothermal theory, according to the prejudices of the reader. It is difficult to understand under any hypothesis why such tremendous amounts of gold are limited to thin conglomerate beds that extend for several hundred square miles. Whichever hypothesis is advocated for the gold, its concentration in the conglomerate beds between quartzites is a function of the original character of the sediment. If the gold is placer, it should be concentrated in the coarser materials near the bottoms of channels (Reinecke, 1927, 1930); if it was introduced in epigenetic fluids, the migration and deposition of gold should have been controlled by the more permeable, coarse-grained beds.

The oldest rocks of the region are Precambrian schists and granitic rocks, which make up a basement complex of involved stratigraphy. Overlying the basement is the Dominion Reef System, a thin sequence of basal conglomerates and lava flows. One gold-bearing zone has been discovered in this unit (Fig. 18-17). Unconformably overlying the Dominion Reef System and basement is the Witwatersrand System, which is divided into lower and upper halves. The Lower Division of the Witwatersrand System consists of shales, quartzites, grits, and conglomerates, with one gold-producing conglomerate bed. These sediments were deposited in a gradually subsiding lake or inland sea.

As the clastic materials accumulated in the center, the edges of the basin became tilted upward and were themselves eroded. Consequently, the basin of deposition was smaller when the Upper Division sediments (mostly quartzites and conglomerates) were being deposited. The Upper Division of the Witwatersrand System contains most of the gold-bearing conglomerates, the most productive horizon being at the base of this unit. Tilting and erosion of the Witwatersrand System, which is 25,000 feet thick, was followed by igneous activity, forming the Ventersdorp System of lavas and interbedded clastic sediments. The Ventersdorp System varies greatly in thickness, exceeding 10,000 feet in places. Following erosion and subsidence of the Ventersdorp System, a thick sequence of clastics and dolomitic limestones, known as the Transvaal System, was deposited in a new basin. The end of this period was marked by further igneous activity, followed by folding and thrusting. In the center of the Witwatersrand basin, the basement was thrust up, thereby doming and overturning the overlying sediments into a prominent structure known as the Vredefort dome, or the Vredefort ring. Erosion stripped the sediments from the top of the dome and left the Witwatersrand System exposed as a pair of concentric circles. At the end of the Paleozoic Era, the Karroo System of

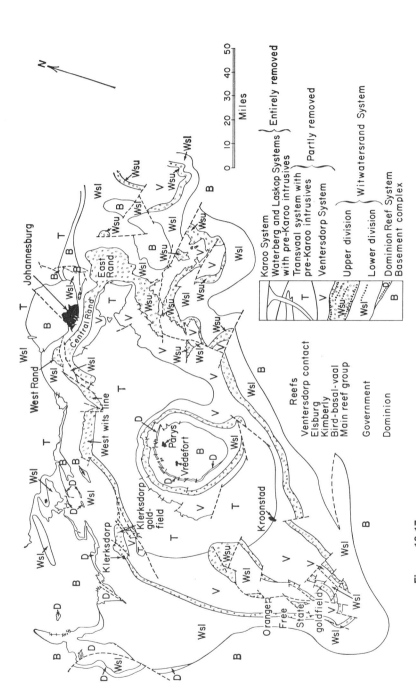

Figure 18-17
Map of the distribution of the Witwatersrand System beneath the cover. (From Borchers, 1961.)

Karoo System
Waterberg and Laskop Systems } Entirely removed
with pre-Karoo intrusives
Transvaal system with } Partly removed
pre-Karoo intrusives
Ventersdorp System
Upper division
Lower division } Witwatersrand System
Dominion Reef System
Basement complex

Reefs

Ventersdorp contact
Elsburg
Kimberly
Bird-basal-vaal
Main reef group

Government

Dominion

Miles
0 10 20 30 40 50

Johannesburg

West Rand
Central Rand
East Rand
West wits line

Klerksdorp
Klerksdorp goldfield

Parys
Vredefort

Kroonstad

Orange Free State goldfield

shallow marine sediments and coal formations was deposited over the entire area, and it still covers half or more of the Witwatersrand System (Borchers, 1961). The sedimentary rocks are intruded by dikes and small masses of mafic and intermediate igneous rocks that are thought to be related to the Ventersdorp volcanics and the Bushveld igneous activity, but no granitic rocks younger than the Witwatersrand System have been found. (Early reports that some plutonic rocks postdate the Rand conglomerates have been discredited.)

The Witwatersrand System forms a large structural trough, which measures about 250 miles in length and 90 miles in width and aligns in a northeast-southwest direction. The principal ore-bearing conglomerates lie not at the outer fringes of this basin, but within, thereby describing a smaller oval roughly 180 by 70 miles (see Fig. 18-17). The mines are concentrated along the northwest side and at both extremities of the synclinal structure. In general, the sediments dip steeply basinward around the outer edges of the synclinal structure, but in the larger mines flatter dips have been encountered at depth. Nearly vertical beds are common, and an overturned sequence is found along the northwest side of the Vredefort ring. The syncline is complicated by many faults. Large-scale thrust and normal faulting are involved, and small normal faults are abundant. Several of the mines were abandoned because faults complicated the mining operations so badly that the gold could not be economically recovered.

The origin of the Witwatersrand System of conglomerates and quartzites has been the subject of almost as much discussion as the origin of the gold. Several environments of deposition have been considered, including a marine shoreline, a large delta, an enclosed basin, a piedmont or flood plain, and a region of alluvial fans (Mellor, 1916; Reinecke, 1930; Gevers, 1961; Brock and Pretorius, 1964–65a, 1964–65b). Most workers agree, however, that the clastic materials were derived from basement rocks to the northwest.

The Witwatersrand rocks have been metamorphosed, and, except for the presence of quartz pebbles, the conglomerate beds have little in common with ordinary conglomerates. They are dense and tightly cemented. Many of the pebbles were stretched during metamorphism, although in places a high degree of sphericity has been retained (see Fig. 18-18). The spaces between pebbles are occupied by pyrite (locally in abundance), grains of quartz, sericite, rutile, chlorite, carbon, and sporadic tourmaline. Locally, many of the quartz pebbles are a bluish, opalescent variety. Coarse-grained gold is rare in these deposits; most of the values come from fine-grained, nearly microscopic disseminations within the coarser, cleaner facies of the conglomerates. As would be expected in either deltaic or flood-plain sediments, the conglomerate layers are not uniform throughout the entire district, though they are surprisingly persistent. The general stratigraphic sequence—a normal sequence for rapidly deposited, coarse, clastic sediments—can be recognized on all sides of the Witwatersrand district.

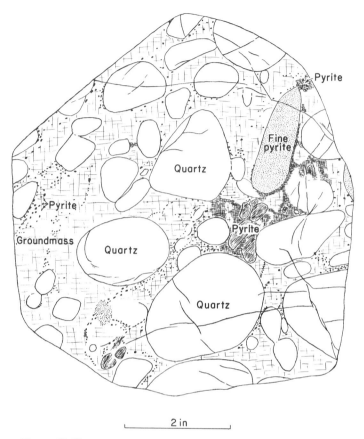

Figure 18-18
Sketch of a hand specimen of conglomerate. Ventersdorp Contact Reef,
Witwatersrand gold district, Union of South Africa.

Ore-bearing conglomerate strata in the Witwatersrand area are known as
reefs or *bankets* (the latter term is Dutch for "almond cake," which the conglomerates resemble). There are six major reef horizons in the Witwatersrand
System, as well as a minor ore zone in the underlying Dominion Reef System.
Most of the gold-bearing reefs are scattered throughout the two-mile thickness
of Upper Division quartzites and conglomerates. Attempts have been made to
unify the nomenclature and correlate the reefs between mining camps (Jones,
1936). The problem is not simple because some of the reefs consist of several
thin beds irregularly distributed throughout a thick sequence of quartzites. For
example, the Main Reef Group, which is the most important zone in the district, includes several ore-bearing conglomerates. In the Central Rand area near
Johannesburg this group includes the Main Reef, the Main Reef Leader, and

the South Reef; about 20 miles to the east, in the East Rand field, only the South Reef persists; along the West Wits line, a mineralized zone 25 to 50 miles southwest of Johannesburg, a rich gold-uranium conglomerate known as the Carbon Leader carries most of the values in the Main Reef Group; and in the Klerksdorp area, about 85 miles southwest of Johannesburg, the Main Reef Group consists of two conglomerates, known as the Commonage Reef and the Ada May Reef (Borchers, 1961). Locally, ore-bearing conglomerates resulted from the erosion of previously deposited Witwatersrand rocks. For example, the Black Reef, in the West Rand area (immediately west of the Central Rand field), is a basal conglomerate of the Transvaal System, overlying the Witwatersrand System and the Ventersdorp lavas unconformably (Pegg, 1950).

The ore-bearing reefs vary in thickness, but they are generally only a few feet in thickness, and productive beds more than 10 feet thick are uncommon; some of the richest reefs are only inches thick. The best gold values are characteristically recovered along the base of the conglomerates, which suggests transgressive shoreline deposits or erosional hiatuses caused by oscillations in the sedimentary basin (Sharpe, 1949).

As early as 1923, the presence of uraninite was noted in the Rand ores (Cooper, 1923). The market for uranium was not large then, and the quantities of uraninite reported were small. Near the end of World War II, however, when interest in fissionable materials was gaining momentum, it was discovered that the frequently mentioned "carbon" of the Rand ores was actually a uraninite-carbon aggregate, known as thucholite (Davidson and Bowie, 1951). The changing economics regarding uranium caused geologists to take a second look at the Rand as a source of supply. It was found to contain much more uranium than previously thought. The average ore contains about 0.03 percent uranium (Burger *et al.*, 1962). Relatively low-grade materials are treated for the uranium content where they can be recovered as by-products of the gold ores; but some uranium is mined from nonauriferous reefs, and in these deposits the uranium content must pay the mining costs. Thus, various grades of uranium ore are recovered.

The thucholite, or carbon, has long been recognized as a close companion of the gold. Throughout much of the development of the district this material was used as a guide to the better ores; where the carbon is richest, the gold values are likely to be best. Some gold is molded on carbon. In the so-called buckshot pyrite ore—a conglomerate containing abundant pyrite in rounded pebblelike forms—pyrite, carbon, and gold appear to have grown in the order listed, forming concentric layers (Macadam, 1931).

The intimate association between uranium and gold has led some workers to insist that the two metals must have a common origin (Davidson and Bowie, 1951; Liebenberg, 1955). Accordingly, the uranium has been studied in detail. As is the case with most lines of evidence in the Rand debate, the uranium has been used as evidence to support both hydrothermal and placer origins.

A wide variety of arguments favor hydrothermal origin of the gold and uranium. Davidson (1953) pointed out that thucholite and uraninite are unstable under surface conditions, making it unlikely that they could have survived erosion at the source and transportation to the site of deposition. Although the thucholite is in rounded grains that appear to be detrital, similar shapes can be found in veins of undoubted hydrothermal origin (Davidson and Bowie, 1951). It has been suggested that the thucholite was formed by the polymerization of hydrocarbon gases under the influence of radioactive bombardment. Methane is abundant in the mine areas, and the association of hydrocarbons with uranium may mean that the thucholite is a secondary product formed where these gases encounted uraninite. Thus the association between gold and carbon may actually be an association between uranium and hydrocarbons, and in turn between uraninite and gold (Davidson and Bowie, 1951).

In addition to these arguments, proponents of a hydrothermal origin emphasize that the gold is seldom found as nuggets. The gold generally is in tiny shreds and flakes within the matrix between pebbles, rather than in both the matrix and the pebbles. Furthermore, the gold is very fine-grained throughout the Witwatersrand area; a placer deposit should show a gradual reduction in grain size away from the source. The gangue includes such typically hydrothermal minerals as chlorite, sericite, and tourmaline. Most placer gold is accompanied by black sands, but magnetite and ilmenite are nearly absent from the Witwatersrand deposits. The hydrothermalists also point out that there was volcanic activity after the Rand sediments were deposited; gold-bearing veins of undoubtedly hypogene origin have been mined in nearby areas, so that the possibility of a genetic association with these igneous rocks is not unreasonable. Some geologists maintain that the presence of abundant pyrite means that a comparable supply of sulfur must have been introduced. The hydrothermal theory explains the restriction of gold to conglomerates as a function of permeability; the most permeable channels for epigenetic fluids would reasonably be the conglomerate beds, where the interstitial openings are larger than in the quartzites and where the clay is at a minimum (Graton, 1930).

Hargraves (1961, 1963) studied many assays of Witwatersrand ore; his reports were based upon random mill samples. He concluded that gold increased in fineness at depth, and that the average silver content of the bullion in individual reefs varies as a function of the relative elevation of the reef. This evidence would seem to support the theory of hydrothermal origin of the ores because the content of silver of many vein deposits increases toward the surface relative to gold, but Hargraves's data can also be used to support the placerist theory of origin. There is general agreement that the conglomerates were laid down by southeasterly flowing streams, so that, as in other placers, the fineness of the gold increases downstream.

Von Rahden (1965) was unable to confirm Hargraves's results by direct sampling of the Ventersdorp Contact Reef and the underlying Witwatersrand Reefs. Von Rahden explained the variations in fineness shown by the ores as a redistribution of silver during the metamorphism of placer gold. His evidence apparently contradicted that obtained by Hargraves.

South African geologists are so united in their defense of a placer origin for the ores that this hypothesis has been referred to as "a national article of faith" by a member of the opposing school (Davidson, 1953). The placerists do not deny the action of hydrothermal fluids on the Witwatersrand sediments, but they ascribe these effects to regional metamorphism rather than to the infiltration of ore-bearing solutions. They point out that the Witwatersrand sediments were derived from the basement, which was known to contain numerous small, gold-bearing quartz veins. The richer ores are concentrated in gravels—generally in the coarser, cleaner gravels—but the quartzites, many of which are also clean and permeable, are essentially gold free. Diamonds, zircons, cassiterite, columbite, chromite, and other heavy residual minerals are present as nuggets that were favorably located—for example, within shaly pockets—and thus escaped destruction during metamorphism. Platinum minerals, especially iridosmium (the natural alloy of iridium and osmium), are also found in the form of nuggets. Davidson (1964-65b, p. 850) points out that the average size of the nuggets is 40 microns—about 10,000 of the nuggets would be needed to equal the value of a British penny. These particles are finer than any known to be recovered from present day platinum placers.

Several of the hydrothermalists' objections to a placer origin can be attributed to metamorphism brought about by temperatures as high as 300°C (Ramdohr, 1953). The replacement textures and shredded appearance of the gold may result from mobilization and recrystallization during metamorphism. The lack of magnetite and ilmenite may be due to a diagenetic reaction (Ramdohr, 1958) or to a metamorphic reaction involving the iron and sulfur, the latter being abundantly supplied by the black shales. Breakdown of the ilmenite by these processes would explain why rutile is associated with the deposits. Careful sampling has shown neither enrichment nor increased gold values of the reefs near dikes, though small amounts of gold have been reworked in the igneous metamorphic aureole, leaving mineralization along cross-cutting dikes and gash veins. Hydrothermal effects can be observed, but they are related to fissures. Although alteration may be shown to grade laterally from the fractures, the gold content does not vary with the degree of hydrothermal alteration. The placerists also point out that no known igneous source for the gold has been found, and at no place does the ore fill strong, persistent, crosscutting fissures.

The presence of uranium minerals with the gold seemed at first to support the epigenetic hypothesis, but recent studies have arrived at the opposite conclusion. The original objection to a placer origin for the uranium was based upon the supposition that uraninite and pitchblende were unstable under surface conditions; for example, Davidson and Bowie (1951) noted that the uranium is leached from the mine dumps by meteoric water. But the placerists argue that the mine waters are abnormally acidic and that the presence of detrital pyrite grains with the uraninite presupposes a low redox environment that might favor the chemical stability of uraninite (Ramdohr, 1958). Laboratory studies have shown that uraninite is physically resistant to the abrasive action of stream transport and should be as persistent in a placer deposit as monazite and other common detrital minerals (Koen, 1958). Furthermore, similar uranium concentrations in the Blind River district of Ontario seem to be of detrital origin (Robertson and Steenland, 1960). In both districts, the uraninite is in rounded grains that show all the signs of attrition and none of the signs of colloidal deposition. Some thucholite grains in the Witwatersrand ores exhibit the relict octahedral cleavage of crystalline uraninite, and in places they have been shattered cataclastically by compaction; it is unlikely that these features would be shown by colloidally deposited, epigenetic pitchblende (Ramdohr, 1958). It has also been pointed out that the gold-"carbon" association may be a product of the metamorphic redistribution of gold. Accordingly, the carbon would be a secondary alteration product of uraninite that acted as a precipitant for gold mobilized during metamorphism. Paragenetic studies corroborate this interpretation (Liebenberg, 1955).

A recent—and devastating—argument against the hydrothermal origin of the uranium minerals is based upon radiometric age determinations of radiogenic lead in galena and detrital monazite. The parent uraninite for the radiogenic lead in galena samples and the detrital grains of monazite are among the oldest materials known (about 3,000 million years old) and match the age of granites in the basement rocks (Allsopp, 1961; Nicolaysen *et al.*, 1962; Burger *et al.*, 1962). The age of the Witwatersrand Series is not known, but it is thought to be about 2,200 million years old (Gevers, 1961).

Whatever the origin of the gold and uranium ores, their localization is a function of sedimentary textures. The ore concentrations are directly related to the conglomerate beds and do not depend upon superimposed structures. Until the advocates of a hydrothermal origin provide unequivocal evidence, the South African geologists will continue to direct exploration according to a placer theory because this approach has proved highly successful in the past.

REFERENCES TO SELECTED ORE DEPOSITS
FORMED BY CHEMICAL PRECIPITATION

Banded Hematite Ores of India

Dunn, J. A., 1941. The origin of banded hematite ores in India, *Econ. Geol.* 36:355-370.

————, 1942. The economic geology and mineral resources of Bihar province, *Geol. Surv. India Mem.* 78.

Jones, H. C., 1934. The iron ore deposits of Bihar and Orissa, *Geol. Surv. India Mem.* 63.

Krishnan, M. S., 1952. The iron ores of India, in *Symposium sur les Gisements de Fer du Monde,* vol. 1, 19th International Geological Congress.

Weld, C. M., 1915. The ancient sedimentary iron ores of British India, *Econ. Geol.* 10: 435-452.

Banded Hematite Ores of Minas Gerais, Brazil

Departamento Nacional da Producao Mineral, 1959. Esbôço Geológico do Quadrilátero Ferrífero de Minas Gerais, Brasil, *Dept. Nac. Prod. Mineral Pub. Especial.*

Derby, O. A., 1910. The iron ores of Brazil, *11th Int. Geol. Congr. Compt. Rend.,* pp. 813-822.

Dorr, J. Van N., II, 1954. Comments on the iron deposits of the Congonhas district, Minas Gerais, Brazil, *Econ. Geol.* 49:659-662.

————, 1965. Nature and origin of the high-grade hematite ores of Minas Gerais, Brazil, *Econ. Geol.* 60:1-46.

————, 1969. Physiographic, stratigraphic, and structural development of the Quadrilátero Ferrifero, Minas Gerais, Brazil, *U.S. Geol. Surv. Prof. Pap.* 641-A.

Dorr, J. Van N., II, and A. L. M. Barbosa, 1963. Geology and ore deposits of the Itabira district, Brazil, *U.S. Geol. Surv. Prof. Pap.* 341-C.

Dorr, J. Van N., II, P. W. Guild, and A. L. M. Barbosa, 1952. Origin of the Brazilian iron ores, in *Symposium sur les Gisements de Fer du Monde,* vol. 1, 19th International Geological Congress.

Guild, P. W., 1953. Iron deposits of the Congonhas district, Minas Gerais, Brazil, *Econ. Geol.* 48:639-676.

————, 1957. Geology and mineral resources of the Congonhas district, Minas Gerais, Brazil, *U.S. Geol. Surv. Prof. Pap.* 290.

Harder, E. C., 1914. The "itabirite" iron ores of Brazil, *Econ. Geol.* 9:101-111.

Harder, E. C., and R. T. Chamberlin, 1915. The geology of central Minas Geraes, Brazil, *J. Geol.* 23:341-378, 385-424.

Leith, C. K., and E. C. Harder, 1911. Hematite ores of Brazil and a comparison with hematite ores of Lake Superior, *Econ. Geol.* 6:670-686.

Scott, H. K., 1902. The iron ores of Brazil, *J. Iron Steel Inst. London.*

Manganese Deposits of Cuba

Burchard, E. F., 1920. Manganese-ore deposits in Cuba, *Amer. Inst. Mining Eng. Trans.* 63:51-104.

Norcross, F. S., Jr., 1940. Development of the low-grade manganese ores of Cuba, *Amer. Inst. Mining Eng. Tech. Pub.* 1188.

Park, C. F., Jr., 1942. Manganese deposits of Cuba, *U.S. Geol. Surv. Bull.* 935-B, pp. 75-97.

Park, C. F., Jr., and M. W. Cox, 1944. Manganese deposits in part of the Sierra Maestra, Cuba, *U.S. Geol. Surv. Bull.* 935-F, pp. 307-355.

Simons, F. S., and J. A. Straczek, 1958. Geology of the manganese deposits of Cuba, *U.S. Geol. Surv. Bull.* 1057.

Straczek, J. A., 1950. Manganeso en Cuba: 1940-45, *Bol. Hist. Nat.* 1:161-168.

Vaughn, T. W., C. W. Hayes, and A. C. Spencer, 1901. *Informe Sobre un Reconocimiento Geológico de Cuba,* Dirección des Montes, Minas y Aguas, pp. 66-74.

Woodring, W. P., and S. N. Daviess, 1944. Geology and manganese deposits of Guisa-Los Negros area, Oriente Province, Cuba, *U.S. Geol. Surv. Bull.* 935-G, pp. 375-386.

REFERENCES TO SELECTED DEPOSITS
OF MECHANICALLY ACCUMULATED ORES

Sierra Nevada, California

Averill, C. V., 1946. Placer mining for gold in California, *Calif. Div. Mines Bull.* 135.

Bowie, A. J., Jr., 1879. Hydraulic mining in California, *Amer. Inst. Mining Eng. Trans.* 6:27-100.

Gilbert, G. K., 1917. Hydraulic mining débris in the Sierra Nevada, *U.S. Geol. Surv. Prof. Pap.* 105.

Hammond, J. H., 1889. Auriferous gravels of California, *Calif. State Mineral. Ann. Rept.* 9:105-138.

Jenkins, O. P., 1932. *Geologic Map of Northern Sierra Nevada Showing Tertiary River Channels and Mother Lode Belt,* California Division of Mines. (Later reprinted with accompanying text on back of map.)

Lindgren, W., 1911. The Tertiary gravels of the Sierra Nevada of California, *U.S. Geol. Surv. Prof. Pap.* 73.

Turner, H. W., 1895. Auriferous gravels of the Sierra Nevada, *Amer. Geol.* 15:371-379.

Malayan Tin Deposits

Adams, F. D., 1929. Tin mining in Malaya, *Can Inst. Mining Metall. Trans.* 31:115-146.

Fitch, F. H., 1952. The geology and mineral resources of the nieghbourhood of Kuantan, Pahang, *Geol. Surv. Dept. Malaya Mem.* 6.

Jones, W. R., 1925. *Tinfields of the World,* London: Mining Publications, pp. 161-207.

Roe, F. W., 1951. The geology and mineral resources of the Fraser's Hill area, Selangor, Perak and Pahang, Federation of Malaya, with an account of the mineral resources, *Geol. Surv. Dept. Malaya Mem.* 5.

———, 1953. The geology and mineral resources of the neighbourhood of Kuala Selangor and Rasa, Selangor, Federation of Malaya, with an account of the geology of Batu Arang coalfield, *Geol. Surv. Dept. Malaya Mem.* 7.

Scrivenor, J. B., 1928. *Geology of the Malayan Ore-Deposits,* London: Macmillan.

REFERENCES CITED

Alexandrov, E. A., 1955. Contribution to studies of origin of Precambrian banded iron ores, *Econ. Geol.* 50:459-468.

Alling, H. L., 1947. Diagenesis of the Clinton hematite ores of New York, *Geol. Soc. Amer. Bull.* 58:991-1018.

Allsopp, H. L., 1961. Rb-Sr age measurements on total rock and separated-mineral fractions from the Old Granite of the central Transvaal, *J. Geophys. Res.* 66: 1499-1508.

Barcza, M., and J. de V. Lambrechts, 1961. Ventilation and air conditioning practice in South African gold mines, *7th Commonw. Mining Metall. Congr. Trans. Pap. Discuss.* 2:725-742.

Bayley, R. W., and H. L. James, 1973. Precambrian iron-formations of the United States, *Econ. Geol.* 68:934-959.

Beerstecher, E., Jr., 1954. *Petroleum Microbiology,* New York: Elsevier Press.

Bender, M. L., T. L. Ku, and W. S. Broecker, 1966. Manganese nodules: their evolution, *Science* 151(3708):325-328.

Benson, W. E., A. F. Trites, Jr., E. P. Beroni, and J. A. Feeger, 1952. Preliminary report on the White Canyon area, San Juan County, Utah, *U.S. Geol. Survey Circular* 217.

Bishopp, D. W., 1964-65. The mode of origin of banket deposits: discussion, *Inst. Mining Metall. London Trans.* 74:658-659.

Bonatti, E., and Y. R. Nayudu, 1965. The origin of manganese nodules on the ocean floor, *Amer. J. Sci.* 263:17-39.

Borchers, R., 1961. Exploration of the Witwatersrand System and its extensions, *7th Commonw. Mining Metall. Congr. Trans. Pap. Discuss.* 2:489-512; also 1962, *Geol. Soc. S. Afr. Trans.* 64.

Bowie, S. H. U., 1964-65. The mode of origin of banket orebodies, discussion, *Inst. Mining Metall. London Trans.* 74:492-497.

Brock, B. B., and D. A. Pretorius, 1964a. Rand Basin sedimentation and tectonics, in *The Geology of Some Ore Deposits of Southern Africa,* vol. 1, ed. S. H. Haughton, Geological Society of South Africa.

———, 1964b. An introduction to the stratigraphy and structure of the Rand goldfield, in *The Geology of Some Ore Deposits of Southern Africa,* vol. 1, ed. S. H. Haughton, Geological Society of South Africa.

Bubenicek, L., 1961. Recherches sur la constitution et la repartition des minerais de fer dans l'aalenien de Lorraine, *Inst. Rech. Siderurgie,* ser. A, no. 262.

Burchard, E. F., and C. Butts, 1910. Iron ores, fuels, and fluxes of the Birmingham district, Alabama, *U.S. Geol. Surv. Bull.* 400.

Burger, A. J., L. O. Nicolaysen, and J. W. L. de Villiers, 1962. Lead isotopic compositions of galenas from the Witwatersrand and Orange Free State, and their relation to the Witwatersrand and Dominion Reef uraninites, *Geochim. Cosmochim. Acta* 26:25-59.

Cannon, H. L., 1957. Description of indicator plants and methods of botanical prospecting for uranium deposits on the Colorado Plateau, *U.S. Geol. Surv. Bull.* 1030-M.

Carpenter, J. H., J. C. Detweiler, J. L. Gillson, E. C. Weichel, Jr., and J. P. Wood, 1953. Mining and concentration of ilmenite and associated minerals at Trail Ridge, Fla., *Amer. Inst. Mining Eng. Trans.* 196:789-795.

Castaño, J. R., and R. M. Garrels, 1950. Experiments on the deposition of iron with special reference to the Clinton iron ore deposits, *Econ. Geol.* 45:755-770.

Cayeux, L., 1909. *Les Minerais de Fer Oolithique de France,* vol. 1, Paris: Imprimerie Nationale.

Coffin, R. C., 1921. Radium, uranium, and vanadium deposits of southwestern Colorado, *Colo. Geol. Surv. Bull.* 16.

Collender, D. F., 1964-65. The mode of origin of banket orebodies, discussion, *Inst. Mining Metall. London Trans.* 74:497-500.

Cooper, R. A., 1923. Mineral constituents of Rand concentrates, *J. Chem. Soc. S. Afr.* 24:90-95, 264-266.

Cornelius, H. P., and B. Plöchinger, 1952. Der Tennengebirge-N-Rand mit seinen Manganerzen und die Berge im Bereich des Lammertales, *Austria Geol. Bunddesanstalt Jahrb.* 95:145-225.

Cousins, C. A., 1956. The value distribution of economic minerals with special reference to the Witwatersrand gold reefs, *Geol. Soc. S. Afr. Trans.* 59:95-113.

Creitz, E. E., and T. N. McVay, 1949. A study of opaque minerals in Trail Ridge, Florida dune sands, *Amer. Inst. Mining Eng. Trans.* 181:417-423.

Davidson, C. F., 1953. The gold-uranium ores of the Witwatersrand, *Mining Mag. London* 88(2):73-85.

———, 1957. On the occurrence of uranium in ancient conglomerates, *Econ. Geol.* 52: 668-693.

———, 1962. On the cobalt-nickel ratio in ore deposits, *Mining Mag. London* 106:78-85.

———, 1964-65a. The mode of origin of banket orebodies, *Inst. Mining Metall. London Trans.* 74(6):319-338.

———, 1964-65b. The mode of origin of banket orebodies, a reply, *Inst. Mining Metall. London Trans.* 74(12):844-857.

———, 1965. A possible mode of origin of strata-bound copper ores, *Econ. Geol.* 60: 942-954.

Davidson, C. F., and S. H. U. Bowie, 1951. On thucholite and related hydrocarbon-uraninite complexes, with a note on the origin of the Witwatersrand gold ores, *Geol. Surv. Gt. Brit. Bull.* 3, pp. 1-19.

Davidson, D. M., Jr., and P.F. Kerr, 1966. Uranium deposits at Kane Creek, Utah, *Amer. Inst. Mining Eng. Trans.* 235:127-132.

———, 1968. Uranium-bearing veins in plateau strata, Kane Creek, Utah, *Geol. Soc. Amer. Bull.* 79:1503-1526.

Davis, G. R., 1954. The origin of the Roan Antelope copper deposit of Northern Rhodesia, *Econ. Geol.* 49:575-615.

Degens, E. T., and D. A. Ross, eds., 1969. *Hot Brines and Recent Heavy Metal Deposits in the Red Sea,* New York: Springer-Verlag.

Denson, N. M., and J. R. Gill, 1956. Uranium-bearing lignite and its relation to volcanic tuffs in eastern Montana and North and South Dakota, *U.S. Geol. Surv. Prof. Pap.* 300, pp. 413-418.

Dessau, G., M. L. Jensen, and N. Nakai, 1962. Geology and isotopic studies of Sicilian sulfur deposits, *Econ. Geol.* 57:410-438.

Dolan, J., 1961. Water problems of the Transvaal and Orange Free State mines, *7th Commonw. Mining Metall. Congr. Trans. Pap. Discuss.* 3:1357-1388.

Douglas, J. K. E., and A. T. Moíre, 1961. A review of South African gold recovery practice, *7th Commonw. Mining Metall. Congr. Trans. Pap. Discuss.* 3:971-1003.

Dowie, D. L., 1964-65. The mode of origin of banket orebodies, discussion, *Inst. Mining Metall. London Trans.* 74:500-503.

Dunham, K. C., 1964. Neptunist concepts in ore genesis, *Econ. Geol.* 59:1-21.

Dunn, J. A., 1935. The origin of iron ores in Singhbhum, India, *Econ. Geol.* 30:643-654.

———, 1941. The origin of banded hematite ores in India, *Econ. Geol.* 36:355-370.

Dutton, C. E., 1952. Memorandum on iron deposits in the United States of America, in *Symposium sur les Gisements de Fer du Monde,* vol. 1, 19th International Geological Congress.

Dutton, C. E., C. F. Park, Jr., and J. R. Balsley, Jr., 1945. General character and succession of tentative divisions in the stratigraphy of the Mineral Hills district, *U.S. Geol. Surv. Prelim. Rept.*

Eugster, H. P., and C. I-Ming, 1973. The depositional environments of Precambrian banded iron-formations, *Econ. Geol.* 68:1144-1168.

Fischer, R. P., 1942. Vanadium deposits of Colorado and Utah, *U.S. Geol. Surv. Bull.* 936-P.

———, 1950. Uranium-bearing sandstone deposits of the Colorado Plateau, *Econ. Geol.* 45:1-11.

———, 1956. Uranium-vanadium-copper deposits on the Colorado Plateau, *U.S. Geol. Surv. Prof. Pap.* 300, pp. 143-154.

———, 1970. Similarities, differences, and some genetic problems of the Wyoming and Colorado Plateau types of uranium deposits in sandstone, *Econ. Geol.* 65:778-784.

———, 1974. Exploration guides to new uranium districts and belts, *Econ. Geol.* 69: 362-376.

———, 1968. The uranium and vanadium deposits of the Colorado Plateau region, in *Ore Deposits of the United States* (Graton-Sales volume) Amer. Inst. Min. Engrs. pp. 735-746.

Frankel, J. J., 1940. Notes on some of the minerals in the Black Reef Series, *Geol. Soc. S. Afr. Trans.* 43:1-8.

French, B. M., 1968. Progressive contact metamorphism of the Biwabik iron-formation, Mesabi Range, Minnesota, *Univ. Minn. Press Bull.* 45.

Gammon, J. B., 1966. Fahlbands in the Precambrian of southern Norway, *Econ. Geol.* 61:174-188.

Garlick, W. G., 1953. Reflections on prospecting and ore genesis in Northern Rhodesia, *Inst. Mining Metall. London Trans.* 63:9-20, 94-106.

———, 1961. Ore genesis: the syngenetic theory, in *The Geology of the Northern Rhodesian Copper Belt*, ed. F. Mendelsohn, London: Macdonald.

Garlick, W. G., and J. J. Brummer, 1951. The age of the granites of the Northern Rhodesian copper belt, *Econ. Geol.* 46:478-497.

Garrels, R. M., E. A. Perry, Jr., and F. T. Mackenzie, 1973. Genesis of Precambrian iron-formations and the development of atmospheric oxygen, *Econ. Geol.* 68:1173-1179.

Gastil, G., and D. M. Knowles, 1960. Geology of the Wabush Lake area, southwestern Labrador and eastern Quebec, Canada, *Geol. Soc. Amer. Bull.* 71:1243-1254.

Gevers, T. W., 1961. Outline of the geology of southern Africa, *7th Commonw. Mining Metall. Congr. Trans. Pap. Discuss.* 1:25-37.

Goldich, S. S., 1973. Ages of Precambrian banded iron-formations, *Econ. Geol.* 68:1126-1134.

Goldschmidt, V. M., 1954. *Geochemistry*, Oxford: Clarendon Press.

Goodwin, A. M., 1956. Facies relations in the Gunflint iron formation, *Econ. Geol.* 51:565-595.

Govett, G. J. S., 1966. Origin of banded iron formations, *Geol. Soc. Amer. Bull.* 77:1191-1212.

Graton, L. C., 1930. Hydrothermal origin of the Rand gold deposits, *Econ. Geol.* 25(supp. to no. 3), pp. 1-185.

Greenberg, R., 1964-65. The mode of origin of banket orebodies, discussion, *Inst. Mining Metall. London Trans.* 74:574-576.

Gross, G. A., 1965. Origin of Precambrian iron formations, discussion, *Econ. Geol.* 60:1063-1065.

Grout, F. F., and J. F. Wolff, Sr., 1955. The geology of the Cuyuna district, Minnesota: a progress report, *Minn. Geol. Surv. Bull.* 36.

Gruner, J. W., 1922. The origin of sedimentary iron formations: the Biwabik Formation of the Mesabi Range, *Econ. Geol.* 17:407-460.

———, 1930. Hydrothermal oxidation and leaching experiments: their bearing on the origin of Lake Superior hematite-limonite ores, *Econ. Geol.* 25:697-719, 837-867.

————, 1946. *Mineralogy and Geology of the Taconites and Iron Ores of the Mesabi Range, Minnesota,* St. Paul, Minn.: Iron Range Resources and Rehabilitation.

Gundersen, J. N., 1960. Lithologic classification of taconite from the type locality, *Econ. Geol.* 55:563-573.

Harder, E. C., 1919. Iron-depositing bacteria and their geologic relations, *U.S. Geol. Surv. Prof. Pap.* 113.

Hargraves, R. B., 1960. A bibliography of the geology of the Witwatersrand System, *Univ. Witwatersrand Econ. Geol. Res. Unit Inform. Circ.* 1.

————, 1961. Silver content of gold in Witwatersrand conglomerates, *Univ. Witwatersrand Econ. Geol. Res. Unit Inform. Circ.* 3.

————, 1963. Silver-gold ratios in some Witwatersrand conglomerates, *Econ. Geol.* 58:952-970.

Hariya, Y., and T. Kikuchi, 1964. Precipitation of manganese by bacteria in mineral springs, *Nature* 202:416-417.

Hatch, F. H., and G. S. Corstorphine, 1909. *The Geology of South Africa,* London: Macmillan, pp. 146-151.

Hawkes, H. E. and J. S. Webb, 1962. *Geochemistry in Mineral Exploration,* New York, Harper and Row.

Hayes, A. O., 1915. Wabana iron ore of Newfoundland, *Can. Geol. Surv. Mem.* 78.

————, 1928. Wabana iron mines and deposits, Newfoundland, *Mining & Metall.* 9:361-366.

————, 1929. Further studies of the origin of the Wabana iron ore of Newfoundland, *Econ. Geol.* 24:687-690.

————, 1931. Structural geology of the Conception Bay region, and of the Wabana iron ore deposits of Newfoundland, *Econ. Geol.* 26:44-64.

Hem, J. D., 1960. Some chemical relationships among sulfur species and dissolved ferrous iron, *U.S. Geol. Surv. Water-Supply Pap.* 1459-C, pp. 57-73.

Hess, F. L., 1914. A hypothesis for the origin of the carnotites of Colorado and Utah, *Econ. Geol.* 9:675-688.

Hewett, D. F., and B. N. Webber, 1931. Bedded deposits of manganese oxides near Las Vegas, Nevada, *Univ. Nev. Bull.* 25(6).

Holland, H. D., 1973. The oceans: a possible source of iron in iron-formations, *Econ. Geol.* 68:1170-1172.

Holmes, A., 1900. In *Nom. Petrol.,* quoted in *Glossary of Geology,* American Geological Institute, 1957.

Horwood, C. B., 1917. *The Gold Deposits of the Rand,* London.

Hostetler, P. B., and R. M. Garrels, 1962. Transportation and precipitation of uranium and vanadium at low temperatures, with special reference to sandstone-type uranium deposits, *Econ. Geol.* 57:137-167.

Hough, J. L., 1958. Fresh-water environment of deposition of Precambrian banded iron formations, *J. Sed. Petrology* 28:414-430.

Huber, N. K., 1958. The environmental control of sedimentary iron minerals, *Econ. Geol.* 53:123-140.

Huff, L. C., and F. G. Lesure, 1962. Diffusion features of uranium-vanadium deposits in Montezuma Canyon, Utah. *Econ. Geol.* 57:226-237.

Hunt, J. M., E. E. Hays, E. T. Degens, and D. A. Ross, 1967. Red Sea: detailed survey of hot-brine areas, *Science* 156:514-516.

Isachsen, Y. W., T. W. Mitcham, and H. B. Wood, 1955. Age and sedimentary environments of uranium host rocks, Colorado Plateau, *Econ. Geol.* 50:127-134.

Jacob, K. D., W. L. Hill, H. L. Marshall, and D. S. Reynolds, 1933. Composition and distribution of phosphate rock with special reference to the United States. *U.S. Dept. Agric. Tech. Bull.* 364.

James, H. L., 1951. Iron formation and associated rocks in the Iron River district, Michigan, *Geol. Soc. Amer. Bull.* 62:251-266.

———, 1954. Sedimentary facies of iron-formation, *Econ. Geol.* 49:235-293.

Jensen, M. L., 1958. Sulfur isotopes and the origin of sandstone-type uranium deposits, *Econ. Geol.* 53:598-616.

Jensen, M. L., and E. Dechow, 1962. The bearing of sulphur isotopes on the origin of the Rhodesian copper deposits, *Geol. Soc. S. Afr. Trans.* 65(pt. 1):1-17.

Jolliffe, A. W., 1955. Geology and iron ores of Steeprock Lake, *Econ. Geol.* 50:373-398.

———, 1966. Stratigraphy of the Steeprock Group, Steeprock Lake, Ontario, in *The Relationship of Mineralization to Precambrian Stratigraphy in Certain Mining Areas of Ontario and Quebec,* Geological Association of Canada.

Jones, G. C., 1936. Correlation and other aspects of the exploited auriferous horizons on the Witwatersrand mining field, *Geol. Soc. S. Afr. Proc.* 39:xxiii-lxi.

Kelley, V. C., 1955. Regional tectonics of the Colorado Plateau and relationship to the origin and distribution of uranium, *Univ. N. Mex. Pub. Geol.* 5.

———, 1956. Influence of regional structure and tectonic history upon the origin and distribution of uranium on the Colorado Plateau, *U.S. Geol. Surv. Prof. Pap.* 300, pp. 171-178.

———, ed., 1963. Geology and technology of the Grants uranium region, *N. Mex. Bur. Mines Mineral Resources Mem.* 15.

Kerr, P. F., 1958. Uranium emplacement in the Colorado Plateau, *Geol. Soc. Amer. Bull.* 69:1075-1111.

Kock, W. P. de, 1940. The Ventersdorp contact reef, *Geol. Soc. S. Afr. Trans.* 43:85-108.

Koen, G. M., 1958. The attrition of uraninite, *Geol. Soc. S. Afr. Trans.* 61:183-196.

Köppel, V. H., and R. Saager, 1974. Lead isotope evidence on the detrital origin of Witwatersrand pyrites and its bearing on the provenance of the Witwatersrand gold, *Econ. Geol.* 69:318-331.

Krauskopf, K. B., 1955. Sedimentary deposits of rare metals, *Econ. Geol. (50th Anniv. Vol.),* pp. 411-463.

———, 1956a. Dissolution and precipitation of silica at low temperatures, *Geochim. Cosmochim. Acta* 10:1-26.

———, 1956b. Factors controlling the concentrations of thirteen rare metals in seawater, *Geochim. Cosmochim. Acta* 9:1-32b.

———, 1957. Separation of manganese from iron in sedimentary processes, *Geochim. Cosmochim. Acta* 12:61-84.

Krumbein, W. C., and R. M. Garrels, 1952. Origin and classification of chemical sediments in terms of pH and oxidation-reduction potentials, *J. Geol.* 60:1-33.

Lake Superior Iron Ore Association, 1952. *Lake Superior Iron Ores,* 2nd ed., Cleveland: Lake Superior Iron Ore Association.

Launay, L. de, 1896. *Les Mines d'Or du Transvaal,* Paris: Baudry, pp. 177-351.

Leith, C. K., R. J. Lund, and A. Leith, 1935. Pre-Cambrian rocks of the Lake Superior region, *U.S. Geol. Surv. Prof. Pap.* 184.

Lepp, H., and S. S. Goldich, 1959. The chemistry and origin of iron formations, *Econ. Geol.* 54:1348-1349.

———, 1964. Origin of the Precambrian iron formations, *Econ. Geol.* 59:1025-1060.

Liebenberg, W. R., 1955. The occurrence and origin of gold and radioactive minerals in the Witwatersrand System, the Dominion Reef, the Ventersdorp Contact Reef and the Black Reef, *Geol. Soc. S. Afr. Trans.* 58:101-254.

Lindgren, W., 1933. *Mineral Deposits,* 4th ed., New York: McGraw-Hill, pp. 415-417.

Ljunggren, P., 1955. Geochemistry and radioactivity of some Mn and Fe bog ores, *Geol. Foren. Stockholm Forh.* 77:33-44.

Lynd, L. E., 1960. Titanium, in *Industrial Minerals and Rocks,* 3rd ed., ed. J. L. Gillson, American Institute of Mining Engineers.

Lyons, J. C., 1957. Wabana iron ore deposits, in *Structural Geology of Canadian Ore Deposits,* vol. 2, Montreal: Canadian Institute of Mining Engineers.

Macadam, P., 1931. The distribution of gold and carbon in the Witwatersrand bankets, *Geol. Soc. S. Afr. Trans.* 34(annex.):81–88.

———, 1936. The heavier metals and minerals in the Witwatersrand banket, *Geol. Soc. S. Afr. Trans.* 39:77–79.

Maclaren, J. M., 1908. *Gold: Its Geological Occurrence and Geographical Distribution,* London: The Mining Journal, pp. 95–98.

Marsden, R. W., 1968. Geology of the iron ores of the Lake Superior region in the United States in *Ore Deposits of the United States* (L. C. Graton-R. Sales Mem. Vol.), ed. J. D. Ridge, American Institute of Mining Engineers.

Martens, J. H. C., 1928. Beach deposits of ilmenite, zircon and rutile in Florida, *Geol. Surv. Fla. Ann. Rept.* 19, pp. 124–154.

———, 1935. Beach sands between Charleston, South Carolina, and Miami, Florida, *Geol. Soc. Amer. Bull.* 46:1563–1596.

Mason, B., 1958. *Principles of Geochemistry,* 2nd ed., New York: Wiley.

McKelvey, V. E., 1950. The field of economic geology of sedimentary mineral deposits, in *Applied Sedimentation,* ed. P. D. Trask, New York: Wiley.

McKelvey, V. E., and J. R. Balsley, Jr., 1948. Distribution of coastal blacksands in North Carolina, South Carolina, and Georgia, as mapped from an airplane, *Econ. Geol.* 43:518–524.

McKelvey, V. E., D. L. Everhart, and R. M. Garrels, 1955. Origin of uranium deposits, *Econ. Geol. (50th Anniv. Vol.),* pp. 464–533.

McKelvey, V. E., and J. M. Nelson, 1950. Characteristics of marine uranium-bearing sedimentary rocks, *Econ. Geol.* 45:35–53.

McKelvey, V. E., J. H. Wiese, and V. H. Johnson, 1949. Preliminary report on the bedded manganese of the Lake Mead region, Nevada and Arizona, *U.S. Geol. Surv. Bull.* 948-D, pp. 83–101.

Mellor, E. T., 1916. The conglomerates of the Witwatersrand, *Inst. Mining Metall. London Trans.* 25:226–348.

———, 1931. The origin of the gold in the Rand banket: discussion of Professor Graton's paper, *Geol. Soc. S. Afr. Trans.* 34(annex.):55–69.

Mendelsohn, F., 1961*a.* Metamorphism, in *The Geology of the Northern Rhodesian Copper Belt,* ed. F. Mendelsohn, London: Macdonald.

———, 1961*b.* Ore genesis: summary of the evidence, in *The Geology of the Northern Rhodesian Copper Belt,* ed. F. Mendelsohn, London: Macdonald.

Mero, J. L., 1964. Mineral resources of the sea, *New York Acad. Sci. Trans.,* ser. 2, 26:525–543.

Miholić, S., 1954. Genesis of the Witwatersrand gold-uranium deposits, *Econ. Geol.* 49:537–540.

Mikhalev, D. N., 1946. On the great and minor epochs of accumulation of manganiferous sediments, *Compt. Rend. (Doklady) Acad. Sci. U.S.S.R.* 54(4):339–341.

Miller, A. R., 1964. High salinity in sea water, *Nature* 203:590.

Miller, D. S., and J. L. Kulp, 1963. Isotopic evidence on the origin of the Colorado Plateau uranium ores, *Geol. Soc. Amer. Bull.* 74:609–630.

Miller, L. J., 1955. Uranium ore controls of the Happy Jack deposit, White Canyon, San Juan county, Utah, *Econ. Geol.* 50:156–169.

Miller, R., III, 1945. The heavy minerals of Florida beach and dune sands, *Amer. Mineral.* 30:65–75.

Moore, E. S., and J. E. Maynard, 1929. Solution, transportation and precipitation of iron and silica, *Econ. Geol.* 24:272-303, 365-402, 506-527.

Nagell, R. H., 1962. Geology of the Serra do Navio manganese district, Brazil, *Econ. Geol.* 57:481-498.

Nicolaysen, L. O., A. J. Burger, and W. R. Liebenberg, 1962. Evidence for the extreme age of certain minerals from the Dominion Reef conglomerates and the underlying granite in the Western Transvaal, *Geochim. Cosmochim. Acta* 26:15-23.

O'Rourke, J. E., 1961. Paleozoic banded iron-formations, *Econ. Geol.* 56:331-361.

Overholt, J. L., G. Vaux, and J. L. Rodda, 1950. The nature of "arizonite," *Amer. Mineral.* 35:117-119.

Owens, J. S., 1965. Discussion, *Econ. Geol.* 60:1731-1734.

Park, C. F., Jr., 1942. Manganese deposits of Cuba, *U.S. Geol. Surv. Bull.* 935-B, pp. 75-97.

————, 1956a. On the origin of manganese, in *Symposium sobre Yacimientos de Manganeso,* vol. 1, 20th International Geological Congress.

————, 1956b. Manganese ore deposits of the Serra do Navio district, Federal Territory of Amapá, Brazil, in *Symposium sobre Yacimientos de Manganeso,* vol. 3, 20th International Geological Congress.

————, 1959. The origin of hard hematite in itabirite, *Econ. Geol.* 54:573-587.

Pegg, W. C., 1950. A contribution to the geology of the West Rand area, *Geol. Soc. S. Afr. Trans.* 53:209-227.

Penning, W. H., 1888. The South African gold-fields, *J. Soc. Arts* 36:433-444.

Pettijohn, F. J., 1946. Geology of the Crystal Falls-Alpha iron-bearing district, Iron County, Michigan, *U.S. Geol. Surv. Strategic Minerals Invest. Prelim. Map* 3-181.

Pirkle, E. C., and W. H. Yoho, 1970. The heavy mineral ore body of Trail Ridge, Florida, *Econ. Geol.* 65:17-30.

Pompeckj, J. F., 1920. Kupferschiefer und Kupferschiefermeer, *Z. Deut. Geol. Ges.* 72:329-339.

Quirke, T. T., 1943. Hydrothermal replacement in deep seated iron ore deposits of the Lake Superior region: a discussion, *Econ. Geol.* 38:662-666.

Ramdohr, P., 1953. Über Metamorphose und sekundäre Mobilisierung, *Geol. Rundsch.* 42(1):11-19.

————, 1958. New observations on the ores of the Witwatersrand in South Africa and their genetic significance, *Geol. Soc. S. Afr. Trans.* 61(annex.):1-51.

Rankama, K., and T. G. Sahama, 1950. *Geochemistry,* Chicago: Univ. of Chicago Press.

Reinecke, L., 1927. The location of the payable ore-bodies in the gold-bearing reefs of the Witwatersrand, *Geol. Soc. S. Afr. Trans.* 30:89-119.

————, 1930. Origin of the Witwatersrand System, *Geol. Soc. S. Afr. Trans.* 33:111-133.

Rice, R., 1964-65. The mode of origin of banket orebodies, discussion, *Inst. Mining Metall. London Trans.* 74:503-504.

Roberts, H. M., and M. W. Bartley, 1943a. Replacement hematite deposits, Steep Rock Lake, Ontario, *Amer. Inst. Mining Eng. Tech. Pub.* 1543.

————, 1943b. Hydrothermal replacement in deep seated iron ore deposits of the Lake Superior region, *Econ. Geol.* 38:1-24.

Robertson, D. S., and N. C. Steenland, 1960. On the Blind River uranium ores and their origin, *Econ. Geol.* 55:659-694.

Royce, S., 1938. Geology of the iron ranges, in *Lake Superior Iron Ores,* Cleveland: Lake Superior Iron Ore Association.

————, 1942. Iron ranges of the Lake Superior district, in *Ore Deposits as Related to Structural Features,* ed. W. H. Newhouse, Princeton, N.J.: Princeton Univ. Press.

Sakamoto, T., 1950. The origin of the Pre-Cambrian banded iron ores, *Amer. J. Sci.* 248:449-474.

Sales, R. H., 1960. Critical remarks on the genesis of ore as applied to future mineral exploration, *Econ. Geol.* 55:805-817.

Sangster, D. F., 1972. Precambrian volcanogenic massive sulfide deposits in Canada: a review, *Geol. Surv. Can. Pap.* 72-22

Sauce, W. B. W. de la, 1926. Beitrage zur Kenntnis der Manganerzlagerstätte von Tschiaturi im Kaukasus, *Abh. Prakt. Geol. Bergwirtsch.* 8.

Schmidt, R. G., and C. E. Dutton, 1957. Bedrock geology of the south-central part of the North range, Cuyuna district, Minnesota, Sheets 1-3, *U.S. Geol. Surv. Mineral Invest. Field Stud. Map* MF 99.

Schneiderhöhn, H., 1923. Chalkographische Untersuchung des Mansfelder Kupferschiefers, *Neues Jahrb. Mineral. Geol. Paläontol.* 47(supp.):1-38.

———, 1937. Die Kupferlagerstätten von Nordrhodesia und Katanga, *Geol. Rundsch.* 28:282-291.

———, 1953. Konvergenzerscheinungen zwischen magmatischen und sedimentären Lagerstätten, *Geol. Rundsch.* 42(1):34-43.

Sharpe, J. W. N., 1949. The economic auriferous bankets of the Upper Witwatersrand beds and their relationship to sedimentation features, *Geol. Soc. S. Afr. Trans.* 52:265-300.

Shawe, D. R., and H. C. Granger, 1965. Uranium ore rolls—an analysis, *Econ. Geol.* 60:240-250.

Shoemaker, E. M., 1956. Structural features of the central Colorado Plateau and their relation to uranium deposits, *U.S. Geol. Surv. Prof. Pap.* 300, pp. 155-170.

Siever, R., 1957. The silica budget in the sedimentary cycle, *Amer. Mineral.* 42:821-841.

Sims, P. K., and G. B. Morey, eds., 1972. *Geology of Minnesota* (Schwartz Vol., 100th Anniv.), Minnesota Geological Survey.

Smyth, C. H., Jr., 1892. On the Clinton iron ores, *Amer. J. Sci.* 143:487-496.

Stanton, R. L., and T. A. Rafter, 1966. The isotopic constitution of sulfur in some stratiform lead-zinc sulfide ores, *Mineralum Deposita* 1:16-29.

Taylor, H. K., 1964-65. The mode of origin of banket orebodies, discussion, *Inst. Mining Metall. London Trans.* 74:801.

Temple, K. L., 1964. Syngenesis of sulfide ores: an evaluation of biochemical aspects, *Econ. Geol.* 59:1473-1491.

Toit, A. L. du, 1940. Developments on and around the Witwatersrand, *Econ. Geol.* 35:98-108.

Trask, P. D., 1925. The origin of the ore of the Mansfeld Kupferschiefer, Germany: a review of the current literature, *Econ. Geol.* 20:746-761.

Trendall, A. F., 1965. Origin of Precambrian iron formations, discussion, *Econ. Geol.* 60:1065-1069.

———, 1968. Three great basins of Precambrian banded iron formation deposition: a systematic comparison, *Geol. Soc. Amer. Bull.* 79:1527-1544.

Trudinger, P. A., I. B. Lambert, and G. W. Skyring, 1972. Biogenic sulfide ores: a feasibility study, *Econ. Geol.* 67:1114-1127.

Twenhofel, W. H., 1950. *Principles of Sedimentation*, 2nd ed., New York: McGraw-Hill, pp. 444-451.

Tyler, S. A., 1949. Development of Lake Superior soft ores from metamorphosed iron formation, *Geol. Soc. Amer. Bull.* 60:1101-1124.

Van Hise, C. R., and C. K. Leith, 1911. The geology of the Lake Superior iron region, *U.S. Geol. Surv. Monogr.* 52, pp. 499-529.

Von Rahden, H. V. R., 1965. Apparent fineness values of gold from two Witwatersrand gold mines, *Econ. Geol.* 60:980-997.

Wade, H. H., and M. R. Alm, 1957. Mining directory of Minnesota, *Univ. Minn. Press Bull.* 40(9).

Waters, A. C., and H. C. Granger, 1953. Volcanic debris in uraniferous sandstones and its possible bearing on the origin and precipitation of uranium, *U.S. Geol. Surv. Circ.* 224.

Way, H. J. R., 1964–65. The mode of origin of banket orebodies, discussion, *Inst. Mining Metall. London Trans.* 74:571–574.

Weeks, A. D., 1956. Mineralogy and oxidation of the Colorado Plateau uranium ores, *U.S. Geol. Surv. Prof. Pap.* 300, pp. 187–193.

Weeks, A. D., and M. E. Thompson, 1954. Identification and occurrence of uranium and vanadium minerals from the Colorado Plateau, *U.S. Geol. Surv. Bull.* 1009-B.

White, D. A., 1954. The stratigraphy and structure of the Mesabi Range, Minnesota, *Minn. Geol. Surv. Bull.* 38.

Wilson, I. F., and M. Veytia, 1949. Geology and manganese deposits of the Lucifer district, northwest of Santa Rosalía, Baja California, Mexico, *U.S. Geol. Surv. Bull.* 960-F, pp. 177–233.

Woolnough, W. G., 1941. Origin of banded iron deposits—a suggestion, *Econ. Geol.* 36:465–489.

Wright, R. J., 1955. Ore controls in sandstone uranium deposits of the Colorado Plateau, *Econ. Geol.* 50:135–155.

Young, R. B., 1931. The genesis of gold in the Rand banket, *Geol. Soc. S. Afr. Trans.* 34(annex.):1–14.

Zapffe, C., 1931. Deposition of manganese, *Econ. Geol.* 26:799–832.

Zealley, A. E. V., 1918. On certain felsitic rocks hitherto called "banded ironstone," in the ancient schists around Gatooma, Rhodesia, *Geol. Soc. S. Afr. Trans.* 21:43–52.

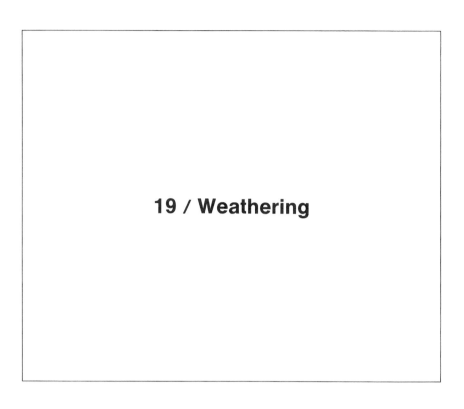

19 / Weathering

Most ore minerals, especially the sulfides and sulfosalts, are formed in reducing environments and at temperatures higher than atmospheric. Where these minerals are exposed to surface conditions, they break down chemically, forming new compounds or going into solution. Similarly, most rock-forming minerals are not stable at the earth's surface and consequently undergo chemical changes as equilibrium is reestablished with the environment. Most geologists regard weathering as an inorganic process related to carbon dioxide and oxygen interchange in the groundwaters; nearly all studies of weathering have been conducted under this premise. However, Ehrlich (1964) has called attention to the activity of bacteria in weathering processes. He reported that oxidation of arsenopyrite and enargite increased markedly in the presence of Thiobacillus-Ferrobacillus group bacteria.

Minor amounts of metals contained in unweathered rock may be concentrated into economic deposits during weathering. Enrichment takes place where the oxidized metallic product is stable and other constituents are selectively leached away. Therefore, depending upon the materials, environments, and products involved, weathering processes may (1) destroy an existing ore deposit; (2) produce one from previously barren or subeconomic rock; or (3) change the mineralogy of a deposit.

The chemical effects of meteoric waters on rocks vary according to the minerals being attacked. Even pure water will dissolve minerals to a limited degree; but meteoric waters contain carbon dioxide, and are slightly acidic. Furthermore, soil water generally contains appreciable amounts of carbon dioxide as well as humic acids, and its pH is lowered to four or five in places. These acidic waters percolate slowly through the zone of aeration on their way to the groundwater table. In transit, or after reaching the zone of saturation, they react with carbonate rocks or feldspars and other silicate minerals to become neutral or slightly alkaline. All minerals are soluble under favorable conditions, and at the earth's surface they eventually break down or are mechanically removed. Mineral solubilities vary widely, and since the rates of mineral chemical reactions vary independently, there is a wide range of effective mineral stabilities. Meteoric waters charged with carbon dioxide and oxygen from the atmosphere will oxidize, hydrate, and carbonatize the rock-forming minerals. The sulfides are converted to sulfates, most of which are soluble, or they are converted to the more stable oxides, native metals, and carbonates. Iron, manganese, and aluminum form oxides and hydroxides that are relatively insoluble at the surface; lead forms a stable sulfate; lead, zinc, and copper are preserved as carbonates in some environments; copper, zinc, nickel, and chromium are retained as silicates and oxides; and native gold, copper, and silver may remain stable in the weathered zone. These elements have all been recovered as residual concentrates and as simple oxidation products near the weathered rock outcrops that contained them (Emmons, 1917).

The depth to which weathering takes place varies greatly and depends upon climate, permeability, structure, and geomorphological history. Where the water table stands close to the surface, and where relief is minor or circulation of groundwaters is low, oxidation may be shallow. Many areas of the north where tundras and swamps prevail are weathered just a few feet below the surface. On the other hand, in arid regions, such as that at the Chuquicamata mine in northern Chile and at the Tsumeb mine in South West Africa, oxidation may be deep. At Tsumeb, copper oxidation products have been encountered along faults at depths of nearly 1,000 meters. At places in the Southern Piedmont of the United States and in the iron formation at Belinga, Gabon, adits and shafts to depths of over 100 meters persist in thoroughly weathered materials.

The mobility of metal ions in the zone of weathering is largely determined by the composition of both the vadose waters and the country rocks (Hawkes and Webb, 1962). Sulfide-free meteoric waters leach such elements as zinc, molybdenum, and uranium from igneous rocks, leaving behind stable oxidation products of iron, aluminum, titanium, chromium, and in places manganese, nickel, cobalt, copper, lead, or antimony. The host-rock environment is especially important in the oxidation of sulfides because some metals that would be leached from a siliceous host are retained in calcareous rocks. Molybdenum, zinc, and silver are especially soluble in sulfate solutions, but under favorable conditions they form stable oxidation products in limestones. Similarly, copper, which is relatively mobile in sulfate waters that circulate through siliceous igneous rocks, forms practically insoluble carbonate minerals in calcareous environments. Iron and lead oxidize to stable compounds in both siliceous and calcareous rocks, and are retained in the zone of weathering. Certain elements are dependent upon the presence or absence of a second element for their stability during weathering. For example, molybdenite oxidizes to the relatively insoluble compound ferrimolybdite [$Fe_2(MoO_4)_3 \cdot 7\frac{1}{2}H_2O$] in siliceous, iron-rich environments, but the molybdenum is readily leached where iron is not plentiful.

The fact that certain elements are concentrated in soils during weathering is useful during geochemical prospecting. Even where the metallic element, such as copper, is comparatively soluble and appears to have been removed, enough remains in the soil so that it may be readily detected by careful geochemical methods. Usually the soil overlying a body of copper mineralization will contain more residual copper than the average soils of the surrounding area. Anomalies can thus be detected. In addition to determinations in soils, four other methods of geochemical study are widely used (Boyle, 1971).

1. Metals are frequently adsorbed in the tissues of plants and show variations similar to those in the soil patterns.

2. They are detectable in stream and lake waters.

3. Metal content in places is increased in stream sediments below mineralized bodies.

4. Bedrock samples may indicate the presence of above average amounts of metal.

In tropical climates, where leaching is especially effective, only the most insoluble oxides remain at the surface. Iron and aluminum form such stable compounds and are so abundant that they are commonly left as residual concentrates. These iron- or aluminum-rich soils are known as "laterites," the Latin word for "brick earth." Laterites are oxides, and thus contrast markedly with the silicate clay soils of the temperate climates. Mafic minerals and feldspars break down and release hydrous iron and aluminum compounds, which are relatively insoluble under oxidizing conditions and consequently remain at the outcrop. During oxidation, the soluble components are dispersed, and the iron and aluminum are oxidized in place or redistributed. The oxidation products may be transported in solution under low Eh and pH conditions; for example, iron is soluble as the simple ferrous ion or in the form of organic, sulfate, and hydroxide complexes (Hem and Cropper, 1959; Hem, 1960; Oborn and Hem, 1961). However, the oxidation products are generally redeposited as they form, because of an immediate increase in the Eh and pH away from oxidizing sulfides in the zone of weathering. Apparently, the mineral structures break down, and the oxidation products go into the colloidal state, as evidenced by the presence of orbicular, concretionary structures known as *pisolites.* Since pisolites form in ways other than from colloidal solutions, their presence alone is not considered definite proof of a colloidal origin. Most, but not all, laterites are pisolitic (see Fig. 19-1).

Iron-rich laterites form over ferromagnesian rocks where the rainfall is heavy and the topography subdued. Flat areas and broad swales are ideal because the water is retained long enough to dissolve the siliceous components of the soil without eroding the residual materials. Apparently, laterites form most readily between the annual high and low positions of the water table; alternating wet and dry seasons are therefore ideal for laterization (Harder, 1952). Laterites have been mined in small amounts as iron ore where the limonitic pellets are abundant and fine-grained soils or clays are nearly absent (Percival, 1966). Very large areas covered with this type of laterite are known in Brazil, Cuba, India, central Africa, the Philippine Islands, and elsewhere in

Figure 19-1
Pellets in laterite, Moanda manganese deposit, Gabon.

the tropics. Many of the iron-rich laterites overlie serpentine bodies or other rocks that originally were rich in iron and deficient in silica. In order to compete successfully with the world's large sedimentary iron-ore deposits and with high-grade magnetite deposits, the iron laterites must be nearly pure and strategically located. A significant factor in the cost of mining laterites is the presence of 10 to 30 percent combined water, which must be transported and then removed during smelting.

In tropical or subtropical environments where the underlying rocks are rich in aluminum and low in other comparatively stable materials, such as iron and silica, concentrations of bauxite are likely to form, especially over syenites and nepheline syenites. Bauxites consist of boehmite [$AlO(OH)$], gibbsite [$Al(OH)_3$], diaspore [$AlO(OH)$], and other hydrous aluminum oxides. The formation of clay minerals may be an intermediate step in the breakdown of some feldspars and feldspathoids to bauxite. Bauxites also form over argillaceous carbonate rocks, in an association with the residual clay known as terra rossa. Although such concentrations of alumina are generally high in iron, they are mined in places, for example, in the Mediterranean region and in Jamaica. Large reserves of aluminous clays are known in many areas, but owing to the difficulty of separating aluminum from silica, the clays are only potential ore.

Manganese oxides commonly form concentrations of relatively pure mineral over bodies of rock that are rich in manganese and are deeply weathered. Many metamorphic rocks contain spessartite garnets and other manganese-bearing silicates and carbonates; in places these minerals are present in large amounts. In India the term "gondite" has been applied to these spessartite-bearing metamorphic rocks. Elsewhere, other names have been applied, but "gondite" has received the widest recognition. Ore is formed where the silica and other valueless materials are removed and the relatively insoluble manganiferous oxides remain at the surface.

Lateritic pellets of manganese oxides are not as common as similar pellets of limonite, but they are found in many places. In Cuba the name "granzon" has been applied to these pellets; this term is now widely used. Economic concentrations of granzon are commonly found on the surface near manganese orebodies.

Besides iron, aluminum, and manganese, several other metals form stable oxidation products. Where they reach economic proportions, these metals are mined as ore, but more often they are merely complicating impurities in iron laterites. Many serpentines contain small amounts of nickel, cobalt, and chromium; these are concentrated in the iron-rich laterites, or in the upper parts of the serpentines; in some districts—for example, at Nicaro, Cuba—these metals are recovered as by-products. The nickel is concentrated in the serpentine, just below the laterite, forming ores containing about 0.3 to 1.5 percent nickel. Exceptional deposits, such as those of New Caledonia, contain as much as 6 to 10 percent nickel. The chromium content of laterites is as much as 1.5 to 2.5 percent; the cobalt content, 0.2 percent.

Most sulfide minerals are less stable in the zone of weathering than are the rock-forming silicate, oxide, and carbonate minerals. Consequently, the sulfides are oxidized, dissolved, or otherwise altered more readily than the surrounding host rocks, leaving only surface exposures of the weathered products. The exploration for ore deposits thus involves the interpretation of altered surface exposures to help determine the subsurface mineralogy.

Thermodynamic relations in supergene processes are illustrated by diagrams such as those of Garrels (see Fig. 19-2). The fundamental process is oxidation, but since the oxidation of sulfides produces hydrogen ion and sulfate, both Eh and pH are controls. The actual mechanisms of oxidation and dissolution of metal sulfides are not well understood. The sulfide minerals no doubt dissolve ionically under the attack of surface waters, and the constituents then react with dissolved oxygen. The oxidation reaction would be much more sluggish if the materials were not first broken down into the ionic state. Sato (1960) suggested that the oxidation agent is H_2O_2, which forms as an intermediate product during the reduction of oxygen. If the oxidation potential of the

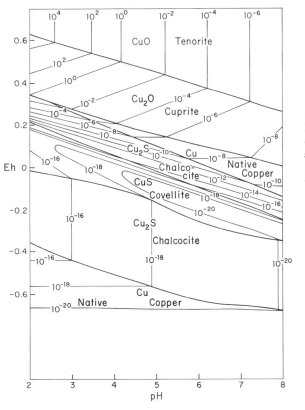

Figure 19-2
Fields of stability of some copper compounds as functions of pH and Eh. Contours are for activity Cu^{+2} + activity Cu^+. (From Garrels, 1954, Fig. 2.)

environment exceeds that of the H_2O_2-O_2 couple, any metal ions present react with the peroxide to form the metal oxide and water. Both theoretical and experimental evidence substantiate this mechanism. Moreover, field measurements of the Eh and pH values in the zone of oxidation fall into a narrow zone just above the standard potential for the H_2O_2-O_2 couple.

Pyrite is the most common sulfide; few sulfide ore deposits lack it. The oxidation of pyrite and other iron sulfides generally leaves limonite or hematite and generates sulfuric acid. The normal reaction involved is

$$2FeS_2 + \tfrac{15}{2}O_2 + 4H_2O \rightarrow Fe_2O_3 + 4SO_4^{-2} + 8H^+ \tag{1}$$

Actually, iron is dissolved as the ferrous ion or as ferrous sulfate, and it is then oxidized to insoluble limonite or hematite:

$$2FeS_2 + 7O_2 + 2H_2O \rightarrow 2Fe^{+2} + 4SO_4^{-2} + 4H^+ \tag{2}$$

$$2Fe^{+2} + \tfrac{1}{2}O_2 + 2H_2O \rightarrow Fe_2O_3 + 4H^+ \tag{3}$$

The iron in massive pyrite bodies is likely to be leached, without leaving much hematite or limonite, because the presence of sulfuric acid keeps the pH low and may form a reducing environment that retains the iron in the soluble ferrous state. Conversely, waters rich in oxygen may alter pyrite directly to ferric sulfate, without going through a ferrous sulfate stage (Eq. 4). The Fe^{+2}/Fe^{+3} ratio may vary widely, depending upon both Eh and pH. The kinds of minerals that form under set conditions of Eh and pH depend upon relative concentrations and the extent to which various solubility products are satisfied.

$$2FeS_2 + \tfrac{15}{2}O_2 + H_2O \rightarrow Fe_2(SO_4)_3 + SO_4^{-2} + 2H^+ \tag{4}$$

Some iron persists at the outcrop as ferric or ferrous sulfate in arid climates, but generally these sulfates are merely temporary transitional states. Iron migrates in the dissolved state where the solutions remain strongly acidic or deficient in oxygen; a low redox potential favors the stability of ferrous ions in solution, whereas a pH below 3 permits ferric ions to remain dissolved. However, such low redox potentials and extremely acidic conditions are unlikely near the surface and above the water table.

The weathering of iron sulfides involves the oxidation of both iron and sulfur. Sulfur in sulfide minerals generally has a valence (oxidation number) of -2. Although the sulfur in FeS_2 would seem to be an exception, it is not, because it occurs as S_2^{-2} "molecules" bound to divalent iron. Oxidation to the sulfate radical increases the valence to $+6$. As an intermediate step, oxidation forms native sulfur, rather than the sulfate radical. For example, ferric sulfate solutions react with chalcocite to release copper ions in solution and produce native sulfur (Sullivan, 1930; Sato, 1960):

$$Cu_2S + 4Fe^{+3} + 6SO_4^{-2} \rightarrow 2Cu^{+2} + 4Fe^{+2} + 6SO_4^{-2} + S \tag{5}$$

This explains the presence of small amounts of native sulfur in many outcrops.

Ferric sulfate and sulfuric acid generated by the oxidation of iron sulfides act as potent solvents for other metallic sulfides. Ferric sulfate oxidizes sulfide minerals to soluble sulfates and is itself reduced to ferrous sulfate. At the expense of hydrogen ions and oxygen, ferrous sulfate probably is oxidized quickly back to ferric sulfate:

$$2Fe^{+2} + 2H^+ + \tfrac{1}{2}O_2 \rightarrow 2Fe^{+3} + H_2O \tag{6}$$

Wherever the acidity of the waters percolating through an ore deposit is maintained, such elements as copper, silver, and zinc are leached. In the absence of iron sulfides, ferric sulfate and sulfuric acid do not readily form, hence the oxidation products of other sulfides tend to remain in place. Similarly, the neutralization of acidic waters by reactive wall rocks, such as carbonates, causes dissolved metals to precipitate as stable minerals. Equations 7 through 21 demonstrate the oxidation of a few common sulfide minerals. Solid reactants and relatively insoluble products are written as stoichiometric compounds; dissolved substances are expressed as ions.

$$\underset{\text{pyrrhotite}}{2FeS} + \tfrac{9}{2}O_2 + 2H_2O \rightarrow Fe_2O_3 + 2SO_4^{-2} + 4H^+ \tag{7}$$

$$\underset{\text{chalcopyrite}}{2CuFeS_2} + \tfrac{17}{2}O_2 + 2H_2O \rightarrow$$

$$Fe_2O_3 + 2Cu^{+2} + 4SO_4^{-2} + 4H^+ \tag{8}$$

$$\underset{\text{chalcopyrite}}{CuFeS_2} + 4Fe^{+3} + 2H_2O + 3O_2 \rightarrow$$

$$Cu^{+2} + 5Fe^{+2} + 2SO_4^{-2} + 4H^+ \tag{9}$$

Equations 8 and 9 show two reactions involving chalcopyrite. It may break down directly to iron oxide and an acidic solution of cuprous sulfate, or it may be taken into solution by ferric sulfate. The sulfuric acid generated in these reactions or generated by oxidizing iron sulfides helps take the copper into solution as cupric sulfate. Bornite behaves similarly:

$$\underset{\text{bornite}}{2Cu_5FeS_4} + \tfrac{27}{2}O_2 + 8H_2O \rightarrow$$

$$Fe_2O_3 + 10Cu^{+2} + 8SO_4^{-2} + 16H^+ \tag{10}$$

$$\underset{\text{bornite}}{Cu_5FeS_4} + 4Fe^{+3} + 4H_2O + 6O_2 \rightarrow$$

$$5Cu^{+2} + 5Fe^{+2} + 4SO_4^{-2} + 8H^+ \tag{11}$$

$$\underset{\text{bornite}}{Cu_5FeS_4} + 4H^+ + 9O_2 \rightarrow$$

$$5Cu^{+2} + Fe^{+2} + 4SO_4^{-2} + 2H_2O \tag{12}$$

The simple sulfides that do not contain iron are oxidized directly or dissolved by ferric sulfate and sulfuric acid produced by the oxidation of associated iron sulfides:

$$Cu_2S + 2H^+ + SO_4^{-2} + \tfrac{5}{2}O_2 \rightarrow$$

chalcocite

$$2Cu^{+2} + 2SO_4^{-2} + H_2O \tag{13}$$

$$Cu_2S + 2Fe^{+3} + 3SO_4^{-2} + \tfrac{3}{2}O_2 + H_2O \rightarrow$$

chalcocite

$$2Cu^{+2} + 2Fe^{+2} + 4SO_4^{-2} + 2H^+ \tag{14}$$

$$PbS + 2O_2 \rightarrow PbSO_4 \tag{15}$$

galena $\qquad$ anglesite

$$PbS + 2Fe^{+3} + 3SO_4^{-2} + \tfrac{3}{2}O_2 + H_2O \rightarrow$$

galena

$$PbSO_4 + 2Fe^{+2} + 2H^+ + 3SO_4^{-2} \tag{16}$$

anglesite

In the presence of calcareous rocks, galena oxidizes to the relatively insoluble carbonate, cerussite:

$$PbS + H_2O + CO_2 + 2O_2 \rightarrow PbCO_3 + SO_4^{-2} + 2H^+ \tag{17}$$

galena $\qquad$ cerussite

The process may involve two steps: oxidation to anglesite (Eq. 15) and conversion of anglesite to cerussite:

$$PbSO_4 + H_2O + CO_2 \rightarrow PbCO_3 + SO_4^{-2} + 2H^+ \tag{18}$$

anglesite $\qquad$ cerussite

The geological evidence of these steps is convincing in cerussite ores, where cores of galena are rimmed in places with transition zones of anglesite. Elsewhere, anglesite is absent and galena appears to alter directly to cerussite. Zinc and silver sulfides are also dissolved in the presence of ferric sulfate:

$$ZnS + 2Fe^{+3} + 3SO_4^{-2} + H_2O + \tfrac{3}{2}O_2 \rightarrow$$

sphalerite

$$Zn^{+2} + 2Fe^{+2} + 2H^+ + 4SO_4^{-2} \tag{19}$$

$$Ag_2S + 2Fe^{+3} + 3SO_4^{-2} + H_2O + \tfrac{3}{2}O_2 \rightarrow$$

acanthite

$$2Ag^+ + 2Fe^{+2} + 2H^+ + 4SO_4^{-2} \tag{20}$$

In a limestone environment the zinc sulfate solution reacts to form smithsonite and gypsum:

$$Zn^{+2} + SO_4^{-2} + CaCO_3 + 2H_2O \rightarrow \underset{\text{gypsum}}{CaSO_4 \cdot 2H_2O} + \underset{\text{smithsonite}}{ZnCO_3} \quad (21)$$

Cupric sulfate solutions react similarly with carbonate rocks to form the hydrous carbonates of copper—malachite and azurite.

Iron oxides and lead sulfate are relatively insoluble and tend to remain at the outcrops, but most oxidation products of other metals are removed in solution and carried downward. Where beds of limestone or other basifying rocks are encountered, the acidic solutions are neutralized, and the more stable carbonates of copper and zinc are formed. In extremely arid countries, such as the Atacama Desert of Chile, water-soluble minerals—sulfates and chlorides of copper and iron—remain in the zone of oxidation.

Native copper is commonly associated with cuprite (Cu_2O) in the upper parts of oxidized copper deposits; in fact, the association is so universal that cuprite without disseminated native copper is rare. The reaction producing native copper and cuprite from chalcocite may be written:

$$2Cu_2S + 8Fe^{+3} + 12SO_4^{-2} + 6H_2O + \tfrac{3}{2}O_2 \rightarrow$$
$$2Cu + Cu_2O + 8Fe^{+2} + 12H^+ + 14SO_4^{-2} \quad (22)$$

The further oxidation of native copper and cuprite produces tenorite (CuO):

$$Cu + Cu_2O + O_2 \rightarrow 3CuO \quad (23)$$

Molybdenum is ordinarily removed almost entirely from soils; only minor amounts can be detected chemically. At places near the base of the weathered zone, or elsewhere if conditions are favorable, molybdic ochre or ferromolybdite may form. This sulfur-yellow mineral may form according to the reaction in Equation 24.

$$6MoS_2 + 4Fe^{+3} + 15H_2O + 36O_2 \rightarrow$$
$$2Fe_2(MoO_4)_3 \cdot 15H_2O + 12SO_4^{-2} \quad (24)$$

Silver behaves much the same as copper in the zone of oxidation but has fewer stable oxidation products. Acanthite, the common silver sulfide, is oxidized by and therefore soluble in ferric sulfate solutions (Eq. 20), and silver sulfosalts are broken down similarly. Consequently, the silver content of sulfide veins is usually carried downward by meteoric waters. Where the climate is somewhat arid, however, the weakly soluble halogen salts, such as cerargyrite (AgCl), bromyrite (AgBr), iodyrite (AgI), and embolite [Ag(Br,Cl)], are left in the oxidized zone. The silver halides in arid or semiarid regions are due to a lack of water sufficient to dissolve them, and usually a windborne supply of salts is available for reaction. Silver carbonates and silver oxides are generally not present, but native silver is common. The native silver forms as a result of reduction of the silver ion, probably by ferrous iron (Stokes, 1906; Cooke, 1913). Equation 25 shows the reaction between ferrous sulfate and silver sulfate.

$$2Ag^+ + 2Fe^{+2} \rightarrow 2Ag + 2Fe^{+3} \tag{25}$$

Zinc sulfates are very soluble; as a result, the zinc content of most oxidized orebodies is dispersed in the groundwater system. However, in arid or semiarid climates zinc may be retained in the oxidized zone as smithsonite, hydrozincite $[Zn_5(CO_3)_2(OH)_6]$, hemimorphite $[Zn_4Si_2O_7(OH)_2 \cdot H_2O]$, or other carbonate and silicate minerals.

Galena is soluble in ferric sulfate solutions; the reaction is so sluggish, however, that unoxidized masses of galena are common in areas where other sulfides have been altered or leached. Leaching of galena is retarded because the common oxidation products—anglesite and cerussite—are also stable in the zone of weathering.

Native gold is inert to most oxidizing environments, but field evidence indicates that solution and short-distance transport of gold do take place under favorable conditions (Emmons, 1912). Gold is somewhat soluble as a chloride complex. In acidic solutions and in the presence of a strong oxidizing agent, such as MnO_2, gold is oxidized to Au^{+3}, which combines with chlorine ions to form the stable $AuCl_4^-$ complex. Manganese, an abundant and widespread element, is present in many gold ores. Enrichments of gold near the surface have been found, suggesting that the gold was dissolved and reprecipitated ahead of mechanical erosion. The solution and reprecipitation of gold results in the presence of coarse gold along cracks and small openings throughout the upper parts of the vein and the adjacent wall rock.

Antimony forms the relatively stable compounds, valentinite (Sb_2O_3), bindheimite, $[Pb_{1-2}Sb_{2-1}(O,OH,H_2O)_{6-7}]$, and stibiconite (also known as cervantite). These minerals are not conspicuous, and, owing to iron oxide staining, are easily mistaken for limonite. Nevertheless, they have been identified in the oxidized zones of many ore deposits. Where antimony is present in complex sulfantimonide minerals, the oxidation products are likely to be complex also. The oxidation of stibnite probably takes place according to the following equations:

$$\underset{\text{stibnite}}{Sb_2S_3} + 3H_2O + 6O_2 \rightarrow \underset{\text{valentinite}}{Sb_2O_3} + 6H^+ + 3SO_4^{-2} \tag{26}$$

$$\underset{\text{stibnite}}{3Sb_2S_3} + 10H_2O + 20O_2 \rightarrow$$
$$\underset{\text{stibiconite}}{2SbSb_2O_6(OH)} + 18H^+ + 9SO_4^{-2} \tag{27}$$

Most arsenic compounds, in contrast with those of antimony, are relatively soluble and consequently are leached from the zone of weathering. Traces of rare arsenates, such as conichalcite [$CaCuAsO_4(OH)$] in oxidized copper ores, reflect the former presence of sulfarsenides, but in humid climates these minerals are dissolved with the sulfates.

Oxide minerals are also susceptible to weathering, but usually at a slower rate than the sulfides. Chromite and ilmenite ores persist at the outcrop until they are mechanically removed. But, given time, they eventually oxidize or dissolve. Martite (Fe_2O_3), hematite, and limonite commonly form in the outcrops of magnetite deposits; maghemite (Fe_2O_3) is stable in many lateritic soils.

Oxidation rates vary over a wide range. Cinnabar oxidizes so slowly that for practical purposes it is considered stable in surface outcrops, even though it is thermodynamically unstable under surface conditions. Galena is commonly protected from oxidation in a jacket of anglesite. By contrast, a few other sulfides oxidize so rapidly that they ignite when exposed to the atmosphere. In deposits of massive sulfides, especially pyrrhotite and marcasite, it is not uncommon for the rock to burn as a result of spontaneous combustion. Experience has taught the miners of Iron River, Michigan, to leave a skin of iron ore on the walls of stopes in order to prevent the highly pyritiferous black slate from igniting. High rock and air temperatures are also produced by oxidizing sulfides, making it necessary in some mines to pump refrigerated air underground so that miners can work comfortably. The processes and end products are the same in these situations as they are on surface outcrops. Acidic waters are produced by the oxidation of sulfides; in some districts the mine waters are so acidic that they dissolve mine rails and clothing, creating a major problem for men and equipment. Similarly, large amounts of noxious and strongly corrosive sulfur dioxide fumes are generated where the sulfides burn, and in extreme cases the mines must be abandoned or parts sealed off until the fire exhausts the oxygen supply or burns all of the combustible materials. The burned country rock is generally bleached and altered as if it had been exposed at the surface for a long time.

The leached, oxidized surface exposures of weathered sulfide deposits are known as *gossans*. A thorough understanding of gossans can be of great value to economic geologists, because gossans may retain distinctive characteristics of the underlying sulfide materials (Blanchard, 1968). It is desirable to know whether sulfides exist at depth and, further, whether the mineralization is simply pyrite or includes appreciable amounts of copper, zinc, silver, and other valuable sulfides. Consequently, many serious efforts have been made to study gossans quantitatively. The criteria by which gossans are evaluated are vague and poorly understood. Those considered include the shapes of the cavities formerly occupied by sulfides; the structure, texture, and color of the limonite; the quantity of kaolinite and other clay minerals; and the presence of metallic oxidation products. Emphasis is usually placed upon the color of the limonite and the configuration of boxworks (the meshwork of porous gossan left after much of the original sulfide is leached). The confident application of gossan studies in exploration, although valuable in some areas, ordinarily requires great skill and a more intimate knowledge of the local ore deposits than most geologists are able to obtain. Extrapolating the characteristics of gossans from one environment to another can be discouraging. Resident geologists, or others familiar through long experience in certain areas, use the color and texture of limonite to indicate whether the underlying rocks are favorable, though seldom is a quantitative approach considered (Locke, 1926). Most geologists agree that gossans should be studied with care; proof of their value has been demonstrated in many mining districts. Lacy (1949) studied the oxidation products at Yauricocha, Peru, and was able to relate individual oxide minerals to their source materials. He divided the oxidation products into residual and transported materials and subdivided these according to the textures developed and the minerals from which they were derived. On the basis of this study, Lacy distinguished among gossans that overlie several kinds of sulfide deposits, including lead-zinc ores, lead-zinc-copper ores, copper-pyrite ores, and massive pyrite bodies.

The presence of gossans does not necessarily mean unaltered sulfides exist at depth. Under relatively low Eh and pH conditions, iron goes into solution in the ferrous state and travels appreciable distances from the oxidation zone. Where these ferrous sulfate solutions encounter limestone or a similar basifying medium, the acid is neutralized, and the iron precipitates as ferric oxide or ferric hydroxide. Displaced iron oxide zones of this type are known as *false gossans*. Clearly, then, it is of critical importance to distinguish between false gossans and indigenous gossans.

Blanchard (1968) attempted to distinguish between the gossan formed by leaching of massive sulfides and the iron-stained material left by weathering of disseminated mineralization. He pointed out that chalcopyrite oxidizing in an inert gangue will theoretically export in solution all of its copper and two-thirds of its iron, leaving one-third of its iron as limonite, in the form of cellular boxwork or sponge. One-to-one mixtures of chalcopyrite and pyrite oxidizing together in an inert gangue will dissolve and export all of the copper and all of the iron in the two minerals. Pyrrhotite yields an excess of acid, but sphalerite yields just enough acid to permit solution of the zinc as zinc sulfate.

Certain sulfides, such as chalcocite, bornite, and tetrahedrite are deficient in sulfur and will not dissolve completely unless external sulfur is introduced. Covellite, sphalerite, and molybdenite contain just enough sulfur to permit solution while pyrite, chalcopyrite, and pyrrhotite yield free acid (Blanchard, 1968).

The depth of oxidation is another significant problem for mining geologists. In tectonically stable regions the oxidation zone generally extends to the water table, especially if the country rocks are permeable. But recent faulting, a fluctuating water table, or impervious wall rocks may modify the pattern of oxidation. The most critical factors are the position and permanence of the water table, because sulfides are generally stable in the slightly alkaline, moderately reducing environment below the groundwater table. In humid climates the sulfide zone may be present a few feet from the surface; in arid environments, where the water table is likely to be deep, the lower limit of oxidation may extend two or three thousand feet below the outcrop. At the other extreme, a rapidly lowered water table may leave sulfides isolated well up in the zone of oxidation. This is especially common in arid environments, where there is not enough water to oxidize and dissolve the sulfides.

The role of permeability in oxidation is strikingly demonstrated at the Tsumeb mine, South-West Africa, where a nearly vertical ore pipe that cuts through steeply dipping sediments is oxidized in the upper and lower parts but is unweathered at intermediate levels. Here, deep oxidation was brought about by groundwaters circulating along a permeable, fractured bed that crops out some distance from the mine but intersects the ore pipe at depth (see Fig. 4-22, p. 89). The unoxidized part of the pipe is protected from groundwater by relatively impermeable strata. Although sulfides remain in the oxidized zones, there are relatively sharp changes in the ratio of oxidized to unoxidized minerals. Between the 1,200- and 2,400-foot levels, the ores are largely sulfides, whereas above and below this, the ores are oxidized. Oxidation along the deep brecciated stratum is so efficient that the lower part of the pipe is more thoroughly oxidized than much of the shallow weathered zone (Söhnge, 1963).

Descending waters charged with dissolved substances tend to seek density levels. This results in a crude stratification of water and explains how concentrations of oxidation products have formed in certain places where deep circulation of water is doubtful; for example, heavy, mineral-laden, and perhaps oxygenated, waters settle below the water table (Brown, 1942; McKnight, 1942).

Many fine-grained oxides, silicates, and carbonates in the zone of oxidation are considered to have traveled as colloids. Typical products of this kind include opal and chalcedony, smithsonite, reniform hematite and limonite, aluminum hydroxides, and manganese oxides. Wad, the amorphous or cryptocrystalline manganese oxide commonly found in residual deposits, is a good example. Some wad gives no pattern with X-ray diffraction equipment and is still colloidal. The composition of wad is indefinite because the colloidal gel tends to adsorb many elements. Unlike most oxide colloids, colloidal manganese oxides carry a negative charge, so that they attract cations of other metals out of solution. Small amounts of nickel are common in many deposits of wad; copper, cobalt, and barite have also been reported in some. The manganese oxides of the Tocantins district, Brazil, contain as much as four percent cobalt (Pecora, 1944). The presence of tungsten, strontium, thallium and a few other metals is considered by Hewett to indicate a hydrothermal, and most probably igneous, history (Hewett and Fleischer, 1960).

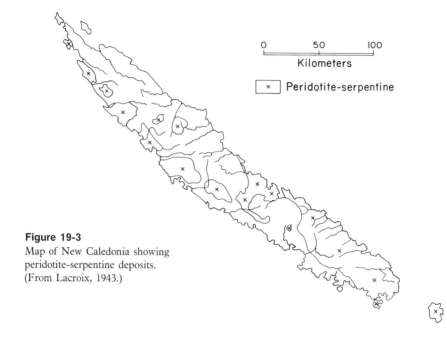

Figure 19-3
Map of New Caledonia showing
peridotite-serpentine deposits.
(From Lacroix, 1943.)

0   50   100
Kilometers

× Peridotite-serpentine

NICKEL DEPOSITS OF NEW CALEDONIA

Surface weathering affects mineral deposits in several ways. It may leach away an older ore deposit; it may oxidize the materials in place without changing the grade of ore; or it may create an ore deposit by the residual concentration of materials that originally were dispersed throughout the fresh rock. The nickel deposits of New Caledonia are a product of the last-mentioned process.

The island of New Caledonia, South Pacific, is about 250 miles long and averages only about 30 miles wide. Nickel deposits were discovered there by Garnier in 1865, and sporadic production has continued since about 1875. Prior to the discovery of the Sudbury district in Ontario, New Caledonia dominated the world's nickel production.

Much of New Caledonia is underlain by ultramafic intrusives—dunite, saxonite, and associated rock types—many of which are partly or entirely ser-pentinized (see Fig. 19-3). The serpentine series, as this igneous complex is known, intrudes Cretaceous and older sedimentary rocks and is consequently thought to be Tertiary in age. Serpentinization essentially involves hydration of the original ultramafic minerals, accompanied by rearrangement and minor loss of constituents; hydration normally results from deuteric or intrusive meta-morphic effects as the hot ultramafics take up water from the intruded country rocks. This mechanism of serpentinization appears to be the one that altered the rocks at New Caledonia because the serpentine is independent of the present topography and of weathering processes.

The many descriptions of the mineral deposits of New Caledonia all agree that the nickel was concentrated during laterization of the serpentine (Glasser, 1904; Ontario Nickel Commission, 1917; Berthelot, 1933; Caillère, 1936; Lacroix, 1943; Chételat, 1947; Routhier, 1952, 1953, 1955, 1963a, 1963b). The serpentine weathers to a dark-reddish laterite, producing soil profiles like those in other tropical areas (see Fig. 19-4). Much of the serpentine series is covered with pisolitic iron oxides. Most of the nodules are about the size of peas, and fine-grained, clayey minerals are nearly absent. Where laterization is complete, the original low-alumina silicates of the ultramafics have been destroyed, and silica, calcium, and magnesium removed. Iron in the laterites is in the form of limonite, carrying 10 to 25 percent water. The serpentine

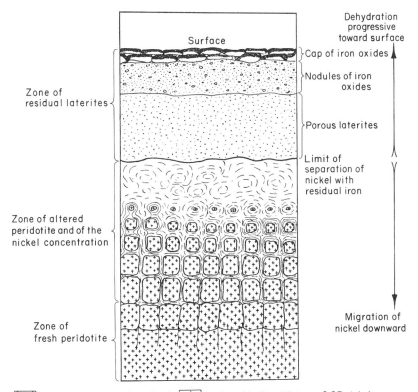

Figure 19-4
Typical section through nickeliferous laterite deposits, New Caledonia. (From Chételat, 1947, Fig. 4.)

series and the laterites are readily distinguished on the surface from other formations by differences in vegetation.

The best nickel ores are concentrated below the laterite near the top of the serpentine series, in nearly horizontal zones, but not all the laterite is underlain by ores. Locally, the minerals are concentrated in minor fissures that extend downward into the rock below, but these fissures are seldom worth mining. The efficiency of the nickel-concentration process was controlled largely by the topography; the best ores are on gentle slopes and on saddles of spurs extending from the main ridges. Figure 19-4 and Table 19-1 clearly show the various stages of laterization and distribution of metals in the New Caledonia deposits. The ores are overlain by soil, by decomposed rocks of the serpentine series, or by pisolitic laterites. In general, the zone of nickel concentration lies one to 20 feet below the surface, but sometimes the cover is 75 to 100 feet thick. Most nickel migration has been downward, though lateral migration has produced a few deposits near the bases of gentle slopes. Small amounts of cobalt and manganese are concentrated in the lower part of the laterites above the more soluble nickel. According to Routhier (1963a), cobalt is present in the porous, high-iron laterites above the nickeliferous horizon.

Garnierite, a nickeliferous variety of serpentine, is the principal ore mineral in New Caledonia. Locally the garnierite is called "nouméite." Another nickeliferous antigorite (népouite and genthite) is also an important ore mineral. Nickel-bearing saponite, known locally as "pimélite," is a third nickel ore. The colors of the minerals range from bright apple green, through light green, to white. Where the iron of the silicates is replaced by nickel, the ore is green, and accordingly is known as green ore; where the magnesium is replaced, the ore is brown, because it retains the color of the iron oxides. Brown ore, known locally as "chocolate ore," is now the common material being mined. Up to

Table 19-1
Zones in the nickeliferous laterite of New Caledonia.

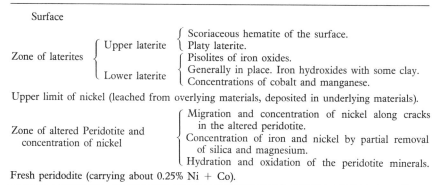

Surface		
		Scoriaceous hematite of the surface.
	Upper laterite	Platy laterite.
Zone of laterites		Pisolites of iron oxides.
	Lower laterite	Generally in place. Iron hydroxides with some clay.
		Concentrations of cobalt and manganese.
Upper limit of nickel (leached from overlying materials, deposited in underlying materials).		
		Migration and concentration of nickel along cracks in the altered peridotite.
Zone of altered Peridotite and concentration of nickel		Concentration of iron and nickel by partial removal of silica and magnesium.
		Hydration and oxidation of the peridotite minerals.
Fresh peridodite (carrying about 0.25% Ni + Co).		

Source: Chételat, 1947, Table 2.

10 percent cobalt is present in the amorphous oxide mineraloid, asbolite (cobaltiferous wad). Black manganese oxides, identified as pyrolusite and psilomelane, have been mined locally in small amounts. Talc and sepiolite $[Mg_4(Si_6O_{15})(OH)_2 \cdot 6H_2O]$ are the most common gangue minerals, and small amounts of chalcedony—representing colloidal silica leached from the laterite zone—are found in fractures cutting the serpentine series.

The ore averages about 3.5 percent nickel, but higher grade ore was mined during the early days, when 10 percent nickel was not unusual. The present grade is still well above the nickel content of laterites in most other districts. Most of the orebodies are small (less than 100,000 tons), the largest deposit containing only about 600,000 tons.

Studies of the unaltered serpentine and ultramafics show that the nickel and cobalt content are relatively uniform throughout the district. There seems to be no direct relationship between the grade of the original serpentine series and the metal content of the ores. Thus, the presence of ore is strictly a function of favorable topography and the effect of weathering conditions on the serpentine series. The effectiveness of the weathering process is more fully appreciated when we realize that it has concentrated the nickel ten- to thirtyfold from fresh ultramafic rock that contained only 0.2–0.3 percent nickel (Chételat, 1947; Blanchard, 1944). Chételat (1947) reported the presence of arsenopyrite and pyrite in the serpentine series and considered the possibility that nickel was a constituent of these sulfides. Other workers think the nickel was contained in the ferromagnesian silicate minerals.

Whatever the original distribution of the nickel, it is clear that weathering caused the enrichment. The best ores are found where chemical weathering was most effective. It has even been noted that the serpentine is less commonly a source material for nickel ores than other rocks of the ultramafic complex, in spite of the fact that the serpentine contains an equal amount of nickel (Chételat, 1947). The difference is a function of permeability because the relative impermeability of serpentine tends to inhibit circulation and downward migration of the mineral-bearing waters.

MORRO DA MINA, BRAZIL

Morro da Mina, in the Lafaiete district, Minas Gerais, Brazil, has been one of the most productive manganese deposits in the Western Hemisphere. The ore is residual and consists of manganese oxides formed by the weathering of silicate and carbonate protores. The protore has been called, by different workers "gondite" (Park, 1956), "queluzite" (Derby, 1901), and "manganese silicate-carbonate protore" (Dorr *et al.*, 1956). Dorr and his co-workers state that the protores are metamorphosed sediments that formed a gradational sequence of cherts, mudstones, and manganiferous carbonates. They are in schists of Precambrian age. The characteristic manganiferous beds have been followed along the strike for more than 40 miles, and probably continue farther. The irregular distribution of the constituents in the sediments helps to explain the present erratic distribution of manganoan carbonates in the protores. The sediments have undergone intense regional metamorphism, and locally are invaded by granitic masses and mafic and pegmatitic dikes (Guimarães, 1935). Igneous metamorphism has been superimposed upon the regional metamorphism, particularly near the northwestern part of the mine—a factor that has led observers in the past to relate the mineralization to igneous metamorphic processes. Igneous metamorphism is now thought to have developed tephroite (Mn_2SiO_4) and other silicates, mainly at the expense of manganese carbonates.

The mineralogy of the protores was studied intensively by Horen (1955), who found spessartite garnets, rhodochrosite and manganoan calcite, rhodonite, tephroite, and a long list of minor minerals, particularly manganese silicates. Pyrite and small amounts of other sulfides, notably alabandite, have been recognized (Park et al., 1951; Ödman, 1955; Horen, 1955). Much protore is a nondescript, sandy looking, gray-brown rock that contains mainly small crystals of spessartite interspersed in a cement of manganoan carbonate and minor amounts of fine-grained sulfides. In the deeper parts of the mine workings the protore is cut by veins and irregular masses of bright red rhodochrosite, rhodonite, and smaller amounts of other minerals. The protore averages about 30 percent manganese. In a few places the carbonates make up as much as 70 percent of the rock (Dorr et al., 1956), but in many places carbonates are sparse or absent, and the rock is massive, dense, and impermeable.

The ore at Morro da Mina consists of manganese oxides—products of the removal of silica and carbonate from the protores by meteoric waters. Only in very few places does the ore grade downward into the protore; rather, the contact is usually a sharp, clearly defined, undulating surface that locally extends downward into the protore along fractures, dikes, and water courses. The presence of carbonates and sulfides that are more easily decomposed than the silicates is thought to aid greatly in the weathering of the protores. Where these materials are present the protore becomes permeable, and decomposition proceeds rapidly and to considerable depths. Ore has been mined to a depth of about 500 feet. Where the carbonates and sulfides are absent in the protores—for example, in those parts of the beds where igneous metamorphism has been active—oxidation is limited, and generally the manganese oxides form only a thin film or at most a few feet of ore.

BAUXITE DEPOSITS OF JAMAICA

The presence of aluminum in the red soil of Jamaica was recognized as early as 1869 (Sawkins et al., 1869), but no attention was paid to this until 1942, when R. F. Innes, an agricultural chemist in Jamaica, said that the aluminum content of the samples he analyzed was high enough to warrant exploitation (Hose, 1950). Today, Jamaica is a principal source of aluminum ore (Blume, 1962; Bramlette, 1947; Hill, 1955; Hose, 1959; Kelly, 1961; Salas, 1959; Schmedeman, 1948, 1950; Vincenz, 1964; Zans, 1952, 1957, 1959).

The oldest rocks of Jamaica are serpentines (thought to be of Cretaceous age) overlain by amphibole schists and marbles of the Upper Cretaceous Metamorphic Series. Above the Cretaceous layer is a sequence of sandstones; tuffaceous shales that weather deep red; and conglomerates, with intercalated layers of limestone. The Wagwater conglomerate marks the base of the Tertiary layer; this is overlain by carbonaceous shales containing sandstone and dark limestone, which in turn are followed by the Tuff Series and the Yellow Limestone Formation. The White Limestone Formation of Middle Eocene-Lower Miocene age is next in the sequence and is the host rock for the bauxite, which is followed by coastal plain beds of limestone and sand.

The central part of Jamaica is marked by the presence of an anticline known as the Central Inlier (see Fig. 19-5). From this central, topographically high, anticline the beds dip toward the ocean. Many sink holes have developed in the White Limestone on the flanks of the anticline, and it is in this karst topography that the bauxitic ores are found. Apparently the ore was formed after uplift and deformation, since it fills the karst depressions and blankets the undulating surface of the White Limestone (Zans, 1952).

The Jamaican aluminum ores are soft and earthy, and sometimes shaley and highly porous. They are commonly dark red or yellow brown, though other colors may be seen locally. At places dark reddish pisolites have been recognized.

The bauxite ores contain mainly gibbsite, boehmite, and diaspore; gibbsite is the dominant ore mineral. The ores are high in iron; generally they carry 18 to 20 percent iron oxides and are known in the industry as *terra rossa*. They contain 48 to 50 percent alumina, two percent titania, and less than three percent silica; combined water makes up about 27 percent. The iron oxide is hematite and goethite; silica is in the clay minerals, largely in the kaolin group.

White Limestone is a chalky, nodular, and well-bedded rock that is dolomitic near the base. This is a pure limestone-dolomite low in silica and iron, but which many geologists have in the past believed to contain enough impurities to make possible the accumulation of the bauxites by weathering since Miocene times (Hose, 1963). The limestone is said to contain on the average no more than 0.2 percent of acid-insoluble material, of which only 0.036 percent is alumina (Chubb, 1963).

The base of the ore deposits is a sharp irregular contact with the underlying and enclosing limestones. The orebodies are irregular and range in size from small pockets to basins comprising hundreds of acres. The average mineable thickness is 10 to 30 feet, but thicknesses range from a few inches to 100 feet or more.

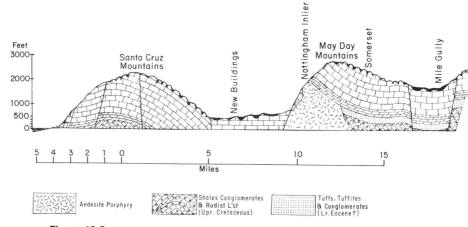

Figure 19-5
Diagrammatic section across central Jamaica, showing the structural relations of the bauxite-bearing White Limestone formation with the older strata. Vertical exaggeration ×7. (From Zans, 1952.)

Ore reserves are estimated to be several hundred million tons, and the presence of such large deposits on relatively pure carbonate rock has led many geologists to question whether or not the ore could have formed by simple weathering of the limestone. Burns (1961), who assumed that no loss of ions occurred during weathering, said that an enrichment factor of 11,600 would be required to produce bauxite of economic grade from the White Limestone, that this factor was unreasonable, and that the White Limestone was not the source of the ore. Waterman (1962) said that a greater thickness than that of the White Limestone must have been eroded to make possible the bauxite deposits. He said that the Eocene Tuff Series may have been the source of the ore. Chubb (1963) adopted the idea advanced by Zans (1956), namely that the source of the bauxites is not the limestone with which they are associated, but the older aluminous silicate rocks of andesitic composition, the Upper Cretaceous, and, in part, the early Eocene andesites and andesitic pyroclastics. Drainage from the central anticline would spread outward through the limestones and would deposit silt and debris in the pools formed in the karst depressions. On the other hand, Sinclair decided from trace element analyses of both the White Limestone and the bauxite that the limestone residue was the source of most of the bauxite (Sinclair, 1967).

Comer (1973) stated that the bauxites formed from the weathering of Miocene volcanic ash (bentonite). A thickness of about five meters of ash would be required to give the known amounts of bauxite. A thickness of this magnitude is known in the Middle and Upper Miocene section of carbonate rocks along the north coast of Jamaica. Rapid and efficient drainage that quickly removed the silica was assured by the karst topography.

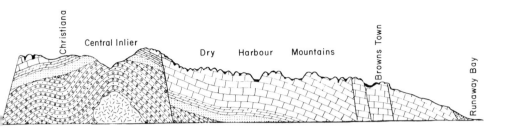

| Yellow L'st Formation (Mid. Eocene) | White L'st Formation with Bauxite (Mid.Eocene-Lr,Miocene) | Coastal Formations (post-Lr.Miocene) | Alluvium |

REFERENCES TO A SELECTED DEPOSIT FORMED BY WEATHERING: CUBAN LATERITES

Cox, J. S., Jr., 1912. The iron-ore deposits of the Moa district, Oriente Province, Island of Cuba, *Amer. Inst. Mining Eng. Trans.* 42:73-90.

Guardiola, R., 1912. Sobre el origen de los criaderos de Mayari, *Rev. Minera Metal. Ingen.* 63:25-27.

Hayes, C. W., 1912. The Mayari and Moa iron-ore deposits in Cuba, *Amer. Inst. Mining Eng. Trans.* 42:109-115.

Kemp, J. F., 1916. The Mayari iron-ore deposits, Cuba, *Amer. Inst. Mining Eng. Trans.* 51:3-30.

Leith, C. K., and W. J. Mead, 1916. Additional data on origin of lateritic iron ores of eastern Cuba, *Amer. Inst. Mining Eng. Trans.* 53:75-78.

Spencer, A. C., 1908. Three deposits of iron ore in Cuba, *U.S. Geol. Surv. Bull.* 340, pp. 318-329.

Weld, C. M., 1910. The residual brown iron-ores of Cuba, *Amer. Inst. Mining Eng. Trans.* 40:299-312.

Woodbridge, D. E., 1912. Exploration of Cuban iron-ore deposits, *Amer. Inst. Mining Eng. Trans.* 42:138-152.

REFERENCES CITED

Berthelot, Ch., 1933. Le nickel et le chrome dans les colonies françaises, *Chimie & Industr.* 29:718-723.

Blanchard, R., 1944. Derivatives of chromite, *Econ. Geol.* 39:448.

———, 1968. Interpretation of leached outcrops, *Nev. Bur. Mines Bull.* 66.

Blume, H., 1962. Der bauxitbergbau auf Jamaika, *Geogr. Rundsch.* 14(6):227-235.

Boyle, R. W., ed., 1971. *Geochemical Exploration,* Canadian Institute of Mining and Metallurgy (spec. vol. 11).

Bramlette, M. N., 1947. Lateritic ore of aluminum in the Greater Antilles, *Geol. Soc. Amer. Bull.* 58:1248-1249.

Brown, J. S., 1942. Differential density of ground water as a factor in circulation, oxidation and ore deposition, *Econ. Geol.* 37:310-317.

Burns, D. J., 1961. Some chemical aspects of bauxite genesis in Jamaica, *Econ. Geol.* 56:1297-1303.

Caillère, S., 1936. Contribution a l'étude des minéraux des serpentines, chap. VII, pt. 1, minéraux de la Nouvelle-Calédonie, *Soc. Franç. Miner. Bull.* 59:289-298.

Chételat, E. de, 1947. La genèse et l'évolution des gisements de nickel de la Nouvelle-Calédonie, *Soc. Géol. France Bull.,* sér. 5, 17:105-160.

Chubb, L. J., 1963. Bauxite genesis in Jamaica, *Econ. Geol.* 58:286-289.

Comer, J. B., 1973. Genesis of Jamaican bauxite; paper presented at Geological Society of America meeting, South Central Section, Little Rock, Arkansas, April 5.

Cooke, H. C., 1913. The secondary enrichment of silver ores, *J. Geol.* 21:1-28.

Derby, O. A., 1901. On the manganese ore deposits of the Queluz (Lafayette) district, Minas Gerais, Brazil, *Amer. J. Sci.* 12:18-32.

Dorr, J. V. N., II, I. S. Coelho, and A. Horen, 1956. The manganese deposits of Minas Gerais, Brazil, in *Symposium sobre Yacimientos de Manganeso,* vol. 3, 20th International Geological Congress.

Ehrlich, H. L., 1964. Bacterial oxidation of arsenopyrite and enargite, *Econ. Geol.* 59:1306-1312.

Emmons, W. H., 1912. The agency of manganese in the superficial alteration and secondary enrichment of gold-deposits in the United States, *Amer. Inst. Mining Eng. Trans.* 42:3-73.

————, 1917. The enrichment of ore deposits, *U.S. Geol. Surv. Bull.* 625.

Garrels, R. M., 1954. Mineral species as function of pH and oxidation-reduction potentials, with special reference to the zone of oxidation and secondary enrichment of sulphide ore deposits, *Geochim. Cosmochim. Acta* 5:153-168.

Glasser, E., 1904. Rapport a M. le Ministre des Colonies sur les richesses minerales de la Nouvelle-Calédonie, *Ann. Mines* 5:29-154, 503-701.

Guimarães, D., 1935. Contribuição ao estudo da origem dos depósitos de minério de ferro e manganês do Centro de Minas Gerais, *Serv. Fom. Prod. Min. Bull.* 8.

Harder, E. C., 1952. Examples of bauxite deposits illustrating variations in origin, in *Problems of Clay and Laterite Genesis,* New York: American Institute of Mining Engineers.

Hartman, J. A., 1955. Origin of heavy minerals in Jamaican bauxite, *Econ. Geol.* 50: 738-747.

Hawkes, H. E., and J. S. Webb, 1962. *Geochemistry in Mineral Exploration,* New York: Harper and Row.

Hem, J. D., 1960. Some chemical relationships among sulfur species and dissolved ferrous iron, *U.S. Geol. Surv. Water-Supply Pap.* 1459-C, pp. 57-73.

Hem, J. D., and W. H. Cropper, 1959. Survey of ferrous-ferric chemical equilibria and redox potentials, *U.S. Geol. Surv. Water-Supply Pap.* 1459-A, pp. 1-31.

Hewett, D. F., and M. Fleischer, 1960. Deposits of the manganese oxides, *Econ. Geol.* 55:1-55.

Hill, V. G., 1955. The mineralogy and genesis of the bauxite deposits of Jamaica, B.W.I., *Amer. Mineral.* 40:676-688.

Horen, A., 1955. Mineralogy and petrology of the manganese protore at the Merid mine, Minas Gerais, Brazil, *Geol. Soc. Amer. Bull.* 66:1575.

Hose, H. R., 1950. The geology and mineral resources of Jamaica, *Colonial Geol. Min. Res.* 1(1):11-36.

———, 1959. The origin of bauxites in British Guiana and Jamaica, *5th Inter-Guiana Geol. Conf.*

———, 1963. Jamaica type bauxites developed on limestone, *Econ. Geol.* 58:62-69.

Kelly, W. C., 1961. Some data bearing on the origin of Jamaican bauxite, *Amer. J. Sci.* 259:288-294.

Lacroix, M. A., 1943. Les péridotites de la Nouvelle-Calédonie, leurs serpentines et leurs gîtes de nickel et de cobalt: les gabbros qui les accompagnent, *Acad. Sci. Paris Mém.* 66.

Lacy, W. C., 1949. Oxidation processes and formation of oxide ore at Yauricocha, *Soc. Geol. Peru (Vol. Jubilar, XXV Aniv.)*, pt. II, fasc. 12.

Locke, A., 1926. *Leached Outcrops as Guides to Copper Ore*, Baltimore: Williams and Wilkins.

McKnight, E. T., 1942. Differential density of ground water in ore deposition, *Econ. Geol.* 37:424-426.

Oborn, E. T., and J. D. Hem, 1961. Microbiologic factors in the solution and transport of iron, *U.S. Geol. Surv. Water-Supply Pap.* 1459-H, pp. 213-235.

Ödman, O. H., 1955. Morro da Mina manganese deposit and its protore, *Eng. Minera. Metal.* (Feb.), p. 57.

Ontario Nickel Commission, 1917. *Nickel Deposits of the World*, Ontario Nickel Commission, pp. 234-264.

Park, C. F., Jr., 1956. On the origin of manganese, in *Symposium sobre Yacimientos de Manganeso*, vol. 1, 20th International Geological Congress.

Park, C. F., Jr., J. V. N. Dorr II, P. W. Guild, and A. L. M. Barbosa, 1951. Notes on the manganese ores of Brazil, *Econ. Geol.* 46:1-22.

Pecora, W. T., 1944. Nickel-silicate and associated nickel-cobalt-manganese-oxide deposits near São José do Tocantins, Goiaz, Brazil, *U.S. Geol. Surv. Bull.* 935-E, pp. 247-305.

Percival, F. G., 1964-65. The lateritic iron deposits of Conakry, with discussion, *Inst. Mining Metall. London Trans.* 74:429-462; and 1966, 75(sec. B):B85-B91.

Routhier, P., 1952. Les gisements de fer de la Nouvelle-Calédonie, in *Symposium sur les Gisements de Fer du Monde*, vol. 2, 19th International Geological Congress.

———, 1953. Observations et ideas nouvelles sur les ressources minerales de al Nouvelle Caledonie, *Echo Minas Metall.* (Sept.-Nov.), pp. 593-597, 725-732.

———, 1955. Geographie de la Nouvelle-Caledonie, IIIe partie, in *Les Ressources du Sous-sol*, Paris: Nouvelles Editiones Latines.

———, 1963a. Les gisements métallifères, in *Geologie et Principes de Recherche*, Paris: Masson.

———, 1963b. Etude geologique de versant occidental de la Nouvelle-Caledonie entre le col de Boghen et la Pointe d'Arama, *Soc. Géol. France Mem.* (n.s.) 32(67):271.

Salas, G. P., 1959. Los depósitos de bauxita en Haití y Jamaica y posibilidades de que exista bauxita en México, *Mex. Univ. Nac. Inst. Geol. Bol.* 59, pp. 9-42.

Sato, M., 1960. Oxidation of sulfide ore bodies, *Econ. Geol.* 55:928-961, 1202-1231.

Sawkins, J. G., and others, 1869. Reports on the geology of Jamaica, pt. II of the West Indian survey, *Geol. Surv. Gt. Brit. Mem.*

Schmedeman, O. C., 1948. Caribbean aluminum ores, *Eng. Mining J.* 149(6):78-82.

———, 1950. First Caribbean bauxite development Reynolds Jamaica Mines, Ltd., *Eng. Mining J.* 151(11):98-100.

Sinclair, I. G. L., 1967. Bauxite genesis in Jamaica: new evidence from trace element distribution, *Econ. Geol.* 67:482-486.

Söhnge, P. G., 1964. The geology of the Tsumeb mine, in *The Geology of Some Ore Deposits of Southern Africa*, vol. 2, ed. S. H. Haughton *Geol. Soc. S. Afr.*

Stokes, H. N., 1906. Experiments on the solution, transportation and deposition of copper, silver and gold, *Econ. Geol.* 1:644–650.

Sullivan, J. D., 1930. Chemistry of leaching chalcocite, *U.S. Bur. Mines Tech. Pap.* 473.

Vincenz, S. A., 1964. A note on the radioactivity of Jamaican bauxite and terra rosa, *Overseas Geol. Mineral Resources* 9:295–301.

Waterman, G. C., 1962. Some chemical aspects of bauxite genesis in Jamaica, *Econ. Geol.* 57:829–830.

Zans, V. A., 1952. Bauxite resources of Jamaica and their development, *Colonial Geol. Mineral Resources* 3(4):307–333; also 1954, *Geol. Surv. Dept. Jamaica Pub.* 12.

———, 1956. The origin of the bauxite deposits of Jamaica, in *Resumenes de los Trabajos Presentados*, 20th International Geological Congress.

———, 1957. Geology and mineral deposits of Jamaica, *Geol. Surv. Dept. Jamaica Pub.* 33.

———, 1959. Classification and genetic types of the bauxite deposits, *5th Inter-Guiana Geol. Conf.; Bt. Guiana Geol. Surv. Proc.*, 1962, pp. 205–211. Reprinted 1962 as *Jamaica Geol. Surv. Pub.* 85.

20 / Supergene Sulfide Enrichment

Chemicals released by the breakdown of minerals in the zone of oxidation either remain in place as stable compounds or are carried in solution with the meteoric waters that migrate downward to the phreatic zone. The dissolved materials may be repeatedly deposited as they descend to the groundwater table, but if they are soluble, they will eventually be carried out of the oxidized zone. Rain waters and solutions that pass through organic-rich soil zones become slightly to moderately acidic, and waters near oxidizing sulfides may be strongly acidic. Reaction of these acidic solutions with carbonates and the rock-forming silicates causes them to become neutral or alkaline. Circulation is very slow below the groundwater table, permitting the basifying process to reach completion; consequently, most groundwaters are alkaline and have a low Eh. As would be expected, many compounds that are soluble in acidic oxidizing waters are precipitated where the solutions become basic and reducing. Accordingly, some of the metals dissolved near the surface are precipitated below the groundwater table (Emmons, 1917; Garrels, 1954). This process is important to economic geology and the mining industry because metals leached from the oxidized upper parts of mineral deposits may be expected at depth. Even more important, this process provides a mechanism by which a small percentage of metal can be leached from a large volume of rock and—provided the conditions below the groundwater table are favorable—can be redeposited as a higher grade deposit in a smaller volume of rock. This process is known as *supergene enrichment*. The primary, subeconomic material underlying the enriched zone is known as *protore*. Theoretically, then, there are three fundamental zones in a near-surface ore deposit: the oxidized zone, the supergene-enriched zone, and the hypogene zone or protore (see Fig. 20-1).

Supergene enrichment is effective in concentrating dispersed metals; the economic value of many disseminated copper deposits is due to this process. Several factors are involved in the development of a supergene-enrichment zone. Where groundwater is actively circulating, the dissolved metals may be widely dispersed. Where the rocks are not permeable, meteoric waters will not be able to leach the oxidation products and carry them down to the groundwater zone. Where erosion is rapid, the groundwater table may be lowered so fast that oxidation of the sulfides above cannot keep pace. Where the water table is not lowered through the disseminated mineralization—that is, where the topography is in old age—the process is at a standstill, and enrichment ceases. Thus, the development of an appreciable concentration requires a balance between the rates of oxidation and erosion as well as a fairly sluggish groundwater system.

Classification of minerals as supergene or hypogene should be avoided as far as possible. Certain minerals are characteristic of high-temperature environments and others of low-temperature environments, but there are exceptions. The availability of the chemical components and the Eh-pH environment may be controlling factors. For example, a certain mineral may be rare or unknown as a product of supergene processes because the proper redox conditions are uncommon to the phreatic zone. But since, at least locally, almost any Eh is possible below the groundwater table, the possibility that the species may form should not be ruled out unconditionally. Although specular hematite

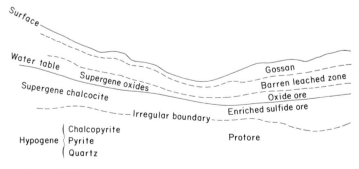

Figure 20-1
Diagrammatic sketch of a deposit containing chalcopyrite, pyrite, and quartz, showing the effects of weathering.

is commonly considered a high-temperature mineral, at the Mountain Home property west of Hanover, New Mexico, plates of specular hematite as much as one-half inch wide are found in limestone above the water table; they are not found at depth and are thought to be of supergene origin. Brown (1936) showed that at Balmat, New York, magnetite forms by the weathering of pyrite; although magnetite had long been considered a high-temperature mineral, this observation was at first difficult for many geologists to accept. In the gold deposits of the South Piedmont there are doubly terminated quartz crystals in the subsoil above the groundwater table. The quartz crystals contain skeleton outlines marked by limonite; similar quartz is not found at depth. Sulfide minerals characteristic of hypothermal and igneous metamorphic zones may indicate hypogene deposition, but again, the possible exception must be anticipated. For example, thermodynamic conditions indicate that pyrrhotite, a typical high-temperature mineral, can form under very restricted (and unlikely) supergene conditions (Garrels, 1960).

Metallic elements have certain definite affinities for sulfur—affinities that are related to the solubilities of their sulfide compounds. Any metal in solution that has a stronger affinity for sulfur than another metal will precipitate as a sulfide at the expense of the more soluble metal sulfide. The sequence of stabilities of the heavy-metal sulfides were first established in 1888 by Schürmann, and accordingly is known as Schürmann's series (Table 20-1). Any metal in the series will replace another that is lower in the series. Thus, copper in solution

Table 20-1
Schürmann's series, stabilities of the heavy-metal sulfides.

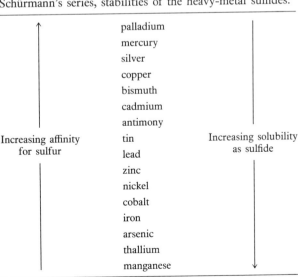

Increasing affinity for sulfur	palladium mercury silver copper bismuth cadmium antimony tin lead zinc nickel cobalt iron arsenic thallium manganese	Increasing solubility as sulfide

Source: Schürmann, 1888.

will replace the iron in pyrite or the zinc in sphalerite. In general, the farther apart the elements are in the series, the more complete the replacement and the greater the rate of reaction. Furthermore, different metals are selectively replaced according to their relative positions in Schürmann's series. For example, silver-bearing solutions react with sphalerite more readily than with covellite or chalcocite.

Schürmann's series has been applied to all ore deposits, both hypogene and supergene. Theoretically, the relationships should be valid in either environment, but hypogene fluids are complicated by so many factors that strict adherence to Schürmann's series should not be expected. The simple relations of the series are especially unlikely in high-temperature, high-pressure systems. Under conditions of supergene deposition, however, where the expected relationships are usually found, Schürmann's series is extremely useful. Since Schürmann's experiments were performed under conditions of standard temperature and pressure, his results would be expected to fit better at low temperatures and pressures than at high ones.

The best examples of supergene leaching and enrichment are found in pyrite-bearing silver and copper deposits. Such deposits, weathered and attacked by meteoric waters, yield dilute solutions rich in sulfuric acid and silver or copper sulfate (see the reactions expressed in Equations 1, 8 through 14, and 20, in Chapter 19). As such a solution descends to the water table, oxygen is gradually exhausted and the solution ceases to be an oxidizing agent. Furthermore, reactions with alkaline constituents neutralize or basify the descending solutions. Thus, the Eh and pH of the meteoric waters change from an oxidizing potential and an acidic environment to a reducing potential and a basic environment that exists below the water table. In the reducing environment the primary sulfide minerals generally remain stable, but they may react according to Schürmann's series with the dissolved metals being carried from the zone of oxidation.

Pyrite is an important mineral both above and below the groundwater table. It weathers to ferric sulfate and sulfuric acid in the zone of oxidation, enabling the meteoric waters to dissolve the ore metals. Below the water table pyrite acts as a host for deposition of the ores because it is relatively soluble and readily relinquishes its sulfur to invading ions of copper and silver. Stokes (1907) showed experimentally how the reaction between copper ions and pyrite may take place in supergene copper enrichment. The reaction, known as Stokes' equation, is

$$5FeS_2 + 14Cu^{+2} + 14SO_4^{-2} + 12H_2O \rightarrow$$
$$\underset{\text{chalcocite}}{7Cu_2S} + 5Fe^{+2} + 24H^+ + 17SO_4^{-2}$$

The excess acid is neutralized, and the chalcocite remains in the enriched zone.

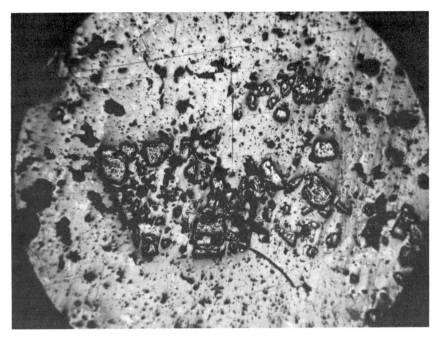

Figure 20-2
Photomicrograph of pyrite replaced by supergene chalcocite. Darwin, California. ×30.
(Photo by Ruperto Laniz.)

Stokes considered this a generalization of the actual process, but the field evidence supports his interpretation (see Fig. 20-2).

Rates of supergene reactions are controlled by many factors. In the presence of abundant calcite or other carbonates, the oxidizing solutions may be neutralized before the groundwater level is reached, and an enriched oxidized zone will result. The character of the sulfides also influences the reaction rate. Pyrrhotite, for example, reacts much faster than pyrite; where pyrrhotite is the primary sulfide below the water table, a supergene chalcocite zone is likely to be thin but high grade. Such a condition was encountered at Ducktown, Tennessee, where the primary ore contains pyrrhotite with subordinate amounts of chalcopyrite. The enriched zone formed just below the groundwater table, in a thickness of two to eight feet; the chalcocite replaced the pyrrhotite almost entirely, and the ore is unusually rich (Emmons and Laney, 1926). By contrast, where the protore contains pyrite and chalcopyrite, as it does in many disseminated copper deposits, it is not uncommon to find enriched zones extending several hundred feet or more beneath the groundwater level.

Copper sulfides also replace the sulfides of zinc and lead, in accordance with Schürmann's series. Many deposits of sphalerite and galena contain small amounts of copper minerals, reworked during weathering and concentrated as supergene products below the water table. The thickness of such zones is generally between the thicknesses of supergene enriched zones in pyrite and pyrrhotite protores.

Supergene chalcocite is commonly soft and powdery. It is referred to as "sooty" chalcocite, to distinguish it from the massive, gray, crystalline chalcocite of hypogene ores. Supergene chalcocite is not necessarily sooty; it may be massive and indistinguishable from hypogene chalcocite. Other copper sulfides also form by supergene processes. Covellite and bornite are not uncommon; chalcopyrite and several of the more complex copper salts of arsenic and antimony also form under special conditions of supergene enrichment. In general, however, supergene sulfides are mineralogically simple compared to hypogene sulfides.

Rates of supergene enrichment may also be controlled by rates of oxidation. Cinnabar thought to be of supergene origin has been reported in some California deposits (Henderson, 1965); supergene cinnabar is not common, however. Low-temperature, sluggish kinetics of the solution and the slow rate of oxidation of cinnabar are believed responsible for the metastable persistence of HgS in the zone of weathering (oxidation), and hence for the limited supergene enrichment.

The chemistry of silver is similar to that of copper, although there are no insoluble silver carbonate compounds to retain silver in the oxidized zone of limestone host rocks. Where halogens are available, as in arid climates, the silver may form insoluble halides above the groundwater table, but silver is ordinarily taken into solution as a sulfate. The supergene minerals, as well as the textures and structures of the deposits, are similar to those formed under epithermal conditions. Consequently, it is sometimes impossible to distinguish between hypogene and supergene deposits of silver. The need for such a distinction is obvious: supergene deposits are likely to be rather shallow, blanket types, whereas hypogene deposits may persist in depth. The most common supergene sulfide compound is Ag_2S. Above 180°C, Ag_2S forms the isometric mineral argentite; at lower temperatures, monoclinic Ag_2S (acanthite) is more stable (Roy *et al.*, 1959). Thus, the presence of isometric crystals of Ag_2S indicates a hypogene origin; but the presence of monoclinic Ag_2S shows only that deposition took place below 180°C, hence acanthite cannot be used to distinguish between supergene and low-temperature hypogene ores. Nor can the conversion temperature of acanthite to argenite be applied to most earlier descriptions of silver deposits, because few geologists in the past attempted to distinguish between the polymorphs of Ag_2S; unless the mineral occurred in elongated crystals of obvious nonisometric habit, it was generally called argentite. The ruby silvers (pyrargyrite and proustite) are more likely to be hypogene than supergene; however, the fact that they can form as supergene minerals means they cannot be considered positive indicators of hypogene deposition. The general observation that supergene assemblages are typically less complex than hypogene assemblages may be of value in the interpretation of silver deposits.

Other minerals, such as those of lead and zinc, have been reported to form enriched sulfide zones, but there are no extensive commercial deposits of these supergene ores. Lead sulfate and lead carbonate are relatively insoluble and they are left in the oxide zone. Zinc also forms a stable carbonate, but the zinc content of sulfide ores is often leached from the zone of oxidation. Many examples of supergene zinc sulfides have been described; perhaps the most noteworthy are the Horn Silver mine in Utah (Butler *et al.*, 1920) and the Balmat-Edwards district of New York (Brown, 1936). Zinc deposition below the water table forms either wurtzite or a light-colored sphalerite. No deposits of supergene zinc sulfides comparable in size to those of supergene copper and silver are known; evidently, most zinc remains in solution to be dispersed in the groundwater system.

CHAÑARCILLO, CHILE

Many examples of supergene-enriched silver deposits could be given, particularly the famous bonanza deposits of the western hemisphere. Today, however, these deposits are nearly exhausted; most of them are abandoned and are largely of historical interest. Deposits of this type extend southward from the United States through Mexico, Central America, and along the western slopes of the Andes in South America. An outstanding district was Chañarcillo, where more than $100 million worth of silver was produced (Moesta, 1928; Whitehead, 1919, 1942; Flores Williams, 1959; Segerstrom, 1962).

Chañarcillo is in the Atacama Desert of Chile, about 32 miles south of Copiapó, in the arid foothills along the western side of the Interior Valley (Fig. 20-3). The district was discovered in 1832 (Miller and Singewald, 1919). During the boom years, from 1860 to 1885, it produced about $2\frac{1}{2}$ million kilograms of silver. Much of the silver was recovered from high-grade masses; one piece of nearly pure native silver weighed over 200 pounds, and another mass of embolite [Ag(Cl,Br)] with native silver weighed 45,000 pounds and contained 75 percent silver.

The rocks in the immediate area of the mine consist of an alternating sequence of Cretaceous limestones and volcanic tuffs intruded by a swarm of diorite dikes and a small granodiorite stock (Segerstrom, 1962). Both the dikes and the stock are highly altered, and parts of the stock consist of diopside, wollastonite, epidote, and similar metamorphic minerals. Narrow copper veins are close to the stock; the silver is farther away.

Although Chañarcillo lies on the axis of a broad regional fold, the dips of the rocks are slight. The rocks are cut by many fractures, which were probably

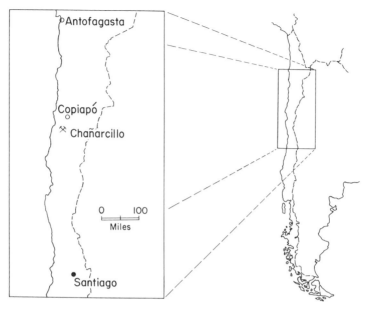

Figure 20-3
Map of Chile showing the location of the Chañarcillo district.

formed during the folding; these fractures contain the ore. Most fractures are parallel to the fold axis, but others cut it at angles of about 45°. The fractures are steeply dipping and of great continuity both in depth and along the strike; at least the upper portions of these fissures were open during metallization (Whitehead, 1919, 1942).

The primary veins vary in width from about one inch to three feet. Hypogene minerals include pyrite, sphalerite, chalcopyrite, galena, arsenopyrite, cobalt arsenides, pearceite ($Ag_{16}As_2S_{11}$), freibergite [$Cu,Ag)_3SbS_3$], proustite (Ag_3AsS_3), polybasite ($Ag_{16}Sb_2S_{11}$), and pyragyrite (Ag_3SbS_3) in a gangue of calcite, barite, quartz, and siderite. The veins extend to the greatest depths explored (about 1,000 meters), but the best ores were concentrated in the purest beds of limestone. Small ore zones were also formed where veins intersect fissures and dikes (see Fig. 20-4). The major veins are aligned along the axis of the anticline; there seems to be a direct correlation between proximity to the crest of this fold and continuity or richness of the parallel veins.

After the hypogene minerals were deposited the rocks were faulted. The major structure—a normal fault with about 150 feet of displacement—divides the Chañarcillo district into northern and southern parts. During this period of faulting the ore-bearing veins were refractured, thereby permitting subsequent erosion and weathering to redistribute the silver minerals in the near-surface parts of the veins.

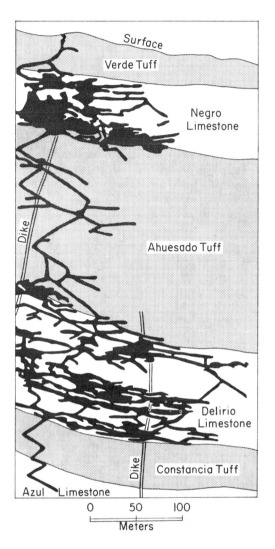

Figure 20-4
Cross section through the Constancia mine, Chañarcillo, Chile, showing the concentration of ore in limestones and the extensions of ore parallel to the bedding. (From Whitehead, 1919, Fig. 2.)

The shapes and sizes of the primary orebodies were not significantly changed by supergene sulfide enrichment; they were merely enriched in silver at the expense of iron, antimony, arsenic, and sulfur. The zone of supergene enrichment ranges in thickness from a minimum of 130 feet in the northern part of the district to a maximum of 500 feet in the south (Whitehead, 1942). Descending meteoric waters precipitated their silver content when they reached the primary ore beneath the water table. Accordingly, the upper parts of the supergene sulfide zones were selectively enriched, and the ore grade diminished from the top to the bottom of the bed. Enrichment was also extensive along faults and water courses. Enrichment obliterated the primary minerals in the richest zones, but pseudomorphous relationships between hypogene and supergene minerals are characteristic in all the enriched ores.

Supergene minerals include stephanite (Ag_5SbS_4), "argentite" (acanthite?), dyscrasite (Ag_3Sb), native silver, stromeyerite ($AgCuS$), and minor amounts of pearceite and polybasite (Whitehead, 1919). Enrichment was by replacement of early sulfides by secondary sulfides, antimonides, and native silver; little replacement of the gangue minerals was noted. As a rule, the supergene minerals were richer in silver than the hypogene minerals.

Studies of the supergene sulfide paragenesis have clarified the processes of enrichment. In the early stages of enrichment ruby silvers (pyrargyrite and proustite) were replaced by "argentite" (acanthite?), stromeyerite, stephanite, and small amounts of polybasite and pearceite. Because pyrargyrite was more susceptible to replacement than proustite, it was the first hypogene mineral attacked. Replacement was clearly recorded in concentric bands of supergene minerals formed around unaltered cores of the ruby silvers. Dyscrasite and native silver were restricted to zones of intense enrichment, forming irregular dendritic masses in pyrargyrite, but also in proustite, pearceite, and polybasite as enrichment progressed. Further supergene enrichment eventually resulted in the complete replacement of primary minerals, until native silver and dyscrasite replaced even the earlier supergene minerals in the upper part of the enriched zone. The most intense enrichment was represented by veins of massive dyscrasite and native silver, which usually replaced sulfides but locally filled open spaces or replaced calcite along cleavage planes.

Above the supergene sulfides, and separated from them by a thick bed of tuff (see Fig. 20-4), is a second bed of limestone, containing oxidized ores. In contrast to the supergene-enriched lodes, the oxidized orebodies were changed from their original configurations. The veins were thickened to as much as 30 feet, and became irregular rather than elongated. Oxidation developed silver halides, which were zoned according to relative solubilities. Cerargyrite ($AgCl$), the least soluble, formed the upper zone, embolite [$Ag(Br,Cl)$] defined the intermediate zone, and iodyrite (AgI) formed the lower zone (Miller and Singewald, 1919, p. 277). Bromyrite ($AgBr$) and iodembolite [$Ag(Cl,Br,I)$] were also common oxidation products. In places the oxidized ores cropped out, but elsewhere they were overlain by tuff. The oxidized ore minerals were deposited both as replacements of calcite and sulfides and as open-cavity fillings, replacement being the dominant process. As the groundwater level was slowly lowered through the silver-bearing rocks, the oxidized zone gradually encroached on the supergene-enriched sulfides. Thus, parts of the oxidized ores (native silver and dyscrasite) were replaced by silver halides. After the halides were formed, there followed a short period during which the oxidation processes were reversed and native silver and small amounts of argentite coated, transected, and replaced some of the silver halides (Whitehead, 1919).

Whitehead (1919) studied the chemistry of the enrichment processes and concluded that sulfuric acid and ferric sulfate, mixed with halides (which were probably wind-blown from the sea), were the active solvents and reagents. The abundant calcite quickly neutralized the acid, but enough ferric sulfate was present to facilitate the leaching process.

Very little hypogene ore was mined in the Chañarcillo district, even though the primary veins contained 60 to 150 ounces of silver per metric ton. Oxidation and supergene enrichment increased the silver content 25 to 80 percent, forming deposits with 100 to 240 ounces of silver per ton. The remoteness of Chañarcillo prohibited mining of anything but high-grade ores, and the primary veins were further excluded because of their depth and the presence of considerable water in the lower levels. Thus, because of both their nearness to the surface and their higher content of silver, the supergene-enriched zones contained the most favorable mineralization for mining.

MIAMI, ARIZONA

Supergene enrichment has been especially important in the history of many disseminated or "porphyry" copper deposits. Several deposits of disseminated copper in the Miami (or Globe-Miami) district, Arizona, have been affected by supergene enrichment (Ransome, 1919; Tenney, 1935; Peterson *et al.,* 1951; Peterson, 1954; 1962). Miami is in southeastern Arizona, near the center of the great southwestern United States copper province (Fig. 20-5).

The principal orebodies at Miami form an irregular, elongated mass of disseminated copper that extends for two miles along a schist-granite contact

Figure 20-5
Map of Arizona showing the location of the Miami district.

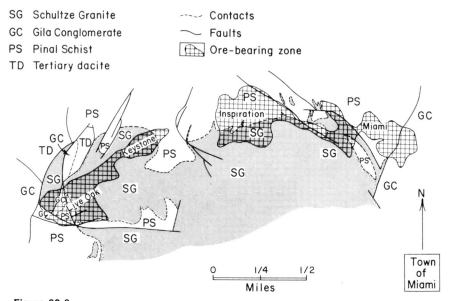

SG   Schultze Granite   ----- Contacts
GC   Gila Conglomerate   ⌒ Faults
PS   Pinal Schist   [▦] Ore-bearing zone
TD   Tertiary dacite

Figure 20-6
Outline of the disseminated copper orebodies at Miami, Arizona, and their relation to the
contact between Pinal Schist and Schultze Granite. (From Ransome, 1919, Plates 31 and 39.)

(Fig. 20-6). Evidently, the mineralizing fluids ascended along the granite contact, where late-magmatic turbulence or post-intrusion tectonics created repeatedly opened veinlets in the contact zone. Most ore is in the Pinal Schist,
of Precambrian age, but some copper is mined from a porphyritic facies of the
granite—the Schultze Granite of Tertiary(?) age. On the western end of the
mineralized zone is the Live Oak orebody; near the center, the Inspiration
orebody; and on the eastern end, the Miami orebody. Live Oak and Inspiration
are largely overlain by a sheet of granite porphyry.

Hypogene mineralization formed an assemblage of ore sulfides and alteration
products that typify the disseminated copper deposit (see Chap. 13). The
unenriched protore at Miami averages about one percent copper (Ransome,
1919) and consists of pyrite, chalcopyrite, and molybdenite distributed along
minute veinlets throughout the schist and granite porphyry. Other orebodies
in the district contain much less copper in the protore than that directly under
the Miami orebody. The alteration products include pyrite, quartz, sericite,
orthoclase, and clay minerals (Schwartz, 1947).

Supergene enrichment increased the grade of ore from one percent or less
to as much as five percent in localized zones. In the early days of mining, when
the upper, high-grade part of the supergene-enriched ore was recovered, the
ore averaged more than two percent copper; today, materials that have been
only weakly enriched are being mined. The Pinal Schist is more permeable

than the Schultze Granite, and it was therefore more amenable to supergene enrichment. Leaching was relatively complete in the oxidized zone of the schist, but the oxidized zone in granite typically retains its copper in the form of malachite and chrysocolla. Accordingly, supergene sulfide enrichment was more thorough in the schist. The supergene-enriched zone starts abruptly beneath the gossan, anywhere from 100 to 600 feet below the surface; it indicates the position of the water table at the time of supergene enrichment (Fig. 20-7). The copper content increases abruptly at the chalcocite zone and tapers off gradually down to the protore (Fig. 20-8). This reflects the fact that supergene sulfide enrichment started at the water table and continued downward to the limit of groundwater circulation or until the supply of copper ions in solution was depleted. Incipient enrichment took place along grain boundaries and tiny cracks in pyrite and chalcopyrite. Further replacement of the primary sulfides by chalcocite increased the copper tenor until the upper zone, the richest, contained about three percent copper. The effects of supergene enrichment are found as far as 1,200 feet below the surface (Lindgen, 1933).

Chalcocite is practically the only supergene sulfide mineral in the Miami district. It replaced both pyrite and chalcopyrite, but chalcopyrite was more susceptible to replacement. In the upper parts of the supergene zone both the chalcopyrite and the pyrite were completely replaced by chalcocite, and intermediate stages of replacement are represented by chalcocite envelopes around remnant cores of pyrite. Small amounts of covellite are found in places; covellite represents the first step in the enrichment of chalcopyrite. In the Copper Cities deposit, $3\frac{1}{2}$ miles north of Miami, supergene enrichment did not progress far because the protore was not exposed to weathering until recently (Peterson, 1954). Here, pyrite is preserved in the enrichment zone; even the chalcopyrite is only partly replaced by chalcocite.

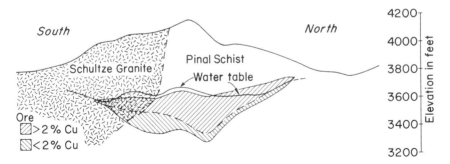

Figure 20-7
North-south cross section through the Inspiration orebody, showing the shape of the ore zone, the relationship of high-grade ore to lower-grade ore, the thickness of the ore-bearing zone in Pinal Schist, and the relationship between the top of the supergene-enriched zone and the groundwater table. (From Ransome, 1919, Plate 41.)

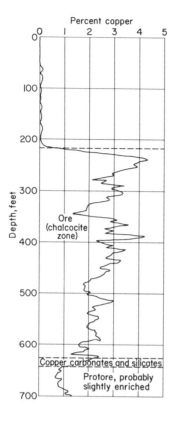

Percent copper

Depth, feet

Ore
(chalcocite zone)

Copper carbonates and silicates

Protore, probably slightly enriched

Figure 20-8
Assay graph of No. 2 shaft, Miami copper mine, Arizona, showing the copper content in the oxidized zone, supergene-enriched zone, and protore. The thin zone of oxidized ore beneath the chalcocite zone is anomalous. (From Ransome, 1919, Plate 47.)

All the ore mined in the Miami district has been taken from supergene-enriched deposits. In the early days of mining, when the cut-off grade was 1.5 percent copper, even the exceptionally rich protore beneath the Miami orebody could not be mined economically. With the development of modern mining and metallurgical techniques, the limits of ore grade have gradually been dropped, so that rock containing only 0.5 percent copper can be profitably mined under favorable conditions. The Copper Cities deposit is now being mined even though it has been enriched only slightly above the tenor of the protore (which averages 0.4 percent copper). In each deposit, then, the copper has been mined as soon as the definition of ore grade has been broadened to include the supergene-enriched part of each deposit. Where leaching and supergene enrichment were thorough, a thick, high-grade chalcocite zone was produced; these deposits were mined first. Deposits that were not favorably weathered and enriched had to await economic changes before becoming classified as ore. Today, some of the marginal materials are leached artificially with acidic solutions, and the copper is recovered from collecting basins in the lower levels of the workings.

REFERENCES TO SELECTED SUPERGENE DEPOSITS

Rio Tinto, Spain

Bateman, A. M., 1927. Ore deposits of the Rio Tinto (Huelva) district, Spain, *Econ. Geol.* 22:569-614.

Finlayson, A. M., 1910. The pyritic deposits of Huelva, Spain, *Econ. Geol.* 5:357-372, 403-437.

Herrero Aleixandre, A., 1947. Mining in Spain's Huelva district, *Eng. Mining J.* 148(10):90-92.

Rutherford, W. P., 1953. El campo de piritas de Huelva (con referencia especial a los macizos de la "Zarza"), *Mineria Metal. Madrid* 151, pp. 11-20.

Williams, D., 1934. The geology of the Rio Tinto mines, Spain, *Inst. Mining Metall. London Trans.* 43:593-678.

———, 1942. Rio Tinto, Spain, in *Ore Deposits as Related to Structural Features*, ed. W. H. Newhouse, Princeton, N.J.: Princeton Univ. Press.

———, 1950. Gossanized breccia-ores, jarosites, and jaspers at Rio Tinto, Spain, *Inst. Mining Metall. London Bull.* 526, pp. 1-12.

Williams, G., 1932. The genesis of the Perrunal-La Zarza pyritic orebody, Spain, *Inst. Mining Metall. London Trans.* 42:3-80.

Ely, Nevada

Bateman, A. M., 1935. The copper deposits of Ely, Nevada, in *Copper Resources of the World*, vol. 1, 16th International Geological Congress.

Bauer, H. L., Jr., R. A. Breitrick, J. J. Cooper, and J. A. Anderson, 1966. Porphyry copper deposits in the Robinson mining district, Nevada, in *Geology of the Porphyry Copper Deposits*, ed. S. R. Titley and C. L. Hicks, Tucson: Univ. Arizona Press.

Fournier, R. O., 1959. Mineralization of a portion of the porphyry copper deposit near Ely, Nevada, *Econ. Geol.* 54:1348.

Lawson, A. C., 1906. The copper deposits of the Robinson mining district, Nevada, *Univ. Calif. Pub. Dept. Geol. Bull.* 4:287-357.

Pennebaker, E. N., 1942. The Robinson mining district, Nevada, in *Ore Deposits as Related to Structural Features*, ed. W. H. Newhouse, Princeton, N.J.: Princeton Univ. Press.

Schwartz, G. M., 1947. Hydrothermal alteration in the "porphyry copper" deposits, Econ. Geol. 42:319-352.

Spencer, A. C., 1913. Chalcocite enrichment, *Econ. Geol.* 8:621-652.

———, 1917. The geology and ore deposits at Ely, Nevada, *Univ. Calif. Pub. Dept. Geol. Bull.* 8:309-318.

Neihart, Montana

Bastin, E. S., 1923. Supergene process at Neihart, Montana, *Econ. Geol.* 18:87-93.

Hurst, M. E., 1922. Supergene process at Neihart, Montana, *Econ. Geol.* 17:383-388.

Schafer, P. A., 1935. Geology and ore deposits of the Neihart mining district, Cascade County, Montana, *Mont. Bur. Mines Geol. Mem.* 13.

Weed, W. H., 1901. The enrichment of gold and silver veins, *Amer. Inst. Mining Eng. Trans.* 30:424-448.

REFERENCES CITED

Brown, J. S., 1936. Supergene sphalerite, galena, and willemite at Balmat, N.Y., *Econ. Geol.* 31:331-354.

Butler, B. S., G. F. Loughlin, V. C. Heikes, et al., 1920. The ore deposits of Utah, *U.S. Geol. Surv. Prof. Pap.* 111, pp. 521-527.

Emmons, W. H., 1917. The enrichment of ore deposits, *U.S. Geol. Surv. Bull.* 625.

Emmons, W. H., and F. B. Laney, 1926. Geology and ore deposits of the Ducktown mining district, Tennessee, *U.S. Geol. Surv. Prof. Pap.* 139.

Flores Williams, H., 1959. *Apuntes de Geología Economica de Yacimientos Minerales,* vol. 2, Santiago: Universidád de Chile Escuela de Geología, pp. 88-91.

Garrels, R. M., 1954. Mineral species as functions of pH and oxidation-reduction potentials, with special reference to the zone of oxidation and secondary enrichment of sulphide ore deposits, *Geochim. Cosmochim. Acta* 5:153-168.

——, 1960. *Mineral Equilibria at Low Temperatures and Pressure,* New York: Harper.

Henderson, F. B., III, 1965. Deposition and oxidation of some mercury ores, *Amer. Inst. Mining Eng. Preprint* 65137.

Lindgren, W., 1933. *Mineral Deposits,* 4th ed., New York: McGraw-Hill, pp. 849-850.

Miller, B. L., and J. T. Singewald, Jr., 1919. *The Mineral Deposits of South America,* New York: McGraw-Hill, pp. 273, 274, 277.

Moesta, F. A., 1928. El mineral de Chañarcillo, *Bol. Minero Santiago Chile* 40:167-182 (tr. of 1870 article).

Olmstead, H. W., and D. W. Johnson, 1966. Inspiration geology, in *Geology of the Porphyry Copper Deposits,* ed. S. R. Titley and C. L. Hicks, Tucson: Univ. Arizona Press.

Peterson, N. P., 1954. Copper Cities copper deposit, Globe-Miami district, Arizona, *Econ. Geol.* 49:362-377.

——, 1962. Geology and ore deposits of the Globe-Miami district, Arizona, *U.S. Geol. Surv. Prof. Pap.* 342.

Peterson, N. P., C. M. Gilbert, and G. L. Quick, 1951. Geology and ore deposits of the Castle Dome area, Gila County, Arizona, *U.S. Geol. Surv. Bull.* 971.

Ransome, F. L., 1919. The copper deposits of Ray and Miami, Arizona, *U.S. Geol. Surv. Prof. Pap.* 115.

Roy, R., A. J. Majumdar, and C. W. Hulbe, 1959. The Ag_2S and Ag_2Se transitions as geologic thermometers, *Econ. Geol.* 54:1278-1280.

Schürmann, E., 1888. Ueber die Verwandtschaft der Schwermetalle zum Schwefel, *Justus Liebig's Ann. Chemie* 249:326-350.

Schwartz, G. M., 1947. Hydrothermal alteration in the "porphyry copper" deposits, *Econ. Geol.* 42:319-352.

Segerstrom, K., 1962. Regional geology of the Chañarcillo silver mining district and adjacent areas, Chile, *Econ. Geol.* 57:1247-1261.

Simmons, W. W., and J. E. Fowells, 1966. Geology of the Copper Cities mine, in *Geology of the Porphyry Copper Deposits,* ed. S. R. Titley and C. L. Hicks, Tucson: Univ. Arizona Press.

Stokes, N. H., 1907. Experiments on the action of various solutions on pyrite and marcasite, *Econ. Geol.* 2:14-23.

Tenney, J. B., 1935. Globe-Miami district (Arizona), in *Copper Resources of the World,* vol. 1, 16th International Geological Congress.

Whitehead, W. L., 1919. The veins of Chañarcillo, Chile, *Econ. Geol.* 14:1-45.

——, 1942. The Chañarcillo silver district, Chile, in *Ore Deposits as Related to Structural Features,* ed. W. H. Newhouse, Princeton, N.J.: Princeton Univ. Press.

21 / Metamorphism of Ores

A great deal of confusion exists in the literature on the metamorphism of ore minerals, especially the sulfides, sulfosalts, and native metals. Seldom are attempts made to distinguish between ores that have been mildly metamorphosed and those that have been subjected to intense, high-grade metamorphism. Large areas of metamorphosed terrane, such as the Precambrian shield of Canada, contain many examples of both cross-cutting and stratiform deposits within the regionally metamorphosed schists and gneisses. It seems unrealistic to assume that none of these deposits is premetamorphic in age; rather, it is logical to assume that at least some of the deposits existed prior to metamorphism, even though they now appear to be normal replacements in foliated rocks (Hanson, 1920). How can these orebodies and their degree of metamorphism be recognized? Until more is known of the behavior of ore minerals at both high temperature and pressure, much of the discussion concerning their metamorphism will be inconclusive.

The characteristics by which metamorphism of sulfide orebodies are recognized are unsatisfactory and inconclusive. Certain textures are thought to be products of dynamothermal metamorphism, but they generally have alternative interpretations (Bastin, 1950; Edwards, 1954; Betekhtin, 1958; Vokes, 1969;

Spry, 1969; McDonald, 1967). Vokes (1969) has published one of the few general reviews of the subject of metamorphism of sulfide ores. He attempts to describe the effects on sulfides of contact (thermal), cataclastic, hydrothermal, and regional metamorphism. Changes in mineralogy and fabric are discussed, as is the mobilization of minerals and elements from existing sulfide bodies.

Vokes states that the wide stability ranges of sulfides, the relatively few phases present in most ores, and the ease with which the high-temperature sulfide phases revert to low-temperature phases on cooling, help to confirm the fact that metamorphism is not reflected in the mineralogical assemblages of sulfide masses. During metamorphism pyrite may be converted to pyrrhotite and magnetite, sphalerite may become enriched in iron, and the exsolution of chalcopyrite in sphalerite is a common phenomenon.

According to Vokes, "A more thorough understanding of the features which metamorphism can produce in ores of all types will greatly assist in the true assessment of their original mode of formation." Such an understanding runs into many difficulties. For example, the banding in many so-called metamorphosed ores actually may be a product of dynamic metamorphism, or it may be relict foliation that was retained through the replacement of a schist or gneiss. According to Kalliokoski (1965), metamorphosed sulfide ores characteristically show granoblastic textures. They have grain boundary relations that are paragenetically inconsistent and complex; the sequential deposition of minerals in the more highly metamorphosed ores cannot be deduced from a study of mineral textures. Some minerals grow porphyroblastically, others show regained equilibrium under metamorphic conditions. Minerals like chalcopyrite may migrate in the stress field (Kalliokoski, 1965).

Arguments about the origin and geologic history of the lead-zinc deposits at Broken Hill, New South Wales (see Chapter 12), illustrate the uncertainty that exists in the study of the metamorphism of sulfide ores. Broken Hill has been described by many careful geologists as old deposits that were rearranged and recrystallized during intense regional metamorphism (Andrews, 1922; Ramdohr, 1950, 1953; King and Thomson, 1953). The ores conform to the contorted bedding of Precambrian sediments. High-grade metamorphic rocks and the associated aplites and pegmatites attest to the severity of deformation during metamorphism. Folding took place under conditions of considerable plasticity and was succeeded by faulting while flowage was still possible. King and Thompson (1953), who advocated a syngenetic origin for the metals, argued that during metamorphism the zinc-lead deposits were recrystallized and structurally concentrated along particular fold axes. Ramdohr (1950, 1953) also thought the ores were involved in the high-grade metamorphism that affected the enclosing sediments, though he suggested that the lead and zinc were deposited as epithermal replacement ores before metamorphism. Other workers have classified the ores as hypothermal and consider them to be postmetamorphism in age (Gustafson *et al.*, 1950; Gustafson, 1954). According to this interpretation, the ore and gangue minerals replaced foliated metamorphic rocks and therefore possess inherited banding. Gustafson (1954) points out that the ore minerals are not deformed; if they are considered to predate the folding, there must have been a subsequent period of recrystallization in the absence of directed stress.

Kinkel and his associates studied the massive sulfide deposits at Ore Knob, North Carolina, and at other deposits between Ducktown, Tennessee, and the Gossan Lead, Virginia. They concluded that ores result from the metamorphism of preexisting deposits (Kinkel, 1962, 1963, 1967; Kinkel *et al.*, 1965; Hutchinson, 1963). Closely following deposition of the massive pyrrhotite-chalcopyrite-pyrite ores, the sulfides and enclosing silicates were recrystallized under a rising temperature gradient to coarse-grained unoriented aggregates that contain pyrite porphyroblasts in pyrrhotite. The common boudinage, dilation, and flow structures are shown in Figure 21-1.

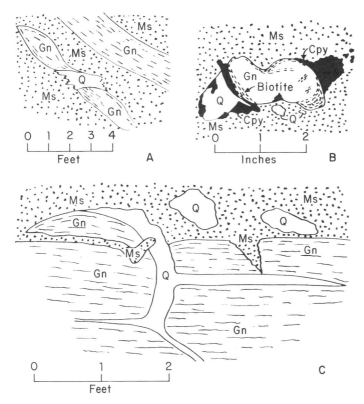

Figure 21-1

Boudinage and dilation structures. (A) Boudinage structure; (B) rotation of fragment with development of pressure shadows; (C) pull-apart structure. *Gn,* gneiss; *Ms,* massive sulfide; *Cpy,* chalcopyrite; *Q,* quartz. (From Kinkel, 1962, Fig. 2.)

Similar differences of opinion have centered around the gold of the Southern Piedmont in the United States. It is not clear whether emplacement of gold predates or postdates the principal metamorphism. Pardee and Park (1948) noted that the ore is broken and shattered in places, but that abundant evidence for postmetamorphic ore deposition can be found. Gold-bearing quartz replaces deformed layers, and mineralization cuts sharply across foliation planes in places. Most of the ore shows little evidence of directional stresses. Other workers have held that the ores were recrystallized completely during metamorphism, and the presence of predeformation quartz veins in the Piedmont rocks of Virginia has been convincingly demonstrated by Cloos and Hietanen (1941).

Much work has been done on the ore deposits in the metamorphic terranes of eastern North America from Quebec to Alabama. Some investigators agree with Kinkel that the present form and distribution of the ores result directly from metamorphic processes (Emmons, 1909; Kalliokoski, 1965), but others say that at least some of the ores are postmetamorphic selective replacements in sericite schists and other metamorphic rocks (Watson, 1954).

Schistose sulfide ores apparently have formed through metamorphism and shearing of sulfide bodies as well as through replacement of schistose rocks (Newhouse and Flaherty, 1930). Flow banding in sulfide masses has been described in Coeur d'Alene, Idaho, where plastically deformed galena contains streaks of sphalerite, tetrahedrite, and pyrite (Waldschmidt, 1925). This "steel galena" formed along fault zones as a result of shearing, but it was not subjected to high-grade metamorphism. Similar sheared galena has been mined from the Slocan district in British Columbia (Uglow, 1917). By contrast, banded ores from the Mammoth mine, Shasta County, California, inherited their foliation from a sheared, schistose alaskite prophyry (Graton, 1909); the relict foliation was replaced by pyrite, sphalerite, and chalcopyrite.

Galena is especially susceptible to plastic flowage under stress, and it is not uncommon to find curved cleavage planes on galena within rocks that show no other signs of deformation. The plastic deformation is readily detected both megascopically and microscopically. Bent cleavage planes apparently reflect only minor stress because continued deformation results in granulation, which produces the fine-grained steel galena (Bastin, 1950). The deformation of galena and other sulfides is primarily due to translation, movement that takes place along crystallographically oriented shear planes without actually rupturing grains (Grigor'yev, 1961; Osborn and Adams, 1931).

The mobilization of sulfides under metamorphic conditions may explain many features of ore deposits, but the behavior and stability of sulfides under high temperatures and pressures are essentially unknown. For example, galena melts at $1,130°C$ (Kracek, 1942), and its presence in furnace slags proves that it will form at high temperatures. Its stability range under conditions of dynamothermal metamorphism is a different matter; there are no quantitative data. Nor are there quantitative data for most other sulfides. If, as many geologists contend, the ore minerals remain in residual hydrothermal fluids until the waning phases of igneous or tectonic activity, then it should follow that where metamorphic conditions approach rheomorphism (partial fusion) or anatexis, or where the volatiles are metamorphically activated, the same ore minerals would be the first to be mobilized for migration. It would be expected that the sulfides would migrate much more readily than most silicates, and, if unable to escape to regions of lower temperatures and pressures, would eventually be reprecipitated.

Roberts (1965) conducted experiments on recrystallization and mobilization of sulfides at 2,000 atmospheres and in the temperature range of 50° to 145°C. The conditions simulated those of a deposit subjected to moderate metamorphism. Textures resembling those of exsolution and replacement were obtained.

A metamorphic origin has often been suggested for the unique deposits of zinc oxides at Franklin Furnace and Sterling Hill, New Jersey. Metsger (1962) emphasized the preferred orientation in the plane of foliation of the C-axis of willemite. He decided that the deposits were involved in regional metamorphism.

Solomon (1965) concluded that the ores of Mount Isa, Australia, were metamorphosed along with the enclosing country rocks. He said that the ores of Mount Isa were local zones of sulfides deposited during sedimentation of the country rock. The present form and distribution of orebodies are thought to have resulted from tectonic deformation.

The behavior of native metals during metamorphism closely parallels that of sulfides, but depends, as would be expected, upon the metal. Under ordinary atmospheric conditions gold is one of the most stable elements; yet the absence of any clearly defined nuggets in Precambrian placer deposits indicates that gold migrates under metamorphic conditions at least for short distances. No characteristic of the gold in Precambrian metamorphic rocks has been found to be applicable in determining whether the gold was mobilized during metamorphism or introduced hydrothermally after the metamorphic processes had ceased.

Metamorphic deposits of native silver and copper are extremely difficult to recognize. Native silver recrystallizes under very mild pressures. Even the grinding of polished sections causes a distortion in crystals of the surface layer— an annoying phenomenon that makes the microscopic study of original silver textures difficult (Bastin, 1950). Native copper begins to recrystallize at 450° to 500°C (Carpenter and Fisher, 1930; Edwards, 1954), although it will not melt until the temperature reaches 1,083°C (Kracek, 1942).

Iron oxides are more stable than the sulfides or native metals under high temperatures and pressures. In the Lake Superior region, and in other areas of iron formation, the rocks have locally been intensely deformed into hematite schists, yet the amount of metal in the formation is not noticeably lowered. In places the oxides form iron silicates, and the minerals are completely re-crystallized; the composition of the minerals has been rearranged but the iron content remains about the same.

LaBerge (1964) concluded from his studies in the Lake Superior district that small amounts of magnetite resulted from the reduction of hematite but that much of the hematite resulted from oxidation of siderite and greenalite during regional metamorphism. Owens (1965) disagreed, pointing out several common field relationships, such as alternating beds of magnetite and siderite, that do not fit LaBerge's hypothesis. Cumberlidge and Stone (1964) also disagreed with LaBerge. They said that the regional metamorphism did not alter the oxidation state of the original minerals.

Many other metal oxides originate in high-temperature, high-pressure environments; consequently, they tend to be stable under conditions of high-grade metamorphism. Chromite is characteristically a product of magmatic segregation. Rutile, zircon, and monazite are also common constituents of igneous rocks. Corundum, ilmenite, and the spinels are found in high-grade metamorphic rocks, where they apparently are stable. By contrast, metamor-phosed manganese oxide appears to be unusual; the manganese combines with silica to form tephroite (Mn_2SiO_4) and other silicates.

DeVore (1955) suggested that ore-forming fluids in general result from the mobility of metals and sulfides under metamorphic conditions. Accordingly, the metals are considered to migrate from centers of metamorphic activity, and the associated plutonic rocks are thought to represent a further product of rheomorphism. The geosyncline becomes the furnace that generates hydro-thermal fluids. This hypothesis is gaining favor among economic geologists, and it is considered the best explanation for certain deposits. For example, in the Michigan copper district native copper is found along the tops of Precam-brian basaltic flows. Stoiber and Davidson (1959) presented field and laboratory data in support of the argument that metamorphism of the lavas at depth altered the basalt to pumpellyite, thereby releasing copper and other chemical constit-uents that formed the native copper deposits associated with albitized wall rock and amygdule fillings of calcite, chlorite, epidote, quartz, prehnite, and laumontite ($CaAl_2Si_4O_{12} \cdot 4H_2O$). If abundant sulfur had been available (for example, from sapropelic rocks in the zone of high-grade metamorphism), perhaps a normal assemblage of sulfide minerals would have resulted.

The concept of a metamorphic origin for all hydrothermal fluids seems a little narrow when applied to deposits of obvious late-magmatic origin. More-over, metamorphism in general is believed to be a process of dispersion rather than concentration. Perhaps it would be more realistic to consider several possibilities in the zone of anatexis of a geosynclinal prism of sediments. If avenues of escape are available for the mobilized fractions, these fluids may be expelled from the high-temperature, high-pressure zone, forming veins or replacement ores in a cooler zone. Dispersed metals would in effect become concentrated along available channels of escape. The complete fusion of geosynclinal sediments, forming a fluid magma, would incorporate all the metals that had not escaped; in a highly plastic environment no open fissure could form. The magma produced by anatexis would contain the metals as dissolved ions to be expelled during the waning stages of crystallization, according to the classic concept of hydrothermal fluids. Conceivably both mechanisms could operate in a single district. Metallogenetic provinces may result from metamorphic or igneous action involving widespread sedimentary concentrations of the metals. Accordingly, the presence of numerous copper-bearing quartz monzonites in the southwestern United States may reflect anatexis that involved a bed of stratiform copper sulfides in the deeper parts of the geosynclinal prism—a bed with copper deposits similar to those in Zambia and Zaire.

A type of metamorphism that has received considerable attention in recent years is thermal metamorphism, related to local intrusions or at places associated with regional metamorphic processes (Izawa and Mukaiyama, 1972; Lawrence, 1967, 1972; Muraro, 1966; Mikkola and Väisänen, 1972; Spry, 1969; McDonald, 1967). Izawa and Mukaiyama described many conformable copper-pyrite orebodies in Japan that were thermally metamorphosed, either with or without regional metamorphism. Such thermal metamorphism is characterized by irreversible desulfurization relations, for example, the formation of pyrrhotite from pyrite, or, at higher temperatures, the formation of intermediate solid solutions $((Cu,Fe)_{1+x}S)$ from pyrrhotite and chalcopyrite. The intermediate solid solution is in turn broken down by retrograde processes and produces such phases as chalcopyrite, cubanite, pyrrhotite, and a small amount of mackinawite. The iron sulfide (FeS) content in sphalerite commonly increases during thermal metamorphism. During the process of thermal metamorphism the heated fluid seems to perform an important role in heat transfer and desulfurization.

In the Mt. Morgan orebody in Australia, pyrite, pyrrhotite, chalcopyrite, and magnetite were thermally metamorphosed by the intrusion of dikes of mafic to intermediate composition (Lawrence, 1972). This resulted in annealing recrystallization of ore along the contact zones, and the textures formed are similar to those obtained in annealed metals.

REFERENCES CITED

Andrews, E. C., 1922. The geology of the Broken Hill district, *Geol. Surv. N.S.W. Mem. Geol.* 8.

Bastin, E. S., 1950. Interpretation of ore textures, *Geol. Soc. Amer. Mem.* 45.

Betekhtin, A. G., 1958. Structures and textures of ores (tr. 1962 E. A. Alexandrov), *Int. Geol. Rev.* 4(8):940–946.

Carpenter, H. C. H., and M. S. Fisher, 1930. A study of the crystal structures of native copper, *Inst. Mining Metall. London Bull.* 306, p. 32.

Cloos, E., and A. Hietanen, 1941. Geology of the "Martic overthrust" and the Glenarm Series in Pennsylvania and Maryland, *Geol. Soc. Amer. Spec. Pap.* 35, pp. 31–32.

Cumberlidge, J. T., and J. G. Stone, 1964. The Vulcan iron formation at the Groveland mine, Iron Mountain, Michigan, *Econ. Geol.* 59:1094–1106.

DeVore, G. W., 1955. The role of adsorption in the fractionation and distribution of elements, *J. Geol.* 63:159–190.

Edwards, A. B., 1954. *Textures of the Ore Minerals and Their Significance,* Melbourne: Australasian Institute of Mining and Metallurgy.

Emmons, W. H., 1909. Some regionally metamorphosed ore deposits and the so-called segregation veins, *Econ. Geol.* 4:755–781.

Graton, L. C., 1906. Gold and tin deposits of the southern Appalachians, *U.S. Geol. Surv. Bull.* 293.

———, 1909. The occurrence of copper in Shasta County, California, *U.S. Geol. Surv. Bull.* 430, pp. 71–111.

Grigor'yev, D. P., 1961. Three types of plastic deformation in galena, *Int. Geol. Rev.* 3:586–597.

Gustafson, J. K., 1954. Discussion: geology of Australian ore deposits, Broken Hill, *Econ. Geol.* 49:783–786.

Gustafson, J. K., H. C. Burrell, and M. D. Garretty, 1950. Geology of the Broken Hill ore deposit, N.S.W., Australia, *Geol. Soc. Amer. Bull.* 61:1369–1437.

Hanson, G., 1920. Some Canadian occurrences of pyritic deposits in metamorphic rocks, *Econ. Geol.* 15:574–609.

Hutchinson, R., 1963. The Ore Knob massive sulfide deposit, North Carolina: a discussion, *Econ. Geol.* 58:997–998.

Izawa, E., and E. Mukaiyama, 1972. Thermally metamorphosed sulfide deposits in Japan, *24th Int. Geol. Congr. Sec. 4 Mineral Deposits,* pp. 455–462.

Kalliokoski, J. O., 1965. Metamorphic features in North American massive sulfide deposits, *Econ. Geol.* 60:485–505.

King, H. F., and B. P. Thomson, 1953. The geology of the Broken Hill district, in *Geology of Australian Ore Deposits,* ed. A. B. Edwards, Melbourne: Australasian Institute of Mining and Metallurgy.

Kinkel, A. R., Jr., 1962. The Ore Knob massive sulfide deposit, North Carolina: an example of recrystallized ore, *Econ. Geol.* 57:1116–1121.

———, 1963. The Ore Knob massive sulfide deposit, North Carolina: a reply, *Econ. Geol.* 58:1159–1160.

———, 1967. The Ore Knob copper deposit, North Carolina, and other massive sulfide deposits of the Appalachians, *U.S. Geol. Surv. Prof. Pap.* 558.

Kinkel, A. R., Jr., H. H. Thomas, R. F. Martin, and F. G. Walthall, 1965. Age and metamorphism of some massive sulfide deposits in Virginia, North Carolina, and Tennessee, *Geochim. Cosmochim. Acta* 29:717–724.

Kracek, F. C., 1942. Melting and transformation temperatures of mineral and allied substances, in *Handbook of Physical Constants,* Geological Society of America (spec. pap. 36), pp. 139–174.

LaBerge, G. L., 1964. Development of magnetite in iron-formation of the Lake Superior region, *Econ. Geol.* 59:1313–1342.

Lawrence, L. J., 1967. Sulfide neomagmas and highly metamorphosed sulfide deposits, *Mineralium Deposita* 2:5–10.

———, 1972. The thermal metamorphism of a pyritic sulfide ore, *Econ. Geol.* 67:487–496.

McDonald, J. A., 1967. Metamorphism and its effect on sulfide assemblages, *Mineralium Deposita* 2:200–220.

Metsger, R. W., 1962. Notes on the Sterling Hill ore body, Ogdensburg, New Jersey, in *Northern Field Excursion Guidebook* International Mineralogy Association.

Mikkola, A. K., and S. E. Väisänen, 1972. Remobilization of sulfides in the Outokumpu and Vihanti ore deposits, Finland, *24th Int. Geol. Congr. Sec. 4 Mineral Deposits,* pp. 488–497.

Muraro, T. W., 1966. Metamorphism of the zinc-lead deposits in southeastern British Columbia, *Can. Inst. Mining Metall. Spec. Vol.* 8, pp. 239–241.

Newhouse, W. H., and G. F. Flaherty, 1930. The texture and origin of some banded or schistose sulphide ores, *Econ. Geol.* 25:600–620.

Osborn, F. F., and F. D. Adams, 1931. Deformation of galena and pyrrhotite, *Econ. Geol.* 26:884–893.

Owens, J. S., 1965. Origin of the Precambrian iron formation, discussion, *Econ. Geol.* 60:1731–1734.

Pardee, J. T., and C. F. Park, Jr., 1948. Gold deposits of the Southern Piedmont, *U.S. Geol. Surv. Prof. Pap.* 213.

Ramdohr, P., 1950. Die Lagerstätte von Broken Hill in New South Wales im Lichte der neuen geologischen Erkenntnisse und erzmikroskopischer Untersuchungen, *Heidelberger Beitr. Min. Pet.* 2:291–333.

———, 1953. Über Metamorphose und sekundäre Mobilisierung, *Geol. Rundsch.* 42(1):11–19.

Roberts, W. M. B., 1965. Recrystallization and mobilization of sulfides at 2000 atmospheres and in the temperature range 50°–145°C, *Econ. Geol.* 60:168–171.

Solomon, P. J., 1965. Investigations into sulfide mineralization at Mt. Isa, Queensland, *Econ. Geol.* 60:737–765.

Spry, A., 1969. *Metamorphic Textures,* Oxford: Pergamon Press.

Stoiber, R. E., and E. S. Davidson, 1959. Amygdule mineral zoning in the Portage Lake lava series, Michigan copper district, *Econ. Geol.* 54:1250–1277, 1444–1460.

Uglow, W. L., 1917. Gneissic galena ore from the Slocan district, British Columbia, *Econ. Geol.* 12:643–662.

Vokes, F. M., 1969. A review of the metamorphism of sulfide deposits, *Earth-Sci. Rev.* 5:99–143.

Waldschmidt, W. A., 1925. Deformation in ores, Coeur d'Alene district, Idaho, *Econ. Geol.* 20:573–586.

Watson, K. D., 1954. Paragenesis of the lead-zinc-copper deposits of the Mindamar mine, Nova Scotia, *Econ. Geol.* 49:389–412.

22 / Metallogenic Provinces and Epochs

A *metallogenic province* is defined as an entity of mineral deposits that is characterized by related mineral compositions, form, and intensity of mineralization (Petroscheck, 1965). Bateman (1950) defined a metallogenic province as a region characterized by relatively abundant mineralization, of which one type predominates. The concept of a metallogenic province was used years ago by de Launay (1913), Spurr (1923), and Lindgren (1909, 1933). In recent years considerable impetus has been given to the study of metallogeny by the establishment of an international commission to prepare a metallogenic map of the world (Guild, 1968, 1974; Petroscheck, 1963; Stoll, 1965). Russian geologists in particular have done much work in this (Smirnov, 1959, 1968; Tvalchrelidze, 1957, 1964, 1967; Belibin, 1955; Radkevich, 1956).

A *metallogenic epoch* is the time interval favorable for the deposition of certain useful substances (Lindgren, 1933). Turneaure (1955) used the term to designate periods during which mineral deposition was most pronounced. Several authors have emphasized the close relationship between metallogenic provinces and metallogenic epochs. Petroscheck (1965) restricts provinces to tectonic intervals within major tectonic units. This is difficult to do because the ages of many deposits are problematical. In most districts the age of the deposits can be fixed only within broad limits, and age is based primarily on indirect evidence. Thus, although the Canadian Precambrian Shield is frequently described as a metallogenic province, the geologists who work with its mineral deposits say that it is composed of many metallogenic epochs; its ore deposits were formed at many different times. At present the correlation between metallogenic provinces and epochs is difficult or at times impossible, but as more information is obtained, closer correlation should be possible.

The size of a metallogenic province ranges from a single mining district to regions that extend hundreds or even thousands of miles. The Canadian Shield has been defined as a province, as has the Climax-Urad-Henderson area of central Colorado, a small province that is characterized by the deposition of molybdenite, which probably occurred during one epoch.

The concepts of province and epoch are useful theoretically, and, as more is learned by detailed mapping, they may become more useful in exploration and in classification. Detailed studies tend to define the time intervals or epochs more closely than the limits of the province. Burbank (1930) described two major epochs of ore mineralization near Ouray, in the San Juan Mountains of Colorado—one of Late Cretaceous-Eocene age, the other of Late Tertiary. The mineralization of the two epochs is similar, and no effort was made to describe or define the provinces.

The term "metallotect" was introduced by Laffitte, Parmingeat, and Routhier (1965) to indicate those geologic features that are believed to play a role in the concentration of one or more elements and hence are thought to have contributed to the formation of ore deposits. This term may be useful in the discussion of both metallogenic provinces and epochs.

A few geologists have tended to equate metallogenic provinces with ore-deposit zoning (Rastall, 1923). Rastall described the zoning at Cornwall as metallogenetic zoning. Clearly the processes that give rise to metallogenic provinces and to zoning must be closely related, but they are not the same thing. Deposits within a province may or may not be zoned; an example is found in the Appalachian province of North America where some, but not all, of the deposits are zoned (Gabelman, 1968; Pardee and Park, 1948).

Three types of metallogenic provinces may be distinguished on a broad scale and with respect to the tectonic environment: the metamorphic Precambrian shield areas, the mountain (or orogenic) belts, and the stable (or platform) areas. Geologists tend with increasing frequency to relate these three types of provinces to periods of geosynclinal developments. The ore-forming fluids are obtained by tapping deep-seated subcrustal layers and by the remobilization of metals in sedimentary rocks during metamorphism and palingenesis.

White (1966) stated that metallogeny during tectonic evolution might be better understood by considering the relation of metalliferous deposits to (1) depositional events; (2) regional metamorphic events; (3) plutonic events, and (4) deformational events expressed by regional folds and fracture patterns.

Probably no ore deposit owes its origin exclusively to any one of these events; some or all may have played a part of greater or lesser importance.

Solomon, Groves, and Klominsky (1972) give an excellent description of the ore deposits in the Tasman Orogenic Zone of eastern Australia. The massive sulfides of the Tasman Geosyncline furnish a clear example of metallogeny.

Smirnov (1968) speaks for many geologists when he groups the sources of ore-bearing materials into three categories: (1) *juvenile,* related to subcrustal basaltic magmas; (2) *assimilated,* related to palingenic magmas of crustal origin;

Figure 22-1

Principal post-Eocene endogenic metallogenic provinces of the world in relation to the major lithospheric plates. Key to ornament: (1) accreting plate margin; (2) transform plate margin; (3) consuming plate margin with dip direction of downgoing plate; (4) margin of uncertain and (3) *filtrational*, related to the circulation of nonmagmatic underground waters.

The juvenile group, supposedly of subcrustal origin, is characterized by the presence of iron, manganese, vanadium, titanium, chromium, the platinoids, copper, and zinc. These are essentially the elements of the siderophile group that have close spatial relations with magmatic rocks, particularly with those of the mafic type. The ores are believed to be concentrated in skarns and sulfide deposits during the earliest stages of geosynclinal development. The ores are

nature and (or) location; (5) relative plate motion; (6) area of mineralization of post-Eocene age; (7) minor or suspected post-Eocene mineralization; (8) major or noteworthy isolated ore deposit of post-Eocene age. (From Guild, 1972, Fig. 1.)

characterized by similarities in composition irrespective of the ore province. In isotopic composition, the ore-forming elements of the deposits are restricted to fairly narrow limits, close to those of meteorites.

Deposits of crustal origin are related to palingenic magmas or magmas formed by the melting of crustal rocks. These magmas generally are of intermediate silicic compositions. The ores commonly contain tin, tungsten, beryllium, lithium, niobium, and tantalum in minerals that are concentrated mostly in the intermediate and late stages of geosynclinal development. The mineral deposits are not so closely related to the ore-forming rocks as are the deposits of the juvenile group, and their compositions are not so uniform from one province to another. Variations in the isotopic composition of the ore-forming elements are more extensive than in the deposits of the juvenile group.

The so-called telethermal deposits, which include the stratiform lead-zinc deposits, are considered by Smirnov to be examples of orebodies derived from nonmagmatic sources. These were formed during either the late stages of geosynclinal development or the platform stage of geologic history. The ores apparently have no direct relationship to magmatic rocks. Their ore-forming elements display maximum isotopic variation; this suggests mobilization of elements from underlying and surrounding rocks (Smirnov, 1968).

Subcrustal basaltic source material is thought to predominate during the early stages of geosynclinal development. As the geosyncline slowly matures, granitic materials become increasingly abundant. During the intermediate or crustal stage of geosynclinal development the granitic materials, or palingenic magmas, predominate. The latest (or platform) stage of geosynclinal development draws its ore-forming materials from subcrustal and nonmagmatic sources alike.

Gehlen (1968) attempted to define precisely the criteria for distinguishing between ores from subcrustal and from crustal sources. He concluded that it was not possible. Nevertheless, he states that ore deposits derived from subcrustal or juvenile sources appear to be genetically related to their ultramafic or mafic host rocks, which he considers to be differentiates of intruded masses of mantle origin. In the crustal deposits that correspond to those referred to by Smirnov as palingenic magmas, the presence of pegmatites and granites is widely recognized. Gehlen states that isotopic chemistry may lead to more distinctive, more diagnostic criteria. The sulfur isotopes in particular appear to deviate from the average ratio wherever deposits are of crustal origin.

Smirnov (1968) described the metallogenic provinces or subdivisions of Russia in an attempt to clarify the relationships of the broad features of the metallogeny of the country. He attributed a polycyclic character to the metallogeny of most mineralized areas, and stated that many metallogenic epochs exist within the limits of a single metallogenic province. In some places, such as south central Siberia, metallogenic epochs have been repeated from Precambrian times to the Tertiary.

In recent years a great deal of thought has been given to the relationships of ore deposits to the processes involving plate tectonics and sea-floor spreading. Metallogenic provinces may well be associated with continental margins and subduction zones, though the older the ore deposits, the less obvious this relationship is likely to be. Guild (1972) published an interesting map showing the principal post-Eocene endogenic metallogenic provinces of the world in relation to the major lithospheric plates (Fig. 22-1). Theoretically, where subduction zones associated with sea-floor spreading dip under continental masses, the rocks at depth are melted and thus create magma, which includes various ore constituents. The magma rises in the continental block overlying the subduction zone, and at certain stages ore deposits may be formed. For example, the Andes Mountains of South America contain many active volcanoes supposedly related to the magma created along the subduction zone at depth beneath the continental mass. The numerous ore deposits of the Andes Mountains and along the coast of South America may be related to this process. Belts, or the linear distribution of various types of ore deposits, may also be explained. For example, the iron ore deposits of Chile and Peru are mostly aligned in a metallogenic province relatively close to the coast, whereas the copper deposits are in a separate province farther inland at considerably higher elevations. The position of lead and zinc deposits is not entirely clear, in places they are farther inland than copper, whereas elsewhere they may lie between the iron and copper ores.

Further study of plate tectonics and its relation to metallogenic provinces may furnish a valuable generalized prospecting tool. However, much more thought and information will be required before the ideas form anything more than an interesting hypothesis.

REFERENCES CITED

Bateman, A. M., 1950. *Economic Mineral Deposits,* New York: Wiley, pp. 316–325.

Belibin, J. A., 1955. *Les Provinces Métallogèniques et les Epoques Métallogèniques,* Moscow: Gosgeolizdat.

Burbank, W. S., 1930. Revision of geologic structure and stratigraphy in the Ouray district of Colorado, and its bearing on ore deposition, *Colo. Sci. Soc.* 12(6):151–232.

Gabelman, J. W., 1968. Metallotectonic zoning in the North American Appalachian region, *23rd Int. Geol. Congr. Proc.* 7:17–33.

Gabelman, J. W., and S. V. Krusiewski, 1972. The metallotectonics of Europe, *24th Int. Geol. Congr. Sec. 4 Mineral Deposits,* pp. 88–97.

Gehlen, K. v., 1968. Potential criteria for distinguishing between ores from crustal and sub-crustal sources, *23rd Int. Geol. Congr. Proc.* 7:179–184.

Guild, P. W., 1968. Metallotects of North America; paper presented at Society for Economic Geology and American Institute of Mining Engineers meetings, New York, February 27.

———, 1972. Metallogeny and the new global tectonics, *24th Int. Geol. Congr. Sec. 4 Mineral Deposits*, pp. 17-24.

Laffite, P., F. Permingeat, and P. Routhier, 1965. Cartographie métallogénique, metallotect et géochimie régionale, *Soc. Franç. Miner. Bull.* 88:3-6.

Launay, L. de, 1913. *Gites Metalliferes*, vol. 1, pp. 241-288.

Lindgren, W., 1909. Metallogenic epochs, *Econ. Geol.* 4:409-420.

———, 1933, *Mineral Deposits*, New York: McGraw-Hill.

Pardee, J. T., and C. F. Park, Jr., 1948. Gold deposits of the Southern Piedmont, *U.S. Geol. Surv. Prof. Pap.* 213.

Petroscheck, W. E., 1963. Die alpin-mediterrane Metallogenese, *Geol. Rundsch.* 53: 376-389.

———, 1965. Typical features of metallogenic provinces, *Econ. Geol.* 60:1620-1634.

Radkevich, E. A., 1956. Les zones métallogéniques du primorie et les particularites de leur development: geologie de la partie meridionale de l'Extreme-Orient et de la Transbaikalie, *Geol. Rud. Mestorozdh. Petrog. Miner. Geokhim. Trans.*, vyp. 3.

———, 1972. The metallogenic zoning in the Pacific ore belt, *24th Int. Geol. Congr. Sec. 4 Mineral Deposits*, pp. 52-59.

Rastall, R. H., 1923. Metallogenic zones, *Econ. Geol.* 18:104-121.

Smirnov, V. I., 1959. Essai de subdivision métallogénique de territoire de l'U.R.S.S., *Soc. Géol. France Bull.*, ser. 7, 1:511-526.

———, 1968. The sources of the ore-forming fluid, *Econ. Geol.* 63:380-389.

Solomon, M., D. I. Groves, and J. Klominsky, 1972. Metallogenic provinces and districts in the Tasman orogenic zone of eastern Australia, *Australas. Inst. Mining Metall. Proc.* 242, pp. 9-24.

Spurr, J. E., 1922. *The Ore Magmas*, New York: McGraw-Hill.

Stoll, W. C., 1965. Metallogenic provinces of magmatic parentage, *Mining Mag. London* 112(5):312 ff; (6):394 ff.

Turneaure, F. S., 1955. Metallogenic provinces and epochs, *Econ. Geol. (50th Anniv. Vol.)*, pp. 39-91.

Tvalchrelidze, G. A., 1957. Les epoques métallogénique du Caucase, *Sov. Geol.*, sb. 59.

———, 1964. On genetic types of deposits in composite parts of geosynclines (Caucasus taken as an example), in *Problems of Genesis of Ores, Reports of Soviet Geologists*, 22nd International Geological Congress, pp. 322-333.

———, 1967. Main metallogenic features of basaltic and granitoid type geosynclines, *Geol. Rud. Mestorozhd. Acad. Sci. U.S.S.R.*, no. 5.

White, W. H., 1966. Problems of metallogeny in the Western Cordillera, *Can. Inst. Mining Metall. Eng. Spec. Vol.* 8, pp. 349-353.

Wilson, H. D. B., and P. Laznicka, 1972. Copper belts, lead belts, and copper-lead lines of the world, *24th Int. Geol. Congr. Sec. 4 Mineral Deposits*, pp. 52-59.

Indexes

Name Index

Dunham, K. C., 392, 425
Dunn, J. A., 396, 397
Dutton, C. E., 405, 412

Edwards, A. B., 45, 46, 47, 70, 102, 108, 158, 180, 181, 199, 268, 307, 497, 501
Ehrlich, H. L., 456
Eitel, W., 196
Ellis, A. J., 19
Emmons, S. F., 8
Emmons, W. H., 9, 21, 70, 161, 181, 262, 466, 481, 485, 500
Engle, A. E. J., 206
Epstein, S., 25, 206
Eskola, P., 30
Ettlinger, I. A., 319, 321, 326
Eugster, H. P., 396
Ewers, W. E., 223

Fairbairn, H. W., 111, 246
Farmin, R., 16
Farquhar, R. M., 306
Febrel, T., 344
Fenner, C. N., 21, 58
Field, C. W., 205
Finucane, K. J., 65
Fischer, R., 220, 237
Fischer, R. P., 234, 419, 421, 423
Fisher, L. W., 223
Fisher, M. S., 501
Flaherty, G. F., 500
Fleischer, M., 469
Fletcher, J. D., 316
Flores Williams, H., 64, 487
Folinsbee, R. E., 204
Forgan, C. B., 82
Forward, P. S., 299
Fountain, R. J., 144
Frankel, J. J., 436
French, B. M., 244, 412
Frenzel, G., 380
Frietsch, R., 235
Froberg, M. H., 153
Frondel, C., 51
Fryklund, V. C., Jr., 315, 316, 317, 318
Fujita, Y., 98
Fukuoka, I., 368

Gabelman, J. W., 507
Gair, J. E., 293
Galkiewiez, T., 364
Gammon, J. B., 428
Gardner, M., 4
Garlick, W. G., 30, 427, 436
Garnett, R. H. T., 66, 168
Garrels, R. M., 44, 112, 180, 396, 398, 403, 422, 423, 461, 481, 483

Gastil, G., 400
Gaudin, A. M., 192
Gehlen, J. W., 510
Geier, B. H., 87
Geijer, P., 15, 73, 224, 235, 236
Gemmill, P., 70, 71
Gerhard, C. A., 6
Gevers, T. W., 261, 439, 444
Geyne, A. R., 348, 349, 350, 351
Gill, J. E., 47
Gill, J. R., 423
Gillson, J. L., 153, 233, 234
Gilluly, J., 17, 128, 143
Gilmour, P. C., 379, 386, 388
Gittins, J., 59
Glasser, E., 471
Goldich, S. S., 395, 396, 397, 398
Goldschmidt, V. M., 429, 430
Goodspeed, G. E., 31
Goodwin, A. M., 380, 388, 396
Goranson, R. W., 17
Govett, G. J. S., 398
Granger, H. C., 421, 423
Graton, L. C., 10, 39, 214, 436, 442, 500
Gray, A., 30
Greenberg, R., 436
Griffitts, W. R., 380, 382
Griggs, A. B., 380
Grigor'yev, D. P., 500
Grout, F. F., 77, 405
Grover, D. I., 508
Gruner, J. W., 402, 405, 423
Gruszczyk, H., 364
Guild, P. W., 224, 229, 230, 506, 509, 511
Guimaraes, D., 30, 31, 474
Gunderson, J. N., 393
Gunning, H. C., 56
Gushkin, G. G., 24
Gustafson, J. K., 299, 302, 305, 306, 307, 498
Guy-Bray, J. V., 244, 246

Hack, J. T., 71
Hagner, A. F., 31
Hall, G., 382
Hammer, D. F., 319, 321, 323
Hammond, P., 233, 234
Hanley, J. B., 257, 260
Hanson, G., 497
Haranczyk, C., 124, 362, 363, 364
Harder, E. C., 102, 280, 402, 459
Harder, J. O., 290
Hargraves, R. B., 435, 442
Hariya, Y., 392
Hashimoto, K., 380
Hatch, F. H., 436

Locality Index

Subject Index

Acidic, definition, 22
Adiabatic expansion, 96
Aeromagnetic studies, 286
Algae, 418
Alkaline, definition, 22
Alteration, 136
  deuteric, 281
  sericitic, 144
Apatite veins, 153

Bacterial action, 418, 456
  anaerobic, 100
Banded iron formation, 393
Banded ironstone, 393
Bankets, 440
Basin-range structure, 103
Belt series, 315
Bernoulli principle, 96
Biochemical processes, 392
Bleaching, 140, 305, 318
Bonanza silver deposit, 348, 487
Boudinage structures, 499
Bowen's reaction series, 260
Breccia pipes, 59, 69, 90
Brunckite ore, 124, 362
Buried stream channels, 421

Canga, 402
Carbonaceous matter in sands, 421
Carbonatite, 59, 237
Chemical precipitates, 392, 428
Chimneys, 69
Chloride ion complexes, 180
Chromite ore, classes, 224
Chromium deposits, 224
Colloidal deposition, 123
  migration of metals, 50
Colloidal systems, 51

Colloids, 362, 375, 381, 469
Complex ions, 20, 180
Complex equilibrium, 80
Connate water. See Waters.
Copper-sulfur system, 195
Copper deposits
  disseminated, 143, 311, 491
  supergene enriched, 491
Crystallization, 57

Darcy's law, 48
Decrepitation method, 190
Deposition of ores, 55
  sequential, 177
Depositional textures. See Textures.
Deposits
  diagenetic, 356, 428
  diplogenetic, 218
  lithogenetic, 218
  epigenetic, 56, 356
  syngenetic, 56, 356, 391
  epithermal, 10, 145, 337
  hydrothermal, 10, 145
  hypogene, 482, 488, 492
  hypothermal, 10, 142, 289
  magmatic segregation, 57, 139, 220
  mesothermal, 10, 142, 311
  placer (see Placers)
  sedimentary, 10, 391
  supergene (see Supergene enrichment)
  telescoped, 162, 368
  telethermal, 10, 145, 355
  volcanogenic, 379
  xenothermal, 10, 367
Deuteric alteration, 281
Deuteric solutions, 22
Diagenesis. See Deposits, diagenetic.
Diatremes, 71